NATURE FROM WITHIN

NATURE
FROM
WITHIN

Gustav Theodor Fechner
and His Psychophysical
Worldview

Michael Heidelberger

Translated by Cynthia Klohr

University of Pittsburgh Press

Published by the University of Pittsburgh Press, Pittsburgh, Pa., 15260

Copyright © 2004 University of Pittsburgh Press

Manufactured in the United States of America

Printed on acid-free paper

10 9 8 7 6 5 4 3 2 1

Originally published as *Die innere Seite der Natur: Gustav Theodor Fechners wissenschaftlich-philosophische Weltauffassung*
© Vittorio Klostermann GmbH, Frankfurt am Main, 1993

Library of Congress Cataloging-in-Publication Data

Heidelberger, Michael.
  [Innere Seite der Natur. English]
  Nature from within : Gustav Theodor Fechner and his psychophysical
worldview / Michael Heidelberger ; translated by Cynthia Klohr.
      p. cm.
Includes bibliographical references (p.   ) and index.
  ISBN 0-8229-4210-0
  1. Fechner, Gustav Theodor, 1801-1887. I. Title.
  B3237 .H4513 2004
  193—dc21
                        2003012375

CONTENTS

Preface   vii

Introduction   1

PART I History

Chapter One: Life and Work   19

1.1 Ancestry and Early Years   19

1.2 Oken's and Schelling's Philosophies of Nature   21

1.3 Turning to Physics and Overcoming Philosophy of Nature   26

1.4 Herbart's Psychology   31

1.5 The Aesthetic and Romantic View of Nature   35

1.6 The Philosophy of Late Idealism   38

1.7 Beginning Philosophical Work   43

1.8 Illness   47

1.9 The Day View's Origins   50

1.10 Written Work after 1851   57

1.11 The Day View as Contrasted with the Night View   62

1.12 Fechner's Life after Recovery   66

PART II Philosophy

Chapter Two: Nonreductive Materialism   73

2.1 Knowing and Believing   75

2.2 Fechner's Mind-Body Theory: "The Identity View"   91

Chapter Three: Philosophy of Nature   116

3.1 Philosophy of Nature and "Belief"   116

3.2 Psychical Phenomena as Functional States   118

3.3 The Day View as Scientific Identity Philosophy   121

3.4 Direct Realism: The Objective Reality of Phenomena   124

3.5 Further Implications of the Day View   127

PART III Day View Science

Chapter Four: Scientific Realism and the Reality of Atoms   137

4.1 Fechner's Early Writing on Atomism   138

4.2 The "Theory of Atoms"   142

4.3 Realism Includes Phenomenalism   149
4.4 Mach Turns to Anti-Atomism   154

Chapter Five: Psychophysical Parallelism:
The Mind-Body Problem   165

5.1 Psychophysical Parallelism Dates Back to the 1850s   167
5.2 Psychophysical Parallelism from Fechner to Feigl   174
5.3 Schlick and Carnap Enter the Scene   183
5.4 Psychophysical Parallelism in the United States: Herbert Feigl   188

Chapter Six: Psychophysics: Measuring the Mental   191

6.1 Basic Concepts   191
6.2 The General Principle of Measurement and Measuring Sensation   193
6.3 Applying the Principle of Measurement to Gauging Sensations   200
6.4 Objections to Quantifying Psychical Phenomena   207
6.5 Ernst Mach's Theory of Measurement   234
6.6 Measurement Theory and the Day View   244

Chapter Seven: Self-Organization and Irreversibility:
Order Originating from Chaos   248

7.1 Life and Organic Development   249
7.2 The Philosophical and the Scientific Context   255
7.3 From Fechner to Freud and Peirce   260
7.4 Self-Organization Today   271

Chapter Eight: Indeterminism: From Freedom
to the Laws of Chance   273

8.1 Fechner's Indeterminism   273
8.2 Excursus I: Freedom and Physiology   285
8.3 Excursus II: Epigenesis and Philosophy of History   293
8.4 "Collective Objects"   296
8.5 From Fechner to Von Mises   309

Conclusion   318

Appendix   321
Notes   325
Bibliography   367
Index   439

PREFACE

This book began with a paper on Fechner's *Theory of Measuring Collectives* contributed to a project at The Center for Interdisciplinary Research at the University of Bielefeld. That article now forms the final chapter of *Nature from Within*. The overall project dealt with what is known as the "probabilistic revolution," or the pattern of how probabilistic and statistical methods gradually became established in empirical science (see Krüger 1987). While working on that paper I developed an interest in other areas of Fechner's thought, not directly connected with issues of probability and statistics. I found links of which I had previously been unaware: I was astonished to see that even the philosophical tradition to which I personally owe the most, namely, logical empiricism, was itself originally part of a significant historical trend within the overall philosophical movement of German idealism. (This is not to deny that logical empiricism in one respect does represent a "new epoch in philosophy," as Moritz Schlick has said.) But a scholar can indeed, whether or not he is conscious of it, remain faithful to a certain underlying tradition, although on the surface he seemingly resists it. Once we know how contemporary philosophy of science of a logical-empirical provenience is related to the philosophy of the early nineteenth century, several other historical and systematic ties become easily discernible. Fascinated by this discovery, I resolved to pursue the matter further. My readers now hold the fruit of these efforts in their hands: a study in Fechner's philosophy of science and nature.

Writing this book helped me to examine my own convictions. I wanted to understand just how some of the motives that shaped contemporary philosophy of science relate to reasons of my own—reasons that led me in a particular direction in the philosophy of science.

An early version of this work was presented as a habilitation thesis during the winter term of 1988–1989 to the philosophical faculty at the University of Göttingen. I am grateful to both my advisors and my critics there. Special thanks is due to Lorenz Krüger for critical and encouraging debate and strong support as this work progressed. I also thank my friends in Göttingen; their enthusiastic friendship helped me through tough moments. My deepest gratitude goes to my wife and children; words cannot express just how decisive our bond has been.

Sections of chapters four through eight appeared in earlier versions as Heidelberger 1988a (chapter 4), 1986, 1993 (chapter 6), 1990, 1990a (chapter 7), 1987 (chapter 8). Chapter 2 slightly coincides with Heidelberger 1988. Chapter 5 appeared as Heidelberger 2003.

I would like to thank Cynthia Miller and her colleagues at the University of Pittsburgh Press for making this edition possible and for guiding me through the process. I thank Cynthia Klohr, too, for her fine rendition and valuable advice. Thanks go to David J. Murray for his suggestions and to Olivier Darrigol for his comments on chapter 6. I am also grateful to the Goethe Institute Inter Nationes for cofunding the translation and to all the reviewers consulted. My university relieved me of duties for one semester in order, among other things, to complete the English edition—a gesture much appreciated.

Bringing forth an English edition allowed me to slightly revise the work and enhance it. Chapter 5, for instance, is entirely new. I also took this opportunity to additionally include most of the more recent literature on Fechner. Since no up-to-date English renditions of Fechner's opus are currently available, the translator has taken the liberty of phrasing the quotations in English. The original appeared as *Die innere Seite der Natur: Gustav Thedor Fechners wissenschaftlich-philosophische Weltauffassung*, Frankfurt am Main: Vittorio Klostermann, 1993.

Tübingen, August 2003

NATURE FROM WITHIN

INTRODUCTION

DURING THE NINETEENTH CENTURY the relation of philosophy to the sciences underwent a fundamental transformation. After Hegel's death, an unprecedented success of the sciences usurped philosophy's formerly dominant position. Science abandoned philosophy to develop its own kind of rationality, a rationality giving rise to widespread change. It was then at conflict with its greatest philosophical competitor, the philosophy of nature. Many men of science feared and fought *Naturphilosophie*, the alleged "plague of the century."[1] The establishment of twentieth-century logical empiricism and subsequent progress in analytic philosophy of science seemingly epitomize this severance and emancipation. Now natural science and mathematics provide themselves—as it were—with their own philosophy, namely a "scientific worldview."[2]

I would like to introduce Gustav Theodor Fechner's thought. On one hand Fechner's notions are thoroughly marked by that liberation of natural science, which he wholeheartedly endorses. On the other, Fechner remains faithful to a philosophical outlook that came from a school surrounding Schelling. Fechner held empiricism and naturalism to be evident and indispensable attitudes underlying all empirical science: They were not to be compromised. Yet, he also yearned to combine this insight with a universal vision from the philosophy of nature. Allied with the philosophy of nature, he felt it impossible to deny that the psychic element of the world—being the source of all animation in the universe—is, just like the physical part of the world, fundamental and not further reducible. Ultimately he elaborated a new kind of philosophy of nature, one that entirely renounces the validity of any special philosophical source of knowledge disrespecting experience, while simultaneously condoning principles for generating reasonable metaphysical hypotheses that reach beyond what we know directly through experience. Fechner viewed his novel concept of nature and science as a prolific heuristic device for gaining scientific entry into hitherto neglected fields. The most significant and renowned outcome of his endeavor was psychophysics, which became the foundation for quantitative empirical psychology. To this day, psychophysics has remained a steadfast discipline in psychology.

Fechner called his outlook "the day view." It combined elements taken from both philosophical idealism and scientific materialism and proposed to be the opposite of what he called "the night view," the mechanistic and reductionist theory of life, and its varieties. The mixture of philosophy of nature and science that Fechner had in mind at first seems improbable. One might take it for delusive reconciliation, a trade-off with a disadvantage for science. Far-flung speculation, so typical of the philosophy of nature, seems much too remote from the sober ways of science for the cleft between them to be bridgeable. As Moritz Schlick once remarked: Fechner's philosophical ruminations "cheer us poetically, but offer no new insights."[3]

I invite you to consider all of this from a different angle. I think that Fechner's attitude toward philosophy of nature actually did emanate new insight and that it can take us further yet. Fechner's philosophy and his kind of science called "the day view" amount to an original and successful accomplishment—in terms of its own ambitions as well as in terms of modern scientific criteria. Borrowing a pertinent idea from Schnädelbach: I would like to see Fechner remembered not as merely a follower nor as a forerunner—those positions most frequently attributed to philosophers positioned between Hegel and Heidegger.[4]

Fechner devised a kind of 'scientific philosophy' that drew much of its impact and many challenges from the philosophy of nature coeval with Schelling. His ambition was to reconcile that impetus with novel scientific rationality and transform it into productive natural science. His project softened the distinctions demarcating philosophy of nature, reflection on science, and ultimately, science itself.

In presenting Fechner's program for the philosophy of nature, I examine this unusual way of thinking about nature and science. It is a method not exhausted by analytically reworking problems suggested by science. Instead, it also absorbs into philosophy some synthetic elements taken from science.[5] Today this sort of scientific philosophy is barely understood and rarely attempted, although one does occasionally happen upon signs of it. Indeed, contemporary philosophy of science will only have a future and be capable of fulfilling its purpose by mustering the courage to embrace 'speculative'—albeit scientifically controllable—conjecture. Philosophy of science should be at liberty to accommodate more elements from the philosophy of nature, even if this involves metaphysical faux pas by the standards set by the Vienna Circle. Straitjacketing ourselves into purely reconstructive analytical and internal positions in the philosophy of science—as important as those positions may seem and as undesirable as it may appear to relinquish or revise them—has created an explanatory vacuum, into which all sorts of unwelcome so-called philosophies pour in from all sides. We are neglecting a wealth of solutions from philosophy and overlooking our own (in places dark) philosophical past. If left unchecked, overstated theoretical demands will uncontrollably promote a faltering, if not the loss, of our critical philosophical capacities.

In themselves, naturally, these reflections do not conclusively argue the case for studying what Fechner writes on philosophy of nature. Fechner's work harbors a wealth of oddities—a trait common to much nineteenth-century writing. Today no one would believe that plants or worlds have souls—one of Fechner's more notorious ideas. Even during Fechner's lifetime these vivid conceptions were rarely accepted as serious insights in the philosophy of nature.

Things look different, though, for Fechner's mind-body theory. Today scholars leave no stone unturned in their search for solutions that are at once materialistic or physicalistic and nonetheless nonreductive.[6] Not only did Fechner pursue this very objective, but beyond that, he intended to transform the consequences that this theory would have for psychology and all other sciences into an entirely new empirical theory. This makes his suggestion for solving the mind-body problem unexpectedly topical.

Paul Feyerabend has shown that we can best test the strength and scope of any particular scientific theory by juxtaposing it with an alternative, plausible, realistically interpretable *theory* that creates as much havoc as possible for the first one.[7] We can detect the flaws in one theory by comparing that theory with another; it is not enough to simply claim that a theory is at odds with reality. Analogously, we might say that the real problems (or what should be considered the real problems) in the philosophy of science do not stem directly from the particular *sciences* themselves, but from interspersed philosophical reflection on the explanatory models that science provides—in other words, from a properly understood (historically enlightened) *philosophy of nature*. This philosophy should not only seriously analyze science, but also examine just how we formulate hypotheses about the fundamental set up of the world. Of course, conjecture of this nature goes beyond what is scientifically certain. But this sort of vision is important for philosophical orientation in the world. If philosophy of science neglects the responsibility of dealing with overall questions of orientation, unwelcome 'experts' will begin 'philosophizing' about them. Philosophy of nature will become ostracized from traditional philosophy and little notice will then be made of the philosophy of science.

In the search for philosophical programs akin to Fechner's, we automatically think of Charles Sanders Peirce. In a letter to William James he wrote: "If you were to call my philosophy Schellingism transformed in the light of modern physics, I should not take it hard."[8] Fechner's philosophy can also be characterized along those lines. We could even say that within German literary culture Fechner parallels Peirce, at least in terms of philosophy of nature. There is striking convergence among their cosmologies, a similarity of views on the development of convictions (Peirce: "belief," Fechner: "*Glaube*"), and an agreement in notions regarding how metaphysics can be cleansed empiristically and controlled scientifically. Chapter seven provides some evidence that there is a historical link between them.

Fechner's work is important for reasons beyond his synthetic philosophy of science and his nonreductive solution to the mind-body problem. Knowing and appreciating his ideas prods us to re-evaluate our most recent philosophical and scientific past. The impact of his "day view" can be traced along several meandering historical routes right up to contemporary philosophy, natural science, and philosophy of science. His ideas were also significant in terms of how the closing nineteenth century gradually turned from mechanistic thinking in the natural sciences and towards positivistic, empiristic tendencies. His notions were passed on predominantly by Ernst Mach, who felt a close personal and scholarly affinity to Fechner. On the whole, Fechner's work was well received in Austrian-Hungarian Empire and helped shape "Modern Vienna." Thus the font of logical empiricism's "scientific worldview" was not so much the *break* of natural science from philosophy of nature, as described above, but the desire to *unite* both ways of thinking.

If this appraisal is correct, then Fechner establishes a link in the historical line beginning with Schelling's philosophy of nature, proceeding to Mach and on to contemporary philosophy of science and natural science. To securely confirm this hypothesis we would need to scrutinize the pre-history of the philosophy of nature itself, which is something I was unable to do within the scope of this book. My emphasis was initially on becoming adequately acquainted with Fechner himself, on gaining a panorama of his thought, on correcting prevalent misinterpretations, and on peering over his shoulder, watching him devise novel empirical theories. By themselves, my detailed studies on the historical effects of Fechner's work provide adequate support for the claim that Fechner's work does mark a significant phase between German idealism and twentieth-century logical empiricism (as represented by Mach et al.) and other related scientific movements.

This book has three parts: history, philosophy, and day view science. The first part, which coincides with chapter one, reviews Fechner's life and work. Along with providing biographical information and a characterization of Fechner's writings, I have taken care to adequately describe the background with respect to the history of ideas. As mentioned above, Fechner's philosophy was shaped by Schelling's, and particularly, by Lorenz Oken's philosophy of nature. But the later phase of idealism and the notion of nature and natural aesthetics propagated by late romanticism were equally important. Fechner modeled his idea of natural science after that of the French *Physique expérimentale* in the manner in which it was exercised particularly at the *École polytechnique* in early nineteenth-century Paris. He was chiefly interested in electricity and magnetism.

The second part of this book deals with Fechner's philosophy. We focus on Fechner's emphasis on philosophy of science, which I call 'nonreductive materialism' (chapter 2). Fechner wanted to demonstrate that even a scientist could accept a view that postulates the identity of the psychical and the physical in an empirically clear and phenomenological interpretation, as a solution to the mind-body problem. We are not forced to question the causal inclusiveness of the material world. Today's terminology might label Fechner's mind-body theory a "double aspect theory" (also known as "property dualism" or "dual perspective theory"). A weaker version of this theory was standard for natural science throughout the late nineteenth century, followed by a notable revival of interactionist substance dualism.

Fechner felt that only "a philosophy of identity" (Schelling's term) adjusted to natural science could appropriately explain the presence of the psychical in the world. We must call Fechner's mind-body theory nonreductive, because in it the psychical is not further reducible to matter; none of its independence is 'interpreted away.' The theory is materialistic, because it considers the psychical a whole part of nature and all events (thus also psychical events), though they may not be physically *caused*, must have physical *conditions*, and are therefore to a certain extent explainable in physical terms.

In order to justify the existence of the psychical beyond that of one's own in-dividual consciousness, Fechner designed a theory of self-stabilizing methods for hypothetical conjecture ("believing"). An assumption or belief becomes more probable, the longer people uphold it, the more useful it is, and the more diverse and widespread is the empirical evidence in support of it. Equipped with this theory of belief quite similar to American pragmatism, Fechner also later tried to lay a foundation for other hypotheses that transcend experience.

Chapter three centers more narrowly on Fechner's philosophy of nature. This is the component in Fechner's philosophy founded on nonreductive materialism and that, being based on his theory of "belief," ultimately provides natural science with inductively justified metaphysics. While natural science deals with the outer side of nature, the philosophy of nature is thematically concerned with nature's "inner side," in other words, the psychical aspect of the world—an aspect that each of us knows directly only via our own consciousness and that otherwise must be tapped into indirectly.

Part III discusses Fechner's concern with natural science, or, to be more pre-cise, the consequences he drew from his philosophical "day view" for branches of science and their respective underlying philosophies. We shall look carefully at five central challenges discussed in Fechner's scientific work.

The first of these studies (chapter 4) deals with Fechner's defense of scientific realism.[9] Fechner does not discuss this type of realism abstractly, but instead defends atomism in physics and philosophy. We must take heed in phrasing what Fechner meant by "realism." After empiristically reworking Schelling's identity doctrine, he went on to advocate a sort of phenomenalism that enhanced his criticism of various metaphysical concepts: the concept of the noumenon (thing-in-itself), the concepts of substance, the ego, and force. Ernst Mach's critique of metaphysics took up right where Fechner's left off. Mach adopted Fechner's definition of mass for his critique of mechanism. This definition later caused quite a stir in physics and philosophy. But Fechner's "phenomenalist realism" also prompted his follower Mach to abandon Fechner's type of atomism (and realism) and establish his own infamous position of anti-atomism and antirealism. Fechner's psychophysics and Herbart's philosophy also significantly inspired Mach's change of mind. For modern philosophy of science this episode in the Fechner-Mach relationship is particularly valuable, pressing, and informative with respect to the ongoing debate between realists and anti-realists.

In the second study (chapter 5), Fechner's mind-body theory and its enormous influence on philosophy and different sciences is treated. It has become known as "psychophysical parallelism" and has dominated German philosophy, psychology, and physiology way into the twentieth century. Three forms of this parallelism are distinguished: The first and foremost is taken as an empirical postulate or methodological rule which does not pretend to explain anything but only to presuppose that any mental occurrence has a physical basis. The second form is more metaphysical and takes a stand for an identity theory of mind and body thus rejecting a causal relationship between them. The third form amounts to panpsychism, which not only assumes a physical side to each mental one, but also the other way around. Especially the first form has been widely influential. It is shown that a version of the second form of parallelism by a neo-Kantian and positivist, A. Riehl, has served as blueprint for Schlick's and Feigl's identity theory. With Feigl, mind-body theory entered its present-day phase.

Chapter six is about psychophysics and empirical scientific psychology seen as quantitative science. In laying a foundation for empirical, quantitative psychology, Fechner elaborates a theory of measurement that includes day view phenomenalism at its crux. As intense debate on psychophysics followed, Fechner's opponents were for the first time compelled to actually state the theory of measurement immanent in the mechanistic worldview and from which they contrived to deduct the immensurability of sensations. Ernst Mach, though, recognized that Fechner's principles of measurement were easily applicable in physics and would terminate in a critique of the mechanistic worldview. Fechner's work on measurement later encouraged the differentiation of different types of scale, a method that has become fundamental for today's theory of measurement.

Beyond Fechner's contribution to the theory of measurement, even today the particular *methods* he devised for gauging and scaling the mental can be found in every textbook on empirical psychology. Although his *theory* of the relation between sensation and stimulus was long overshadowed by Stanley S. Stevens's "new psychophysics," his psychophysical law recently experienced a surprising renaissance within the context of fundamental issues involving Stevens' approach.

Chapter seven deals with Fechner's theory of self-organization. Fechner considers self-organization a characteristic that distinguishes organisms from purely mechanical systems. It is caused by a special type of movement of material particles within a system. The movements of these particles are subject to the "principle of a tendency towards stability." Put simply, this principle states that due to the effects of internal forces, a closed or relatively closed system will, in the long run, tend toward a stable state, meaning that it periodically approximates a state in which it has already once been. Fechner believed to have good reasons for assuming that this principle is also true for the entire universe. The world was originally in a completely irregular state, that, conforming to the principle of stability, gradually consolidated itself, achieving an ever-increasing degree of regularity. All that is inorganic is a mature final product of this process, while organisms are seen as the remains of the spontaneous state of the still young universe. Fechner tried to make this theory tangible for biology by accommodating Darwin's theory of evolution. This amounts to an early version of biological systems theory.

After thoroughly analyzing Fechner's ideas on self-organization, I turn to Oken's philosophy of nature as their source and to the effect they had on Mach, Freud, Peirce, and gestalt theory. Not long ago, Ilya Prigogine pointed out that Fechner's theory of self-organization led directly to his own theory as well as to that of Manfred Eigen. This shows once again the contemporary relevance of Fechner's ideas.

In chapter eight, finally, I investigate indeterminism, a notion that is implicit in Fechner's theory of self-organization and therefore already apparent in the previous chapter. Fechner was one of the first thinkers, if not even the first, to take the idea seriously that we, as Peirce later was to say, live in a "universe of chance." The world's development is, to a great extent, indeterminate. We cannot entirely explain the existence of irregular events by taking recourse to limited human knowledge of the conditions and laws of the universe, we must also acknowledge that there are objective factors supporting it.

Fechner demonstrated how science can be done even under the assumption that indeterminism is true—done even better and more appropriately than under the opposite assumption of determinism, which, in the mechanistic system, had advanced to the status of being the absolute postulate for scientific understanding. The theory of freedom, as it had been developed by the philosophy of late idealism in critique of Hegel's logicism, proved crucial for Fechner. It led him to a special concept of causality, which later prompted Mach to think that there is no more to a concept of causality exempt of metaphysics than the functional dependence of two variables.

Fechner initially phrased his notion of indeterminism in philosophical terms and later made this the basis for a mathematically formulated empirical theory of statistic distribution. His posthumously discovered *Theory of Measuring Collectives* already defines probability as the limit of relative frequency. Thus, Fechner established one of the most important branches in modern theory of probability. Richard von Mises, a logical empiricist, sharpened Fechner's notions and further developed them systematically. Twentieth-century empiricism held its interpretation of probability as frequency for one of its most noteworthy accomplishments. So we see that Fechner was significantly involved in the "probabilistic revolution" in natural science.[10]

All of this shows the importance of Fechner's day view and how a number of developments—right down to our own interests—were influenced by it. These examples also make the sort of philosophical natural science and scientific philosophy embodied in Fechner's thought tangible and let it seem plausible. Compared to Fechner's profuse and prolific philosophy of nature, present-day philosophy of science often looks quite boring and barren. These five samples of Fechner's philosophic-scientific thinking exemplify how philosophy of science can be successfully interlaced with philosophy of nature.

NOW I would like to provide a survey of the source material and secondary literature on Fechner.[11] The major source for information on Fechner's life is a comprehensive biography published by Johannes Emil Kuntze (1824–1894) just five years after Fechner's death.[12] Kuntze was the oldest son of Fechner's sister Emilie and lived in the Fechner household as a foster-son from 1834 until after 1864.[13] He later became professor for jurisprudence at the University of Leipzig and was temporarily the dean of the faculty for law. This biography, undertaken at the behest of Fechner's widow, was written, as Windelband noted, "with tenderness and vivid portrayal, but not without a touch of biting criticism."[14] Not only does the bigoted tone render this biography unsatisfactory, but also the author's lack of expertise in philosophy and natural science. Nevertheless, Kuntze's biography of Fechner is indispensable as a source of information on the circumstances of the scholar's life, particularly because he had access to Fechner's unpublished writings that were lost in World War II. He included some excerpts word for word. In contrast, Wilhelm Wundt, coeval with Fechner and a witness of the latter's work, provided a pleasingly impartial and informed report on him.[15]

In 1890 the physiologist William Th. Preyer edited a selection from his exchange of letters with Fechner and Fechner's correspondence with Carl von

Vierordt, a physiologist from Tübingen.[16] Finally, in 1905 Lipps published an index of Fechner's unpublished writings in the archive of the Royal Saxon Society of the Sciences, furnished for the most part with tables of content and text samples of varying length.[17] Most of the Fechner archive was destroyed when Leipzig University was bombed on December 4, 1943. The section for handwritten manuscripts at the university library still contains Fechner's diaries (for 1828, 1831, 1842–1850, 1860, 1862, 1864–1875, 1877–1879—in total 7282 pages), a few more than two hundred letters to and from Fechner and a small number of other manuscripts. Döring and Plätzsch provided a description of these assorted papers in 1987.[18]

Among contemporary contributors to Fechner studies, Marilyn E. Marshall is at the top of the list. She offers a clear, precise, informed, and reliable overview of Fechner with a very scientific inclination; I warmly recommend it as a primary source of orientation.[19] In additional essays Marshall also writes on William James's relationship to Fechner as well as on Fechner's 1823 habilitation thesis in philosophy of nature.[20] For other published letters to and from Fechner see Weber 1850, P. 1887, Fechner 1890, Preyer 1890a, Kiesewetter 1891, Koenigsberger 1903, Thiele 1966 and 1978, and Hoffmann, Dieter 1991, 338–42. See also Lennig 1994. A chronological bibliography of Fechner's writings can be found in Müller, Rudolph 1892 and in Altmann 1995, which is based on Müller. Altmann 1995 also has a chronological bibliography of the secondary literature up to 1994. The Gustav-Theodor-Fechner Gesellschaft, founded in Leipzig in 1990, has produced a CD-ROM with many of Fechner's works, including rare items, unpublished letters, and excerpts from the *Tagebücher* (diaries). See http://www.uni-leipzig. de/~fechner/, or contact gustav@uni-leipzig.de.

Among the more recent literature the following papers are also recommendable: Adler writes on the fate of Fechner's psychophysics in America; Bringmann paints a colorful portrait of Fechner, adding new facts; Ellenberger explores the relationship between Freud and Fechner; Scheerer provides valuable information about Fechner's work, particularly written for psychologists and Fechner's concept of inner psychophysics and its meaning; Mattenklott's work is a brief but intense (psychological) study on Fechner; Schreier writes on the roots of psychophysics in physical concepts; Sprung makes an attempt to fully honor Fechner's role in the history of psychology; and Oelze basically provides an interpretation of Fechner's *Zend-Avesta*.[21]

Particular mention is due to the writings that Brozek and Gundlach collected in 1988 for a conference commemorating the one-hundredth anniversary of

Fechner's death: Allesch and Ebrecht on Fechner's aesthetics, Jäger on Fechner's worldview, Marshall on Fechner's desire to be in two places at once, Murray and Ross on Fechner's relationship to Ernst Heinrich Weber, Scheerer and Hildebrandt on the fair evaluation of Fechner's accomplishments in psychology, Tögel on the relationship between Fechner and Freud, Wolters on Fechner and Mach, and papers by other authors on many intriguing topics. This volume marks a new phase in Fechner research.

The first dissertation on Fechner to appear in a long while, and that includes much hitherto neglected material, was presented to the faculty for social sciences at the Humboldt University in Berlin in 1990 by Petra Lennig.[22] Her work is remarkably careful and well balanced.

In a paper written in 1993 I attempted to very briefly explain the merits of the fundamental theses of both Fechner's philosophy of science and philosophy of nature.[23]

The renaissance of Fechner's psychophysics within contemporary empirical psychology is extremely well documented in the journal *Behavioral and Brain Sciences*.[24] Fresh appreciation for Fechner's psychophysics has gradually altered the way psychologists think about Fechner.

The previous picture of Fechner upheld by nearly two generations of psychologists (especially in English-speaking countries), was sketched almost exclusively by the psychologist and historian of psychology Edwin G. Boring.[25] Boring's portrayal of Fechner, while written in a vivid and understandable style that fulfilled its purpose at the time, is meanwhile outdated in many details and in general. This becomes particularly evident below in chapter six, although I hardly make direct reference to Boring there. Boring frequently measures Fechner's philosophical and scientific theories against his own operational and verificationist criteria—which we today find overly restrictive and exaggerated. In addition, Boring is frequently a victim of those very handy and plastic categories with which he himself tries to organize the material.

Although Boring greatly admired Fechner's endeavors in psychophysics, like his follower Stevens,[26] he actually denied them any scientific worth. By doing so, Boring and Stevens encouraged an attitude at least partially aroused by William James. After being initially skeptical about Fechner's philosophy, James later (unlike Boring and Stevens) became very enthusiastic about Fechner's *Zend-Avesta* and theory of souls.[27] But he found Fechner's psychophysics worthless. In *Principles of Psychology* he joked about "patient whimsies" with which the "dear old man" Fechner pestered his colleagues.[28] Even a widely read summary of Fechner's psychophysics published by Titchener in 1905 can be counted among the American writings tinged by this ambivalent view of Fechner. In France, Fechner's work gained stormy and controversial attention, as related in chapter six, and was given the decisive blow by Bergson's critique in 1889, only to come to a near halt after being reworked by Foucault.[29] Later, in 1925, Séailles wrote a remarkable piece expressing appreciation for Fechner's philosophy.

It may be perhaps almost too subtle to say that older German writings (and not only these!) on Fechner fall into one of two groups: Authors of the first group present Fechner's pan-psychical worldview as a pleasant, otherworldly—yet freakish—Biedermeier idyll, each of them quite glad not to have been the one to have come up with it. The second group goes to the other extreme and sees in Fechner's doctrine of the world-soul, enriched with some of Schopenhauer's philosophical themes, the solution to all the world's problems. Many of these writings contain valuable insights and startling details, but none of them examine the special link that Fechner saw between philosophy of nature and science.

In various ways the following pieces are exceptions to that rule: Seydel's work on Fechner's antithesis to neo-Kantianism; Bölsche's enthusiastic, very poetic yet keen evaluation of Fechner by one of Haeckel's followers; Hartung's work on Fechner's roots in Schelling's philosophy of nature; the piece by Siegel, which provides valuable facts on Fechner's doctrine of atoms; and the article by Wundt mentioned above, which, while it praised Fechner's thoughts on science, also coined the pejorative label "philosophy as poetry," a derogation that stuck for a long time, right up to today.[30] Summaries of Fechner's psychophysics were not as significant for German scholars as they were for French and English readers. Of the German overviews, Gutberlet's work published in 1905 and Wilhelm Wundt's *Physiological Psychology* may be the most comparable to the Fechner studies completed by Titchener and Foucault.[31]

Among early comprehensive works on Fechner's life and work, particular merit is due to Lasswitz—Bölsche writes that solely Kurd Lasswitz is responsible for reviving interest in Fechner at the turn of the century.[32] In the first part of his work, Lasswitz (by the way one of the first German science fiction authors) provides the most readable and stylistically mature description of Fechner's "Life and Work." Although Lasswitz deserves credit for providing the first balanced and suitable depiction of Fechner's philosophic and scientific thought in the second part of his book ("The World Picture"), he does sporadically distort Fechner's philosophy. Lasswitz was an enthusiastic follower of late nineteenth-century neo-Kantian theory of knowledge and consequently this served as his standard for evaluating Fechner's ideas. He tended to attribute covert Kantianism to Fechner and then criticize it as unsatisfactory.

Lasswitz also deserves credit for vehemently fighting the "masters of new gnosticism" such as Willy Pastor and others, who, around the turn of the century adopted Fechner as an authority for their anti-scientific "emotional philosophy," replete with "subjective fantasy." [33] May this be a warning to all of today's postmodern new age philosophers hoping to exploit Fechner's work for their own purposes.

It is surprising that although in Germany Fechner's philosophical writings were indeed popular from the turn of the century on, up at least until World War I and somewhat longer than that no one, after Lasswitz, produced a truly serious study and interpretation of his work. (Adolph is perhaps the laudable exception.[34]) Instead, Fechner was misused—in Eduard von Hartmann style—for fashionable and mostly shallow idealism.[35] To my knowledge, to this day no truly comprehensive study of Fechner's thought exists that does adequate justice to both aspects: Fechner's scientific work and his philosophy of nature (without dismissing it as an oddity), and the reciprocal relationship between both.

Much relevant literature has appeared in the last ten years, after the German original of this book came out. We can note some kind of Fechner revival which has taken place since then. Gundlach 1993 gives many new details on Fechner's descent and family and presents a new evaluation of Fechner's *Elements of Psychophysics*. Lennig 1994 treats the relation of metaphysics and science in Fechner in a new and thoughtful way. Arendt 1999 is an attempt by a general historian to place Fechner in his time. Many of the well-known Fechner scholars (including myself) have continued their work on Fechner and expanded it into new areas in the meantime (see works by Adler, Arendt, Brauns, Gundlach, Lennig, Marshall, Meischner-Metge, Murray, Scheerer, the Sprungs, and others).

Some recent works relevant to Fechner are general accounts of the relationship of psychology and philosophy in the nineteenth century. They include Sachs-Hombach 1993, Rath 1994, Reed 1994, Ash 1995, Hatfield 1995, Kusch 1995, Murray 1995, Schmidt, Nicole 1995, Danziger 1997, Hatfield 1997, and Borgard 1999.

Works that are more focused on the history of philosophy and on philosophers relevant to Fechner's context are Adler 1992 (on William James's relation to Fechner), Kruck 1994 (on Weisse), Sieg 1994 (on the neo-Kantian school of Hermann Cohen), Pester 1997 (on Lotze), Briese 1999 (on leading philosophers between 1830 and 1850), Gödde 1999 (on the history of the concept of the unconscious), Stephan 1999 (on the development of the concept of emergence),

Schneider 2001 (on I. H. Fichte and Weisse). Köhnke 1986 (history of neo-Kantianism) remains the seminal study of this period, however. New relevant literature on Ernst Mach includes Blackmore 1992, Stöltzner 1999, Heidelberger 2000a, 2000b, Blackmore 2001, Hatfield 2003 and several works by Ley. Works on other scientists closely related to Fechner are Cahan 1993 (Helmholtz), Caneva 1993 (Julius Robert Mayer), Heidelberger 1993 (Helmholtz), Turner 1994 (Helmholtz and Hering), Ross and Murray 1996 (E. H. Weber), Haupt 1998 (G. E. Müller), Favrholdt in Bohr 1999 (on Bohr), and Baumann 2002 (Hering).

The last years have also seen a vast increase of work on romantic *Naturphilosophie* and its relation to and impact on science which is, of course, important for assessing Fechner's roots in this movement. See especially Gloy/Burger 1993, Grün 1993, Poggi/Bossi 1994, Baumgartner 1994, Bonsiepen 1994, Kanz 1994, Snelders 1994, Strack 1994, Poggi 1996, Schmied-Kowarzik 1996, Caneva 1997, Mischer 1997, Wilson 1997, Warnke 1998, Heidelberger 1998, Breidbach/Ziehe 2001, Breidbach et al. 2001, and Köchy 1995–98.

Another notable increase in interest concerns Fechner's aesthetics and its context. Works in this vein include Pauen 1995, Machotka 1995, Höge 1995–97, Breidbach 1997, and Benjafield 2001. Batschmann 1996 deals with the Holbein exhibition of 1871 for which Fechner seems to have developed the first statistically evaluated psychological questionnaire in history.

For recent and authoritative treatises on psychophysics, which increasingly include a chapter or two on the historical development of their discipline, see Link 1992 (essay reviewed by Smith 1994), Mausfeld 1994, Baird 1997, Gescheider 1997, Laming 1997, Narens 2002, and Falmagne 2002. See also further historical work in this respect by Laming, Link, Murray, and Ross. Michell 1999 deals with the history of the concept of measurement (not only) in psychology.

It is highly welcome that, after a long absence in the debate, French scholars seem to have developed interest in the French connection to psychophysics. See the work by Nicolas and also Dupéron 2000 (which, however, is strangely ignorant of almost all secondary literature). See also Laming and Laming 1996. Literature that expands on the topic of the last chapter of this book (indeterminism, probability, and statistics) includes Plato 1994, Porter 1994, Toyoda 1997, Stigler 1999, Hochkirchen 1999, and Stöltzner 1999–2003.

The last years have also seen an increase of interest among historians of German literature in Fechner. See Czaja 1993, Fick 1993, Borgard 1999, and Fix 2003. This latter work is the outcome of a celebration of Fechner's two hundredth birthday, as is Sommerfeld et al. 2001.

Veritable discoveries (at least for me) were the following two books: Hofer 1996 and Barham 1998. Hofer discovered that the work of the well-known systems theorist Ludwig von Bertalanffy is ultimately rooted in Fechner (see also Hofer 2002). Barham was able to put together interesting documentary evidence of Gustav Mahler's engagement with Fechner's works. It seems that Mahler learned about Fechner from S. Lipiner (whom I discuss in chapter 1, section 1.11), with whom he discussed Fechner's works.

PART I
HISTORY

1

LIFE AND WORK

1.1 Ancestry and Early Years

GUSTAV THEODOR FECHNER was born on April 19, 1801 in Gross-Särchen,
a village situated on the Neisse river in the southeastern region of Lower Lusatia.[1]
His father, Samuel Traugott Fechner (1765–1806), had been a pastor there (probably Lutheran) since 1793, as his father before him had also been. Fechner's
mother Dorothea Fechner (1774–1806), née Fischer, descended likewise from a
regional pastoral family. From childhood on Fechner himself, the second of five
children, was meant to join the clergy, too. His older brother Eduard Clemens
(1799–1861) became an artist and moved to Paris in 1825, where he later died. The
three younger siblings were Fechner's sisters Emilie, Clementine, and Mathilde.

Fechner's father has been described as a typical pastor of enlightened times:
of a zealous nature, yet open-minded for progress. He was the first in his region
to have a lightning rod mounted on the church; he upset the congregation by
not wearing a wig during sermons; he had his children vaccinated, and he was
a passionate fruit-grower. His young children were taught Latin—at the age of
three, little Theo (Fechner's nickname) spoke Latin as fluently as he did German.
Fechner's mother was affectionate, cheerful, friendly, and poetic, a woman who
gathered a social circle around herself in all of life's situations.

Following their father's premature death in 1806, both sons were sent for a few years to their maternal uncle, also a pastor, in Wurzen and Ranis in Thuringia. In 1814 Gustav Theodor was enrolled in secondary school in Sorau (now called Zary), a town near the village where he was born; later he spent two years at the School of the Cross in Dresden, where the Fechner children were reunited with their mother. He continued his education first by attending the medical academy for surgery for six months in Dresden and then—as a penniless sixteen-year-old student—registering to study medicine at the University of Leipzig. He maintained lifelong membership with his alma mater.[2]

In Leipzig Fechner also attended lectures on logic (held by the philosopher Wilhelm Traugott Krug [1770–1842]), botany, zoology, physics, chemistry, pharmacy, anatomy, physiology, obstetrics, and algebra. The eager student was not satisfied with the quality of teaching; he preferred learning from books. Yet, two particular lecture series caught his attention: "Having endured Kühn's boring lessons, Weber's lectures on physiology and Mollweide's lectures encouraging me to study mathematics were the only two exceptions [to the otherwise tiring curricula] and provided me with ideas that promised to become extraordinarily fruitful. I even became Mollweide's famulus for a few years. But despite all my effort, a lack of talent prevented me from being successful at mathematics."[3] (In due time we shall see that this self-assessment regarding mathematics is questionable.)

During Fechner's early university years, Ernst Heinrich Weber (1795–1878) was just completing his habilitation work in Leipzig. Ernst Heinrich was the oldest of three brothers who all were to become well known scientists and remain Fechner's lifelong friends.[4] In 1818 Ernst Heinrich became a lecturer on anatomy and three years later a professor for anatomy and physiology in Leipzig. A prolific author, he gained renown for "The Sense of Touch and Common Feeling," an article in Wagner's *Pocket Dictionary for Physiology* (1846).[5] This vade mecum entry examines "the slightest differences in weight discernable by touch, the lengths of lines discernable by sight, and sounds distinguished acoustically," all features that subsequently were to become fundamental for Fechner's law of psychophysics.[6] "Weber's Law"—as Fechner dubbed Weber's scientific research results—states that the smallest discernable distinction between two weights, two lengths, or two sounds is expressible as an invariable ratio between those weights, etc., regardless of the sizes of the weights and lengths or the intensity of the sounds themselves. The difference between two stimuli is thus always perceived as having the same intensity, as long as the ratio holding between the stimuli remains unchanged.

A professor for mathematics at the University of Leipzig, Karl Brandan Mollweide (1774–1825), who also taught astronomy there, was equally interested in color perception. He gained renown for his critique of Goethe's color theory (1810). He was most likely the one to arouse the student's interest in subjective optical phenomena.[7] Fechner became his assistant for a while.

As his studies progressed, Fechner realized that medicine had been an unfortunate choice. Training was so neglected that he had no opportunity to practice blood-letting (considered a common cure-all), let alone a chance to perform an operation of even the simplest kind. Fechner himself admitted that he lacked the hands-on talent required for becoming a physician. He finished his studies by taking the necessary exams, but he did not fulfill the requirements for acquiring a doctoral title. He received a bachelor in medicine, passed the practical physician's examination and received a master's degree paired with the license to teach at the faculty of medicine.[8] While still a student, he subtly demonstrated a dislike for medicine by writing acerbic satires, published under the pseudonym "Dr. Mises" and titled *Proof that the Moon is Made of Iodine* (chiding the fashionable administration of iodide as a panacea) and *A Panegyric for Today's Medicine and Natural History.*[9]

This aversion to medicine, however, did not impair Fechner's fascination with natural science. On the contrary: He concluded that he was built for scientific theory, although he was not always pleased by this insight. The study of medicine also contributed to a loss of religious faith and to becoming atheist.

At the university, interest in science was only one of several factors shaping Fechner. Other impressions and influences, which at times conflicted with his scientific interests, were to become significant for his subsequent life and work. These crucial factors include:

- acquaintance with Oken's and Schelling's philosophies of nature,
- discovering Herbart's philosophical psychology,
- developing a romantic, aesthetic attitude toward nature and life, and
- being influenced by the philosophy of post-Hegel late idealism.

1.2 Oken's and Schelling's Philosophies of Nature

J. E. Kuntze's biography on Fechner includes two notes written by Fechner expressing his delight in Lorenz Oken's philosophy of nature:

In February 1820 I discovered Oken's philosophy of nature. I was so fascinated by the first chapter that although I did not thoroughly understand it and continued reading without real clarity, it busied my mind for years afterward.[10]

My studies in medicine had convinced me to become an atheist, estranged from religious notions; I now saw the world as a set of mechanical workings. Then I discovered Oken's philosophy of nature and began reading it together with my friend Spielberg, a student of theology. It suddenly shed new light on the whole world, including science, and I was dazzled. Naturally, I understood little of it—as one would expect—and I admit not getting much further than the first chapter. But in a nutshell, all at once I found a perspective for a comprehensive and unified worldview and set out to study Schelling, Steffens and other philosophers of nature. None of them provided much real clarity, but I felt that I myself might contribute to it. Some of the papers among my *Stapelia mixta* (1824) attest to that attitude.[11]

It is important to keep in mind that—at least in hindsight—Fechner interpreted his conversion to philosophy of nature indirectly as alienation from inanimate mechanism and materialism and returning to religious notions, perhaps even as recapturing the religion of his youth on a higher level.

Before portraying how Fechner came to terms with the philosophy of nature, some general comments on the philosophical tradition in question are in order.[12] The beginning of all philosophy of nature throughout the early nineteenth century was Kant's *Critique of Pure Reason* (1781), which presented the forms of the world of appearances as functions of the human capacity for knowledge. Philosophy of nature aimed to subdue the dualism that results when appearances are separated from noumena; it intended to discover the connection between the world of noumena and human consciousness. Schelling said:

> It is undesirable that nature converge *by happenstance* with the laws of our mind (as would be the case if it were mediated by a *third party*). We prefer that *nature herself* necessarily and originally not only *express* the mind, but also *realize it herself;* and that she can only be nature and be called nature by doing just this.
>
> Nature is to be visible mind, the mind is to be invisible nature. *Here* then, in the absolute identity of the mind *within* us with nature *outside* of us, must lie the solution to the problem of how nature can be possible without us. The final goal of all our further research is therefore this notion of nature.
>
> Nature's system is simultaneously the system of our mind.[13]

Philosophy of nature's task, then, is to portray nature's unfolding as the development of the mind: "*All* of nature, not merely a *portion* of it, is an ever-*evolving* product. Nature in its entirety is constantly being created and everything is involved in this process of creation."[14]

In its time, philosophy of nature appeared to be a serious scientific alternative to the widespread Newtonianism propagated by eighteenth-century French philosophers. That doctrine took the universe for a clockwork, ticking away mechanically. Life and consciousness are of minimal significance. In contrast, philosophy of nature exchanged "soulless mechanism" for an "animated organism," trying to understand and explain the phenomena and the progress of life and consciousness within the context of the organic world.

An early, radical version of this kind of philosophy of nature was most consistently staked out and elaborated by Lorenz Oken (1779–1851). He lacked all restraint in exploiting the famous and infamous vernacular of the philosophy of nature. Although he had no real theory of organic evolution to offer, such as one that would allow for the transmutation of one species into another, he did take his theory of development one step further than most of his coevals and taught that higher order organisms spontaneously originated from organic "primeval slime." Schelling, in contrast, denied all theory of organic evolution and Hegel thought of development as a conceptual change, but not as the metamorphism of nature.[15]

It helps to sketch the basic ideas of Oken's theory. (Further evidence of Oken's influence on Fechner is presented in 7.2). Oken's *Textbook* [*Lehrbuch*], that immensely impressed Fechner, begins as follows:

> *The philosophy of nature is the science of God's own eternal transformation within the world.*
>
> It must show the stages of development of the world from its beginning in primeval nothingness; it must show how the heavenly bodies and elements originated, how these rose to a higher level and eventually became organic and developed into reason in mankind.[16]

This clearly characterizes the objective of philosophy of nature. Its purpose is to expound the development of the universe, starting with God's original ideas and leading up to its highest form, the human being.

Oken's philosophy of nature includes three parts: The Mathesis, a doctrine of the whole, deals with God and his activity. The Ontology is a doctrine of particulars, dealing with individual appearances in the world, or, the individualization of matter. The Pneumatology (later called Biology) is a doctrine of how the whole is also part of individuals. It deals with the continued effects of divine activity in individual things, in other words, with what is organic.

The whole has a real, material side (ether) and an ideal, immaterial side (God). The ideal side takes the form of pure oneness, the material side is diverse. The idea and reality, however, are identical and differ merely in their form.[17] Reality is created when God juxtaposes himself:

> By postulating itself, the real, or diversity, or the world, comes into being. The creation of the world is nothing other than an act of self-consciousness, God himself appearing.
>
> What we find as thoughts in our consciousness are the individual appearances of the world in God's consciousness. The things of the world are no more real for God than our thoughts are for our own minds. We carry a world within ourselves and create one each time we think or postulate ourselves; in the same way, God created by becoming self-conscious and he continues to create for eternity because he is continually becoming self-conscious; he is eternal self-consciousness, and nothing else.[18]

Oken makes use of the process of apprehending oneself (of becoming self-conscious) as a paradigm for all processes of nature. One could say that this turns a theory of self-consciousness into a theory regarding the whole world. Three "ideas": the postulator, that which is postulated, and self-consciousness become the source of all of the world's diversity. God's activity when he postulates himself is what Oken calls primeval activity, or the entelechy of God. One result of God postulating himself is the creation of polarity. And polarity is the cause of movement in the world:

> All motion originates in duplicity, thus from the idea, and in a dynamic, not in a mechanic manner. The idea of a mechanical movement set off by other merely mechanical movements, or impulses, and continuing on through eternity is nonsense.
>
> Nowhere do there exist purely mechanical movements, nothing in the world has become what it is through impulse; at the source of all motion is an inner act, an entelechy.[19]

"Polarity" became Oken's magic word for describing individual phenomena of the empirical world. Polarities occur in nature's universal substrate, namely ether. They set it in motion, allowing chemical elements to develop. New polarities arise and encourage the development of ores, salts, earth(s), and heavenly bodies, progressing further to galvanism, and further yet to the organic, plant, and animal kingdom. Extensive studies in embryology led Oken to a recapitulation theory stating that the development of the embryo repeats the pattern of developmental history of life on earth.[20] Further, his doctrine of "primeval slime" with its infusion-like "bubbles" representing the smallest units of life of which organic life is comprised paved the way for the theory of protoplasm and for cell theory by Schwann and Schleiden.

Oken's system is an example of how idealism in the philosophy of nature turns into stalwart materialistic monism. The distinction between pantheism and pure materialism gets hazy. At bottom, Oken's philosophy of nature deals exclusively with matter. God is no longer the idea of the universe, but rather straightforwardly identical with matter, namely with ether: "Ether is the first instance of God becoming real and simultaneously his eternal position. God and ether are identical. Ether is the foundational matter of creation, everything arises from it, it is the ultimate divine element, the divine body, ousia, or substance."[21] Reacting to this, Johann Eduard Erdmann notes in 1853 that "philosophy's tendency to become heathen"—latent in Kantianism and surfacing in the early works of Schelling—established itself permanently through Oken's work. He comments dryly: "For Oken the word 'God' merely means the universe and when he uses the word 'spirit' he does so using the widest possible meaning of the word so that if necessary, we could even subsume mint liqueur and such under it. Here a system of identity is transformed into simple philosophy of nature; what is ideal does not correlate to, but is merely a continuation of what is real."[22]

This "heathen" aspect of the philosophy of nature had a profound effect on mid-nineteenth-century philosophical materialism, particularly influencing Moleschott, David Friedrich Strauss, and probably also Feuerbach and Engels. This fact is often overlooked. In 1856, for example, the philosopher Gotthilf Heinrich Schubert noted (in his memoirs): "Moleschott, Vogt, and others wholly endorsed the same thing openly stated or latently inherent in Oken's philosophy of nature."[23]

The momentum that Darwinism gave German scholarly thinking in general cannot be understood without explaining Oken's impact. Darwin's theory was welcomed as unexpected empirical evidence and a logical furtherance of the notion of progress given by philosophy of nature, in all its materialistic varieties. Wilhelm Lütgert claims that German materialism did not stem from French materialism, but was instead an immediate and direct sequel to idealistic philosophy of nature, united by Oken.[24] Even Ernst Haeckel's theory of evolution—as Carl Guettler emphasizes—concurred with Oken's.[25]

The coming chapters will show just how Oken's "heathendom" survived in Fechner's philosophy. In any case, Fechner was so impressed and captivated by the philosophy of nature that he believed he had finally found the right profession. He began preparing himself for a career in the philosophy of nature and achieved a master's degree in 1823 (comparable to today's doctoral degree). Within a year he completed a habilitation thesis on *Praemissae ad theoriam organismi generalem*, was given permission to teach at the university (*venia legendi*), and planned to give lectures on "Schelling's and Oken's ideas in philosophy of nature."[26]

His *Praemissae* deal abstractly with the nature of ideas, with qualitatively and quantitatively defining objects in the world, with the concepts of singleness and plurality, and with how units are made up of their members.[27] He tried to sketch a typology for the constitution of parts and how they must fit together if they are to make up a qualitative or quantitative unit. He hoped that these reflections would help him find general metaphysical categories relevant for all objects of nature and enable a general theory of organisms.

Guided by his insights in the philosophy of nature Fechner felt that he was on his way "towards discovering the secrets of the world and creation, and in tune with the philosophy of nature popular among scientists of the time, namely Schelling's and Oken's philosophies, [he could] lay a foundation for the entirety of human knowledge." But this conviction did not last for long. Soon after his habilitation he felt serious doubts about the Schelling-Oken philosophy. Working in philosophy of nature gradually turned into a nightmare and "a struggle I had always contained within myself that denied me satisfaction in my endeavors. I believed myself to be headed in the right direction, but never reached a sure goal. I racked my brains from dawn to dusk and sometimes on into the night searching for solid ground, but I was never happy with what I accomplished."[28] Eventually Fechner mentally overworked himself to the point of exhaustion. He was obsessed by the problems pursued in the philosophy of nature.

1.3 Turning to Physics and Overcoming Philosophy of Nature

Fechner's frustration led to such anguish that he finally ceased working in the philosophy of nature altogether. The decision to do so was particularly supported by hearing physiology lectures given by Ernst Heinrich Weber. Fechner claims to have learned the "correct conception of science" for the first time from those lectures. They aroused in him first doubts about philosophy of nature.[29] Also, his financial standing compelled Fechner to seek out a new area of work. To prevent drowning economically he dabbled in "literature."[30] With a style both lively and vividly terse he wrote two semi-popular science booklets; one was an outline for logic as it was taught by Krug, the other a manual for physiology.[31] He also flung himself with vigor into the translation of French science books, which he began editing and rewriting on his own. The most important of these are his translations of *Précis élémentaire de physique expérimentale* by Jean-Baptiste Biot (1774–1862)[32] and *Traité de chimie élémentaire* by Louis-Jacques Thénard (1777–1857).[33] Fechner wrote additional volumes and extra chapters supplementing both works. Between 1822 and 1838 he produced between fifteen hundred and two thousand printed pages of text yearly as a source of income. After 1827 he added numerous scientific articles and books of his own.

Beginning in the winter of 1823–1824 Fechner held lectures in general and specific physiology for the faculty of medicine.[34] As an acknowledgment of his successful translation and congenial editing work on Biot, the second edition of which appeared as soon as 1828–1829, Fechner was awarded a temporary chair at the university of Leipzig following the death of the physicist Ludwig Wilhelm Gilbert (1769–1824) and continuing until it was taken over by Heinrich Wilhelm Brandes (1777–1834).[35] Originally there had been some intention to give him the chair permanently, but the idea was dismissed because Fechner was too young.[36] Translating Biot's work, however, helped Fechner because, as he wrote, it guided him "down a more exact path and I became aware that this was the only avenue to clear, certain and productive findings in science . . . I recall asking myself: could any of all those orderly and law-like arrangements of optical phenomena, that Biot expounds with such clarity, have been discovered with the Oken-Schelling approach? It is hardly a scientific method."[37]

Fechner applied himself zealously to his new task, experimenting thoroughly with what was known about electricity. After completing his stint for the chair position, he continued teaching unpaid lessons on electricity,[38] galvanism, electrical chemistry, magnetism, electro-magnetism, and very generally on "the latest progress in physics and chemistry."[39] During the winter of 1831–1832 and the following summer term he also gave lectures on meteorology. Having just fled from the philosophy of nature, he hurled himself wholeheartedly at "Cauchy's most difficult doctrines," prodded by the conviction that without mathematics nothing can be achieved in physics.[40]

At the time, this attitude towards physics was certainly not common in German scholarly circles. A distinction was usually made between "natural history" (*Naturlehre*), in which mathematics played a subordinate part, and applied mathematics (mechanics, geometry, geometric optics, hydrodynamics, astronomy . . .). Not until the later half of the nineteenth century were methods introduced and enhanced for using mathematics in physics[41]—mostly further developing the *Physique éxpérimentale* from the *École polytechnique* in Paris.

Thomas Kuhn has described this general process by which knowledge becomes more theoretical as a fusion between two separate scientific traditions, namely Baconian experimental science and classical mathematical science.[42] The more axiomatic deductive sciences practiced since antiquity (astronomy, optics, mechanics) were augmented in the seventeenth century by experimental, math-free areas of research (magnetism, theories of heat and electricity, chemistry) resulting from an increased appreciation of knowledge gathered by craftsmen, pharmacists, and alchemists. The dawn of the nineteenth century in France then witnessed the mathematization of Baconian science, a development that did not occur in England and Germany until later.

Fechner's translating accomplishments and his editing of the newest French publications on physics and chemistry such as those of Biot and Thénard were, next to Gilbert's *Annalen der Physik* [Annals for Physics], the main source for disseminating French scientific knowledge; they contributed decisively to the commencement of similar theorizing in Germany, thereby reforming physics. The importance of Fechner's endeavors for establishing a scientific community in physics in Germany should not be underestimated.[43]

Subsequently, Fechner's fresh interest in modern physics and chemistry effected his career choice. In 1827 he received a grant from the government of Saxony and visited Paris for three months, meeting Biot, Thénard, and Ampère.[44] In 1831 and 1832 he was a non-salaried university lecturer and finally in 1834 he became professor for physics, taking over the chair he had applied for following the death of physics professor and rector Heinrich Wilhelm Brandes.[45] In Winter 1834 he held his last lectures in experimental physics for the faculty of medicine.[46] In 1835 he executed his predecessor's plan and set up the first institute for physics in Germany in the newly built Augusteum.[47]

Fechner's research in physics was chiefly on the theory of electricity, electromagnetism, and electrical chemistry. His studies in transition resistance became fundamental for later research on the polarization effects of electric current. He was among the first to recognize the importance of Ohm's law and tested it extensively in experiments, publishing the results in 1831 under the title "Quantitative Determinations of the Galvanic Chain" [*Maaßbestimmungen über die galvanische Kette*].[48] Fechner's work contrasts sharply with other research of his time, especially because he uses a strict quantifying method and because of the great amount of data which he produced in experiments and then discussed and evaluated.

In 1827 Ohm discovered the law later to be named after him. As trivial as it may seem to us today, at the time this law induced a profound change of meaning for the concept of electricity; it did not become common knowledge until the 1840s.[49] Fechner revised the third volume of the textbook on Biot in 1828, already basing it on Ohm's law. He said that for the first time this law "made sense of the causal relations in the galvanic chain."[50]

At that time Fechner was not only active in physics research, but endeavored to keep abreast of advances made in chemistry as well. In 1830 he founded the journal *Pharmaceutisches Centralblatt* [Central Pharmaceutical News], which is still in publication today; he himself contributed articles to it throughout the first five years of its history.[51] In 1832 he published a *Repertorium der Experimental-Physik* [Repertory of Experimental Physics] that surpassed the revision of Biot's textbook in acquainting German physicists with research done in France by Fresnel, Ampère, Poisson, Laplace, Navier, Cauchy, and others.[52] We may say that this was the first journal for physics in Germany, a forerunner of the yearly journal *Fortschritte der Physik* [Progress in Physics], established in 1845 with Gustav Karsten as editor.[53]

Besides these more experimental studies in physics, Fechner also fostered his inclination for general theoretical contexts. At a very early stage he expressed the opinion that material bodies are nothing other than "a system of atoms or molecules" and that "all material appearances . . . can be traced back to attraction, according to the law of gravity."[54] Here Fechner anticipated Helmholtz's notion that later was to become very influential, an attempt undertaken in *Erhaltung der Kraft* [Conservation of Force], published in 1847, in which he tried to show "that all effects in nature can be reduced to forces of attraction and repulsion, the intensity of which depends on the distance between the reciprocally effective points."[55]

In order to demonstrate how physical phenomena could be reduced to reciprocal gravitational forces of atoms, Fechner devised a planetary model of atoms in which imponderables (electricity, heat), as small particles, orbit around a larger ponderable at the center. It was probably the first time in the history of physics that atoms were thought of in this way. (Chapter 4 discusses Fechner's atomism in more detail.)

Alongside of all this work, Fechner still found time to continue studying subjective optical phenomena, particularly contrasting colors.[56] In articles written in 1838 and 1840 he attempted to prove experimentally that the simultaneous contrast of collateral images depends on the subject; it is not part of the physical properties of light. He also found that the successive contrast accompanying after-images is due to retina fatigue under prolonged stimulation, and not due to an inner development of light, as Joseph Plateau had taught. Both of Fechner's theories influenced future research. Helmholtz, when writing about contrasting colors in his book *Handbuch der Physiologischen Optik* [Handbook of Physiological Optics] published in 1867, relied greatly on Fechner's findings.[57]

What became of Fechner's philosophy of nature after he turned to natural science and became an avant-garde physicist? No doubt, several motives rooted in his philosophy survived to be overtly and covertly incorporated into Fechner's later work, as we shall discover from case to case. What is important at this point is how Fechner himself saw his flight from philosophy of nature and his new attitude, and how both were evaluated by his coevals. In the work entitled *Tagesansicht* [Day View], published in 1879, Fechner once again remarked how as a student of medicine he had initially been fascinated by materialism, from which Schelling's philosophy distracted him for quite a while. He continued: "only to then be drawn all the deeper into it [into materialism]."[58] Fechner's "conversion" in 1820 was, then, contrary to claims frequently made in secondary literature, not enduring and ended quickly (at least according to his own *reflected* evaluation in retrospect). Deserting philosophy of nature was thus equivalent to returning to materialism. Proof that at the time Fechner's befriended peers shared his own judgement on this is given by a letter written by the painter Wilhelm von Kügelgen. He reports on a conversation between the two of them in July 1844 after playful boxing at the castle ruins of Falkenstein. His vivid and amusing portrait of Fechner reads: "Our talk was very interesting. He has a composed, sober mind . . . We didn't mention religious issues, but spoke instead exclusively about matters in philosophy of nature and I had to contend with his materialist bent. He seems to see all of life as merely mechanisms, therefore he sees not much difference between a locomotive and a lion or horse, except that the latter produce offspring. He is the first person of this sort that I have ever met and I cannot deny finding his toughness quite Saxon."[59] In 1849 he also wrote that Fechner's sentiments "tend mightily toward the coarsest sort of materialism."[60]

In spite of returning to materialism, Fechner hardly considered his time spent pondering philosophy of nature to have been wasted. His benefit from that phase was that he learned to find and formulate comprehensive principles and general contexts in nature: "Yet what remained important to me from that period is the idea of uniformly culminating and mentally penetrating nature, an idea that I expressed in later writings, although the Schelling-Oken notion no longer held authority for me."[61] In the following sections we shall see what Fechner particularly meant by these vaguely formulated concepts of unity and where he thought a guarantee for the unity of nature can be found.

It is interesting to note once again that Fechner composed satirical and poetic texts, writing under the pseudonym Dr. Mises, as a means of easing the personal tension created by the relation of natural science to philosophy of nature, or that of materialism to religion. As early as 1821 in the aforementioned satire (*Proof that the Moon is Made of Iodine*) he pokes fun at Oken's *Textbook for Philosophy of Nature*; and in *A Comparative Anatomy of Angels* [*Vergleichenden Anatomie der Engel*]—a capriccio swinging between satire and seriousness—he cleverly imitates the thought patterns and persuasion habits of philosophers of nature. He later said that parts of this last treatise, which was once dubbed "the best scholarly satire" written in the German language,[62] "playfully suggested" the idea that plants have souls.[63]

1.4 Herbart's Psychology

While writing on Fechner's *Theory of Atoms* one of Herbart's devotees, Moritz W. Drobisch, noted in 1856 that he had first heard of Herbart from Fechner in 1824.[64] Obviously Fechner had early acquaintance with Herbart's theories. He never acknowledged Herbart's psychology and philosophy though; on the contrary, he exploited every opportunity to demonstrate how his own position deviated from Herbart's. Still we can assume that the conflict growing within him while reading Herbart's works encouraged him to articulate his own views. The difference between their views was not insurmountable, however. It did not prevent Fechner from adopting some of Herbart's particular concepts and approaches.

In his time, the philosopher Johann Friedrich Herbart (1776–1841) was considered the first scholar to not merely have spoken of the necessity of psychological laws, but to also have attempted to lay out and precisely define them.[65] He was also considered the one who deemed faculty psychology obsolete. Instead of tracing psychological phenomena back to any natural predispositions, to a "mere aggregate of the soul's faculties," Herbart sought to describe a model of mental life in which various phenomena could be explained uniformly by "psychological laws linked by necessity." His procedure for psychology must have effected his readership similarly to Descartes's rejecting *qualitates occultae* and replacing them with mechanic explanations in early modern physics.

Thus in *Psychology as a Science* Herbart sets out to "research the soul in the same way that we research nature; inasmuch as this implies presupposing an ubiquitous, completely regular relationship of phenomena, and investigating this by surveying facts, making careful deductions, by using novel, tested and corrected hypotheses, and finally, whenever possible, by contemplating values and calculation."[66]

The laws of psychology are to be found in the "facts of consciousness." But since self-observation and observation by others always remain unsatisfactory, psychology must try to discover the connections of inner perceptions via something that is not directly given in perception. "Psychology on the whole can be nothing but an addendum to inwardly perceived facts."[67]

The most important annex that is not part of the facts—but that Herbart believed it was necessary to presuppose—is the existence of a substance (a substantial being) whose states we are aware of as the facts of consciousness. Here Herbart was guided by his desire to define the nature of ego or the self. Whenever I want to perceive myself, I am aware of myself as thinking, acting, being passive, or feeling. I am never aware of myself as merely "I," but always only as something thought of. Since thoughts always require something that thinks them, there must be some immutable, simple and unified bearer of these attributes that are given to us as thoughts. Herbart calls this substantial bearer of mental phenomena the "soul" and defines it as a monad, or a "simple real being" (they have come to be called "reals").

> It has been proven in detail that observing one's own self inevitably involves contradictions, inasmuch as we want to understand ourselves directly using the concept of ego, as if being an ego were the ground for our whole being. Being an ego must depend on something. And the carrier that supports it is, as always, a substance. We call it in particular a *soul*; because according to general principles of metaphysics, a substance is foremost incapable of modification in order to protect itself against being disturbed *by other beings* . . . and because in our case self-protection takes the form of *thought*, in such a constellation and context that it results in consciousness of the self, or being an ego."[68]

This amounts to a compact basic model of Herbart's psychology. All mental phenomena—in fact, all appearances whatsoever—are the results of the modification of one being by another. Unlike his teacher Johann Gottlieb Fichte (1762–1814), who defined "self" as a self-positing activity, Herbart suspends the idea that consciousness autonomously posits thoughts. Thoughts, rather, are self-sustaining elements of the soul whose function is to prevent chaos: "Given the impulse by the objects, and guided by them, we come back to ourselves, because *without* objects self-consciousness would be nonsense. It is not a matter of freedom at all."[69] Since the soul is a simple entity, whatever disturbs it cannot merely coexist barring all connection. Opposing thoughts hinder one another, similar thoughts tend to coalesce. Each thought must have a certain intensity in order to survive in consciousness. An obstructed thought is obscure, an unhindered thought is lucid.

> Each of us himself notices that at any given moment an incomparably smaller amount of knowledge, thoughts, and desires actually occupies our mind, than could be available if prompted. This absent, albeit not distant knowledge, this persistent knowledge that we have—in what state is it within us? . . . *Two* thoughts are sufficient to entirely expel a *third* from consciousness, and to produce an entirely different state of mind. . . . Just as we are accustomed to speaking of thoughts entering consciousness, I call the boundary that a thought seems to cross when it changes from being an entirely restrained state to the state of manifesting a degree of real thought: *the threshold of consciousness.*[70]

Beginning with the simple mechanisms of obstructing and connecting thoughts, which is the soul's tool for preserving its integrity, Herbart splits his investigations into two parts. The first is mathematical. Here he explains the laws governing the connection and repulsion of thoughts (he calls it the "statics and mechanics of the mind") as a function of individual thought intensity and the degree to which thoughts differ. The second part is applied psychology. Here he explains various mental phenomena using the mathematical laws. While psychology presupposes the monadic soul, the soul cannot be its own object of study, because its true essence remains forever hidden. Granted, Herbart's mathematical excursus is difficult to follow: In order to determine the equilibrium or movement of thoughts, he calculates the sum of obstacles. Nonetheless, his explanations are replete with interesting studies, for example on the development of self-consciousness in children. A child observes how a dog evades being hit with a stick and thus learns that the animal understands some idea of pain and sticks. "The child now forms a thought; this is important . . . progress and indispensable preparation for developing self-consciousness."[71] The concept of the self or "I" is evoked by the conflation of two sets of ideas.[72] Notions of space also result from the conflux of ideas.

Fechner's main objection to Herbart's philosophy is one of method. Like every other science, psychology must set out using what is given, making no metaphysical assumptions. What is given must be acknowledged as real and not considered a mere appearance relying on something that exists independently; something that alone is real, beyond what is given. If it makes any sense at all to ask what exists non-relatively, the question must apply to all of reality; what is given is not relative to individual, non-given substances, but to the enduring greater entirety of all reality. Speaking to a follower of Herbart, Fechner said, "To me the final or absolute reason for everything is not an X to be found beyond what is given, as are your simple qualities, but rather it is the whole, it includes everything given anywhere and anytime, and the laws and relationships that can be found to apply."[73]

If one is inclined to barge into metaphysics, one must first use "unpretentious empiricism" and once science has matured, one may then engage in conjecture about what is absolute, or about the entirety of reality, which is compatible with the given. But by beginning with metaphysics, as Herbart does, we "estrange science from life."[74]

Fechner's second objection follows from that. Herbart's notion of being, or substance, is a general metaphysical concept lacking validity for science, unless it applies to everything. Science gets along fine without any notion of substance." Since we are comfortable with what is relative, the concept of being also serves as a relative concept."[75]

Fechner derives his third strategy for criticizing Herbart from his training in physics. He confronts Herbart's use of concepts with comparable concepts in physics, while simultaneously reflecting on how to realize Herbart's hypotheses in terms of physics. The mental events that Herbart contrived must also always be physically realized. This insight changed Fechner's appreciation for mathematical psychology. To measure means to use a materially given standard. Herbart's psychology is unsuccessful because it lacks this connection to the physical world. New mathematical psychology must first produce a spatio-temporally valid standard. In *Zend-Avesta* he writes: "For reasons too lengthy to discuss here, Herbart's principles of mathematical psychology are wrong. If mathematical psychology is possible at all—and I believe it is—it must be founded on measuring and calculating the material phenomena to which mental phenomena are attached, because these allow precise measurement."[76]

Although Fechner found Herbart's mathematical psychology erroneous, he retained the latter's notion of dealing with the mental mathematically. His criticism on Herbart's theory did not prevent him from adopting two extremely important concepts for psychophysics, the concept of threshold and the notion of psychical intensity.

Herbart remained a role model for Fechner and his psychology counted many followers until at least the 1870s. Fechner often encountered Herbart's theory because Leipzig was the center of the Herbart School. The scholars there included professor Moritz Wilhelm Drobisch (1802–1896), the editor of Herbart's works Gustav Hartenstein (1808–1890), and Ludwig Strümpell (1812–1899).[77] Many scientists, for whom the years between 1820–1850 were formative, had come to terms with Herbart's theory in one way or another. These included, for example, Johannes Müller and even Helmholtz. For philosophy and psychology it is important to note that while Hermann Cohen and Wilhelm Wundt later became his heftiest critics, they originally worked on theories from Herbart.

Herbart's approach was obsolete by 1879, at the latest. His contemporary Benno Erdmann characterized Herbart's psychology as "transitory to contemporary precise psychology." Herbart, the Fichte-Ulrici group and some other congenial scholars continued to ground their sort of psychology metaphysically, preferring "exclusively methods of introspection." In contrast, new or experimental psychology is depicted as liberating psychological issues from the reign of metaphysics. Psychologists from the Herbart School, however, had the advantage over Fichte and Ulrici that at least "precise psychology . . . profited from their work."[78] Nowadays Herbart's psychology in the narrower sense is all but forgotten. But his pedagogical theories are still quite popular and these, in turn, include several elements of his psychological theory.[79]

1.5 The Aesthetic and Romantic View of Nature

Besides being engrossed in the philosophy of nature and the analysis of Herbart's theories, another epochal factor shaping Fechner's life and work was the instillment of a general life attitude permeated by an aesthetic and romantic view of nature. This gradual change was inspired by Martin Gottlieb Schulze, a friend two years his senior. Fechner wrote that the "acquaintance and long close association with a kind of depraved genius throughout the 1820s" was one of the events that "altered my mental life" and "remained a sustained influence."[80]

Around 1823 Fechner was introduced to Schulze by Christian Hermann Weisse (whom we shall meet again later). Schulze, too, was the son of a pastor and a student of medicine who had interrupted his studies after his theoretical examination and led the life of a restless Bohemian and penniless author, full of ambitious plans that failed to reach fruition. He abhorred what he called sterile science and tradition; he reveled in nature and poetry and had a fascinating, demonic allure for the young people around him, including Fechner. It was perhaps in an attempt to disengage Fechner from Schulze that Fechner's mother moved from Dresden to Leipzig in 1824, taking charge of her son's household. Fechner had more or less regular contact with Schulze until at least 1846. Schulze himself eventually ended up in a lunatic asylum.[81]

Inspired by Schulze, Fechner absorbed the underlying impulse of the romantic worldview, which Wetzel has aptly described as desiring "the union of man and nature, the equality of knowledge and imagination, science and poetry, critical and ironical reflection and intuitive immediacy of feeling."[82] According to this romantic attitude, nature is whole. Any adequate description of nature, therefore, must also encompass the emotional, aesthetic, and ethical meaning that it has for us and consider these elements just as important as theoretical content. Man himself and all that defines man is part of nature; to find his way, man must be aware of his far-reaching and full roots in nature. Humanities and natural science are one—or at least they meet at an ideal vanishing point.

An early expression of this view of nature can be found in Johann Wolfgang Goethe's famous reaction to Baron Paul Thiry d'Holbach's *Système de la nature* published in 1770, which at that time was considered a paradigm for materialism and determinism in both moral and cosmological terms. Goethe did not worry that Holbach's theory is wrong or absurd, but that it is "gray," "dark," "lifeless," "quintessentially senile," "unpalatable, indeed tasteless."

Goethe writes, "How hollow and empty we felt in this tristful atheistic twilight, where earth and all its creatures and heaven with all its stars disappeared."[83] For Goethe a system like that could never do justice to the vital, vivid character of life and the splendor of the "great embellished world" and therefore is not to be taken seriously.

Even Alexander von Humboldt's writings contain a trace of this attitude toward nature, combined with a religious twist. For Humboldt, both natural science and poetry on nature intend to portray the world as a whole: "Nature is the living expression of God's omnipresence in the works of the sensual world."[84] Clearly, art is much better equipped to portray nature than is science. A serious believer in the aesthetic-romantic attitude would feel compelled to become an artist or poet, rather than a scientist.

Evidence that Fechner saw himself in this predicament is expressed in a letter he sent as a young man to the poet Jean Paul, whom he revered, accompanied by a packet of his attempts at literature (probably the *Stapelia mixta*, appearing

in thought" (Löwith) and on this point was of one opinion with Kierkegaard and Feuerbach's critique of Hegel and the many types of materialism that originated therein.[97]

On the other hand late idealism strives to preserve what it considers the accomplishment and intention of German idealism by eliminating the shortcomings it saw in its early phase. Within the highly diverse spectrum of post-Hegel criticism of idealism, late idealism attained its specific identity through 1) the challenge of freedom and personality that arose out of the critique and 2) the method, disdainful of abstract categories, that was applied to solve the problems.

Weisse and I. H. Fichte considered the fundamental challenge of post-Hegel philosophy to be how to defend, redefine, and understand the real and finite in comparison to what is merely an idea, and individual freedom as opposed to what is general and absolute. Whatever is tangibly individual is more than a mere abstract moment in the course of the world's development, more than a mere "mask of a universal spirit,"[98] as Hegel had taught. In order to correct Hegel's conceptual errors on these matters, it would take a new concept of personality, for both God and the finite individual.[99]

So late idealists want to overcome Hegel's pantheism by fusing it with speculative theism founded on a personal God. God, or The Absolute, is not to be thought of as an impersonal spirit, but as a personal and free, absolute "I." This fusion is justified by an ontological proof of God's existence conjoined with the cosmological. Within himself, God unites the idea of the Truth as the logical form of the conceivable process of the world with the idea of Beauty as the principle of vital individuation, meaning that it is the principle of all purposes that free individual persons autonomously devise in their pursuit of happiness. "The true concept of divine personhood results from what is particular penetrating what is spiritually absolutely universal."[100]

Since idealism as it was known had neglected the topics of freedom and what is special about individual reality, within his philosophy Weisse attributes fundamental significance to aesthetics, "the science of the idea of beauty." Of all his writings, his coevals valued his *System of Aesthetics* published in 1830 the greatest.[101]

All that is real originates in acts of freedom, originates in "voluntary action in general," either acts of God or acts of his creation. These acts are voluntary because they totally lack subjection to necessity. In creating the world, God willingly limits his own power and thereby continuously establishes the spontaneity and freedom of the beings he created. Spontaneity and freedom of action occur within time and thus make God a historical being, unfolding himself over time. These two traits also restrict his capacity to entirely foresee the future.

In the process of reconstructing and reordering speculative thinking the need became obvious to insert several steps in order to mediate and make relative the transition from thinking to being, a transition that Hegel makes without much ado, and to rethink the conceivability of objective knowledge. Beginning around 1831, I. H. Fichte and Weisse call this endeavor "Theory of Knowledge" [*Erkenntnistheorie*]. It was intended to move forward in the direction Kant had indicated and investigate afresh the conceivability of speculative knowledge.[102] In 1847 Weisse published an oratory discussing *How contemporary German philosophy should follow Kant's lead.*

Weisse's and Fichte's efforts eventually bring forth a peculiar synthesis of speculative idealism and Christian theology, simultaneously emphasizing contingent, empirical experience in nature and history. In this spirit Fichte founded the journal *Zeitschrift für Philosophie und spekulative Theologie* [Journal for Philosophy and Speculative Theology] in 1837, which also served as an organ of opinion for Anti-Hegelians.[103] This publication was meant to promote speculation, while incorporating what is real in nature and history. It encouraged appreciation for empiristic thinking, which at the time had a poor reputation in academic philosophy.[104]

The objective of philosophy was no longer viewed as merely understanding the universal form of the logical idea in terms of abstract and dialectical categories, but to describe "the idea's unique way of being," as it actually appears in the real, contingent world, and to further develop it into a philosophy of nature and anthropology. The periodical mentioned above (and renamed in 1847 *Zeitschrift für Philosophie und philosophische Kritik* [Journal for Philosophy and Philosophical Critique]) took this task so seriously that until it was discontinued in 1918 it was among the nineteenth century's most important organs for what we today call "philosophy of science." Alongside Fechner, Ernst Mach and Moritz Schlick also published here; even Hans Reichenbach's dissertation was reprinted therein.[105]

Weisse's philosophy influenced not only Fechner, but also the physician and philosopher Rudolf Hermann Lotze (1817–1881), who likewise lived in Leipzig.[106] Until 1844, when Lotze became Herbart's successor at the university in Göttingen, Weisse, Lotze, and Fechner comprised the heart of a "little circle" that met once weekly.[107] Late idealism, however, effected Lotze differently than it did Fechner. The main difference relates to their views on the mental. While Lotze saw the unity of consciousness, similar to Herbart's psychology, as proof of a uniform substance of the soul, Fechner assumed a synechiology, in which the mental is an emergent property of a system. Granted, both wanted to reconcile the concept of teleology with natural science's concept of causality. But to achieve this Lotze succumbed to the existing mechanistic worldview, while Fechner— without becoming unscientific—was able to productively reinterpret the prevalent philosophy of science. Over the years Lotze critically analyzed Fechner's writings repeatedly.[108]

But let's get back to Fechner. How is his philosophy related to Weisse's philosophy? Basically both agree on what to demand of philosophy: namely, that it rekindle inquiry into how being and thinking are related and employ new methods in searching for ways and restrictions for a solution—methods resulting from logical, empirical thought.

As much as they may have agreed on objectives, their way of thinking was hardly similar: In spite of all his criticism on Hegel's philosophy, Weisse did conform to the speculative, Hegelian way of thinking, while Fechner vehemently rejects even the smallest residue of speculative method and favors solely those methods allowed in natural science. Kuntze writes: "It is true that on the question of *what* we know, i. e. what the thinking mind should learn, their opinions [Fechner and Weisse's] did not diverge, for Fechner also sought (not merely what is particular, but also) what is universal and whole—what is universal in the world, the world's wholeness, but on the philosopher's second question of *how* we recognize the universal and the whole and try to conceptually grasp the universe, they had basically and characteristically different views."[109]

Nonetheless, an entry in Fechner's diary testifies to their deep friendship, despite all their differences: "Although I share none of his [Weisse's] opinions on science and art, he is my most valuable friend here and with him I enjoy the most intellectual discussions."[110]

The exchange between Fechner and Weisse must have been marked mainly by debate over the aforementioned issue of the acceptable means for establishing knowledge. Kuntze reports that their conversations were vigorous: "Whenever they met, their minds collided and explosions took place . . . It was a contest, where no one was spared, arms clashed intensely, and a knight had to sit firm in his saddle or else roll in the dust."[111] These debates apparently reflected the whole range of conflict between German idealism and modern natural science. Yet the rigorous arguments seem not to have harmed, but to have invigorated their friendship.

Of all the philosophical ideas that Fechner, in spite of his wariness, adopted from Weisse (or conceived jointly with him), the most important was the idea of spontaneity in nature. Fechner's ideas on non-determinism, his view of the role of mathematics in science, and his theoretical statistics depend directly on the particular philosophy of freedom and individuality cultivated in late idealism. We shall return to this context in detail in chapter 8.

All of his life, Fechner was critical of academic philosophy. He often made sarcastic and ironic remarks about Schelling, Herbart, and Hegel, particularly regarding how they constructed concepts a priori and their rationales. Hegel's philosophy, he wrote, is "in a certain sense the art of how to unlearn correct inference."[112]

THIS look at late idealism completes our overview of the most significant factors at work in shaping Fechner's scientific personality. Reviewing all the influences, we notice that on the one hand Fechner's development is typical for many physicists and natural scientists of his day. Like so many scientists, he was born into a protestant pastor's family, he matured early, was intelligent, and educated himself more through his own studies and experiments than in lectures at the university. Like many others of his generation, he was more receptive for physics as they were taught in France than for the ideals of domestic "*Naturlehre*" (natural history). Quite typically, he observed the necessity for enhancing the role of mathematics in science and demanded it, although his own insufficient knowledge of mathematics made it difficult for him to fulfill that ideal. With other men of his times, such as Johannes Müller, Justus Liebig, Hermann Helmholtz and Matthias Schleiden, he shared an interest in combining physics with state of the art philosophy and toyed with philosophy of nature.

On the other hand Fechner differs uniquely from his contemporaries in science: First, he has a literary and artistic bent, unusually strong religious struggles and an immense passion for work. And while others gradually settled for either philosophy or science, religion or materialism, a literary existence or the life of a professor, and romanticism or realism, Fechner remained undecided for a long time, seeking general ways of thinking that would enable all these contradictory facets to be united. After eventually deciding upon a career as a physicist, he wasn't really content with that choice. His illness, which shall be discussed in section 1.8, quasi-enabled him to be an outsider who combined contradictory realms in concepts. While it is true that indecision and inner strife generally are characteristic of the German literary epoch called Biedermeier, it was not representative of scientists of that time. Fechner also embodies independent and courageous thinking, remaining uncompromising, even if it cost him his career or fame. Wilhelm Wundt's opinion is typical and touches upon a theme varied by many who knew Fechner personally: "I know no other general expression for this trait than to say that he *absolutely lacks intellectual prejudice and is fearless in his own convictions.* I cannot recall ever having observed this quality developed to a similar degree in any other person, a character for which I count it one of the most unforgettable rewards of my life to have met him."[113]

1.7 Beginning Philosophical Work

We interrupted the story of outer circumstances in Fechner's life just when he had been granted a professorship for physics. In 1833 he married Clara Volkmann (1809–1900), whom he had met through her brother Alfred Wilhelm Volkmann (1801–1877), one of his university friends who later became a physiologist and anatomist. Clara Volkmann came from a respected bourgeois family. Her father Johannes Wilhelm Volkmann (1772–1856) was the City of Leipzig's senator for architecture.[114] Fechner and Clara's marriage has been described as happy; they remained childless.

Prior to inauguration as professor, Fechner had taken on the job of editing a *Home Encyclopedia* (a guide for organizing the household) totaling eight volumes published between 1830–1834 by Breitkopf and Härtel containing eight hundred to nine hundred pages per volume.[115] This task cost Fechner so much time and effort that for a while he felt unable to accept the chair in physics. The university and the government agreed to allow him to postpone taking office until December 1835.[116] He wrote nearly a third of the encyclopedia contents himself, including such prosaic entries as "Carving Meat and Setting the Table."[117]

Fechner embarked upon this project as a means of achieving financial success and independence, at least temporarily. The idea was realistic, since all sorts of conversational dictionaries established at that time by Brockhaus, Meyer, and Pierer sold with enormous success. And Fechner's encyclopedia does seem to have been profitable to a certain extent. It was printed twice without any changes.

But the third edition, revised by one Dr. Hirzel, failed utterly and brought a huge loss for the publishing house Breitkopf and Härtel, which belonged to a friend, the lawyer and art patron Dr. Hermann Härtel (1803–1875).[118]

Besides working on the encyclopedia, Fechner also had work to complete for the yearly volumes of the pharmaceutical news mentioned above, another of the obligations he had entered before attaining professorship. In spite of this burden, he found time for thorough experimental investigations into the theory of electricity and physiological optics. His experiments in the theory of electricity dealt mainly with the ongoing debate over whether the process of generating galvanic current should be explained physically in terms of contact or by chemistry.[119] As for publications, Fechner was one of the most prolific German physicists of his time.[120]

In 1835 Fechner wrote a book differing from all his other writings and which in many respects lay the seed for his later philosophical development. He published it as the *Little Book on Life after Death*, using the name Dr. Mises, a pseudonym he had previously used for satires.[121] At that time the topic of immortality was of general interest, a subject of heated debate.[122] Philosophers educated in Hegel's philosophy were the main participants in these debates; among them the late idealists. As they had taught that man and that being in general is unique and individual, it was up to them to also inquire whether individuality is limited temporally or whether the self is enduring and indestructible. If it were transitory, this would mean that in the end Hegel is right and history is nothing but the brain child of an impersonal absolute spirit, indifferent to the individual and personal freedom.

Thus the late idealists were called upon to explain how the individual's continued existence after death could be deduced from the notion of the absolute as a person. In 1834 I. H. Fichte wrote on *The Idea of the Person and Individual Duration* (second edition appearing in 1855), followed by *Continuance of the Soul and Man's Place in the World* in 1867. Weisse also wrote *The Secret Philosophical Doctrine of the Human Individual's Immortality*, published in 1834. But not only Hegelians and former Hegelians wrote on the topic. Bernard Bolzano (1782–1848), for example, had written as early as 1827 *Athanasia, or Reasons for Immortality*, in which he carefully critically analyzed various attempts of his contemporaries to deal with the notion of immortality.

Fechner's treatment of this theme can be viewed as an effort to show that the late idealist notion of personal continuation is compatible with a naturalistic concept of man. As he relates in the preface, Fechner got the main idea for the little book during a discussion with a theologian named Billroth, one of Weisse's followers. Billroth thought that spirits of the deceased continue to exist in the living.

According to Fechner, humans pass through three phases of life: a prenatal phase, life on earth, and life hereafter. During the transition from the second to the third phase humans are destroyed physically, but consequences of one's previous expressions of life continue to exist. "Each cause retains its consequences as an eternal possession."[123] So whichever material system comprises all the causal effects of prior conditions, that material system also carries on the individual spiritual property belonging to a person during his life on earth: "Whatever permits the body of an old man to continue the same consciousness that was in his body as a child, of which he has not one atom anymore, will also let the body in the hereafter continue the consciousness of the old man, of which he has not one additional atom." Here Fechner views the property of having consciousness as a functional property of systems. One and the same consciousness can be realized by entirely different systems. In death man awakes in a "self-made organism, a unity of innumerous mental creations, effects, and moments; it can be larger or smaller and more or less have the strength to continue developing, depending on the extent and degree with which the mind of the person grasped his world during his life."[124] This new "organism" no longer requires sense organs for receiving information from the outside world. External light and sound waves that the organism previously used to determine the nature of its surroundings now themselves make up a part of his new organism, insofar as they are connected to traces that the physical organism left behind during his life. The deceased, therefore, has the same relationship to the processes of nature that the living person had to the functions of his body.

The organism that supports consciousness after death is connected in various ways to the organisms of the living and the deceased. The new physical bearers of consciousness are so interwoven that it is no longer possible to identify the particular carrier of one individual consciousness. One and the same physical system can support several consciousnesses (or at least parts of them). Casting several stones into a pond simultaneously, after a while we can no longer tell which stones caused which ripples, although the distinction undoubtedly could be made.

Yet even a living body supports consciousness that is influenced by other consciousness. Our mental life consist of more than what it itself creates, it includes the effects of other consciousness, as can be observed most clearly when in abnormal states (such as somnambulism or mental illness) the real reciprocal relation of interdependence between those states is disturbed in favor of another consciousness.

The inner schism so often found in a person is nothing but the struggle of foreign minds seeking to conquer that person's will, reason, or concisely, his entire internal nature, for themselves. Just as a person perceives the agreement of other minds within him as peace, clarity, harmony and security, he perceives their struggle as discord, doubt, faltering, confusion and dissension within himself. . . . A person's self therefore remains out of danger amidst all this struggle, as long as he preserves his innate freedom and strength and never tires of exercising them.[125]

This resembles Sigmund Freud's constellation of the ego, superego, and id.

Fechner's notion of survival after death is entirely compatible with the materialism he endorsed at the time. He mentions no tenet for the ontological status of consciousness or its relationship to the body. He simply commences from the fact that consciousness exists in the world and that our experience associates it with certain physical systems, namely (living) bodies. It is important to realize that when Fechner uses the term "hereafter" he is not speaking of a transcendental, empirically inaccessible realm, but instead, he means that part of *this* world that functions as a new bearer of the former consciousness.

> This reflects the immense justice of creation, namely, that each person himself creates the conditions for his future being. One's actions are not requited by reward or punishment; there is neither heaven nor hell in the normal Christian, Jewish, and Heathen sense of the word, where a soul goes after death; the soul neither ascends nor descends, nor does it remain idle; it neither bursts nor does it flow into the universal; instead, after surviving the transitional illness called death, it continues to grow calmly according to the permanent logical consistency of nature on earth that erects each phase on the foundation of an earlier phase, and leads to a higher form of being.[126]

In this booklet Fechner anticipates a number of ideas that come to fruition in his later writings and which enable him to eventually advocate animation of the whole universe. It is interesting to note that two years after Fechner published this piece, Weisse, using the pseudonym Nicodemus, also wrote a "booklet"— partly to refute and partly to amend Fechner's booklet—calling it the *Little Book on Resurrection*. He opined therein that life after death is to be understood as purely mental. The only souls granted life after death are those possessing original minds. Mindless persons "melt into nothing and burst like empty bubbles."[127]

Today, Fechner's ideas may seem foreign to us. One reason is that contemporary philosophy shuns the question of life after death. Nevertheless, Gebhard is wrong in claiming that Fechner's *Little Book* preludes "the development of irrationalism in Fechner's works," pursuing a "glorification of death."[128] For Fechner "the world beyond" is precisely *not* a region remote from the earthly sphere, but instead, it is the realm that each person creates for himself during his life on earth. Fechner should be read conversely as *criticizing* the exaggerated unrealistic fantasies about life after death so typical of the epoch: If we must have life after death, says Fechner, then please provide a naturalistic explanation for it. And with his booklet Fechner intended just that. In doing so he came much closer to Feuerbach and David Friedrich Strauss than one might initially imagine; both of whom denied immortality, but thought that a person survives through his or her works.

Fechner's view is no more speculative than science fiction fantasies like those devised in artificial intelligence by Hans Moravec and Marvin Minsky. His ideas can be easily translated into contemporary computer jargon: Death represents a point in time at which the program executed by one's mind is taken over by other hardware: namely by the causal chains which we have started through our actions during our lives. Moravec thinks it possible that in the near future we will be able to program a computer with our mind (from our brain) in the course of our biological life. "When you die, this program will be installed in a mechanical body, which then without interruption easily takes over the responsibility for your life."[129]

1.8 Illness

Gradually Fechner's load of obligations and emotional struggles ruined his health. It exhausted him to continue lecturing in physics; he had felt not up to it from the start. Above all, he believed that a "lack of talent in mathematics and practical things" disqualified him for professorship in physics.[130] He soon suffered headaches and insomnia, a lack of motivation and long stretches of ennui. Lack of concentration and disruption entailed by these symptoms made working a nightmare; even two medical treatments in curative bath sanatoriums brought no relief.

After Christmas of 1839 this torment grew into an acute and severe crisis, forcing him to discontinue lecturing.[131] Using himself as a test person in numerous experiments while investigating after-images and contrast phenomena, experiments in which, for example, he looked at the sun through colored glass, Fechner induced massive eye trouble, including constant flickering. The after-images of his observations persisted unusually long: "Through a two-hour long observation of five to five minutes the image of the scale viewed through a tele-scope imprinted itself onto my eye so strongly that even twenty-four hours later I still saw the same image every time I closed my eyes or directed my glance to a dark or merely dusky place."[132] His hearing also suffered. Eventually Fechner be-came so sensitive to light that he blindfolded himself and diagnosed himself as blind. He tried all kinds of harmless and harmful treatments. In 1841 he lost all ap-petite and emaciated himself to the point that he could no longer stand upright.

In Leipzig the scholar's condition became the talk of the town. One day a woman he did not know sent him a meal, about which she had dreamt that it would deliver him from his ordeal.[133] The recipe actually helped Fechner to slowly take in food once more.

In addition to the anguish of starvation and blindness, Fechner was afflicted throughout the following year by worry about a mental disturbance, namely, the "total destruction" of his "mental energy"—as he put it. A severe loss of thought caused him to break off all social contact. His main activity consisted in trying to control his thoughts. The scenario he had depicted in the *Little Book* now happened to his own mind. It became the arena for a struggle among foreign influences, threatening his autonomy:

> My inner self split up as it were into two parts, my self and my thoughts. Both fought with each other; my thoughts sought to conquer my self and go an inde-pendent way, destroying my self's freedom and wellbeing, and my self used all the power at its will trying to command my thoughts, and as soon as a thought at-tempted to settle and develop, my self tried to exile it and drag in another remote thought. Thus I was mentally occupied, not with thinking, but with banishing and bridling thoughts. I sometimes felt like a rider on a wild horse that has taken off with him, trying to tame it, or like a prince who has lost the support of his people and who tries slowly to gather strength and aid in order to regain his kingdom.[134]

For a while he tried to alleviate his agony by inventing riddles or somehow keep-ing himself busy. Using auto-suggestive techniques he was gradually able to read-just himself to light.

Meanwhile, everyone assumed that Fechner would no longer fulfill his duties for a long time.[135] In 1840 his public lectures were taken over by the older son of his precursor in office, professor for mathematics and physics at the Nikolai

School, K. W. Brandes. After much consideration, finally Wilhelm Weber (1804–1891), who, as one of the Göttinger Seven had lost his professorship, was nominated to take up Fechner's chair.[136] Weber sent a statement to the government that he would reinstate Fechner as director at the institute for physics, if he should convalesce.[137] In that case Weber would be made the director of a new laboratory and observatory for magnetism.

Following several swings and a peak in August 1843, Fechner's illness abruptly changed for the better in October of the same year, accompanied by a manic phase and megalomania:

> The quick and favorable transformation that happened in the course of my psychical and physical life, and the way it happened, pushed me in October and part of November into a strange, crazy state of mind that I was unable to describe, particularly since once it was over, a clear memory of it all but vanished. I am certain that I believed that God himself called me to do extraordinary things and that my suffering had prepared me for it, that I felt that I in part possessed extraordinary psychic and physical powers, and in part was on my way to achieving them, that the whole world now appeared to me in another light, than it had before and does now; the riddles of the world seemed to reveal themselves; my earlier life had been extinguished and the present crisis seemed to be a new birth. Obviously my state was close to that of mental disorder; nevertheless, gradually everything settled into symmetry.[138]

One of his sisters describes Fechner as being "in a state of remoteness," "full of ecstatic excitement" throughout that phase.[139]

In 1845 Fechner wrote up a report on his illness, which Kuntze later printed in the biography that he wrote on Fechner.[140] Fechner's own history of his illness is sober and aloof; he describes in detail the procedures he used to heal himself.[141] As the manic phase gradually faded it was also the end of the crisis, although visual difficulties and periodic headaches never entirely ceased.[142] In 1846 Fechner reported to the ministry for culture that while his health had been restored, he still felt unable to return to his former profession "because neither can my eyes tolerate keen observation, nor my mind tolerate mathematical thinking." He requested permission to lecture on philosophical topics.[143]

Several attempts have been made to explain Fechner's period of illness. In Leipzig in 1894 the neurologist Paul Julius Möbius (1853–1907), a friend of Fechner's, spoke of "akinesia algera" or "lack of mobility due to the painfulness of movement without tangible cause of pain" with a neurasthenic pre-history. In 1925 the analyst Imre Hermann (1889–1984) interpreted Fechner's illness psychoanalytically. He suspected an uncured Oedipus conflict and interpreted the problems as an "intrauterine regression." That seems superficial and questionable. In 1970 the physician and historian of psychiatry Henri F. Ellenberger tied Fechner's illness to "heavy neurotic depression with hypochondriac symptoms, possibly complicated by damage of the retina." In 1976 the psychologists Bringmann and Balance came to the conclusion that Fechner suffered from a complex psychoneurosis with compulsive, depressive, and hypochondriac facets. Finally, in 1991 the historian of medicine Christina Schröder and the clinical psychologist and physician Harry Schröder diagnosed a "depressive psychosis with hypomanic to manic post-variation," encouraged and initiated by chronic exorbitant demand and exhaustion.[144]

1.9 The Day View's Origins

After overcoming the crisis, Fechner, as a private scholar, occupied himself with philosophical and scientific topics, glad to be rid of lecture obligations in physics.[145] Although he retained the title of professor for physics, he considered himself merely a supernumerary at the university, though it continued to pay him a modest salary. Without being obligated to do so, in the summer of 1846 he began once again to give lectures on topics that presently interested him. He first lectured "on 'the greatest good,' later on philosophy of nature, on 'the final things' [i.e., on life after death], on anthropology, on the seat of the soul, on the relationship between body and soul, on psychophysics and on aesthetics."[146] He held one lecture series on the "fundamental relationship between the material and the physical principle" fourteen times between 1846–1874. The success of these endeavors was limited, few students came to hear him lecture.[147]

In 1845 Fechner once again published some writing. It began with a short article in Poggendorff's *Annalen der Physik* [Annals of Physics] "On the Link between Faraday Induction Phenomena and Ampère's Electro-Dynamic Phenomena." It is Fechner's last piece dealing exclusively with physics, but it is also his most important in that field.[148] Just as Gauss, Weber, and Franz Neumann later also tried to do, Fechner attempted to link the two different laws of electricity that were known at the time. This paper written in 1845 put physics in Germany on a new course until Maxwell's Field Theory took its place. In chapter 4, we will discuss these events in some detail.

After publishing this piece in physics, Fechner proceeded to publish on entirely different topics. First evidence of this new creative period appeared in 1846 as "On the Greatest Good," in which he sketched a naturalistic ethics and tried to draw some ethical conclusions from his doctrine of immortality. Again using the pseudonym Dr. Mises he published a satire titled *Four Paradoxes*.

In the first paradox he advocates the thesis that shadows are alive. The second paradox suggests, half in jest and half in seriousness "to consider the variable t (time) a fourth coordinate in space."[149] This appears to be the first time in history that anyone has suggested, in a thought experiment, the idea of a two-dimensional being on a flat surface. In the third satire Fechner makes fun of how some of his contemporaries increasingly banish the belief in miracles from religion, yet defend all kinds of adventurous nonsense as well-founded convictions in science, medicine, and philosophy. Their arguments could just as well prove that witchcraft works. Here Fechner heaps a fair amount of ridicule on Hegel, as he continues to do in the fourth paradox,[150] where he uses the dialectic method to "prove" that beginning with the self-motion of a concept, the world cannot have resulted from an originally creative principle, but from a destructive one.

In his work on the greatest good, which followed from the first lecture he gave after recovering, Fechner believed to have found the source of a worldview that would unite all the various elements that seemed so incompatible to him before his crisis: Oken's philosophy of nature, the romantic-aesthetic attitude, physics, late idealism. He came to call this worldview the "day view" in contrast to the "night view" of the mechanistic world concept.[151] All of his writing to follow is for the most part mosaic pieces filling in this view, be they of a philosophical or a scientific nature. Once he had formulated it, he did not really further develop his day view except in *Ideas on the History of Creation and Development of Organisms* (1873).

By "the greatest good" Fechner means "the final purpose, to which all human thinking and action, devising and planning should be directed, not only for the individual, but in terms of it, it should unite all people."[152] Once one has determined what "greatest good" means, one can also know what the greatest "moral principle" is. According to Fechner, the greatest good for individual humans and for man in general is "pleasure." Every moral rule is guided "on the whole by a gain in pleasure." The general principle of this rule, the "pleasure principle" states that: "Man should, as much as he can, seek to bring the greatest pleasure, the greatest happiness whatsoever into the world; seek to bring it into the whole of time and space. Reducing dullness means the same as increasing pleasure."

This highest normative principle rests, according to Fechner, on a fundamental fact: "All of a person's subjective and objective motives for action, whatever they may be called and whichever part of nature to which they are related, all of his motives and purposes include an aspect of pleasure, either openly or covertly, consciously or unconsciously, and easily recognizable for an analyzing mind." All beings have a "desire for pleasure" within them, pleasure is "vitally, causally connected to everything that exists and is effective in the world." Fechner goes on to say that "No motive exists that is not directed towards creating or maintaining pleasure, or eliminating or preventing displeasure."[153] God himself, as a spirit, finds pleasure in bringing each individual and the entire universe closer to the ultimate goal of the greatest pleasure. Evil in the world is always only temporary. Taking up ideas from the *Little Book on Life after Death*, Fechner views hell and paradise as states which each individual creates for himself. The world is set up such that displeasure—or evil in the world—is always transitory. Whoever acts contrary to the pleasure principle punishes himself and everything else with displeasure. But "God lets what is harmful be devoured by its consequences, and what is good is multiplied by its seeds."[154] Although there is apparently so much disaster, evil, and pain in the world and although it may increase for certain periods of time, in the long run within the world, pleasure will always increase.

Fechner saw the enormous merit of the pleasure principle in the fact that it is established on experience of the "empirical nature of man and things." Accordingly, he disagrees with Kant's verdict that eudemonism is unsatisfactory because it turns morals into an empirical theory. The categorical imperative is justified, but it must be filled with empirical content. Fechner asks, "How can a theory of action, that must prove itself empirically, be itself independent of what is empirical? To me that would be as if physics were abstracted from the empirical nature of bodies and movement, or as if it were developed solely in the mind—something that has, granted, been attempted, but with which success?"[155] The effect of an unempirical attitude in ethics has always been that "the empirical side of life in turn disregards the morals of science." With this decidedly naturalistic view of ethics, Fechner adheres (whether or not he is aware of it) to one of Oken's dictums, found in the latter's philosophy of nature: "Philosophy must grow out of the philosophy of nature, just as the blossom grows out of the stem . . . Philosophy or ethics without a philosophy of nature is nonsense, a complete contradiction, just as a flower without a stem is nonsense. . . . The reason why thinkers in the philosophy of mind still run around without a foundation and without a compass lies solely in the lack of knowledge of nature of those who write and teach philosophy."[156] The fact that man's existence is determined by pleasure and pain is the most important "insight in nature" for Fechner's ethics.

Fechner's writing was echoed not only within internal discussion among university philosophers, but also within the political events of the on-going revolution of 1848. He reports, for example, on being accused that his ethics condone "the most dangerous communist, emancipatory and egalitarian tendencies."[157] This is not surprising because in the final chapter he analyzes Christian aversion to pleasure in clear words: "Individual lust of the flesh" is only "reprehensible" if a conflict arises between the individual's pleasure and the "principle of pleasure," namely "divine commandments." It is not conclusive from this that generally every "pleasure is meaningless and despicable; and this has brought forth monks, mortification, and sermons condemning the pleasure of this world."

His appeal to pleasure eventually culminates in the vision of pleasure-attuned religion and morals: "One day morals and religion will come, not as the destroyer of what has been, but as the flowering of what has been, returning the word 'pleasure' to its right honor. It will close monasteries, open up life, and revere art, and yet consider the Good more holy than the Beautiful, namely not just what generates present pleasure, but also future pleasure and all that goes with it."[158]

In a subsequent essay of 1848, in reply to a review written by Hermann Ulrici, Fechner forms his empirically established ethics into a philosophically significant doctrine. Here he more precisely explains how pleasure and pain determine man's actions and direct the will. Our actions are "always and inevitably determined by the pleasant and unpleasant character of a thought about the planned action or its omission, or its consequences."[159] He takes this for a psychologically proven fact. But the pleasant character of a future event is never important for our actions. It is not the expectation of pleasure, not the *idea of pleasure*, but the *pleasure of the idea*, the pleasure of expectation, that determines action. Even love is merely a motive created by the pleasure of an idea. Fechner gives a poignant illustration involving the case of Curtius, who supposedly leaped to death for his country. He did not do so believing that he would thereby win future pleasure by inducing a pleasurable event, he leapt solely because he *presently* found the thought very pleasant that through this act he might achieve fame and rescue the country.

The fact that human actions are often not compelled by pleasant ideas does not contradict the claim that our actions are guided by pleasant ideas—just as the law of gravity is not contradicted by a balloon ascending in the sky. Fechner thinks that cases like this merely demonstrate the existence of various layers of motives for behavior and that pleasant ideas can be rendered ineffective by opposing motives. Usually these opposing motives are of a moral nature.

The fact that we actually do experience pleasure caused by some ideas and a lack of it because of others and that this happens in degrees is in part based on instinct and in part a result of individual development. All of our "experience, instruction, and thought related to pleasure and displeasure and which we have experienced consciously" either consciously or subconsciously causally effects all subsequent motives and determination of will: "The more one studies how other people and oneself have been instructed by God and other persons, the more one discovers that precisely this after-effect [of prior pleasure and pain and thoughts about them] is crucial and confirmed."[160] This outline of Fechner's teachings on the principle of pleasure sufficiently shows that here Fechner is speaking as a psychoanalyst. We shall return to the question of how Fechner influenced Freud in chapter 7.

Within the year 1848 Fechner also presented a book called *Nanna—or On the Soul Life of Plants*. The direct occasion had been a mystic experience he had had in October 1843, towards the end of his crisis. For the first time in a long while he was able to take a walk through the garden without covering his eyes and he literally soaked-in the beauty of the flowers.[161] He saw everything in exaggerated clarity and believed to perceive that the plants' souls were "glowing."

In *Nanna* Fechner tries to prove that plants have souls, using scientifically sober, albeit sometimes long-winded explanations. He meant that plants also have their own psychic side, which only they and no one else can apprehend. He rarely concentrates on positive reasons *supporting* the theory of plant animation, but instead tries to refute the opposing *skeptical* arguments. The main arguments included the following: Plants have no nerves, they possess no central organ, they are incapable of voluntary movement, they serve no purpose of their own, one cannot imagine the life of their souls.

Fechner's counter-move is to outline a functionalistic mind-body theory. Long before Putnam and Fodor, the American philosophers generally heralded as the founders of functionalism in the contemporary mind-body debate, Fechner imagined that mental states could be realized somehow other than by the brain.[162] He often borrowed illustrations from music: Just as we can play music on the violin as well as on the flute, feelings can be manifested by something other than nerve tissue and brain mass.[163] Fechner's subsequent reflections on the mind-body problem all rest on this functionalistic notion.

Two of Fechner's rare specimens of brief, clear explanation are lectures given in 1849 at public meetings of the Royal Saxon Society of the Sciences—one on "The Mathematical Treatment of Organic Shapes and Processes" on the occasion of the king's birthday on May 18; the other "On the Law of Causality" commemorating the anniversary of Leibniz' death on November 14. Fechner had been a founding member of the Society and became the vice secretary for the section on mathematics and physics in 1848.[164]

Both speeches aim to show that nature can be subsumed under mathematically formulated laws even if in the world these phenomena are subject to indeterministic variation. Natural science loses none of its universal nor its necessary character, even if the world is really as Weisse described it in his system of freedom. These two speeches laid the foundation for Fechner's mathematical statistics, a field which he worked in for the rest of his life. Both papers are highly significant for the history of indeterministic thinking. We shall analyze them in detail in chapter 8.

In 1851 Fechner wrote the book he considered to be his major philosophical work: *Zend-Avesta or On the Things of Heaven and the Afterlife: From the Standpoint of Meditating on Nature.* Therein he compiled all the topics hitherto elaborated in his philosophy and cultivated them further. He enhanced the functionalistic doctrine of the animation of plants to include the stars and the whole universe.

This work has three parts. The first two deal with the "Things of Heaven," the third with the "Things of the Afterlife." In the first part, which we could call a sort of religious cosmology, Fechner attempts to demonstrate the plausibility that the earth, heavenly bodies in general, and the entire universe is animated. The problem arises of how an individual can be part of a greater psychic being without losing his individuality. Fechner solves it by presenting a theory of the psychic levels of the world. Consciousness of living individuals belongs to the consciousness of the earth, which in turn belongs to the divine consciousness of all things.

The second part consists almost entirely of appendixes to the first. Topics are repeated, widened, studied in depth and supplemented by reflections on method. The third part, finally, further details the subjects of the *Little Book on Life After Death,* equating future life by analogy to life of memory.

The most important passage of the *Zend-Avesta* can be found in an appendix to the second part. Here Fechner develops his "fundamental view" of the relationship between body and soul: "Behind all of this writing" says Fechner, "lies a basic notion of the relationship between the body and soul or body and mind, which seems to include the primary foundation for the harmonious link of otherwise very heterogeneous, actually contradictory appearing ways of viewing the world or basic tendencies in philosophy."[165]

Fechner's mind-body theory is an empiristic and phenomenalistic continuation of both Oken's and Schelling's Spinozism, and thus an "identity view," as Fechner calls it. It later came to be called the "dual perspective theory," "psychophysical parallelism" or the "double aspect theory."[166] Hidden in an addition to this appendix is also a "new principle of mathematical psychology," in which slumbered the first germ for the basic idea of psychophysics which were to unfold later.[167] In his own words, on "October 22, 1850 at dawn in bed" Fechner came upon the idea of making "the *proportionate* increase in living energy ... be the measure of the *increase* of pertinent mental intensity."[168] At Harvard University this date was later celebrated as "Fechner Day," the day when psychophysics began. Whether or not this was justified is questionable.[169]

On the whole, the *Zend-Avesta* (which means "living word") is tedious reading. Long-winded boring descriptions alternate with carefully explained details mixed with terse to-the-point and profound analyses; edifying observations on the verge of bigotry and piousness, written in pietistic pamphlet style, follow keen methodological discussions. The book lacks systematization and the main points are often stuck somewhere in additions to appendixes, while themes of lesser importance are discussed at length in almost eccentric reverie. The style ranges from a careless conversational tone to polished rhetoric. It also includes very personal remarks, impressionistic pictures, poems and long quotations from books in every imaginable field of science.

No wonder the book was not well received. Julius Schaller, a *Naturphilosoph* of Hegelian provenience, complained that the work lacked "concise determination in thought." While he called it "aphoristic metaphysics," at the same time he deemed the fundamental tendency of the book valuable, a tendency toward the same direction exhibited by Humboldt's *Cosmos*: "The idea of the Whole, the unity of all appearances in nature is to be developed. And this unity, as Fechner describes it, is the most penetrant, intensive unity imaginable, it is the unity of the Soul, the Mind conscious of Itself."[170]

Naturally, the reasons for *Zend-Avesta's* failure were not purely stylistic. In the 1850s, materialism (which should preferably be called 'philosophical naturalism') of the type advocated by Vogt, Büchner, Moleschott, and Czolbe dominated the scene. Traditional philosophy suffered a severe identity crisis following the decline of German idealism. Academic philosophy and philosophical endeavors in general enjoyed little esteem. Add to this the post-revolution repression in Germany, which university philosophy strove to escape by imposing far-reaching limitations on itself.[171] Natural science and humanities began to go noticeably separate ways.

The leading science at that time, physiology, saw within grasp the explanation of nature and man using principles from physics and chemistry; in research it followed the maxim that organisms are commanded by the same forces and laws valid for unanimated matter. Consider Helmholtz's *Erhaltung der Kraft* [Conservation of Force] (1847), the source of origin for the influential bio-physical school.[172] This setting did not tolerate ruminations about plant-souls and a world-soul. Fechner's claim that his approach was compatible with the most recent work in physiology, and that it was ideologically neutral, could not change that. Five years later, in a sarcastic and self-ironic tone, Fechner described the fate of his theory of souls: The idea of the plant-soul found undivided applause among the ladies, "spoken and written laudation, from friends and strangers." The chief reward for his writing and the symbol of its success was a crooked carrot that had fallen to his lot; a lady from Altenburg, with whom he was not acquainted, had sent it to him from her garden "as a sign of her interest in my book."

"Almost the same undivided rejection" reverberated among scientists, philosophers, and even theologians. No one took sides for the plant-soul. The first group saw "philosophy of nature barging in with a world-soul, . . . the others saw complete pantheism barging in. It does not help to protest that one is not a philosopher of nature, or a pantheist in this or that sense."

Fechner sympathized with his publisher: "Poor Voss! He printed 1000 copies of the Zend-Avesta, but sold only 200! Those few were probably mistaken for a novel like Nanna. . . . And if a natural scientist happens to grab it, only to find that the Zend-Avesta is a doctrine of the souls of the stars, he drops it immediately in dismay: 'Plant-soul, world-soul, fool's soul! If only the author had stuck to what he does best; he has done better in the past; this must reflect some remnant of his illness.'"[173]

1.10 Written Work after 1851

In 1855 Fechner published a book on the principles of physics called *On the Physical and Philosophical Theory of Atoms*. Besides the *Little Book on Life After Death* it was to become the only publication in the "day view" manner to see two editions within Fechner's lifetime. The book appears to have met with some approval among physicists and philosophers alike. It would seem that after the catastrophe of the *Zend-Avesta* Fechner was determined to prove that he could stick to what he did best, namely physics, but also that he wanted to show that the teachings of *Zend-Avesta* were fertile for physics and had remarkable consequences in store.

In this book, Fechner advocates a decidedly scientific realism regarding atoms. He surveys evidence in physics and chemistry that in his opinion supports the thesis that the fundamental structure of matter is atomic, and he vehemently defends this against philosophical objections. He augments scientific arguments with general philosophical reasons, ultimately tying them in with the "day view." Affronting the tradition of all the philosophical schools existing at the time, he claims that these reasons are not of an a priori nature, but that they are generalizations and extrapolations based on scientifically accepted facts. And with the same vehemence he had applied to defending atomism, Fechner tries to show that no theory of monads can be deducted therefrom, on the contrary, one gets "Synechiology." He disagrees with Herbart, I. H. Fichte and—without mentioning them by name—with many natural scientists such as Helmholtz, who (at least implicitly) assumed that 'the soul is situated at one point' and does not fill out the entire animated body.

In his sometimes quite satirical and highly witty book *Professor Schleiden and the Moon*, of 1856, Fechner presents two entirely different topics that only go together in his ingenious conceptual acrobatics. The first chapter is a very recommendable, dense, self-ironic, and poetic survey of all his previously published works expounding his new worldview. But the real subject (of the first part) is a rhetorically brilliant refutation directed at Matthias Jakob Schleiden, a botanist and founder of cell theory. Schleiden, an advocate of a Jakob-Friedrich-Fries sort of Kantianism, had derided Fechner's doctrine of plant-souls: "There exists almost no scientific absurdity which Fechner has not ridiculed under the name Dr. Mises . . . , of which he is not just as guilty of committing under his own name, or even worse than those whom he scourged."[174] Schleiden was so impressed by Fechner's reply to his banter that he later visited him in Leipzig and they resolved their dispute.[175]

In the second part Fechner moves on to questions of the moon's effect on the weather, plant growth, illness, and so on. He uses statistical methods for the first time and surveys an enormous amount of data about the moon's influence—data he has gleaned from scientific journals. This work seems to refute the idea that minimal factors (secondary causes) balance each other out over time and are therefore negligible.

The year 1860 finally brought forth the great work that made Fechner famous: *Elements of Psychophysics*.[176] In two volumes totaling 907 pages, Fechner establishes experimental psychology based on quantitative measuring methods. The *Elements*, preceded and followed by a series of shorter pieces written on related topics, are of an entirely different character than most of Fechner's previous books. These volumes are thoroughly scientific, packed with formulas, collected data, reflections on method, minute descriptions, and discussions of difficult experiments. A tie to the "day view" is made with just a few lines, but made unmistakably. The work was addressed above all to physiologists and philosophers.

Fechner's purpose with the *Elements* was twofold: For one, he intended to establish a new branch of science—psychophysics, the science "of the functional or dependency relationships between the body and the soul, or more generally, between the bodily and mental, or the physical and psychical world." Beyond this, he also wished to demonstrate that his philosophical "fundamental view" of the relationship between the body and the soul is not absurd after all; it can be founded on an "exact doctrine," which, like physics, is based on "experience and the mathematical linking of facts."[177] One gathers, as Wilhelm Wundt did, that since Fechner's theory of souls either went unnoticed or met with hostility, he changed his "tactics" and now attempted to establish his worldview—as it were—'from a different angle'.[178]

Since in Fechner's opinion every "exact doctrine" must first measure its objects of investigation, psychophysics too must demonstrate that psychical dimensions are measurable. (This task can be neglected for physical dimensions, since physics already studies and performs this kind of measurement.) Fechner draws several conclusions from the principles of gauging sensations, including things that are not directly observable. He calls this the field of "inner psychophysics." It deals with the relation between mental and bodily activity that is directly tied to sensation. Many of the findings that Fechner presented in this work resulted from joint experiments done with his brother-in-law Alfred Volkmann, the physiologist mentioned above.

At first only a few, albeit important scientists took note of the *Elements of Psychophysics*. Boring gives us a list of scholars discussing Fechner's *Elements* throughout the 1860s: Helmholtz, Mach, Wundt, Volkmann, Aubert, Delboeuf, Vierordt, and Bernstein.[179] Contemporary periodicals carried apparently only one review, actually only a report of the contents, rather than an analysis of the book.[180] The first edition of the *Elements* had a print-run of 750 copies; 277 were sold within a year.[181] In 1875 Fechner's publisher suggested terms for a second edition, but the project was not accomplished during Fechner's lifetime.[182]

In the *Elements of Psychophysics* Fechner announces a forthcoming supplement promising far-reaching consequences for religion and the philosophy of nature, consequences resulting from the "most general form of psychophysics." His conclusions were meant to "anticipate the future goal of psychophysics, erected on the fundamental principles provided in this work [namely, the *Elements*]."[183] The announced essay was soon published in 1861 titled *Concerning the Soul. Passage through the Visible World, in Search of the Invisible World.* It exhibits Fechner's clearest treatment of his "fundamental view" and evidences his effort to overcome the stylistic imperfections permeating *Zend-Avesta*. Here Fechner's line of thought is inclusive and systematic. He recapitulates the reasons supporting the notion of plant-souls and the earth-soul, presenting them more precisely and abstractly than before.[184]

The same work also contains a detailed exposition of "empirical principles of belief," in other words: an epistemology. By "belief" [*Glauben*] Fechner means all assumptions (of religious and nonreligious kinds) for which, while they cannot be "proven exactly"—meaning that they cannot be directly demonstrated in observation—there nonetheless exist empirical reasons making those assumptions at least empirically probable. Fechner further developed this epistemology, which shares some important features with pragmatism, in 1863 in the book *Three Motives and Reasons for Belief*. Therein he sketches various principles for judging the probability of beliefs.

Beginning around 1865, Fechner devoted himself to another area that he envisioned exploiting for the purpose of enhancing the "day view," namely: aesthetics. Between 1865 and 1872 he wrote fourteen articles on aesthetic topics, and in 1876 he published two volumes called *An Elementary Course in Aesthetics* [*Vorschule der Aesthetik*], crowning those endeavors.[185] Starting with the claim he had expressed in *On the Greatest Good*, stating that all of man's behavior is a function of the pleasure of his ideas, he takes this notion a step further to claim that man's aesthetic judgments are also functions of the pleasure of such ideas and sensations, as these are elicited by aesthetic objects. Once again, Fechner strives to chart facts, or empirical laws, that underscore empirical aesthetics, instead of "supervening" a priori aesthetics, namely aesthetics that proceed from general ideas to particulars. For him the whole purpose and method of aesthetics relates to psychophysics.

Fechner's aesthetics begins with Zeising's investigations into the rule of the golden mean. An aesthetic impression must be subject to various conditions, if it is to arouse desire or aversion. Fechner distinguishes here between a direct and an associative factor. An aesthetic impression's direct factor is determined by a large number of varying principles, among others, the "principle of uniformly linking diversity," the "principle of lack of contradiction, of unanimity, or of truth" and the "principle of clarity."[186] The associative factor concerns the cultural background which in part determines our aesthetic reaction. Our aesthetic judgment is connected to a variety of ideas via memory and habit, ideas of which we are aware to varying degrees while experiencing an aesthetic impression.

The entire theory is interwoven with prolonged ruminations on method. Fechner would have liked to discover a way to directly measure the degree to which something affords us satisfaction or dissatisfaction. But since this is not testable, we must be content with information on how many people prefer one particular aesthetic impression over another. Here for the first time Fechner employs the concept of "collective object," a term on which he later established a system of mathematical statistics.

Fechner sought underpinnings for his aesthetic theory by designing an experiment for an art exhibition featuring two versions of a Madonna portrait painted by Hans Holbein, The Younger. He wrote up lengthy questionnaires and distributed them among the visitors at the exhibition, in an attempt to discover which of the two paintings was considered more beautiful. Unfortunately, a mere 113 of the 11,842 visitors completed the questionnaire, rendering the findings negligible. But Fechner has the honor of being the first researcher to have contrived a questionnaire survey for psychology and to have interpreted the results statistically.

Besides being absorbed in aesthetics, Fechner contributed to the raging debate on Darwinism and the evolution of species. In 1873 he published *Some Ideas on the History of the Creation and Development of Organisms*, to which we shall return in chapter 7. In the foreword to this publication Fechner admits that "after resisting the doctrine of descent for quite some time" he was finally convinced. He goes on to praise Haeckel's *Natural History of Creation* for presenting Darwinian theory with such lucidity that it affords "clear insight" into it.[187] The purpose of Fechner's book is to examine very general principles of development which would encompass both Darwinian theory of evolution and a theory of development of a philosophic nature, thereby reconciling them. The so-called principle of the tendency toward stability, which is closely related to the second principle of thermodynamics, plays a fundamental part in it. An irreversible direction of progress is immanent to nature; in the long run, it moves from absolute spontaneity and irregularity to extreme stability. The whole concept is linked in a grand manner to psychophysics, eudemonism, and the "day view": The physical tendency towards stability serves as a bearer of the psychological tendency to increase pleasure in the world. By the end of his life Fechner had done much to prepare a second edition of the *Ideas*, but was no longer able to complete it.[188]

After Fechner had basically completed his work on aesthetics, he returned to psychophysics and elaborated his position in answer to the controversies that his ruminations had elicited. None of his prolific ideas aroused as much interest as the notion of measuring mental dimensions. Belated public awareness of the *Elements of Psychophysics* turned into stormy debate throughout the 1870s. We shall discuss this in chapter 6.

In 1877 Fechner analyzed the objections brought forth by his critics and published his defense: *The Case for Psychophysics*. A thick *Review of the Main Points of Psychophysics* followed in 1882, in which he not only replies to his most acerbic adversary Georg Elias Müller from Göttingen, but also presents an updated substitute for the *Elements* of 1860, which by then were out of stock.

1.11 The Day View as Contrasted with the Night View

Between the issue of both publications in defense of psychophysics, Fechner published *The Day View as Contrasted with the Night View* (1879)—a work that perhaps is even more Fechner's prize piece than the *Zend-Avesta*. Here he summed up his own philosophy and astutely discussed the intellectual tendencies of his time. For the first time he engaged himself in comprehensive philosophical polemics.

The *Day View* goes back to a manuscript written in 1871 that originally was planned to refute Eduard von Hartmann's *Philosophy of the Unconscious* and bore the title of "The Day View in Contrast with the Night View of Consciousness." It seems that Fechner rewrote it several times, before finally completing it.[189] As we learn from a letter written by an enthusiastic Fechner admirer, the Viennese physician Josef Breuer, and addressed to the philosopher Franz Brentano, the crucial impulse for finishing the book appears to have come from a Viennese student and author originally from Galicia, Siegfried (Salomon) Lipiner (1856–1911). Breuer reports: "Lipiner can pride himself on being the one to have urged Fechner to condense his otherwise quite diffuse exposition, oscillating as it does between poetry and logical explanation, into a compendious, readable, and comprehensible book; this is how the "day view" was produced."[190]

Lipiner was something of a precocious genius, who maintained personal friendship with Richard Wagner, Friedrich Nietzsche, Nietzsche's friend Erwin Rhode, and particularly with Paul Natorp and Gustav Mahler.[191] For a while he enjoyed fame for his poetry *Prometheus Unchained*, published by Breitkopf and Härtel in Leipzig in 1876. Lipiner studied philosophy under Brentano in Vienna, changing to the university at Leipzig for the winter semester 1875–1876 and the following summer term. He became friends with Fechner, as a former student of Fechner's relates: "Usually Lipiner had very few acquaintances, at least among fellow-students. I observed him more often accompanying the elderly, admirable Professor Fechner, the honorable psychophysicist and aesthetician, which surprised everyone, since the aged gentleman left his house only unwillingly . . . And here this queer fellow got the old man to stroll in Rosental Park just like everyone else."[192]

Lipiner had also joined the "academic-philosophical club" in Leipzig, where he swaggered and drew attention with his considerable eloquence, just as he had done in the "Vienna German Students' Reading Club."[193] In a lecture given to the Vienna Reading Club in 1878 *On the Elements of Renewing Religious Ideas Today*, which caused quite a commotion, Lipiner mentioned Fechner and the astrophysicist Karl Friedrich Zöllner as authorities for the contention that "the main doctrines of all true religion" must not contradict science and that science itself inspires us to "idealistic and even theistic notions."[194]

One of Lipiner's friends was young Sigmund Freud (1856–1939). Together they coedited a philosophical periodical from 1874–1875, vigorously exchanging ideas until at least 1877.[195] Certainly Freud and Lipiner discussed Fechner, for their philosophical ideas were similar. In 1874 Freud wrote twice to his childhood friend Eduard Silberstein, studying in Leipzig at the time, asking him for information about Fechner and his teaching.[196] He also probably heard of Fechner's ideas in Brentano's philosophy courses, particularly since Brentano corresponded with Fechner in 1874 on his most recent publication *Psychology from an Empirical Standpoint*, and on how to gauge sensations.[197]

The Lipiner episode is characteristic of how at that time intellectual circles of the Danube monarchy were linked to those of Leipzig. The bond is reflected by the fact that the cultural circles of Austria-Hungary took more note of Fechner's work than did those of Prussian Germany. From 1872–1878 the University of Leipzig counted more students than any other German university;[198] it was apparently quite attractive for Austrian students and the top choice for studies outside the country. Theodor Gomperz, Heinrich Braun, Thomas Masaryk, Eduard Silberstein and Edmund Husserl all studied for some time in Leipzig. Professors were also exchanged between the universities in Leipzig and Vienna; the physiologists Ewald Hering and Karl Ludwig held chairs at both institutions.

The perhaps pivotal factor tying the intellectuals of the Danube monarchy to many scholars in Leipzig was that both were deeply influenced by Johann Friedrich Herbart's philosophy. In the pedagogical reform of 1848 Franz Exner (1802–1853), a philosophy professor who had been summoned from the University at Prague to come to the Vienna Ministry for Education, saw to it that Herbart's philosophy and pedagogical doctrine were adopted throughout Austria —a feat that equally prevented German idealism from flourishing there.[199] By the turn of the century Vienna and Prague had become strongholds for Herbartian philosophy. As noted in 1.4, Leipzig was the heart of the German Herbart school. Herbartianism in Austria (which notwithstanding its own metaphysical underpinnings was critical of metaphysics, abhorred idealism and admired empiricism) was open for Fechner's psychophysics. One of Exner's students, the Herbartian Gustav Adolph Lindner wrote a *Textbook for Empirical Psychology* used as a course text in philosophy in upper grades of Austrian high school and in which he mentions Fechner.[200] Even Freud read this book in school.[201] The common background that favored Herbart's philosophy created a kind of immunity to neo-Kantianism after 1878 in the academic life of Leipzig and the Danube monarchy.

The "basic points" of the *Day View* run as follows: Physical appearances are not sensation states belonging to perceiving beings; they exist objectively in the world, they are outside of subjective consciousness. In terms of perceptual theory, then, Fechner advocates a variety of "direct realism" and rejects the doctrine of secondary qualities: Perception is immediate acquaintance with an external object and its properties. All appearances are interconnected within the highest consciousness. The psychical part of human nature also belongs to divine consciousness, and this divine consciousness is the inner side of the divine body, namely, the outer material world. We can draw conclusions about the nature of the afterworld from our knowledge of the constitution of this world.[202]

For Fechner this starting point suggests solutions to all kinds of philosophical and scientific challenges: Teleology, pleasure and pain, determinism and indeterminism, the mind-body problem, causal law, evolution theory, and so forth. His "day view," he says, is an "equal opponent" for the "two major tendencies exhibited by the prevalent night view"—one (tendency) which tries to "generate the entire contents of the world from the a priori emptiness of abstract concepts," as well as the other, that limits "human knowledge of the world to knowledge of our own subjectivity."[203] Philosophically, then, the day view contradicts both Schelling/Hegel theory and Kant/Schopenhauer theory.

Fechner begins with a dramatic charge against both mechanistic materialism's and philosophical pessimism's night view. These two approaches leave the world cold and barren and with all the despair of Hades; all vitality, every color, sound, and fragrance is merely subjective illusion. Fechner contrasts this devastation with the redeeming, liberating bright beauty of the "day view," where violins and flutes do not "pretend" to make sounds, and butterflies do not "pretend" to have colors.[204]

Fechner's *Day View* is often misunderstood as a backwards, antimodern, and irrational case for spiritual idealists warding off materialistic natural science. One audacious claim even says that Fechner confused the realism of traditional common sense with an erroneous nocturnal outlook on life and the world—only to find a truly meaningful worldview in insanity, mystery, obscurantism, and romanticism.[205] He is sometimes even associated with spiritualism. Such interpretations are nonsense. In contrast to the official mechanistic worldview preferred by his contemporaries, Fechner's work actually sketches a new sort of epistemology, explaining the reality of the mental and the organic, bridging the cleft that separates nature and consciousness, reality and perceptual appearance, and combining science with direct human experience.

Compared to Eduard von Hartmann's irrational *Philosophy of the Unconscious*, an enormously popular book at the time, Fechner's day view is very rational; its tenets always allow critical empirical scrutiny. While we can't deny that some of Fechner's motives sprang from (late) romanticism and religion, this alone does not render his opinion irrational and cannot cancel any significance it may have for the future. Fechner's day view is an attempt to understand science in a way that reunites science with the real world of people, with all the ethical and aesthetic implications involved, instead of excluding them from it, as mechanistic materialism does.

We may question the worth of Lotze's claim in a review of Fechner's *Day View*, that David Friedrich Strauss's very influential and popular creed for materialism, *Old and New Belief. An Avowal* (1872), is "in every respect the perfect embodiment and depiction of the worldview that Theodor Fechner calls nocturnal and hopes to see expelled by the revival of a day view."[206]

Naturally, it may apply to some of Strauss's blunt opinions that do indeed conform to the mechanistic concept of the world. But it does not apply to this theologian's "heathen religious" side, namely his pantheism, nor to his opinion that the difference between materialism and idealism is merely a matter of words.[207] Notwithstanding a heart-felt sympathy for Fechner's book, Lotze disqualifies Fechner's literary soaring and notes that "the analogy's bridges between this world and the next are not stable enough to entice one to set foot upon them."[208]

In the year that he died, 1887, a longer article by Fechner given the title "On the Principles of Measuring the Mental and Weber's Law" appeared in a periodical edited by Wilhelm Wundt, *Philosophical Studies* [*Philosophische Studien*]. With youthful enthusiasm Fechner once again dove into the debate on psychophysics. Wundt found this contribution "the clearest and most complete exposition of the matter given at all throughout the nearly forty years Fechner labored at it."[209]

From among the huge selection of papers Fechner left for posterity, a posthumous book called *Theory of Measuring Collectives* was edited in 1897 by the psychologist and psychophysicist Gottlob Friedrich Lipps. Here Fechner—as it were—freehandedly sketches a new sort of mathematical statistics that is closely connected to the notion of freedom cradled by late idealism. This outlook can be traced back to ideas on indeterminism that Fechner had already once expressed in 1849. This book is also the source for interpreting probability as frequency, a method that was to become highly significant for the twentieth century.

1.12 Fechner's Life after Recovery

After Fechner abandoned the university position in physics his finances were a source of endless worry.[210] The ministry for education gave him half-pay, at first 850 talers, then 600, later more.[211] He remained a member of the faculty for philosophy at the University of Leipzig, holding the title of professor for physics. But he no longer held any office; he was not given a chair for the philosophy of nature, as secondary literature often suggests. Except for lecturing, Fechner was not active in university life. He did participate regularly in meetings of the Royal Saxon Society of the Sciences. After the demise of his friends Weisse (1866), Hermann Härtel (1875), Ernst Heinrich Weber (1877) and Karl Friedrich Zöllner (1882), there were few left to keep him academic company.

There is little to report on the outward circumstances of Fechner's life after he recovered from illness and psychological crisis. He was a bookworm, spending his days absorbed at his desk. He lived in the same quarters (No. 2 Blumengasse, nowadays called Scherlstrasse), a stone's throw from the center of Leipzig's old downtown area from 1850 until his death. Kuntze and Wundt describe a typical Fechner workday: He labored from early morning until one o'clock in a small, austere chamber, sitting on a stool or standing at the high desk.[212] After a scanty lunch he strolled to Leipzig's Rosental Park (now the zoo), had an afternoon coffee at Café Kintschy and read the newspaper. Occasionally he played a game of chess there. At home again, he continued to work until late in the evening. He also often sought distraction by visiting friends, relatives, or acquaintances in Leipzig's society. Once a week he gave a two-hour long lecture at the university.

The only extravagances he allowed himself were afternoon coffee and an enormous amount of stationery. As he aged he read less and concentrated on writing down every thought that occurred to him. At the invitation of a friend he often took summer vacations with his wife, without interrupting his work. In 1875 the ministry relieved him of his teaching duties at his own request. His eyesight faltered and he underwent various cataract operations in Halle, performed by the ophthalmic surgeon Gräfe. But otherwise, he maintained good health until his last day.

In 1877–1878 Fechner was "almost involuntarily"[213] involved in spiritual meetings with the American medium Henry Slade (1840–1904) and the Danish mesmerist Hansen, having been brought along by one of his friends, the astrophysicist Karl Friedrich Zöllner (1834–1882). At the time, Slade's seances were a sensation all over Europe.[214] In 1872 Zöllner had begun a campaign against the alleged signs of decline in the established natural sciences by writing a piece called *On the Nature of Comets*. His polemic had been particularly directed at leading representatives of natural science in Berlin. Zöllner became acquainted with spiritualism in 1875 in England and imagined therein a great way to prove his theory of a four-dimensional, positively warped universe. He needed four-dimensional space in order to prove that the law of gravity was reducible to Wilhelm Weber's basic law of electrodynamics; a project he had already been laboring at for quite some time. It was related to his idea that from the darkness of the sky and other facts, one can deduce that the universe is non-Euclidean.[215] After 1878 Zöllner's polemics became manic and pathological, sometimes ending in wild anti-Semitic tirades.

It severely damaged Fechner's scientific reputation at the time to have been caught up in the ruckus surrounding Slade and Zöllner.[216] One chapter of the *Day View* cautiously, skeptically, almost reluctantly admits the possibility of spiritualistic phenomena. It is understandable that someone like Fechner, who himself had so often pondered the notion of life after death and advocated mental functionalism would be curious about parapsychology.

The ultimate reason for believing that Slade's experiments were not trickery and swindle after all, as he had originally thought to be the case, were testimonies given by the Leipziger mathematician Wilhelm Scheibner (1826–1908) and particularly by his friend Wilhelm Weber, who, in Fechner's opinion, "embodied the spirit of exact observation and logical methods."[217] But Ernst Mach, who personally witnessed the spiritualist scene in Leipzig, reports that Fechner was not alone in not relying solely on his own experience as proof and initially needing affirmation by another in order to be fully convinced. In the end, each member of this spiritualist circle named one of the others as a decisive, competent authority.[218]

In evaluating the episode of spiritualism and Fechner's role therein, we should not overlook the fact that although Fechner liked Zöllner, their opinions diverged at significant points. Zöllner was a Kantian, zealously lecturing on Kant's work,[219] as well as a devotee of Eduard von Hartmann and Schopenhauer—all philosophers that Fechner disdained. Zöllner taught a monadical, hylozoic theory of souls that attributed sensibility to matter itself, while Fechner advocated synechiology, in which the property of having a soul is an emergent property of complex systems.[220] Although Fechner thought that spiritualism might perhaps support his theory of life after death, he desired that the day view be judged independently of any link with it. If spiritualism had nothing to offer philosophy, Fechner was convinced that it could hardly be used for proof in scientific conjecture, as Zöllner had done for the notion of the fourth dimension of space.[221] Fechner's contemporaries did not see these distinctions clearly and lumped Leipzig's spiritualistic professors together, particularly when viewed from distant Berlin, the seat of the professors most often attacked in Zöllner's polemics. The commotion caused by Zöllner temporarily damaged the reputation of non-Euclidean geometry, also bringing disrepute to Helmholtz's work on the subject.

When evaluating the conflict sparked between scientists in Leipzig and Berlin by scholars dabbling in spiritualism we must also keep in mind that as of 1870, Helmholtz and Wilhelm Weber were involved in an ongoing dispute on whether or not Weber's basic law of electrodynamics fulfilled the law of the conservation of energy.[222] It bothered Helmholtz that according to Weber's principle the force between two electric "masses" (today called "charges") depends not only on the inverse square of the distance between them, but also on their relative velocity and relative acceleration. In *Conservation of Force* (1847) Helmholtz denied the existence of velocity-dependent forces. In 1872 Zöllner joined the dispute with his book on comets. His criticism eventually resulted in an all-around attack on the inductive method of the British physicists William Thomson, Peter Guthrie Tait, and John Tyndall—all friends of Helmholtz. Two years later, Helmholtz retorted that these reproaches represented a relapse into metaphysical speculation, disparaging "Zöllner and his metaphysical friends."[223]

As far as we know, the last eye-witness account on Fechner is that given by the chemist Wilhelm Ostwald. In his memoirs he relates that on the occasion of being called for a chair at the University of Leipzig he made an inaugural visit to Fechner just a few weeks before the elderly scholar passed away. The report testifies to Fechner's spryness right up to his last day:

> I consider it a special deed of fortune that I was able to personally meet Gustav Theodor Fechner, the founder of quantitative psychology. I had read much of his work and had long revered this rare personality . . . Just entering his house felt like coming home—the floor in the foyer was strewn with white sand, the way it was done where I came from. In spite of his advanced age he was as nimble as a youngster. He had heard of me, probably through Wundt, and inquired immediately whether all the gaugings I had undertaken included any in which one and the same value had been measured repeatedly. For he was preoccupied with the theory of measuring collectives and sought as much diverse data of this kind as possible. Unfortunately I had no such data to offer, otherwise I would have gladly seized the opportunity to see him again. He quickly involved me in vigorous conversation and I was unhappy to break it off, when the time came for me to leave.[224]

Despite all of his scintillating, vivacious ideas, Fechner's work habits had something compulsive about them. He feared nothing as much as boredom, and without intellectual tasks he suffered ennui. Even in the final diary entry written prior to his death he complained that his faltering eyesight "often forced him to be embarrassingly bored" so that he took walks "to kill time."[225] Neither were the topics he covered really matters of his own choice; he felt compelled: "He was enslaved to them."[226] Bringmann and Balance calculate that for the period between 1843 and 1887 Fechner wrote twenty-six books and sixty-one articles totaling approximately eight thousand pages of print.

According to Kuntze, Fechner was not prone to lengthy correspondence, but he was visited by numerous scientists from all over the world.[227] One diary entry is a spirited and detailed report on a particular visitor, Dr. Thomas Masaryk, who later became the president of the Republic of Czechoslovakia.[228] Masaryk was twenty-six years old at the time and sought contact with professors and publishers in Leipzig. It is also known that Franz Brentano visited Fechner in Leipzig,[229] as did Ernst Mach, Carl Stumpf, and the American psychologist Granville Stanley Hall.[230]

Fechner had no devotees in the narrower sense of the word. Kuntze mentions Wundt and Preyer as members of the younger generation closest to him. Beginning in 1876 Fechner corresponded regularly with the Swabian rural physician, advisor for medicine in Stuttgart, and honorary doctor of the University of Tübingen, Wilhelm Camerer (1842–1910), who, upon recommendation from his former teacher, the physiologist Carl Vierordt from Tübingen, had approached Fechner with questions about the psychophysics of the tactile sense. He eventually became Fechner's closest coworker. In the 1880s Camerer published articles on the tactile and gustatory senses in the *Journal for Biology* [*Zeitschrift für Biologie*]. Fechner himself noted in one of his articles on the spatial sense that he wrote it in conjunction with Camerer.[231] Fechner perused all of these publications prior to printing and commented on them in correspondence. Camerer visited Fechner in 1883.[232]

Fechner died in Leipzig on November 18, 1887 after suffering a stroke twelve days earlier. Three days later, Wilhelm Wundt read a funeral oration at the burial in the Johannis-Cemetery. Ten years thereafter, Paul Julius Möbius donated a Fechner monument (renovated in 1983) to the Rosental Park in Leipzig.

PART II
PHILOSOPHY

2 NONREDUCTIVE MATERIALISM

THIS PART OF the book presents Fechner's philosophy as a systematic position. Two things characterize Fechner's philosophical endeavors: in terms of method he had an inclination to natural science, and the subject he worked on was a special solution to the mind-body problem. He elaborated a position centered on what I call *nonreductive materialism*.[1] The position is nonreductive, because it describes life and consciousness as having an independent, original nature that cannot be further reduced to physical phenomena, or—as the nineteenth century would have it—reduced to "the mechanics of atoms." Yet, at the same time, Fechner's position is materialistic. He sees every change in the physical world as wholly explicable by laws of nature, and he sees for every mental change some change in the physical world that precedes it. The intriguing thing about Fechner's philosophy is just how he gets these two seemingly contrary tendencies of non-reducibility and materialism to harmonize.

Guided by natural science, Fechner is a radical empiricist with a phenomenalistic outlook. He tries to show how the world is made of phenomena [*Erscheinungen*], and how to refute philosophical views by proving that they lack any relevance for real phenomena. His theory relies on verificationism: he adheres as far as possible to the maxim that claims about the world only make sense and are justified if they are linked to actual observations. The only methods permitted are those that "allow no conclusions other than those based on experience."[2] Yet from the outset, he also insists that in quotidian life and science, we cannot avoid underpinning our activities with claims that we do not, or cannot, verify. If we take metaphysics to be beliefs about the state of the world that cannot be derived empirically (which is how Fechner defines metaphysics), then we cannot do without a certain amount of metaphysical presumption, both in everyday life and in science.

It would be hasty to conclude from this that Fechner's philosophy is vain metaphysics and speculation. If that were correct, the same would hold for today's philosophy of science, which some time ago abandoned the exaggerated verificationism demanded by the Vienna Circle. Fechner's search is for the very criteria needed to test the plausibility of an assumption that we cannot examine directly in terms of phenomena, in light of the other facts we know. These criteria would limit any deviation from our empiristic ideal to a required minimum. The point is that Fechner finds the very reasons that permit deviation from verification in a "harmless" quotidian way to be equally sufficient for justifying metaphysical conclusions in "less harmless" areas.

Fechner's philosophy has two parts: one is a reconstructive, analytic part that by contemporary standards can be classified as non-metaphysical; the other is a synthetic-constructive part, which applies the means developed in the analytic part to draw conclusions about phenomena that are (still) outside the realm of experience. The non-metaphysical philosophy of the first part, which is committed to phenomenalistic empiricism, is the heart of Fechner's philosophy and—as I mentioned—can be characterized as nonreductive materialism. The philosophy of nature proper contained in the second part, itself based on the non-metaphysical first part, deserves the description "inductive metaphysics."[3] Fechner himself considers it "objective idealism."

Dividing Fechner's philosophy into a non-metaphysical, fundamental part and an appended supplementary philosophy of nature is not the result of restructuring it in hindsight; the division is immanent to Fechner's work itself.[4] Granted, Fechner did not always make this distinction as clearly as one might desire. When he philosophized, his focus was usually the philosophy of nature, or—as he also called it—"a philosophical closure" for natural science. It is also true that Fechner's contemporaries were more interested in his philosophy of nature and usually took little notice of the materialistic and empiristic prerequisites supporting his philosophy of nature. The non-metaphysical part was also quite sober and complex. This elicited the deep-rooted prejudice that is still valid today; Fechner is often considered one of the nineteenth-century's purely metaphysical and eccentric "conceptual poets," who produced untestable, meaningless, and worthless philosophical sayings.

2.1 Knowing and Believing

Knowledge of Phenomena

The concept of "phenomenon" is fundamental for Fechner's non-metaphysical philosophy; he gives this term approximately the same meaning as "idea" had in John Locke's time: It indicates anything that can be an object of consciousness.[5] Phenomena are given through either outward or inner experience:

> Normally, all we know about what exists is what enters consciousness, only our feelings, sentiments, thinking, willing . . .
> Accepting consciousness in its unity and with its contents, as it is, is the purest, direct experience we can have of what exists. Consciousness is a being that knows its own constitution and is exactly as it thinks that it is. There is no other being for which we can say the same.[6]

Because of this directness, phenomena form an appropriate foundation for knowledge. They are "the Given," that is, they are "whatever is shown." All that we know about the world in the strictest sense of the word, all that is "simply or objectively *what is certain*" within the realm of experience, is either given through immediate experience or derived by "cogent logical conclusion" (meaning: deduction):

> I know—in this sense of the word 'knowledge'—that a sensation of red, green, or yellow is in the world, when I experience it myself; there is nothing to criticize; what's there, is there.

> Within the realm of what is experienced, certainty of knowledge does not extend beyond what is directly experienced, nor beyond its logically analyzed, combined, or further developed contents.[7]
> When my mind . . . itself appears immediately to me, that is the end of doubt.[8]

So Fechner thinks: 1) that our empirical knowledge consists of propositions about immediately given phenomena and their logical consequences (and nothing further); 2) that such propositions cannot be refuted by propositions about other phenomena; and 3) that these propositions are indubitable.

Beyond the empirical certainty secured by the phenomena, Fechner envisions even mathematical certainty—also worth mentioning for the sake of completeness. Compared to his contemporaries, Fechner entertains an atypical and surprisingly lucid idea of the analytic character of mathematics, demonstrated in his pronouncement that mathematical truths "by themselves say nothing about existence [meaning what empirically exists], they only claim: if this is so, then such is such. Mathematics cannot prove that space has three dimensions, only that insofar as space, triangles, circles exist as they are defined, then this or that follows from what is given."[9] Mathematically certain is "whatever, according to the law of identity, cannot possibly be imagined or conceived otherwise," in other words, whatever is a logically true proposition. Accordingly, I know "that every triangle includes *in total* two right angles; for I cannot think it otherwise, without entertaining contradiction in the pre-thought conditions."[10]

Fechner's doctrine of phenomena and the certainty founded upon them fits fairly well within the framework of traditional philosophical theories of perception stretching from early modern times well into the twentieth century, and found particularly in English empiricism. Fechner's variation differs fundamentally by distinguishing *psychical* from *physical* phenomena. In order to explain this distinction we must look at another difference. For Fechner all phenomena fall into one of two categories: Some appear only to the person who perceives them, others are or can also be perceived by other persons. Fechner calls the first group "self-phenomena," in analogy we can designate the others "extraneous phenomena." Examples of self-phenomena are "common feelings, sensations, ideas, intentions, etc.,"[11] examples of extraneous phenomena are houses, humans, trees, my arm, etc. (This last example indicates that we should not be misled by the expression "extraneous." The phenomenon that I experience when I see my arm is by definition extraneous because another person could also perceive it.[12])

Fechner uses the distinction between self-phenomena and extraneous phenomena to carve a criterion for distinguishing physical from psychical phenomena. He arrives at the following fundamental assumptions:

(1)   All and only physical phenomena are foreign phenomena.
(2)   All and only psychical phenomena are self-phenomena.

For John Locke the difference between ideas originating in external objects and ideas originating in inner operations of the mind is fixed—as it were—by the "sense organs" through which the ideas are perceived: Ideas from *sensation* are conveyed by the external senses; the source of ideas resulting from *reflections* "might properly enough be called internal sense."[13] David Hume also speaks of "outward or inward sentiment"[14] from which all "material of thought" is derived. Here the concept of inner sense is understood as the capacity for perceiving that is not linked to external sense organs. For George Berkeley the difference between ideas impressed by the senses and self-generated ideas lies in the independence of the former from personal will.[15] Other than this, the impressed ideas are stronger, more vivacious, and clearer than those generated by the self—a notion also found in Hume. All of these approaches assume that there are two kinds of perception and that—at least in normal cases—it is possible for consciousness, using its own means, to distinguish ideas coming from "outside" from those created by the mind itself. In the end, global certainty is established on the subject's own self-certainty.

In contrast, Fechner postulates only one type of perception. Whether (speaking with Locke) we are aware of a *sensation* or a *reflection*, is, according to Fechner, discernable neither by knowing the phenomenon itself (its strength, vivacity, clearness, and so on), nor by the sense involved in perceiving it, nor by its dependence on the will. The distinction rests simply and solely on whether or not what appears to the percipient is also accessible to other perceivers. The distinction between self-phenomena and extraneous phenomena therefore provides a criterion for distinguishing between physical and psychical phenomena.

In order to evaluate the plausibility of establishing the distinction between physical and psychical phenomena on a distinction made between self-phenomena and extraneous phenomena we need to ask how axioms (1) and (2) are to be understood. Are they definitions? Analytic propositions? Empirical claims? Explanations? Or something else? Fechner's analysis is as follows:

> Ultimately, we can view the whole aforementioned thesis [that physical phenomena are extraneous phenomena and psychical phenomena are self-phenomena] as a generalized expression of our experience; but from a certain viewpoint, it can be seen as a mere clarification of how we customarily use speech.
>
> Of the former we say: it is a common fact of experience that when we think of something as bodily, material, corporeal, or physical, we find ourselves either wholly, or in terms of a particular organ, capable of perception, really or imaginably, from an external standpoint; but when we think of it as mental, or psychical, we find ourselves at a standpoint of inner self-phenomena.
>
> If the latter we say: we *call* something bodily, material, corporeal, physical, or mental, or psychical, depending on whether it appears to someone else or to one-

self, but we call it that in such a way, that even these last expressions '*appear* to one-self' and '*appear* to others,' according to linguistic custom refer to experience, whereby, naturally, a certain fixation of usage that allows for various idiomatic expressions is necessary for scientific stringency.[16]

It might seem that Fechner is confused: (1) and (2) are characterized both as "expressions of experience," thus as empirical statements, as well as "explanation of linguistic custom," in other words, as conventions of everyday language, and these seem to exclude one another. But reflecting for a moment, we discover that there truly are statements that can be simultaneously definitions and empirical—and they enjoy a fairly prominent status in scientific theories. They are what Hilary Putnam calls *framework principles*.[17] Within an empirical theory they play a quasi-analytical part, meaning that they are "as analytic as any non-analytic statements ever get."[18] But they are also of a synthetic nature, because under certain circumstances we can have good empirical reasons for dismissing such statements. Such a circumstance could be, for instance, the availability of an alternative theory that comes to conclusions contradicting the framework principles and that is nonetheless empirically more successful than the original theory. Framework principles, then, are both definitions (conventions of language) and empirical laws. Putnam's examples are how energy was defined before and after Einstein, Newton's second law, and the laws of Euclidean and non-Euclidean geometry. Prior to Einstein's general theory of relativity, "kinetic energy" was *defined* as $\frac{1}{2}$ mv^2, but Einstein's theory gave *empirical* reasons for changing this definition, so that it now reads $mc^2 + \frac{1}{2}$ $mv^2 + \ldots$. Such reasons make the claim that 'kinetic energy equals $\frac{1}{2}$ mv^2' a nonanalytic proposition.

How can we determine whether a definition that fixes a concept within a theory is nonanalytic after all? Putnam claims that the decisive factor is the role that the concept plays within the theory as a whole. If it is a law-cluster concept, meaning that it essentially appears in empirical laws within the theory (and the conditions expressed in the definition are generally accepted as criteria for using the term), then the concept's definition is nonanalytic. For it is then possible to replace the laws containing the concept with improved laws and thereby "force" a change of definition for empirical reasons.

Let us make an excursion into the theory of analytic and synthetic propositions in order to test the plausibility that the definitions of "psychical" and "physical" given in (1) and (2) are to be understood as being nonanalytic in the sense discussed here. The aforementioned quotes from Fechner claim that ignoring stylistic variations our linguistic usage of the terms "psychical" and "physical" can be construed to imply that on the one hand, definitions (1) and (2) are also framework principles and on the other, that the concepts "psychical" and "physical" are also law-cluster concepts as Putnam defines them. Later we shall deal with Fechner's suggestion for the relevant laws in which these law-cluster concepts occur. (In 2.2, when discussing Fechner's mind-body theory.)

In the passage quoted above and in many other places, Fechner—like traditional theories of perception—writes of the psychical as something that is "inside" and about "phenomena collected from the inner standpoint," juxtaposing this to what is "outside" and "phenomena gained from an outer standpoint." For example: "What from an inner standpoint appears to be mental, or psychical, may, for another person—and based on the outer standpoint—appear in a different form, one which is just the form of bodily-material expression."[19] What does this mean? We might think that statements like this prove that Fechner wants to *explain* what is psychical as that which appears from the inner standpoint, and *explain* what is physical as that which appears from the outer standpoint. However, since—in contrast to the British empiricists—he nowhere indicates just exactly what he means by "inner" and "outer" and only provides metaphoric idioms, he actually does not *explain* these concepts as the empiricists had done, by referring to the senses, to vivacity of thought, and so on, he simply defines *obscurum per obscurius*. But if this interpretation were correct, it would rob Fechner's theory of every right to being a productive explanation of the mental straight from the onset. Not long ago, Stubenberg disparaged Fechner along just these lines: "Merely giving the unknown a name does not mean one has expounded it."[20]

Fortunately, another interpretation better suits our discussion thus far. The most obvious guess is that Fechner uses "inner phenomenon" and "outer phenomenon" as *synonymous* with "psychical phenomenon" and "physical phenomenon."[21] On this interpretation, we could substitute the metaphoric of "inner" and "outer" everywhere with the terms "psychical phenomenon" and "physical phenomenon." This would answer Stubenberg's query of what it is supposed to mean that psychical phenomena are inner and physical phenomena are outer.[22] Statements such as: "Inner phenomena are psychical phenomena" and "outer phenomena are physical phenomena" would then be real analytic propositions, and should not be read as claiming to enhance our knowledge about the world. In contrast, the following statements do make this claim, if we understand them in the given meaning as framework principles: "Physical (outer) phenomena are extraneous phenomena" and "psychical (inner) phenomena are self-phenomena." Read this way, axioms (1) and (2) make sense.

In contrast to Descartes and the British empiricists then, Fechner's suggested criterion does not rest solely on capacities of individual consciousness. The distinction between physical and mental, inner and outer—a distinction that is always fallible and revisable—can only be made by interacting with other people. Whatever is inside or outside is *defined socially*. Restricting the phenomena to how they are perceived by individual consciousness does not allow differentiation between subjective and objective phenomena. The thinking subject's self-certainty does not provide an indubitable foundation. How the Given is related to others is what distinguishes "inner" from "outer."

Belief in Phenomena

Fechner understands the difference between physical and psychical phenomena as the difference between self-phenomena and extraneous phenomena. Taking a step further back, we ask which prerequisites determine the existence of extraneous and self-phenomena. Obviously this distinction depends on concepts of the self and other minds. This in turn seems to depend on a concept of the psychical. Fechner's theory looks circular.

But this relationship can be seen as non-circular if we start—as Fechner does—with the existence of a natural human capacity for *belief* (for the purpose of gaining convictions) and view it as fundamental. We discover ourselves and the world initially through the belief in the existence of other subjects. To assume the existence of other persons (and thus one's own existence) means to believe that there are phenomena that are not and cannot be given to oneself. Thus belief is a central concept in Fechner's philosophy.[23]

Fechner elaborated a general theory of belief with great care, following the example of believing in the existence of minds in other persons. In order to understand his philosophy we must make a slight detour and consider Fechner's theory of belief and then return to our main train of thought. (We shall then notice how Fechner's theory of belief is not only necessary for the concept of self and others, but also for other important issues.)

What does Fechner mean by belief? "[I]n the widest sense we and I understand by belief the holding true of things for which we have no certainty through experience or by logical conclusion, including mathematical certainty."[24] Remember that for Fechner the only things that are certain are those given immediately in experience. What persuades us then to enter such uncertain territory of things not secured by immediate experience? Fechner sees two kinds of factors ("motives") at work: for one, "*motives* which *force* us to believe, and *reasons,* which *justify* belief."[25] (This distinction anticipates our contemporary distinction between *context of discovery* and *context of justification.*) Fechner understands motives as meaning the causes subject to laws of nature that allow humans to believe or at least influence this belief to a certain extent. It is the task of empirical psychology to discover them. Reasons for believing a claim, on the other hand, are empirical statements, the occurrence of which makes it seem rational to accept the claim as true. Reasons and motives taken together can be so strong that belief in something not directly experienced can seem certain. But this type of certainty always remains subjective certainty, a certainty that "is equal in importance" to objective certainty,[26] although it is not the same.

One result of the definition of "belief" is that there is no experience from which belief could be deduced. Belief has no reasons "sufficient for proof, only such as are sufficient for conviction; otherwise it would be knowledge and not belief. Thus whatever may establish and support belief, doubts always remain possible in terms of knowledge, without contradicting logic and experience; that is part of the concept itself."[27] From this we cannot conclude, however, that every claim lacking sufficient reasons for proof is a justified belief. Not every "imperfect reason for knowledge" is a reason for belief. Additional specific reasons are needed to prevent a belief from being a mere superstition. One belief is (in a subjective sense) more secure than another, if the number of secure experiences supporting it is greater than the number supporting the other. Fechner explains, "What elicits the belief in a wild savage, who has no knowledge of astronomy, that the sun will rise tomorrow, just as it did today? Only his knowledge that it rose today, yesterday, two days ago, and every day from time immemorial. Induction of this kind can never be perfect; but the imperfect knowledge established on it supplements itself according to a psychological law and becomes a more secure and stronger belief, the less it requires such amendment."[28]

Now, there are two kinds of belief. First, there is the type of belief that under certain circumstances can become knowledge, and second, "belief in the narrower sense" which cannot be converted into knowledge (at least not in a world in which our world's laws of nature hold). If I have reasons for believing that someone is hiding behind a door, I can go and check it and perhaps change my belief into objective certainty. If I have reasons for believing that other people have minds, however, I have no way of investigating whether it is true. This type of belief will "always remain a belief . . . , since every individual in this world can only know about his own soul in this world; while there is another kind of belief that is belief today and knowledge tomorrow, or where an effort to turn it into knowledge looks sufficiently promising. But in this matter [concerning other minds], that type of success is not possible."[29]

For Fechner the acceptance of belief in the narrower sense is where we gain insight into the necessity of metaphysics in our lives, meaning that it exceeds all factually given phenomena. We cannot cooperate normally with one another, we cannot explain social behavior, and we cannot lead normal human lives if we do not believe that others have minds or souls. Demanding empirical evidence in this case and waiving belief would be confusing "a lack of opportunity for absolute certainty of knowledge with an absolute lack of opportunity for knowledge."[30] Fechner goes on: "For us, belief in the existence of our fellow-humans' souls is a necessity, belief in the existence of animal souls comes naturally, and a worldview that includes a universal answer to the question of souls is of utmost necessity. Exact science can dismiss the question as long as it is unable to provide a precise answer; but we cannot dismiss the question altogether; and what we must do when no exact proof is to be found, is to seek a substitute for exact proof."[31]

There are other areas in which Fechner thinks we normally place our trust—our belief in the narrower sense—without profound philosophizing: natural science and the question of "the highest and ultimate things," as he calls it (the belief in God and in immortality). In natural science our belief in the narrower sense of the word is manifest in how we trust our principles of generalization and the way that we extrapolate our own limited knowledge, although we have no proof supporting these habits. We believe in the existence of atoms, without ever having seen them and although we may never see them. We project the validity of natural laws to cover cases that are still unknown (and that we perhaps might never be able to examine) and we allow these beliefs to guide our actions. Further, "Who can say, whether by experience or by mathematics or both taken together, that it is proven or provable that the law of gravity is valid for all space and time? Yet it has proven true thus far, as far and as long as we have applied it to the heavens and time. This justifies belief that is almost as solid as our sturdiest knowledge."[32] It is equally clear that we will never have direct knowledge of God's existence. But for Fechner we can only lead our lives reasonably and explain reality by believing in God's existence.

Three Motives and Reasons for Belief

Belief in the narrower sense would be superstition if we had no reasons to justify it. Which reasons function as "substitutes for exact proof"? Once again, Fechner starts with belief in the minds of others. The sort of argumentation that justifies belief in "fellow minds" can be applied to every other belief in the narrower sense, even to belief overall.

The empiricist that he is, Fechner first examines *motives*, or the causal factors that naturally compel us to believe. This makes his theory of belief a naturalistic enterprise: "If we have a belief of something that is not directly subjected to experience, then there must be some kind of motives whether man is aware of them or not, that are effective from parts of existence, and that generate this belief, or there must be human necessities compelling us to this belief."[33]

There are three kinds of motives: *historical, practical,* and *theoretical*. From historical motives we infer that others have minds because that belief was taught to us as children, we were told that it is so, all the world believes it, and has always believed it. But it is also useful and satisfying to believe in other minds, and this reflects the practical motive compelling us to believe. "We believe what we want to believe, what serves our purposes, or is beneficial."[34] Practical everyday life cannot do without belief in the narrower sense. The theoretical motive, finally, actually convinces us that other people have minds based on inference from experience and reasoning. For Fechner, inference from experience is inference from analogy and induction. He finds all three motives "as far away as possible from all metaphysics on which the philosopher desires to establish belief . . ."[35]

These three motives correspond to three *reasons* that *justify* belief. Each of the motives can be "promoted to a true reason expressed in an argument."[36] The justification to believe in something that we cannot experience directly is greater, the more widespread this belief is in the world, the more useful it is to us, and the more induction and analogy seem to indicate it. A belief is then completely founded, when all three reasons support it. If only one reason supports a belief, then that belief is improbable. The three reasons for a well-founded belief are only sufficient when they are jointly effective and reciprocally supplement one another. Purely theoretical reasons are insufficient. The only "theoretical inference permissible is one that does not contradict practical demands, just as conversely, practical demands must be compatible with theoretical inference."[37] Without a practical reason supporting it, a belief has low probability.

Today Fechner's *historical* motive may appear foreign to us. It seems absurd to think that duration and intensity of a belief are crucial for probability. But this reasoning rested on Fechner's naturalism. He refers (as Peirce later also does) to how, following laws of nature, what nature lets man believe is related to what is actually the case. In the long run, nature cannot "afford" to let man believe things lacking reality: "It would be absurd if nature had made man such that by necessity and in general he is compelled to believe in things for which no real reasons exist."[38] Of course, error can be widespread and persistent. But the dissemination of truth will continue "indefinitely," because it does not contradict other truths and human needs. The spread of error sooner or later meets with resistance because it competes with "the nature of man and things" and competition prevents errors from being circulated further: "Agreement [of belief with the nature of things and humans] must encourage dissemination and persistence, and contradiction [to the nature of things and humans] must have the opposite effect, hindering further spread and stability . . . The ultimate goal being the compatibility of everything in every relation; and this can ultimately only happen to true and good belief."[39] The less our belief corresponds to reality, the quicker we will find ourselves at conflict with our experiences and other beliefs, and the sooner we will dismiss the erroneous belief. And since the belief in other minds is perhaps the most widespread belief of all, each of us has the best reasons for believing it.

Even the *practical* motive for a belief becomes stronger over increasing time for which this belief has proven useful. Every belief influences our thinking, feeling, and action. "Knowledge, belief, action" are related in the end.[40] If a belief is erroneous, in the long run it will be disadvantageous for our thinking, feeling, and actions, by "entangling us in disgusting moods and preposterous actions that in part involve immediate displeasure and dissatisfaction, in part are accompanied by or induce painful consequences."[41] In mastering our lives we cannot find support in an erroneous belief for long. If a belief is true, it will only be to our advantage, secure the success of our endeavors and the satisfaction our actions provide us. Thus we could say that a certain belief is the truest, if it serves mankind the most. We can conclude the truth of a belief from its goodness. Fechner calls this an *argumentum a consensu boni et veri.*[42]

Now, clearly one gets much more satisfaction from believing that one's fellow-men have minds than from relying solely on immediate knowledge and finding oneself the only being in the world possessing a mind.

> We could hardly exist and thrive without believing that our fellowmen have souls . . . the most valuable feelings and efforts are attached to this belief.[43]
>
> No one can stop at the belief that only his consciousness of what exists, exists.[44]

All intercourse with fellow humans in everyday life rests on the assumption that we acknowledge each other as beings with minds. We make inferences about the minds of others for (predominantly) practical reasons.

Yet some reasons are given by "experience and intellect" and being *theoretical* reasons they cause or confirm our beliefs in the narrower sense. We feel that induction and analogy lead us to "extend," "augment," and "generalize" our own limited experience: "Generalization by induction and analogy and reasonable combinations of generalities gleaned from various sides are—in my opinion—the only *theoretical* ways and means that can lead us on the issue of mental and material reality to tenable fundaments of knowledge that are productive for experience, beyond what can be taken for granted and is given directly."[45] At the age of twenty-two Fechner had already begun to scrutinize the nature of inference made through induction and analogy, and had achieved—for his time—a fairly clear idea of them. He thought that empirical knowledge never possesses general and necessary validity because it always refers to single things. But some kinds of reasoning, namely induction and analogy, try to deduct generality and necessity from individual experience. These, however "provide merely probability and not strict certainty, because they proceed from experience. They rest on the assumption that nature adheres exclusively to law, which the intellect feels compelled to accept."[46] We infer by induction that the properties we have discovered belonging to many things of a kind are also attributable to other, not yet examined specimens of that kind. And by analogy we infer that things that exhibit conformity in their known characteristics will also be similar in their unknown features. Reasoning by analogy actually rests on causal inference. We begin with the premise that things that are similar have been caused in a similar way. Something that has occurred because of the same cause as something else will also be similar in its hidden properties.

Our theoretical belief in other minds rests on reasoning by analogy that in fact becomes effective because of a theoretical motive. It has three phases:

Phase One: I notice that certain phenomena that I experience regularly occur together and accompany one another. I find "the outward phenomena provided by my body and my actions are jointly connected to inner phenomena of the soul."[47] If I prick my finger on a rose thorn (outer phenomenon), I feel pain (inner phenomenon). If I take in food, my hunger is quieted, and so on. (Take no offense at the fact that here Fechner prematurely introduces "body," "actions," "outer" and "inner" phenomena, although the distinction of inner world and outer world cannot be made *prior* to belief in the existence of other minds. The argument is valid for every regular occurrence of series of phenomena. This way he curtails an otherwise lengthy explanation.)

Phase Two: I often experience that phenomenal processes occur in which it is not the case that the phenomena characteristically regularly accompany one another. Some phenomena are similar to other phenomena with which I am familiar by my own bodily and behavioral experience, but which are not accompanied by those concomitant symptoms later to be called inner phenomena. For example, I watch a person close his eyes. This observation lacks the (inner) phenomenon that I myself experience when I close my eyes.

Phase Three: The experiences described above allow the inference by analogy that in this case there also exist phenomena which otherwise are regularly given concomitant to other phenomena; but they do not occur in me, they are in someone else, for whom I am only aware of the outer appearances. I conclude from the similarity in appearance and effect of other bodies to the appearance and effect of mine, that the body of another person has just as much consciousness as I myself am aware of. Fechner writes: "After discovering that outer phenomena provided by my body and actions are jointly connected to inner phenomena of my soul, I assume that analogous soul phenomena are also jointly connected to analogous outer phenomena produced by the bodies and bodily expressions of other people, which, however, do not coincide with my inner soul phenomena, meaning that they do not enter my consciousness, because the bodily phenomena of us both do not coincide within me."[48] Thus we can also draw on theory for establishing our belief in other minds, even if our reasons are insufficient for objective certainty and merely suggest a degree of probability.

The relation between motives and reasons for belief in Fechner's theory of belief is not as clearly defined as one might wish. We sometimes get the impression that Fechner assumes that reasons for believing are not themselves sufficient for acceptance of what is justified. To understand this, we must include the theorem that people only accept reasons that comply to their motives. Thus (as Hume said), we must examine man's natural tendencies, if we want to explain why he appreciates inductive arguments. This suggests that whatever allegedly justifies a belief is really only subsequent rationalization for a motive (as Sigmund

Freud said). Dismissing its sporadic fuzziness, Fechner's theory of belief suggests a promising naturalistic approach already headed in the direction of the theory of knowledge later developed in pragmatism.

Self-Phenomenon and Foreign Phenomenon

The belief that there are phenomena of which one is not oneself conscious provides more than just the concept of the other's mind. Belief in the mentality of others is also a prerequisite for the concept of self and for that of the world outside of oneself. I cannot think of myself as *myself* before believing in the existence of certain phenomena (my self-phenomena) that are accessible only by me and not by others, and believing in the existence of other (closely connected) phenomena that could also be accessible by others (the extraneous phenomena of my body). To believe in the existence of another person would mean to acknowledge the existence of particular non-given phenomena (extraneous phenomena) connected to certain given phenomena (the body of the other person). To identify oneself as oneself would then mean to make the converse inference: I, myself, am something analogous to my fellowman. I am both the phenomena not given to anyone else, as well as those phenomena that occur in a regular context and could be given to another person.

Were I able to speak of myself *as* myself *prior* to having belief in another subject, I would have no criterion for sorting what is given into that which is given to me and that which is given to others: I would have to consider *everything* given as belonging to me. "Everything" would be "me." That would be senseless and would not mean identifying "me" or "I." All phenomena might just as well be given to no one, or to everyone. A person with this standpoint would be unable to distinguish between mental and material things, between self-phenomena and extraneous phenomena, between himself and the rest of the world. Fechner thinks that originally each of us passes through this sort of phase "void of differentiation."[49]

If I believe in the other person (in the aforementioned sense) and she believes in me, then the prerequisite is fulfilled that allows us both to make reference to something that we have in common. What we have in common is the realm of phenomena given to us all (or, that under certain circumstances could be given to us all) as extraneous phenomena. We call this realm the external world. Our experience teaches us that those phenomena that are given to me and to others create a lawful context in themselves. In cases of doubt, those contexts also serve as criteria for distinguishing between self-phenomena and extraneous phenomena.

Fechner discusses the question of the right to distinguish hallucinations (alleged extraneous phenomena) from real extraneous phenomena. True extraneous phenomena are embedded in a causal order on which we explicitly or implicitly rely in all our activity in the world and social intercourse with our fellowmen. A context like causal order, worked out within the commonly given realm of self-phenomena—meaning that it is "public"—and is equally binding for all, is not at the mercy of whim. It is externally conditioned and objective. Phenomena that refer to objects in the outside world are those by which

> in part corresponding, in part lawfully continued and thus interconnected [phenomena] can occur in other peoples' minds, of which we can take notice . . . The context that makes a particular phenomena . . . for example, that of a tree, possible for a mental subject solely in the proportions in which it would also be possible for others according to natural laws, justifies the assumption that there is a common cause for it that extends beyond each individual mind; this justification is missing in the case of hallucinations. Therefore we do not allow the phenomena that suggest external things to the person hallucinating to depend on real external things, because they do not submit to the lawful context of ideas that we all have concerning the external world. If there were no such context valid for everyone, we would have no theoretical justification for taking the phenomena of the external world as anything more than the sum of subjective hallucinations.
>
> We must first pass through belief in other minds . . . before finding a theoretical *justification* for belief as contrasted with hallucinations, so that for each one of us things really objectively correspond to the appearance of material external things, meaning that they (also) exist for others.[50]

Just as in the case of hallucination, where we could mistake self-phenomena for extraneous phenomena, there are also some cases where we might mistake extraneous phenomena for mere self-phenomena. This might be the case when one says, "I don't believe my eyes." Beyond this, we may also consider every phenomenon correctly classified as an extraneous phenomenon to also be a self-phenomenon, if we ignore the object of experience and view the experience itself as a special state of consciousness. The perception of a tree and the modification of the state of self induced by being in the state of perceiving a tree are one and the same phenomenon, seen from different viewpoints. The difference becomes obvious when I grasp how one phenomenon is *related* to other phenomena: "And what we call objective about sense experience in comparison to subjective perception is never something existing outside of perception; instead both, that which is objective and that which is subjective are the same phenomenon, merely understood in different relations . . . What is objective is an aspect of a sense phenomenon directed towards our fellowman, what is subjective is an aspect directed towards one's own mind. Yet both are merely two sides, not two things."[51]

It seems clear that Fechner is trying to provide a "logical construction of the world," a "constitutional system" like Carnap's in the early twentieth century.[52] It begins with what is immediately given, or what can be directly shown. This Given is at once without subject or object. Neither a relation to an ego, nor relation to an object is an original property of a phenomenon. The Given, in itself, is neither internal nor external. Our natural inclination is to allow three motives to cause within us a belief in the existence of non-given phenomena. The most important instance of such a belief (and simultaneously for Fechner the first case through which other, similar beliefs first become possible) is the belief in the existence of other minds. This belief enables us in human communication to determine and distinguish two areas of the Given: the realm which is never given to the other mind, being the realm of self-phenomena, and the realm which (at least in principle and under certain circumstances) is given to us all, namely the realm of extraneous phenomena. Extraneous phenomena are what we call the physical world, while my self-phenomena, which the other mind cannot have, and phenomena of which I believe but cannot know that another mind has them as its self-phenomena, make up the mental world. The distinction between physical and mental world, between outer reality and inner phenomenon is not made once and for all for the individual; depending on experience, it can shift and is always revisable.

In conclusion we can evaluate all this as follows: Fechner's phenomenalistic design of the world reflects *neutral monism*, probably even the first variety of it.[53] The phenomena which constitute the world are originally neither physical nor mental, they first become physical or mental by their specific position in the total constellation of relations holding among all phenomena. In terms of neutral monism, the ego, as Manfred Frank once keenly observed, does not owe "its knowledge of itself to a foregoing of 'familiarity with itself,'" but instead must learn it by studying phenomena and their linkages.[54] Fechner's neutral monism thus implies—so to say—a *relational, non-ego-logical* theory of consciousness: Mind, consciousness, self-consciousness—all these are structures of relations existing among experiential data. Likewise, identifying phenomena as external and objective by making reference to other conscious beings and by communicating with them is also something that must be learned.

Endorsing this kind of neutral monism, Fechner broke remarkably with early modern philosophy of mind. The traditional tenet was that self-certainty of the thinking subject, the knowledge of the mental and physical, is prior to all else and occupies a privileged rank in two respects: 1) It is more primitive (being more direct, familiar, perceptible, acquainted, and so on) than knowledge of the body and other external things; 2) its ego-relatedness, or subjectivity, belongs to it from the onset and can be recognized without any further experience. Fechner denies both claims: Knowledge of one's own mind is neither superior to knowledge of the external world, nor is it primordial to it. Without the "other," without what is "external" we cannot meaningfully say that something is "internal."

Even Carnap, in *The Logical Construction of the World* (1928), assumes that an "auto-psychological" basis is fundamental. The physical world is built from this auto-psychological basis, and then in a further step, the psychical world of others is constituted from the physical world. The physical world is prior to the psychical world of others.[55] Fechner reverses this order: The belief in the psyches of others constitutes itself from the Given as a process following laws of nature; from this we constitute our own psyches and then, finally, the physical world. In a sense, then, Fechner overturns the Cartesian *cogito*. He does not say: I think, therefore I am. Instead, he claims: The *other* (mind) thinks (perceives, feels . . .), therefore I am, and therefore there is a world outside of myself.

I would like to emphasize once more, that for Fechner the ego does not identify itself by reference to its own experiences (self-phenomena). Instead, it is those phenomena that later will be identifiable as extraneous phenomena of my own body that are decisive. To believe in other minds means to acknowledge the existence of certain non-given phenomena (the extraneous phenomena) in relation to certain given phenomena (the other person's body). If phenomena are linked such that they constitute an ego, then this is because some relation is established between given phenomena, which experience shows to also be given for others (my body) and other given phenomena that exist in a normal context and are not given for others. Were the ego of my consciousness merely a special property of my own psychical phenomena (and not a special *relation* between psychical and physical phenomena), no belief in the minds of others would occur. Without this belief there would be no way to distinguish between "me" and "the world"; the two concepts would be co-extensive. Using the designation "I" would not imply any real identification. But, this sort identification does in fact happen.

Thus "I" does not refer to a special property of what is psychical, it refers to a psychophysical context.

In contrast, early modern tradition teaches that knowledge of one's own human mind is primordial to knowledge of the body and that the ego is something whose identity is determined by thinking; we imagine it to lack bodily properties.[56] Fechner broke with this conception.

2.2 Fechner's Mind-Body Theory: The "Identity View"

The Phenomenal Contexts "Thing" and "Soul"

We looked at Fechner's reasons for distinguishing psychical from physical phenomena. Since we normally view perceptions as something different from the objects of perceptions, we must now examine what extraneous phenomena have to do with objects and what inner perceptions have to do with consciousness, or the soul. Once again, Fechner is guided entirely by his empiristic attitude. We could say that for him a thing (a soul) is what is perceived. Things (souls) are "collections of phenomena." One thing (a soul) is distinguished from another thing (another soul) by the context in which it collects the Given.

The most refreshing and detailed, and probably also the earliest treatment of this thesis is provided in Fechner's *Theory of Atoms* (1855):

> This orange that I see: I can also touch it, smell it, taste it, peel it, hear a sound by striking it; I can do this not only now, I can repeat it; not only I can do it, but innumerous others can also; and the sum of these possibilities, interconnected and yet limited in certain respects, this possibility of innumerous perceptions represents in us an objective thing, which consequently naturally consists of more than the momentary individual perception or a limited number of such perceptions. On the contrary, in addition to all the individual perceptions of a thing there always still remains something that can provide even more innumerous perceptions; and we easily hypostatize this as some unrecognizable thing, hidden behind them.
>
> But this dark something is nothing other than an unclear interconnected possibility of perceptions itself, which can be linked to the Given.[57]

Here we have an attempt to formulate a phenomenalistic theory of objects very reminiscent of similar, better-known theories later suggested by John Stuart Mill, William James, and Bertrand Russell. Just as Mill in *Examination of Sir William Hamilton's Philosophy* (1865) conceives of matter as the "permanent possibility of sensation," so also for Fechner is the material object nothing other than a lawful context of real and possible perceptions:

> Due to the fact that a real object cannot be described solely by the perceptions it actually arouses, but also by those that could be given under other circumstances (circumstances which themselves in the last instance can always be characterized by the context and relations of phenomena) it becomes necessary, if we want to perfectly describe an object, that we know what it would be like under all *imaginable* circumstances, right down to the most extreme case.[58]
>
> Matter—in the meaning of the term as used by physicists—is only the fixed point of lawful links among phenomena . . . , which enter many minds.[59]

Fechner even considers it an "identical proposition"[60]—in other words: an analytical claim—that objects consist of the relation of their perceptions.

The tenets of *the antithesis* (which Fechner considers responsible for poor philosophy) expounded by coeval philosophers (and this angers him) rely on the existence of the "noumenon." The antithesis refers to two sets of theories: First, those theories that assume an existence of things-in-themselves "beyond" our perceptions of things; but also philosophies that assume something immutable behind what is changeable, something that does not itself appear, but that is the bearer of perceptible properties. Fechner, then, uses the term "noumenon" not only in Kant's sense, but also subsumes thereunder the concept of substance (e.g., in Spinoza's sense), the concept of the absolute as found in Schelling's and Hegel's philosophies and the concept of the "real being" (i.e. the monads, the "reals") used by Johann Friedrich Herbart.

By scrutinizing those conceptions Fechner advances his own empiristic theory of belief. Neither can the thing-in-itself be shown in experience, nor is there the slightest—ever so weak—reason for believing in it. Absolutely no consequences ensue for the explanation of physical reality from assuming the existence of some thing-in-itself. Therefore, talk about things-in-themselves is nonsense: "Every noumenon sought behind the phenomena is nothing, it is an absurdity, an essence lacking essence. . . . I *call* that nothing which does not appear, which cannot appear, and whose being cannot be inferred from appearances using rules that stand the test in the mind's and nature's world of phenomena."[61] The noumenon is superfluous and suggests theory that ultimately

> leads us into the dark, past all experience and everything that is imaginable based on our experience or can be proven in experience; because we have neither an experience, nor an idea based on experience, nor prospects of explanation by experience

for a thing-in-itself or things-in-themselves that are behind consciousness, that
affect consciousness, or that affect actions through their consciousness-generating
effects.[62]

The laws of things, i.e., the law of the relationship holding between the appearances of things is . . . fixed for eternity. For what, then, do we need a solid thing
lurking behind the phenomena? Does it make the law, or make it indispensable, or
explain it? . . . Why do we want a dark, solid thing lingering in the background?
Whatever is already sturdy does not need to be stuck to some other immobile
thing.[63]

Fechner's work is replete with similar sharp remarks about noumena. Fechner
also wrote a radical critique of the traditional concept of substance and essence.
The 'essence' of a thing is nothing but the lawful link between phenomena:

Actually the concept of essence . . . is merely an auxiliary term which can be
clarified and eliminated by tracing it back to its original, demonstrative meaning
and performance; but suing it has the advantage of abridgment and representation
consistent with the general usage of the term 'essence'.[64]

There is no *essence* of things other than their unalterable conditions.

Whatever phenomena are unalterably or lawfully related in such a way that we
can have one in its correct proportion, only inasmuch as we have or can have the
other, are things of the same essence; and this relationship is the very center, it is
what is essential about essence.[65]

But if the essence of things is not something "behind" their appearances,
why is it that we speak of things at all, instead of restricting all our discourse to
talk about appearances? Fechner answers this question by pointing out man's
natural need to "conceptually hypostatize" the relation holding between appearances.[66] When we speak of things instead of appearances we are expressing
our belief that the phenomena which *would* occur, *if* the right conditions pre-
vailed, would be lawfully connected to the phenomena now given to us. The
concept of "thing" is—just like the concept of laws of nature—a device for tran-
scending what is given, allowing us to simultaneously also conceive the non-given
as something that—based on our experience—*would occur* if certain other phenomena were given as the conditions. Since it is never the case that all the aspects
of a thing are perceivable at the same time, such that only parts of it appear to us,
the concept of "thing" designates a "higher form of reality" than can a concept
related only to one single appearance. A thing is "whatever coheres individual
phenomena, or what must be conceived of as the origin, the basis, the end of an
idea and final point of a conclusion; whatever it takes to link ideas in our minds

and infer the unknown from what is known, although it never appears itself, and indeed, is only conceivable as something abstract in itself."[67] The concept of "thing" therefore combines appearances to form a unit and enables inference about what is unknown. Just as David Hume's suggestion that a singular causal claim implicitly contains a general statement about regularity, so also for Fechner does the concept of thing imply that we assume regularity. He sometimes implies this by saying: Not only do phenomena exist, but also the relations holding among the phenomena.[68]

What is true for the physical world applies to the world of consciousness. A mind is "a complex of real and imagined phenomena linked by the unity of consciousness, such as are recognized internally as sensations, thoughts, and so on."[69] Here again, the unity underlying the concept of consciousness or mind is only an expression of the law-like linking of phenomena. A transcendental essence *behind* the phenomena, such as a transcendental ego or a substance that supports the phenomena but does not itself appear, is just not plausible:

> The connection and sequence of appearances in the soul or self is subject to psychological laws. How can an obscure thing, called a monad, help a real being explain the unity that forms the laws governing the soul's phenomena?[70]
>
> We have just as little need for some obscure thing underlying one's own soul as we do for an obscure thing supporting bodies, holding their diverse and changing appearances together.[71]

We will return later to investigate whether these terms allow the unity of consciousness.

It should be clear by now that when reading Fechner we should not be mislead by the mystical aura surrounding the German concept of soul that tends to conjure suspicion of metaphysics. Fechner's concept of the soul—at any rate—is a clear concept and no less metaphysical than his concept of "thing." (Whether it is tenable is another matter.)

Definitions given by Fechner in *On the Soul* [*Die Seelenfrage*] summarize this discussion: "*Mind*, or *soul*, means something that is comprehensible only in its appearance for itself, a unified being characterized only by self-phenomena and the rules of self-appearance; *corpus*, or *body*, is merely a system of external appearance, grasped via the senses, and characterized by relations and rules of external phenomena. *Nature* is the whole system of physical things, of which our bodies are only a small part."[72]

The Body-Soul Relation

Now that we have dealt with Fechner's concepts of "body" and "soul" it is time to examine his view of the *relation* between them. The theory of phenomena developed thus far must be evaluated in terms of whether it can more precisely specify and explain the mind-body relation.

Fechner's own solution to the problem is to interpret the body and soul as "identical," calling his mind-body theory an "identity view."[73] His phenomenalistic approach includes two different ways of saying that two phenomena are identical to each other. In one sense, it can mean that the context of appearances that constitutes one phenomenon is numerically identical to the context of appearances constituting the other. An identity view would then claim that the concepts "body" and "soul" refer to one and the same context. But in another sense it can also mean that the context of appearances captured by the concept of something is so closely connected to the appearances of something else that we can only speak meaningfully of one single context, or, of one and the same identical thing. Fechner's identity view means the latter: "The whole world involves such examples; they prove that what appears to be a single thing seems— when viewed from two different standpoints—to be two separate things, and one does not make the same observations from one standpoint as one does from another."[74] Applied to the body and soul, means that the phenomenon "body" and the phenomenon "soul" are to such an extent "one single thing"—their appearances being so tightly interwoven—that we cannot really say they are two separate things, it only makes sense to say they are *one* thing, as it were: a 'body-soul thing'.

What kind of a relation holds between body and soul, compelling us to say they are one thing? It is not simply one system comprised of two parts. We say a 'machine' is one single thing, although all of its parts are themselves things. But the relation that Fechner sees between body and soul is much closer. It is the relationship between the *ways of appearing* (classes of characteristics) that make up a single thing. We never simultaneously see both sides of a coin on the table (without using auxiliary equipment), no matter how hard we try. We have only one view from *one* limited class of possible views (i.e., one view from the class of views of the heads side, or one view from the class of views of the tails side). The views of one side of the coin are views from *one* way of appearing (or from one context). If we turn the coin over, another way of appearing (or, another context) is given, namely the class of views of the coin's other side. Although we have never seen both sides at once (except indirectly in a mirror), we still say it is *one single* coin. We do not claim to see at two different things, even though every aspect we observe about the coin can always only belong to one of two possible classes of perspective (or, ways of appearing).

Why is the coin one single thing for us, and not two things? Well, we know from *experience* that both classes of appearances are closely related. If I see from one side that the coin is bent, then I also think of the other side as changed; the diameter is the same when seen from both sides; there may be many experiences like this.[75] Fechner expresses this as follows: The possibility of one contextual appearance is "jointly" related ["solidarisch"], or factually related following the laws of nature, to the possibility of the other: "We attribute appearances, properties, mutations, and definitions to one and the same being, when these (characteristics) are connected according to laws of nature in such a manner that the possibility or reality of one by itself elicits the possibility or reality of the other. This is all that can be shown about their common existence."[76] We can apply the coin analogy to the body and soul. The body is 'one side of the coin,' the soul, the other. The contextual appearance of 'body' is as closely connected to that of 'soul' as is normally the case within the lawful context of one single thing. If I view another human being as a unity of body and soul, thus as a person, what I mean by this is that to the perceptible appearances of her body also belong other (for me imperceptible) phenomena so closely connected to that body that a change in one context results in a change in the other.

Although people are nothing more than their phenomena, there nevertheless exists a fundamental difference between the appearance of physical objects and the appearance of persons: For physical objects I can, in principle, discover every possible way in which they appear by altering my perspective (Fechner calls this "changing my standpoint")—although these ways of appearing may be too numerous to exhaust within one lifetime. I can do this, for example, by altering my position relative to the object, by wearing eyeglasses to improve the view, by experimenting with it, or by creating some conditions under which another (up to this point merely potential) aspect of the object manifests itself. In the case of viewing another person, however, even an indefinitely long life would not enable me to exhaust all of that person's possible ways of appearing. For each person there exists namely a context of physical phenomena ϕ as well as another context of phenomena M ($M \neq \phi$), which, together with ϕ constitutes the context T of one single thing, such that this M is given only to this T (thus, M is a self-phenomenon of T). The inner phenomenon of another person is not something that can be given for me.

How does Fechner express this? Fechner's most intent analyses of the mind-body problem can be found in the *Zend-Avesta* and in *On the Soul*. The *Zend-Avesta* contains this summary of the identity view:

> Body and mind or body and soul or matter and ideas or the physical and the psychical (these antitheses are understood here as being synonymous in the *widest* sense) do not differ in ultimate reason and essence, they differ merely in terms of the standpoint from which they are interpreted or observed. What seems to oneself, perceived from the inner standpoint, to be something mental or psychical, appears in a different way to someone outside of you; from his external standpoint, it may seem to him to be a bodily, material expression. The difference in appearance depends on the difference of standpoint of observation (perspective) and the person standing there. The same being has insofar two sides, one mental, or psychical, inasmuch as it appears to itself, and one material, or physical, inasmuch as it appears to someone else. The body and mind or body and soul are not stuck together as two essentially different beings.[77]

Each of us gains this insight from personal experience: our own bodily and psychological experiences are bound together. In *On the Soul* Fechner writes:

> Experience teaches us that each existing soul has one body made for its external appearance, through which one soul is in contact with others . . .
>
> In other words: Self-phenomena can only be interconnected, if they can be perceived by others as connected phenomena. . . .
>
> The common identical essence of body and soul is nothing other than a mutual reciprocality between self-phenomena of the soul and the external phenomena of the body. You cannot have one without the other, there is nothing to be shown or found posterior to it. . . .
>
> This view is wholly an *identity view*, in that it holds both, the body and the soul, to be merely two different manners in which *one and the same being* appears, one way is as perceived from the inside, the other from the outside, and the essence underlying both ways of appearance is nothing but their inseparable mutual interdependence.[78]

But if body and soul are one single thing, then obviously we must also view as identical those physical and mental *processes* to which persons are subjected:

> They are basically the same processes that are interpreted from one perspective as bodily or organic processes and from the other as mental, or psychological

processes. They appear as physical processes to someone viewing these very processes from an external position, or as something that can be studied in the external form in which it is perceptible, in the manner of the anatomist, physiologist,
or physicist. Try as he will, the scientist cannot directly perceive even the tiniest bit
of psychical phenomena in another person; yet these processes are perceptible as
mental processes, namely as feelings, sensations, ideas, desires, and so forth, as soon
as self-perception occurs within them.[79]

Fechner's theory would be worth only half its weight, had he not more carefully specified the connection between body and soul. It is not enough to merely
claim that there is a close, "mutual" [solidarisch] relationship based on laws of
nature. We need to explicitly formulate a law that factually holds for this special
relation. The challenge is to pinpoint the circumstances under which it is appropriate to consider two ways of appearing as two sides of one and the same
thing, and to isolate these from the circumstances under which it is inappropriate. On Fechner's interpretation two ways of appearing can be counted as belonging to one and the same thing as long as for *every* change in one way of
appearing there also occurs a change in the other. That is Fechner's central criterion. Now, quotidian experience and science show that psychological changes
are accompanied by changes in physical events. Nothing is wrong with generalizing finite experience and formulating a universal law: Whenever the soul changes,
the body changes also. The law does not contradict our experience and progress
in medical science confirms it.

Fechner calls this law "the basic law of psychophysics" or "the most general
law" of psychophysics, and sometimes also the "functional principle":[80]

> The most general law is this: Nothing can exist, develop, or move within the
> mind, without there being something in the body that exists, develops, or moves,
> whose effects and consequences reach into the present and future physical world.
> In short: All that is mental is borne by or expressed in something physical and by
> this means has physical effects and consequences.[81]
>
> Proportionate to the degree of similarity among mental circumstances, the same
> degree of similarity will be evident for the corresponding material phenomena. In
> other words: For whatever is the same or different in the mental realm, there is al-
> ways something equally similar or different in the physical realm.[82]

It is important to note that Fechner's functional principle expresses an *asymmetrical*, conditional mind-body relation: Whenever a person changes psychologically, something about him also changes physically. The reverse is not necessarily
true.

It is remarkable how Fechner anticipated what is known today in analytical philosophy as the "supervening" of psychological properties over physical properties. As Donald Davidson, the most eminent advocate of this position, says: The dependence of mental properties on physical properties can be understood as supervenience in the sense that "there cannot be two events alike in all physical respects but differing in some mental respect, or that an object cannot alter in some mental respect without altering in some physical respect."[83] For Davidson, as for Fechner, the supervenience relationship makes a nonreductive version of materialism possible.[84]

It is now time to return to the hypothesis sketched in the first section of this chapter that within Fechner's theory the terms "psychical" and "physical" act as law cluster-concepts (in Hilary Putnam's sense). At that point, we were unable to state a law including "psychical" and "physical" as essential parts. But now we have Fechner's basic law, which is just that. To demonstrate that Fechner's definitions of "physical" and "psychical" (meaning the world of extraneous phenomena and self-phenomena) may, under certain circumstances, be of a non-analytical nature, it must be plausible that we—in light of new experiences suggesting that we revise Fechner's basic law—would also consider Fechner's definition obsolete.

Let us assume that in the future we discover cases of changing psychologically, without altering physically. This would prove that mental and physical groups of phenomena are two different things, which—in one direction or another—can affect one another. It would also show that we can no longer reconstruct the constitution of the self and the physical world using the belief in the psyche of the other (persons) as a basis. In this case we would also have to change the definition of "physical phenomenon" and "mental phenomenon." The physical appearances of a person would now be one isolated object and her mental phenomena would be another. A definition determining when an object should be interpreted as physical or mental would have to rest on special properties of these *objects*, not on the *phenomena* constituting them or on the relationships among those phenomena. In a dualistic world it could be *empirically* true that, while in fact, as a sort of cosmic coincidence, all physical objects are intersubjectively accessible, nonetheless every mental object is accessible only to one subject. But (at least for a phenomenalistic conception), this cannot be used as a *definitional* criterion for the objects 'body' and 'mind'. For the dualist, one object is distinguished from another solely by the specific lawful connection joining their appearances, but not by being a special kind of appearance or way in which these may be given. An object remains identical to itself, whether it is given for you or for me. Fechner would say, in contrast, that phenomena change, depending on whether they are given to one or to us all.

Put the other way around: Is a new instance of experience conceivable that would compel a dualist to alter his definitions of 'mental' and 'physical' objects? At least not in a world that entirely corroborates the identity view. All experiences that confirm the identity view also substantiate dualism. We have seen, however, that it is theoretically possible for some experience to confirm dualism while refuting the identity view. Naturally, this makes it much more difficult to defend the identity view than to endorse dualism. Since (at least for Fechner) no experiences are known that refute the identity view, the dualist relies on what is unknown. He believes that we may someday discover that mental changes are possible without being accompanied by physical changes, thus rendering the identity view false. Therefore, Fechner's definitions stating that the physical is what appears to others and the mental is what appears to oneself are based on the validity of his principle of psychophysical function.

Fechner's identity tenet is an attempt to make the subjectivity, irreducibility, privileged status, and intangibility of the inner phenomena of *other* persons empirically acceptable by assigning them an epistemological status equal to that of *potential* appearances of things. We required the category of potential phenomena for defining the physical object, without considering it problematic. The fact that the inner phenomena of other people always remain concealed to me is no more problematic than the case in which external appearances actually belonging to the contextual phenomena of a thing remain invisible for me because I cannot create conditions under which those phenomena would reveal themselves. In such cases I can refer to previously known phenomena and try to *infer* with probability that one or the other phenomena would occur, if the given condition were fulfilled. The same method applies to the inner phenomena of other people. Combining the external phenomena that I perceive as belonging to the other person with the inner phenomena that I myself experience when I am in that same physical state allows me—with varying probability—to infer the existence and the kind of inner phenomena that the other person is experiencing. If I could get my body to be in the exact same physical state as that of the other person, I would be qualitatively identical to him for the duration of that state and the same inner phenomena would be also given for me. The fact that I am not qualitatively identical to another human being (and we are thus not given the same inner phenomena) is not true a priori; it is merely a contingent fact.

Evidence and Conclusions

The double aspect theory supports what Fechner calls "the parallelism of the mental and the physical."[85] The physical and the mental are strictly *parallel*, but neither is *causally dependent* on the other. This follows from the assumption that body and soul together constitute one single thing. We speak of causation when one thing effects a change in another. It would be nonsense to say that there is a cause-effect relation between the two ways of appearing of one and the same thing. Changes in the world are changes of things. Therefore, if the 'mind-body thing' changes, then by analytical truth its ways of appearing must also change: "Change in the physical and the mental is factually parallel. In fact it *must* be, if body and soul are two ways of appearing of one and the same being, because proportionate to how that being changes, both interwoven ways of appearing must also change."[86] Take again the example with the coin: One side of the coin is not the cause of the other. The reason that one side has a dent is not that the other side was bent; it was caused by the effect of the pliers that I used to bend the coin. Using the pliers is the mutual cause for the changes to both sides of the coin.

Contrasting this to causal dependency, Fechner calls the mind-body relation a "functional relationship." A functional relationship, expressed in modern terms, represents a *law of coexistence*, not a law covering interaction or change.[87] But the change to which the mind-body as a whole is subjected is different. Here, of course, we can speak of causality and if we so desire, we can formulate the present law separately, once for the physical and once for the mental realm—depending on the context we are examining. Since body and mind functionally depend on one another, every psychological change can also be viewed as a physical change, and many physical changes can be considered mental changes: "Common opinion is that the physical influences the mental and alternately the mental influences the physical . . . Our interpretation is not that two heterogeneous beings effect one another, but instead that basically there is only *one* being that appears differently from different perspectives [standpoints]; not that two different causal contexts occasionally interact: there is only *one* causal context, one that is prior to and present in *a single* substance in two different ways, from two different perspectives."[88]

Fechner also sets up a general criterion for functional dependency: "Generally we call the mental a *function* of the physical, saying that it *depends on it* and vice versa, inasmuch as a constant or lawful relation exists between the two, such that we can infer the existence of and changes in one from the existence of and changes in the other."[89] Now, if that causal relationship existed, we would be able to infer a change in one element from a change in the other, granting that we know the relevant governing laws. This means that the causal relation is a sub-relation of the functional relation: If one set of phenomena is causally dependent on another, then it is also functionally dependent on it (but not necessarily vice versa).

Fechner's mind-body theory stands and falls with the issue of the criterion that determines whether a real causal relation holds between two given sets of phenomena, or whether it is a functional relation that is not simultaneously a causal relation. In other words: How can we clearly distinguish the relations between phenomena that together make up one single thing from the relations *among* those single things, or more precisely: the relations between the appearances of different sets of phenomena?

Unfortunately Fechner neglected this question, so we're on our own at interpreting it. It would probably suit our previous interpretation the best to say this: Between two phenomena there exists neither a functional nor a causal relation, if and only if it is not possible to infer a change in one phenomenon from a change in the other. A phenomenon X is functionally dependent on another phenomenon Y (is parallel to it), if, from *every* alteration of the phenomenon X we can *always* infer an alteration of the phenomenon Y. Two phenomena are then causally dependent upon one another, if from an alteration of one we can infer an alteration of the other *only under specific contingent circumstances*. If two phenomena are reciprocally functionally dependent, but neither is causally dependent on the other, then this type of dependency can be taken for a sufficient condition for two appearances belonging to one and the same thing.

For example: If a billiard ball rolls across the table, the change in the location of a dot on one half of the ball's surface is always accompanied by a change in location of a dot on the other half of the ball. Both phenomena are thus functionally dependent on one another and belong to the same thing. However, if the change in location of one dot leads to a change in the location of another dot only under certain circumstances, then the phenomena in question are causally dependent upon one another. This would be the case, for example, if I make one ball collide with another and thereby set a dot on the other ball in motion. That movement occurs, however, only in the special case where a collision has actually taken place. If the first ball passes the second, then the change of location of a dot on the first ball cannot set a dot on the second ball into motion.

Applied to the mind-body relation this means we are justified in considering the body and the mind two appearances of one and the same thing, as long as a change in physical phenomena accompanies *every* psychological change.[90] It implies that the mind is functionally dependent on the body, but not causally dependent on it.

By way of explanation, Fechner compares his mind-body theory to that of Leibniz. For Leibniz the physical also corresponds to the mental, without the one being evoked, effected, or caused by the other. In this sense, Fechner's "parallelism of the physical and mental" does indeed remind us of Leibniz's pre-established harmony. But instead of viewing the body and soul as two synchronized clocks that otherwise have nothing in common, as Leibniz saw them, Fechner joins the body and soul to make one single "clock," "which is aware of itself as a mentally active being, while it appears to others to be material gears' and wheel motion. What mortises the phenomena is not pre-established harmony, but the fact that their basic essence is identical."[91]

Fechner believes that someday his double aspect theory will be understood as an explanation for the unity of consciousness. He thinks that the unity of self-phenomena, roughly speaking, rests on a special relationship holding among those parts of the body that are connected to those self-phenomena. If the individual parts of an animated body are related to each other by a special kind of reciprocal effect and organization, this leads to unified consciousness. The unity of self-phenomena is thus the inner aspect of a particular kind of unity and organization in a physical system. Individual instances of consciousness are connected and unified by the unity of the physical system on which they depend. Fechner asks, "Why . . . should the soul—a unified being that bundles diversity within itself—not be tied to a physical system that itself is a network of diversity?"[92]

The empirical question of which material property on the psychological side of a system actually leads to consciousness of the self cannot be answered a priori. Fechner admits that he does not yet know

> *why* the appearance of inner things that seems unified and relatively simple (the mental) becomes diverse when becoming an outward appearance . . . I do not know *why* this is so, I only know *that* it is so.[93]
>
> How something (physical) can appear much more complex to another person than it does to *itself* (when seen from the mental side) cannot be explained—being just a fundamental fact—but it can be described like this: Juxtapose a system containing five points with another system containing five points, such that each system perceives the total connection of all its points to be one whole, in such a way that

variations in the number and arrangement of the points create merely varying intensities and qualities of simple sensations. Now, one system is not linked to the other system in the same way that each is connected internally, for we do assume that they are two different systems. Thus the interconnectedness of another system will not be felt as strongly as one's own interconnectedness, but rather, we are affected by each point of the other as if it were singular.[94]

However, Fechner does find the "fundamental fact" that "in comparison to the physical, the mental has a relatively unified and simple nature, that can be considered its expression" very important. In generalizing this fact he formulates a general "synechiological principle" stating that "the mind is the linking principle of the physical constitution [organization] and sequence [development over time]."[95] In another passage he uses the term "resultants": Psychological unity and simplicity are resultants of physical diversity, physical diversity brings forth unified or simple resultants.[96] By "resultant" he meant what we today call "emergent property," i.e. a property of a complex system that cannot be reduced to the properties of the individual components of the system and the way they are arranged.[97]

Fechner finds his synechiological principle as important as the "fundamental law of psychophysics." He mentions Herbart and Lotze as earlier attempts of monad doctrines to "make the unity of the soul depend on the metaphysical simplicity of a fundamental being underlying the diversity of soul-phenomena."[98]

We could dismiss Fechner's identity view by saying that it is more natural and appropriate to consider the soul as being something separate from the body, as we do in everyday conversation. Fechner himself concedes that the phenomenal contexts of material things follow different laws than the phenomenal contexts of the soul. Thus it seems utopian to want to prove that the body and the mind-body, but not the "soul," can be thought of in isolation in any meaningful way. Fechner's theory may be compatible with our experience, but it lacks every merit that a two-substance doctrine has. It questions our normal way of speaking, without replacing it with anything really new.

Yet certain criteria for simplicity are fulfilled by the identity view, making it better than other theories. The first minimal requirement that any mind-body theory must meet is that it provide a *description* of those relations between physical and psychological phenomena that we already know to exist—a description of "empirically known relations," that "without involving the metaphysical aspects of this debate [on mind-body doctrines] more related to the so-called essence [underlying the appearance] than to the appearance, . . . uncovers as precisely as possible the factual functional relationships holding between the realms of appearances of body and soul."[99] But a mere description of all previous sequences of phenomena does not constitute a theory. It first becomes a theory when we can also say that certain *law-like* relationships exist between individual phenomena (or between classes of them), or when we can speak of things and the relationships among them, instead of speaking of phenomena. But since claims about law-like relationships always entail more than a mere description of previous experiences (for instance, speaking about an object's noumena exceeds direct experience), every theory is more or less metaphysical.[100]

Now, the identity view has the great advantage that its metaphysical baggage, if any, is less than that of any other mind-body doctrine, making its explanation much simpler. We could even call it a view lacking metaphysics because, as Fechner sees it, it contains no more metaphysics than an ordinary law of physics contains when it is applied to cases that have not yet been observed. The identity view is the single theory that is—as it were—a purely descriptive theory, and which "comprises nothing other than what is factual to one unified point of view, and allows us to concentrate on the facts."[101] A theory that orders phenomena by assuming *one single* kind of lawful relation naturally rests on less assumptions and is simpler than a theory that assumes *two (or more) different* relations. It is weaker to claim that only a *functional* relation holds among phenomena (a certain "relation of appearing"), than to begin with a "relation of dependency," in other words: to begin with the *causal relation* between (sets of) phenomena in the defined field. Thus the theory that body and mind are aspects of *one* thing is, indeed, in a harmless way, not entirely free of metaphysics, inasmuch as it implies that even in the future we will not be able to discover relations holding between the psychical and the corresponding physical phenomena of conscious beings, which would make it necessary to "split" this being into two different "things" (and thereby slip into substance dualism). As long as no such experiences are known, however, any choice from among all the various metaphysical options (materialism, idealism, Spinozism, monism, substance dualism, interactionism, monadical doctrine, metaphysical parallelism of the occasionalists) is permissible. Although it may contain some residuum of metaphysics, the identity view obliges us to less of it than any other option.

If we assume that in fact no known empirical refutation of the identity view is available, we can say that the double aspect theory is *empirically equivalent* to all other metaphysical positions, in terms of all proven "basic facts" about how psychical and physical phenomena are related.[102] Every other theory additionally assumes a cause-effect relation at precisely the point where the identity view already (and only) says there is a functional relation. For materialism, interactionism, and dualism changes in physical things are the *causes* of mental phenomena, while the converse is true for idealism and monadical doctrine. But since every causal relationship is also at least a functional relationship (not always vice versa), the dual aspect theory constitutes—as it were—the 'phenomenological core' of every metaphysically stronger theory. Lacking any fundamentally new experiences, it is most reasonable to accept the double aspect theory. It has the least metaphysical ballast.[103]

The Identity View and Modern Identity Theory

As we have seen, Fechner called his mind-body theory an "identity view." This does not mean that we could directly classify his tenets as "identity theory" in our contemporary meaning of the term.[104] It is more appropriately considered a "*double aspect theory.*"[105] Of course, this is not entirely satisfactory, because "double" can be misleading; one thinks of *ontological* dualism or (as Leibniz does) that there is no connection between the two aspects. This is precisely not what the double aspect theory is about; instead, it is about the *dualism of properties*.

Although double aspect theory and identity theory differ, they are not entirely separate. Identity theory is actually a *special case* of double aspect theory. Popper correctly claims that psychophysical parallelism was the "beginning of the modern physicalistic identity thesis." D. M. Armstrong, a clear advocate of identity theory also considers the "dual attribute theory" (his name for double aspect theory) a variation of identity theory.[106]

Before more closely examining the relation of double aspect theory to modern identity theory we should return to the claim made at the outset of this chapter that Fechner's non-metaphysical philosophy is a form of *nonreductive materialism*. We can now better understand how Fechner's identity view as the heart of his philosophy is simultaneously materialistic and yet nonreductive.

We can distinguish between two kinds of materialism: *ontological materialism* and *explanatory materialism*.[107] *Ontological materialism* claims that all states of affairs existing in our world are realized *in virtue of physical states of affairs*. Imagine, for example, a porcelain cup. A statement saying that this cup is breakable is true in virtue of the fact that it is made of porcelain. The fragility of the cup is manifested by the fact that the cup is made of porcelain. But not everything is fragile because it is made of porcelain; some things are fragile because they are made of glass. While we use the word "because" to express the connection between fragility and material constitution, it is clear that we do not mean a causal relation in this case: being made of porcelain is not the *cause* of being fragile.

Applied to the mind-body relation this means: The fact that some physical systems exhibit mental properties is true *in virtue of the properties* attributed to them. Certain physical properties are sufficient for the existence of certain mental properties (but not unconditionally necessary). The fact that humans possess mental properties is manifested by human biological organization.

This is precisely Fechner's opinion. Whenever he deals with the mind-body relation he emphasizes that no psychological thing is conceivable that does not have a material "carrier." Matter is the "basis" of the mental, the "direct" and "immanent condition" of the mind.[108] He explicitly condemns all idealistic conceptions claiming that matter is "a posthumous product of the mind, or unilaterally dependent on it."[109]

> In one respect the identity view is wholly materialistic, for in it the mental must change at every time and in proportion to how the physical changes, the way it expresses itself appears insofar entirely dependent on that, as a function of the same, yes, it can be entirely translated into it.[110]

> The identity view is wholly *materialistic* in that it allows no possibility of human thought without a brain and movement within that brain, and it is actually supermaterialistic, in that it allows no divine thought without a physical world and movements within that world.[111]

Obviously Fechner's ontological materialism is true not only for properties, but for *processes* as well: For every change in the world there is a physical change in virtue of which it occurs. Applied specifically to the body and mind: If something mental in the world changes, it does so in virtue of a physical change. Mental changes are realized by physical changes. In Fechner's words: Every psychical change depends functionally (but not causally) on a physical change.[112]

Explanatory materialism claims that the connection of physical events to their causes can be completely explained using only causal laws. This condition excludes the option of the mental or the mind being allowed—as it were—to escape subjugation to laws of nature, or as Fechner writes, "to disturb" the material context. An explanatory materialist assumes that the realm of the physical is causally closed: All causes of physical events are themselves physical. Fechner saw clearly that his view implies this type of explanatory materialism:

In terms of the identity view, a natural scientist must no longer put up with the intervention of mental principles in the area he is researching, nor must he advance into the realm of the mental [a habit of some physiologists to fill in the gaps of their observation of the brain with the mind, as if it were a real gap in the body]. Natural science is now pleased to cover the whole context; it is authorized to coincide with pure materialism, the direction it has always preferred, without having had the self-confidence and the permission to head that way; and it never would have been approved, as long as the mind and the body were still fighting over the same territory, about which they are now—in our opinion—in agreement.[113]

Explanatory materialism does not rule out that there may be a psychological explanation for the same event that is materially explicable, i.e. there may be a mental explanation that links a psychological cause with a psychological effect. What it does bar is the possibility that phenomena, in virtue of their mental properties, also have physical effects.

With all its emphasis on materialism, it seems improbable that Fechner's theory could maintain any kind of independence of the mental sufficient for securing nonreduction. But the impression is misleading. Let us look closer at what "reduction" means and then check Fechner's theory for it.

A mind-body theory is reductive, if it implies at least two of the following:

(1)   Mental events are *products* or effects of material processes or events.
(2)   Descriptions of mental phenomena can be completely and intransitively translated into expressions about physical phenomena, such that all circumstances are describable without using terms for mental phenomena.
(3)   Mental states are *nothing but* physical states.

None of the above apply to Fechner's theory:

Regarding (1): A product is causally dependent on its producer. But the realm of the physical is causally closed. Thus all products of the physical are themselves physical, never mental. To this Fechner replies: "Thoughts are not one-sided *products*, or *results* of material events. Instead: material processes capable of carrying thoughts can themselves only result from those which are capable of bearing the same."[114]

Regarding (2): Terms for mental phenomena always refer to mental properties attributed to a (materially describable) bearer. Neither ontological nor explanatory materialism comes to the conclusion that mental predicates mean the same thing as material predicates. Our talk about mental predicates is just as capable or incapable of translation into the terminology of material expressions as our talk about the pressure of gas can be translated into talk of temperature and volume. Temperature and volume do *determine* pressure in the sense that (in an ideal case) my knowledge of them suffices for calculating the pressure present in a system. But this does not imply that "pressure" means the same thing as the quotient of temperature and volume. Similarly, according to Fechner's theory it is in principle possible to derive a mental state from the material state on which the mental state *directly* depends. However: To be in an inner state is not the same as to be in an outer state (it is something different for a gas to be under pressure P than to have the temperature T and volume V), although the two are closely related.

Regarding (3): This is the crucial 'condition for reduction' for the (reductive) identity theorist in the modern meaning of the term. It says that the class of mental states is (contingently) identical to the class of material states. If we ignore the fact that because it includes classes this formulation is actually unacceptable to nominalists, representing the majority of the identity theory advocates, we can say the following.

The identity of body and mind is reductive only if we view mental properties as properties of *states* and talk about states as if they were objects. If we can demonstrate that this itself is a linguistic error, we can restate claim 3 in such a way that the reductive nature of it disappears entirely. Fechner himself did not present an argument of this kind. But he always uses mental predicates to describe *physically identifiable systems*, never the *states* of systems. After all he knew, as the following passage reveals, that identifying the physical with the mental (as current identity theory does) springs from linguistic confusion:

> A primarily essential or overly extreme separation of matter and mind, as is prevalent in common opinion, is confronted with the other extreme, an almost untenable identification or blending of both, as is frequently done in science. In fact, the essential identity (which is also acknowledged by philosophers, if only for other reasons than our own) of *that which underlies* both mind and matter should not mislead us to desire to make identical *the mind* and *matter* themselves, because the identical unity—as mind and matter—occurs in the opposite relation anyway, and therefore should be called the one or the other; the outcome is otherwise a irremediable confusion of language and terminology. And the confusion is complete when we add the principle of identifying everything in nature with objective thoughts, whatever it is and just because it is conceivable.[115]

Why do we say that identity theorists are in error when they use the term 'state' to identify what is physical or mental? It is a case of fallacy similar to that of "sense data." "K exhibits sense data of a giraffe" (meaning that whatever he sees appears to him to be a giraffe) does not permit the conclusion that the sense data are long-necked, but simply that K exhibits the property of possessing a sense data in a particular way. If K has a toothache, feels sad, has a sensation of red, or is in a mental state, it does not follow that the *state* exhibits the property of being an ache, a color, or a mental condition. Equally: If neurons in K's brain in the section 739 always and only fire when he has a toothache, it does not follow that one and the same state of K is both "pain" (mental) and "firing in area 739" (physical), thus making mental and physical states numerically identical; we can only conclude that K has the property pain if and only if he has the property of neurons firing in area 739 of his brain.

We can avoid this mistake by always thinking of states as a structure consisting of a property, an object x as the bearer of that property, and a time interval. The claim that a mental state is always the same thing as a physical state can only mean that the *bearer* of the mental property is identical to the *bearer* of the physical property. (For all x it is true that if x has a particular mental property, then there is a y with a particular physical property and x = y, and the reverse holds for this particular property.)

Put this way, claim (3) is no longer reductive. In the normal sense of the word, we here no longer claim identity between a physical and a mental *state*, except perhaps figuratively, like when we say that expanding mercury is identical to rising room temperature. Although it discloses that the *bearer* of the physical and the mental properties is identical, this type of identity is not reductive. For if all things possessing a specific mental property also have some specific physical property (and vice versa for this property), then this would still not imply that having the mental property is identical to having the physical property. If all living creatures possessing a heart also have kidneys, and vice versa, this does not imply that having a heart is the same thing as having kidneys. At any rate, we cannot derive the identity of mental and physical states from Fechner's materialism.

One might object that Fechner's kind of nonreductive materialism is had by simply denying causal dependency between the mental and the physical and postulating an ominous functional dependency instead. If we allow causal dependency between physical and mental phenomena, then it follows from explanatory materialism, that mental states—since they can only be results of physical events or states—are themselves nothing but physical states. Everything would be reduced to the physical.

To refute this objection I would like to more carefully explain the concept of functional dependency using an example taken from physics. The ideal gas law $PV = nRT$ links pressure P, volume V, and temperature T for a given quantity of gas and describes how these dimensions are functionally dependent on each other. But it says nothing about potential causal dependencies. We can change the temperature of a gas by changing the pressure (for a constant volume). But it would be false to say that change in pressure *causes* a change in temperature. Instead, we say: Whatever caused the pressure to change for the gas also caused a change in temperature.

We can interpret the functional dependency of the mind on the body similarly: While it is true that the physical state changes when the mental state changes, and often vice versa, and these changes conform to laws of nature, this does not mean that a change in one state changes the other; it simply means that whatever event prompted the physical change also prompted a mental change.

Even if the law for gases does not describe causal dependency, it does not bar the possibility that the regularity it describes cannot be *causally explained* by an additional theory, for example, a theory of statistical mechanics, which in turn breaks down the object called "gas" into interacting molecules. Similarly, it also remains logically possible to explain the functional dependency of the mind on the body as causal dependency by thinking of a person as of a system having causally interactive components. But this explanatory option does not make it necessary. Since on the identity view the body and soul constitute one single thing and this suffices as an explanation, any further sort of explanation adds only inadequate, meaningless, superfluous metaphysics.

Once again this nicely illustrates how all experiences supporting Fechner's identity view also confirm interactionism and reductive materialism, but not vice versa. We may anticipate that this will lead to a comprehensive critique of the universality of explanation using mechanical laws of nature. In fact, Ernst Mach elaborated this very idea, beginning with Fechner's theory.

In summary let us recapitulate the crucial features of the *double aspect theory*:

(1) *Ontological materialism:* A system has (or changes) a mental property in virtue of having (or changing) a physical property. Mental properties and mental changes in systems are realized by physical properties or physical changes in systems. The formula is: For all systems x it is true that if x has a mental property M or this property changes, there must be a y with a particular specific physical property P*, which also changes along with the mental

property, and x = y. And the same holds conversely for this particular physical property.

(2) *Explanatory materialism:* The material realm is causally closed, i.e., if a physical event depends causally on another event, then this relationship can only be explained materially (at least in principle). The relationship of a psychical state (x having M at time t) with a physical state (the fact that y has P at time t'; t ≤ t') cannot be explained by material laws. There is therefore no causal relation between the mental and the physical state. (Material laws deal solely with material events.)

(3) *The mental as a self-phenomenon:* Only someone who himself has a mental property can know about it.[116] Formalized: For all mental properties, the necessary condition for a system x to know that a y has an M, is: x = y. Modern identity theory further claims that the class of mental states is identical to the class of physical states.

How this Relates to Schelling's Identity Philosophy

In chapter 1 we discussed factors that shaped Fechner's work. To which tradition did Fechner's mind-body theory belong?

Fechner's identity view is rooted in Schelling's philosophy of identity. His whole philosophy, as Fechner himself put it, "dropped from Schelling's tree."[117] But he did not simply adopt Schelling's doctrine of the identity of subject and object (of mind and nature, of the ideal and the real, of infinity and the definite); he changed it in two respects:

(1) Schelling's reason for the identity of mind and nature (namely the absolute, or pure identity—or whatever he calls it) is interpreted by Fechner as a lawful relation of (both mental and physical) *phenomena,* in other words: a phenomenalistically interpreted interrelation.

(2) Fechner replaces Schelling's definitions of the mind and nature, the ideal and the real, with those of 'self-phenomenon' and 'extraneous phenomenon': self-phenomena being mental, and extraneous phenomena being physical.

The double aspect theory "coincides . . . with the identity views in identifying the substantial foundation of the physical and the mental, but it envisions the relation of the physical and the mental to one another and to that one substance differently than all previous views."[118] For Fechner, "The identity view complies with Schelling's idea of identity inasmuch as Schelling's proposal traces the difference between the mental and the physical to a relative loss of balance during the internal evolution of the absolute; but nothing prevents it from interpreting the inner and outer perspectives of observation, from which—in our opinion—something appears to be mental or physical, as something subjective or objective."[119] Fechner's improvements are an attempt to cleanse Schelling's "doctrine of The Real and The Ideal" of all metaphysical content and give it an empiristic form, making it acceptable for natural science. Fechner believed that these corrections ultimately achieved a clarity woefully absent in Schelling's and Oken's work.

Fechner's mind-body theory only vaguely resembles Spinozism. He shares Spinoza's (and Leibniz's, and Schelling's) opinion that mind and body correspond to one another without being causally related. In contrast to Spinoza (and thereby improving Schelling's position), Fechner sees a nomological relation between self-phenomena and extraneous phenomena, and it is a relation that can be tested empirically. The difference of attributes (thinking or being extended) neglected by Spinoza and obscured in Schelling, is explained by Fechner as the change of perspective that necessarily ensues from the transition from one attribute to the other. Spinoza cannot infer the nature of the physical from the mental, or vice versa. For him, what links extension and thinking is a common substance relevant only to God—it is insignificant for empirical phenomena: "What this [the double aspect] theory has in common with Spinoza's is the parallel process of physical and mental events, but it differs from Spinoza in not seeking the difference of the physical and mental realm within an inexplicably different interpretation of the same substance from *the same* standpoint of a perceiving subject; but traces it to a different way of appearing, depending on a different standpoint of interpretation."[120]

Fechner finds that Schelling does justice to the close relationship between the attributes:

> Schelling's interpretation seems . . . more thoughtful and vigorous than Spinoza's because according to Spinoza, attributes move parallel to each other, unmotivated by anything other than an empty concept of substance; no reciprocal intervention is permitted, and it is even doubtful whether he seeks the difference solely in the external conception of man. According to Schelling, in contrast, the absolute Self makes itself into both subject and object in a lively process where the loss of balance of the one or the other creates the physical and the mental; it transforms itself therein and thus develops its life. Everything here is connected in an animated and active relationship, just as he wrote in one of his explanations, making the Absolute and Life identical.[121]

Spinoza says, "Not only *can* the causal process in each realm be followed up in itself, it *must* be pursued for itself."[122] But in Fechner's teachings we can, by "changing our perspective," describe one and the same causal relation as either purely physical, purely mental, or 'mixed' (meaning that the cause is physical and the effect mental, or vice versa), without claiming that there is an interaction between the body and mind.[123]

Schelling also pondered the question of the causal relation between the physical and the mental, arriving at an answer similar to Fechner's. The solution also demonstrates that by all means he could have stood his ground in today's mind-body debate:

> No causal relation is possible between the real and the ideal, or between being and thinking; thinking can never be the cause of a distinction in being, or conversely, being can never be the cause of a distinction in thinking. For what is real and what is ideal are only different views of one and the same substance; they can effect as little in each other, as a substance can effect something within itself. They do not match, as two different things can match, for which the cause of harmony is something outside of themselves, as Leibniz' harmony has been understood and explained using the example of two clocks; but instead, they match because they are not two different things, they are only one and the same substance. Just as (to use a convenient example) a person who had two names is still one and the same person, and the person named A is the same as the person named B, and does the same things, not because they are somehow linked or one of them causes the other, but because the person called A and the person called B are, in fact, one and the same person.[124]

If Fechner's and Schelling's ideas are similar, why does Fechner not wish to be associated with Schelling? One plausible explanation is that a natural scientist of Fechner's generation would have totally discredited himself by openly acknowledging the merits of Schelling's philosophy of nature.[125] Still, it was not merely a tactical maneuver: Fechner was much too independent and straightforward for that. On the contrary, he was convinced (and rightly so) that the identity view is changed so fundamentally by being empirically established "from beneath"— as Fechner calls it (in contrast to speculative justification "from above" as provided by Schelling) that it has entirely different consequences for natural science than what Schelling envisioned. Thus Fechner could not identify himself with Schelling's philosophy. At any rate, Fechner's coevals attest that he eschewed German idealism and was openly dismissive of academic philosophy in general. Few were aware that Schelling's identity philosophy survived in a new guise as part of Fechner's identity view.

One rare instance to the contrary occurred in 1875 on Schelling's one hundredth birthday when Rudolf Seydel, a follower of Weisse, dared to give a revealing salutatory, hailing Schelling's philosophy as the "common origin" of various intellectual tendencies of his time. Schelling's identity philosophy had lain "the new, truly monistic fundaments . . . , upon which in turn also Herbart and Lotze and the author of the Zend-Avesta and Nanna now living among us could each post himself in a novel and unique way." Despite the sincere avowal that Schelling's philosophy of nature is useless for science, Seydel's oration peaks in a vision of united natural science and identity philosophies:

> What if—in spite of their fundamental difference—Schelling's volition monism should be found combinable with the results and demands of modern scientific monism, and if his path of structural thinking should meet with the research avenues of exact empirical science, like the paths of two miners, who, having started at two opposite ends are busy breaking a tunnel through and finally fall into each other's arms and greet each other brotherly as the authors of one common work? I believe that on this very day for an observant ear, calls and hammering from one side to the other are already discernible.[126]

Fechner was certainly one of the few scientific "miners" aware of the pounding on the other side.

3

PHILOSOPHY
OF NATURE

3.1 Philosophy of Nature and "Belief"

BY ENDORSING NONREDUCTIVE materialism Fechner covered his retreat, as it were, and was then able to devote himself to what mattered most to him, namely, philosophy of nature.[1] Science is for observing nature from the outside. The philosophy of nature, in contrast, studies nature from within, or, from the side of nature visible only to nature itself.

Since we ourselves are part of nature, our psychical phenomena put us directly in touch with nature's inner side, albeit only to that part which is functionally dependent on our bodies. That, of course, cannot suffice as grounds for a philosophy of nature. Since psychical phenomena are accessible solely to the person experiencing them, it would seem as if one could do "philosophy of nature" in only one case, namely one's own. It is a tautology that no one has access to one's own psychical phenomena except oneself. A theory applicable to exactly one single case, which, withal, is itself not intersubjectively accessible, does not deserve the title of 'theory.' This way philosophy of nature could never get off the ground.

So Fechner resorts to his theory of belief. In everyday life and in natural science we constantly use explanations that rely on a belief in things that are not directly empirically accessible to us. Our natural human status often forces us to augment direct experience "by imagination and inference."[2] Many quotidian activities rest on the assumption that our fellowmen have minds, although we actually only observe physical changes in them. No physicist has ever seen an atom, yet physicists are convinced that atoms exist: Working out the explanation of all physical phenomena in terms of the movements of atoms is physics' alleged central research program.[3] If these considerations are not entirely unfounded, we must have some effective "ersatz" (as Fechner calls it) for exact proof, one that is capable of instigating strong belief. Fechner surveys motives and reasons for belief that are perhaps sufficient to be seen as this type of substitute; more than any others, theoretical reasons enhance belief—if not justifiably, then at least plausibly—when they are inferences from the known to the unknown, based on probability.

Why should it be impermissible or empirically disreputable to exploit in philosophy of nature the same method that is used in day-to-day social intercourse and in physics, where it instills such compelling "strong confidence in the belief" of non-observable phenomena? We cannot preclude from the outset that the inferences we make about the inner lives of others, lives of which we have no direct evidence, are not equally applicable to other nonhuman beings (or systems). Whatever suggests that others have minds is something that results from external observation. Philosophy of nature must stay open-minded and investigate whether and to what extent we can conclude from external observation that nature has an inner side (and thus a subjective character): "The methods we use to infer things beyond ourselves are basically the same as those with which we infer from here to there, from today to tomorrow, and which all empirical science employs to infer from the given to the non-given."[4]

In contrast to natural science, such a philosophy of nature must unfortunately rely on surmise. Since the psyche appears only to itself, the philosopher of nature can gain no *knowledge* in this field of study (except, as we have said, in his own case). So we cannot expect philosophy of nature to offer accurate evidence of the kind that natural science provides. But this does not mean that we are free to contrive inference methods arbitrarily as a substitute for accurate evidence, for the sole purpose of getting the desired results, as earlier philosophers of nature had done. Instead, Fechner acknowledges exclusively those methods of inference that compel us in normal everyday life to believe in other minds. Using these we can gain for philosophy of nature exactly the same kind of certainty about the minds of our fellowmen as that which we exercise everyday. "In fact," according to Fechner, "this is the fundamental point from which I proceed in the following: Regarding the entire question of the mind, do not demand any certainty about another character other than the certainty we have about the existence of the minds closest to ourselves, but do use the means necessary to do so."[5]

3.2 Psychical Phenomena as Functional States

Now, what are these external, material signs that can provide us with the inference we need for philosophy of nature? What Fechner finds most crucial is *functional similarity*. The notion that a particular system has a soul is all the more likely to be true, the more the functions exercised by the system's (externally visible) components in serving that system resemble the functions of (material) components in human beings, that—as we know from experience—are essential for psychical life. Fechner here combines the identity view with *functionalism*; he takes psychical phenomena for functional states of material systems, such that they could also be realized in some other way, namely, other than by the human nervous system and brain. Fechner contends that souls can be organized in many different ways.[6]

What sound reasons can Fechner offer in support of the claim that a given material system has a soul? We must discover and list the empirical features from which, when they occur, we can conclude with greater or lesser probability that they are functionally similar to the human body and allow us to infer the existence of a soul. Fechner wrote up long lists of such features, of which I shall name a few. One system is functionally similar (in terms of having a soul) to another system that has a soul, and therefore probably also has a soul, if it fulfills the following criteria:[7]

- It must form a unified whole, relatively closed in relation to its environment, and organized according to a uniform plan; it must be a "closed system with self-regard."
- It must be distinguishable from other systems of the same type, meaning it must be individual and autonomous.
- It must, when supplied with energy coming from outside itself, be capable of self-development and self-production; it must be a system of "self-regulation, self-development."

- It must be capable of innumerable effects, some of which are "unpredictable innovations."
- Its individual components must more or less promote system-self-preservation.

Fechner calls any argument based on criteria such as these an "argument of similarity or analogy." He elaborated five other types of argument involving the biological relation of a system having a soul to another system, in hopes of demonstrating the probability that the second system in question also has a soul.[8]

Fechner tries, using every empirical and rational means at his command, to prove that these criteria apply not only to animals, but also to plants, to the earth, indeed—to the entire world. In his opinion, all these systems fulfill the criteria. He feels confident in assuming that the universe is animated, meaning that it has a side of which only it is aware. Since every animated system is furnished with unified consciousness (following the principle of synechiology, cf. chapter 2), so also does the universe, and we call this God's Spirit. This spirit is embodied by the physical part of the universe. The world's soul and God's existence are thus plausibly extrapolated from experience: "It is not a premised concept of God that determines God's essence, but rather, what is divine in the world and perceivable in us determines our concept of Him."[9] This remark once again shows how Oken's thought influenced Fechner. Fechner's "proof of God's existence" includes some surprising insights that turn up later in twentieth-century systems theory, for instance in ecology.

Naturally, Fechner's provocative opinion that plants, the earth, and the entire universe have souls, and that the world is God's body, met with enormous resistance in science, in the humanities, and in theology: "My unfortunate little books dealing with the souls of stars and plants—materialists tearing at one end, idealists at the other, tossed away by disapproving natural scientists, sold cheaply by the book shops, and scrapped—have suffered plenty."[10] The major objection raised time and again was that neither plants, nor the earth, nor the universe have nervous systems—which is reason enough for pantheism and panpsychism to collapse. In a famous oration on "The Limits to Knowledge of Nature" (1872) Emil Du Bois-Reymond protested:

> A scientist cannot concede that there is animated life where there are no material conditions for mental activity in the form of a nervous system, such as is the case for plants. And he will rarely encounter disagreement on that. But what would we reply if he, prior to accepting the idea of a world-soul, were to demand to be shown somewhere in the universe a convolution of ganglia cells and neural fibers

appropriate for the mental activity of such a soul, embedded in neuroglia, provided with warm arterial blood under the right pressure, and attached to appropriate perceptive nerves and organs?[11]

Between 1848 and 1882 Fechner fought this attack (calling it the "nerve issue") offering arguments that could have originated in present-day functionalism.[12] He finds no reason to believe a priori that "only neural coils are capable of psychical phenomena."[13] And he also considers the empirical evidence brought forth either illogical or irrelevant. He calls the belief that neuroglia are mandatory for animation an "obsolete superstition." If the "nerve argument" were not a fallacy, we would also have to conclude that: "Since violins need strings to sound, then flutes also need strings; but since flutes have no strings, they cannot sound; candles and petroleum lamps need wicks in order to burn, so gas lamps also need wicks, but they have none; thus they cannot burn. Yet flutes sound without strings and gas lamps burn without wicks." Fechner goes on to say that mammals and birds breathe with lungs, but this does not mean that all animals must have lungs: "But if fish and worms can breathe without having lungs, while mammals and birds can breathe only if they have lungs, why cannot plants without nerves experience perception, while animals can only perceive when they do have nerves?"[14]

In general Fechner appeals to the fact that nature varies the "forms of means for fulfilling functions in many ways," meaning that one and the same organic function can be fulfilled by different organisms in different ways.[15] In order for an organ to function as an "organ of the soul" (i.e., a physical organ subtending sensation), it probably must be capable of generating ordered oscillating processes. (Here Fechner is following the resonance theory laid out by the physicists Herschel, Melloni, and Seebeck, stating that psychophysical processes are oscillatory movements.[16]) But as many experiments show, sound and light waves can be produced in numerous ways; therefore, it is conceivable that yet other organs can manifest sensations, and so on, although they are built differently than the brain. Thus, "There is no way that we can conclude from the nature of intellectual movements what the nature of the subtending physical movements must be like, in other words, we cannot conclude which substrate and which type accommodate these movements, but we can conclude . . . that the psychical context corresponds to a psychophysical context."[17]

Another critical theme expressed by Schleiden and others said that Fechner's arguments are teleological, an explanatory method having no use in modern science, but purely didactical value for practical philosophy. Fechner protests that a teleological inference from external similarity to functional similarity can be just as correct as causal inference. We must simply be careful not to assume that nature has any purposes a priori; instead we must infer from the factually observed fulfillment of purposes in an organism to similar functions in analogously organized organisms.[18]

Notice that throughout the entire discussion Fechner struggles to avoid the slightest deviation from scientific principles and methods. In the programmatic exordium to *Zend-Avesta* he writes: "Anyone who closely examines my writing will find that the exalted vitality I attribute to nature in no way limits the exacting demands of the natural scientist. I do not restrict scientific procedures in the way that philosophies of nature, since Schelling and Hegel, more or less restrict them. I do not debase scientific findings, I make use of them. No law of nature is less binding for me than for the most accurate natural scientist."[19]

Unfortunately Fechner's arguments and claims seem not to have convinced his coevals. In a typical quip he remarks in *On the Soul* that he won't complain about a lack of response, for he already has three devotees for the doctrine of star-souls: a Viennese police chief, a German colonist in America, and a West-Indian businessman who immigrated to Paris.[20]

3.3 The Day View as Scientific Identity Philosophy

Adopting a pantheistic concept of God[21] enabled Fechner to take the identity view, which he had hitherto applied solely to solving the mind-body problem, and apply it to the entire world—making his philosophy (once again) an "identity *philosophy*." He writes: "God's Spirit and Corpus are themselves simply two ways or views of appearing, the inner and the outer ways of the One Being, God. Two sides, not two things."[22] But this entirely alters the status of Fechner's philosophy. Up to this point we emphasized its materialistic bent. While this intellectual cornerstone of his theory remains intact within its internal context, unmoved by the philosophy of nature, overall it is elevated to become, as Fechner calls it, "objective idealism."[23] This idealism encompasses nonreductive materialism like one Russian Matroschka doll encloses another.

Nonetheless, we should be careful in interpreting Fechner's notion of "objective idealism" from our own historical standpoint. He was always rather independent and willful when it came to terminology. The term "objective idealism" actually first turns up in Fechner's late work, that in its somewhat affected pious tone differs entirely from the "heathen" tendency of his early work influenced by Oken. Not to mention that it is indeed a rather strange "idealism" since it holds that the idea of a world spirit can only be materialized by a material bearer.[24]

But for Fechner it is this very conversion to pantheism that breaks the spell of science lacking belief and thus banishes—borrowing a phrase from Goethe—the "atheistic twilight." By interpreting Oken's identity philosophy in terms of empiric phenomenalism, Fechner is at last able to conceptually unite accurate natural science with a romantic-aesthetic bond to nature and late idealism's freedom and Divinity doctrines. The resulting "basic view" immediately became an existential personal necessity for him. He writes about himself: "The author cannot drop this basic view that has come to engulf his entire knowledge, belief, and thought, without falling himself."[25] He calls this new worldview the "day view" and contrasts it in general with the dualistic-pessimistic "night view" of mechanistic materialism and all other important philosophic tendencies of his time.

Exploiting external experience in order to correctly infer the probability of an inner side of nature, a side that pinnacles in the utmost unity of consciousness, Fechner arrives at the following typical conclusions for his worldview:

(1)   All that is physical and conceptual appears in God;

(2)   The world objectively has the qualities that we perceive it to have (meaning that no secondary qualities exist). Thus the day view is a more natural outlook than the night view;

(3)   We are subordinate parts of the Person of God, our bodies and souls are parts ("individual partial beings or partial spheres") of God's body and soul;

(4)   Immortality can be explained in naturalistic terms;

(5)   The view enables a kind of teleology compatible with the strict causality found in natural science;

(6)   The world's development gains an ethical meaning;

(7)   Freedom is consistent with "absolute lawfulness."

Let us briefly examine these points one at a time.

Just as physical and mental phenomena come together in the unity of a person, so also do all phenomena come together in God—those that we perceive as being internal, as well as those we perceive as being external:

> The world of appearances is a vault, held together by utmost consciousness. This is why we can rightly say that God, the highest form of consciousness, is

The Highest Being; because within the unity of His consciousness, everything exists and is a mutual condition of everything else and all is interrelated, for all the appearances of things, and thus also for all things, which themselves are only contexts of appearances.[26]

Obviously, Fechner's critique of the concept of noumena and his critique of the concept of substance were first made possible by the way in which he united "the quality of being a noumenon" and the substantiality of things and subjects *in* the world with one another to become one single great world-noumenon (in other words: God). God becomes—as it were—the residue of substance lost in the world. Natural science, being the doctrine of external phenomena and their relations, can do without substances and noumena; indeed, its methods forbid the acceptance of such assumptions. But a "worldview," being—as Fechner calls it—a "theory about the entire realm of existence," cannot desist assuming a highest unity of consciousness. The passage quoted above continues: "But since God is the Highest Being above all things, we no longer need dark beings hiding behind the appearances of things."[27]

Fechner does not actually use the expressions "substance" or "noumenon " for God. He finds it appropriate to view God, just as he does the things of the world, as a lawful context of phenomena. I find, however, that his concept of God must be seen in terms of a substantial subject. He writes, for instance:

> Perceptual phenomena extending beyond humans and animals cannot float freely, they require a subject, an overarching consciousness.[28]
>
> The body and the soul, each of these represents in itself a context of appearances that is only possible and real within a higher, universal unity of consciousness, and the same is true of the relation holding between the two contexts.[29]

This twist in Fechner's thought seems to revive of one of Oken's opinions, expressed in 1808 in a typically rapturous manner, calling the belief in noumena a "belief in ghosts" and claiming that correctly understood, God is the only real subject:

> The world is not apposed to man, it is only his body, it is *not* divided into mind and matter, split up within man himself. There is no opposition within the universe, instead there are various tiers and subordination. The world of plants is not apposed to the animal kingdom, it is subordinate to it. Femininity is not apposed to masculinity, it is subordinate. There exists neither noumena, nor egos, and even less a thing such as unconsciousness. I cannot imagine how someone could think it up. Only a total ignorance of mathematics, physics, botany, zoology, compara-

tive anatomy, and physiology could devise such belief in ghosts. It is the universe itself that appears to itself and is called The Ego. But in order to understand this, one must have been born on The *Holy* Night."[30]

3.4 Direct Realism: The Objective Reality of Phenomena

In denying that the concept of substance is applicable to the things of the world, Fechner's philosophy slightly resembles that of Berkeley.[31] Berkeley spurned applying the concept of substance to matter, saying that only "spirits" (human persons and the Person of God) are substances. His ontology contains only two kinds of things, namely spirits and ideas. For Fechner there are only two ways of expressing ontology. In terms of our inner world, we humans would be incapable of "clear and practical communication" if we were to assume that there are only "individual spirits" with private phenomena. This is why both common sense and science proceed from the assumption that there is an "external world" that is "prior and exterior to these spirits."[32] But judged from God's perspective, Fechner's (and Oken's) ontology is even more radical than Berkeley's. For in the end there is solely one single Being, namely an "idea spirit" (we would have to call it), namely God, whose Being is identical with the constant production of ideas, i.e., phenomena perceived internally by Him, while the being of the world and all of its creatures exist in being perceived, or produced, by God. "The way God sees things is the way they really are, and God's view of them belongs to the definition of their existence."[33] As this passage suggests, today we would say that Fechner's concept of God results in "direct realism." A direct realist thinks that that the qualities we perceive in the world are objective properties of the things themselves. The redness of an apple is not something originating in our brains as a result of stimulation by light waves; instead it is seen as a property that is objectively attributable to the apple. In other words: A direct realist rejects the distinction between primary and secondary qualities that has commanded thought in natural science since John Locke. The direct realist accepts only the existence of primary qualities.

Fechner can easily argue for direct realism: If there is an overarching divine consciousness that is the focal point of all phenomena, then the world *is* as it appears to us. It is not the case, as the night view would have us believe, that everything "that we see, hear, and opine of the world . . . is merely an inner deception, an illusion."[34] Instead, the world is illuminated by God's sight and rung through and through by his listening; what we ourselves see and hear of the world is only the last offshoot of His vision and audition.

Perceptible phenomena that extend beyond us are not a mere illusion; they are objectively distributed throughout the world.[35]

For Fechner this is by far the most significant implication of his day view. It guarantees that an uncompromising natural scientist, an adherent of scientific principles and methods, need not feel condemned to supporting the "barrenness" and "dreariness" of the world as it is presented by mechanistic materialism and other varieties of the night view. The world of "nocturnal philosophy" is "dismal, mute, and desolate," emptied of all purpose, all beauty, and all meaning. Psychoanalytically we would say that it just demands too much denial from us. The day view, in contrast, truly 're-enchants' the world. The day view provides a "direct and much more pleasurable picture" of the world than the night view can offer. It holds that colors, sounds, fragrances, tastes, and sensations of warmth and touch exist objectively and that beauty and vitality are not empty illusion, but material reality!

Fechner's *Day View* published in 1879 demonstrates impressively how he struggled to (re)create the bond between the scientific worldview and the normal, natural way of seeing the world. It is indeed possible that Fechner designed the grand scenario that introduces his *Day View* as a reply to Emil DuBois-Reymond's famous oratory of 1872, mentioned above. In that speech DuBois-Reymond, the self-designated administrator of scientific knowledge on nature of his time, had proclaimed without much ado:

> The fact that in reality there are no such things as qualities results from dissecting our capacities for sense perception . . . A perceptual sensation as such originates . . . in fact in the sense substances, as Johannes Müller has named the brain regions belonging to the sense nerves, and for which now Hermann Munk has distinguished a part of the cerebral membrane as the visual sphere, the auditory sphere, and so on. The sense substances translate the excitement that is the same to all nerves into sensations, and being the real carriers of what Johannes Müller calls 'specific energies,' these substances then proceed to generate the various qualities, each depending on its particular nature. Moses' announcement "And then there was light" is wrong, in physiological terms. There was no light until the first red eye spot of an infusorian for the first time made a distinction between light and dark. Without substance for vision and audition, this colorful, resounding world around us would be dark and silent.

And just as silent and dark in itself, i.e., without properties, as results from subjective dissection, *is* the world as it is found through objective observation in the mechanistic view, that in place of sound and light knows only the oscillations of a primordial stuff lacking properties, that has become matter that is sometimes weighable, sometimes not.[36]

Fechner found that this opinion reflects an entirely unjustified belief in a "causal leap," through which "the clump of protein called brain" "conjures up" sensations and "shining, sounding vibrations" from "blind, mute waves." The day view proves that we are not forced to accept the "spiritualist magic" of that leap and can leave or return a qualitative definition to external perceptions.[37]

With his own type of direct realism Fechner—as we suggested—also claimed above all to draw science closer to the natural way of seeing the world:

Natural man resists the wisdom of the night view. He believes that he sees the objects surrounding him because it is really light all around him, not that the sun begins to shine somewhere behind his eyeball; he believes that flowers, butterflies, and violins are as colorful as they appear to him . . . , in short, he believes that throughout the world there is light and sound outside of himself and it pervades him. But he lets science teach him, and now he believes to be cleverer, and believes that he has one less illusion. But this illusion remains and mocks, as far as he knows, how his knowledge scoffs at his illusion. . . .

Proud of this wisdom full of folly we observe with sympathetic condescension the simple and modest silliness of Africans and Turks and believe ourselves much farther advanced than previous centuries, because they had even fewer of these follies. But we should be prouder of our little matches; they will continue to glow, even when all the sparkling will-o'-the-wisps of the night view are extinguished and sunken.[38]

Viktor von Weizsäcker interpreted Fechner's day view as a mission "to rescue humanity from science."[39] That is correct. But on a closer look it means even more: He wants to rescue respectable science from restricting itself to the mechanistic worldview.

We have yet to deal with the difficulty arising for Fechner from the fact that appearances of the same things vary from person to person. It would seem as if consistent direct realism forces us to accept that the world's phenomena simultaneously possess differing qualities. If a certain substance seems red to you and green to me, it would have to be both, simultaneously. Or if I find a painting beautiful and you find it repugnant, then the picture would have to be both beautiful and repulsive. Since it cannot be both at once, we would have to conclude that an individual's perceptions of the world cannot necessarily be considered the true qualities of the world. Fechner solves this problem, in his own opinion, by saying that we must pay attention to our own individual peculiarities of perceiving. Since the "way the soul is furnished" depends on bodily organs and these differ from person to person, it will also be true that all the shining, sounding, fragrant movements that originate outside of us and effect the material system of our bodies "will be modified in a different way according to the individual make-up of each body. They will experience different associations and thereby also be connected to variously modified sensations, and enter into a system of different psychological associations; and yet there can remain something common in the quality of sensation between inside and outside."[40] We must think of how a quality perceived by an individual relates to the way this quality appears to God as similar to how a light beam is modified by variously cut glass and curved mirrors. The quality of the light remains the same throughout all of these modifications, even though it appears modified in individual cases. We can approximate the true quality by discovering our personal deviations and taking them into account. As close as the day view otherwise comes to the "natural view," it does depart from it by making room for the "particular make-up of each individual." At this point we do recognize Fechner's debt to Leibniz; that is apparent despite all his opposition to the latter's "monadical worldview."

3.5 Further Implications of the Day View

Fechner's pantheism particularly needs to explain what it means for a person, i.e. her whole mind and soul, to be part of a superior subject. For this purpose he developed the doctrine of the "psychophysical tiers of the world"; expounded in the greatest detail in *Psychophysics*, chapter 45. If we think of the oscillations of the physical bearer of psychical phenomena as exhibiting the form of a wave, whereby the psychical phenomena first enter awareness after the waves crossover a certain level, namely the threshold of consciousness, this gives us a model for explaining how one consciousness is subordinate to another. God's consciousness is represented as a main wave. It carries smaller surface waves with shorter vibration periods and thresholds higher than that of the divine wave. There is discontinuity in the transition of one consciousness of surface waves to another, but there is continuity in God's main consciousness, whose threshold for awareness lies much deeper.

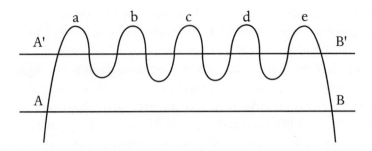

Waves. Taken from Fechner, *Elemente der Psychophysik*. Leipzig: Breitkopf & Härtel 1860, vol. II, 540.

Fechner attempts to support this idea by demonstrating an analogy holding for our brain and its parts. He refers to a series of experiments in which animal cerebral hemispheres were severed and also mentions experiences of amnesia in patients with brain injuries. If parts of the brain normally connected above the level of the threshold of consciousness are severed, the result is discontinuity. The separated segments now achieve, each on its own, albeit weaker, the same "unifying psychological performance" that was previously attained by the whole brain. If we could split a person into two parts such that the one half connected to the left hemisphere and other the half connected to the right hemisphere were separate, without the person dying, then we would have two specimens of the same person, with the same frame of mind, the same talents, knowledge, memories, consciousness. Just as some individual sections of the brain can take on the functions of other sections when these fail, in order to maintain a unity of consciousness, so do we, as humans, have within us the capacity to take on functions of the entire cosmic consciousness in order to maintain its unity.

From the day view Fechner also gleans a simple way of presenting a naturalistic notion of life after death. This is certainly one of his most unusual doctrines.[41] Through metabolism our bodies continuously change, such that after a while entirely new matter continues to bear one's former mind. Psychological continuity is thus possible, in spite of physical change. It is therefore also conceivable that our psychical side survives death by making use of the physical effects we have generated during the course of our lives. During her "earthly" life a person creates a "body of deeds" to which her soul will be transferred. The effects of our actions are—like all effects—indestructible, because it is the nature of causes to continue themselves "unaltered in their consequences," even though other effects might possibly join them in determining the results. Survival after death is thus the survival of the soul with a new physical body. The "afterlife" takes place in the same world as our "present existence," we experience merely an actual, not substantial immortality. Fechner explains: "If there is future continued existence, it can only be founded on the fact that the main wave of our psychophysical system [meaning the brain as the direct bearer of our psychological phenomena], on which our main consciousness depends, leaves that part of the earthly system on which it depends or in which it now crosses over the threshold and is transferred *in continuo* to another part or to supplement of this system."[42]

In order to more precisely see how psychological phenomena are transferred to material effects, we need to research the laws that are already valid for "changes of consciousness" during our earthly life. We know from experience that we sometimes—under certain circumstances—remember sense perceptions from earlier times in our lives—although these are no longer given. This experience gives us a foothold for imagining psychological life after the decay of the body. We will probably lead a life of memories in which we become aware of all the causal consequences our life has left behind. We can have no perceptive life, since we no longer posses any sense organs:

> If we ask what it is that in this life continuously supports the continuity of identical consciousness from childhood to old age in the same body, although the material and the shape of the body have changed, it is the circumstance that the later body grew out of the earlier one, and the bearer of the earlier mind generated consequences to which consciousness continued to adhere. And if we ask what the conscious relation is between a memory and the perception from which it grew, once again we find the circumstance that whatever bears memories within us is a consequence of whatever bore the perception. Therefore, a principle of continuity is generally valid for consciousness, as far as we can follow it up in this life, and we will be able to follow it from this life to the afterlife.[43]

Life after death will continue to be personal, although it will be impossible for the living to identify the "body" of the deceased. After death our bodies become part of "one and the same huge body," but our souls do not "intermingle," although they are borne in this life and in the afterlife by God's consciousness. Even now the ideas left behind by our forefathers influence us, without implying that we loose our identity by it. We can also recall the most diverse memories individually, although the material effects are all mixed up within the brain. "The same world is common property for all of us, but it is different for each one, has a different meaning for each one of us."[44]

Fechner's philosophy of nature also renders feasible a teleological worldview that is quite harmless in terms of its scientific implications. One and the same process that represents a causal consequence in physics can—when seen from the inside—appear to be a final process. For this reason, from the inside, the physical development of the world as a whole can be understood as a teleological process, without implying an interruption or reversal of causal order in the outer world process. Fechner's teleology is therefore not vitalistic and harmonizes cleverly with materialism and mechanism of the external observation of nature: "All material impulses or lawful increases in velocity in nature . . . are essentially—depending on how they were created—bound to mental or psychological impulses, endeavors of a managing world soul."[45] In a chapter to come we will see how these "endeavors," when undertaken by the self-phenomena, are reflected in a lawful "tendency" of physical systems towards "stability."

In order to render it plausible that in philosophy of nature the world seems purposeful, purpose must be psychological. One must name the inner objective towards which the development strives. We can only discover this objective by inferring from cases in which direct insight into the motives and purposes of the mind is given (thus, from our own minds) to the inner side of the world as a whole. Now, according to everything we know directly about our own psyches, "all our motives and the purposes of our actions by nature characteristically are related to . . . pleasure and pain."[46] All our actions serve to prevent pain and increase pleasure. This gives us good reason to believe that the development of the entire world aims to achieve the greatest state of pleasure.

In terms of ethics, this increase in the state of pleasure for everything becomes the ultimate standard for action. Fechner calls this type of ethics "determinism."[47] This contradicts neither his "indeterminism" nor his idea of free will. Perhaps he should have coined it "equifinalism." His choice of words misleads all those writing commentaries on this topic[48] and Fechner himself seems to have not been totally clear about using the term.

The idea is probably this: No matter what an individual may decide, the (causal) consequences of his actions will—in the long run—be positive. Just as every system for which the second axiom of thermodynamics holds will, over either a short term or long term, make a transition into a homogenous, ordered final state independent of what the original state had been, the causal consequences of human actions will also—in the long run—turn for the better, no matter what an individual person in an individual case may decide. If they turn to evil, this can merely postpone the victory of the Good, but not make it impossible. It pleases God to "lead the whole world and every individual in it to a pleasant final goal, or, inasmuch as there is no end within the world, to lead them to an ever greater approximation to pure happiness, such as he himself enjoys."[49]

Fechner's "equifinalistic ethics" is tenable only if the normal laws of process that rule the movements of individual atoms are supplemented by 'dispositional laws,' i.e. statistical laws, that cannot be reduced to lower level laws of process. Of course, this kind of ethical "determinism" has no resemblance to mechanistic determinism as classically described by Laplace. Fechner's 'equifinalism' is compatible with both a deterministic and an indeterministic notion of individual acts of volition. We shall return to the question of whether the freedom of will that Fechner says is entirely naturalistically undetermined is compatible with the causal inclusiveness of the physical world, to be explained in detail in chapter 8.

PART III
DAY VIEW SCIENCE

Do nonreductive materialism and the day view have real consequences for our concept and practice of natural science? Fechner definitely thinks so. But those consequences neither limit nor substitute standard scientific methods of reasoning. This is no surprise, for by no means does Fechner want to cripple either the demanding scientist or scientific rights. He has more subtle consequences in mind. He finds two ways in which whether a scientist embraces or rejects the day view makes a difference. The first is how the scientist views the relationship between phenomena and the quantitative laws of nature (for Fechner, this equals mechanics). The second concerns the scientist's chosen field of research.

Regarding the first way: Natural science inspired by the day view "surpasses and supplements" the abstract, or quantitative scientific view by also leaving nature a "qualitative definition," a description lost under the ruling night view. The day view corrects the prevailing idea that mechanistic science alone "covers reality entirely," being in principle a perfect theory for all phenomena. The day view resists the notion that a quantitative scientific description can exhaustively describe any object: "Just as there is an outward quantitative cause there is also equally an outward qualitative definition . . . the same as there is an inward quantitative cause within us. And this is how our day view surpasses and supplements the view of natural science, without questioning its exact form in the least."[1] This reasoning is buttressed by the business of physics: "The task of physics is, put prosaically, nothing less than this: to discover the most general and accessible interrelation of experiences in the physical domain *such that beginning with what is given we can predict and produce what is not given*."[2] Thus, a purely instrumental notion of science follows from Fechner's day view, making predictability a criterion for scientific acceptance.

The day view understands "nature" as the epitome of outward phenomena and their interrelatedness. If we go along with this definition of nature, then the "interrelation of experiences" means the links among external phenomena or the discovery of relationships inherent in nature. Of utmost importance to Fechner are relationships connecting tactile phenomena: "By the term 'matter' the physicist understands—quite in line with common usage, to which he is always indebted—that which is noticeable to the sense of touch, or, whatever is tangible." But for valid reasons, physics does not halt at the sense of touch. The scientist finds "that according to experience (for how else could he discover it), other properties capable of being demonstrated exhibit themselves jointly with the property of being tangible; these are phenomena of balance and motion; they are more easily understood by vision than by touch. The scientist proceeds to count these among the definitions of matter, along with the laws he has discovered through experience (including the auxiliary concept of forces). (This is probably the origin of the claim 'matter is what moves in space.')"[3]

Gradually scientists discovered that there exists a thoroughly reliable connection between the qualities of a body and its phenomena of balance and motion (its relations in terms of mechanics). This led to an understanding of qualities as functions of mechanical dimensions: "Physics can . . . list rules . . . such has how certain qualities are modified after a relational change in the composition of physical parts; in other words: physics tells us which functions the modifications express. Subsuming an individual instance under such a general rule is to explain something *physically*."[4]

In all of this, the physicist is not interested in whether or not something exists (and if so, what it is) 'behind' the experience of touch, something that could be considered its cause. That does not imply that the physicist must reject the notion of noumena (the concept of substance), but rather that from his point of view, noumena are of no interest:

> The physicist knows matter only as something tangible, as something that is bound to touch by laws of nature.[5]
> Within the physical world it all boils down to the sense of touch and vision as the final elements into which physical things can be qualitatively decomposed.[6]

But science has discovered that expressing natural laws *quantitatively* is most useful for its objective of "inferring with utmost accuracy what is not given from the given circumstances in the external world of phenomena, or to predict the success of modified relations in contexts of outer phenomena."

Science thus abstracts from sensation and deals solely with the quantitative aspects of natural phenomena. (For Fechner this always implies that the sciences of external phenomena perform only "spatio-temporal measurement," they deal with the mechanically defined side.) This sort of abstraction is recommended for "physics, chemistry, astronomy, physiology . . . and is sanctioned in its purest form in the science that commands and pervades them all—mechanics."[7]

But we cannot conclude from the victories of the procedure of abstraction, as the night view does conclude, that an abstract description and the quality of a thing are identical, thereby "draining" the world of quality: "It is strange that the materialist—and not just the materialist, but actually the whole world of science infected by the night view—believes that no qualitative definition is attributable to nature outside of ourselves, just because the scientist *abstracts* from it."[8]

The materialist overlooks the fact that the method of abstracting serves a certain *objective*, namely, that of making optimal predictions. We know that this is the motive behind the method because science quickly reverts from abstraction back to perceivable qualities, as soon as it no longer pursues the purpose intended by abstracting. It is a fundamental mistake to identify the real object as exclusively that aspect of the object described in physics by abstracting from sensation.[9] At its best, physics may disclose the universal correlation between the quantitative (mechanical) relations of phenomena and our sensations; it deals with one aspect in lieu of the other. But we must always be prepared that perhaps someday, for one or the other kind of phenomenon, we will discover that sensations are no longer representable as functions of mechanical circumstances. We have no reason to assume a priori that feeling will be universally eliminated from physics via abstraction.

Fechner finds another argument for his view in the analysis of scientific measurement. The basic idea is that when we measure, that is, when we observe objective equality between what we have measured and our standard of measurement, we still cannot waive our subjective impression. So once again, the advocate of the night view is thrown back to sensation at a crucial point. Sensations can simply not be eliminated from science. When discussing psychophysics in chapter 6 we shall take a closer look at this and other conclusions that Fechner derives for the process of measurement when analyzed under the day view perspective.

On Fechner's second main consequence of the day view for science: Another benefit that the day view offers programs of scientific research is a certain *heuristic method*, a method that cannot be founded on the night view. If the world has a psychical side and every psychical process corresponds to a physical causal process, then it must be possible to somehow find traces of the psyche within the physical domain and vice versa. What is physical must "also contain" an expression of the mental, and the mental must contain an expression of the physical. By keen attention to the "other" side of nature, a user of the day view can discover otherwise unseen aspects that further his research. Knowledge about one realm also provides a heuristic device for the other realm. Change and development in the physical world manifests itself as change in the psychical world, and change in the psychical world manifests itself in the physical world.

Fechner makes impressive use of this heuristic method when dealing with indeterminism. Our acts of volition, which seem to us to be self-determined acts, correspond in external physical nature to new, uncaused beginnings. We shall follow up this theory in detail in chapter 8.

Fechner envisions that when psychophysics is highly developed and mature it will provide empirical criteria for refuting other mind-body theories and buttress the identity view of mind and body. He felt that even his own rudimentary psychophysics offer arguments that refute monadical theory with its postulate of a point-like seat of the soul. He called these sorts of preliminary arguments "starting points" for the day view. One day we will have discovered sufficient empirical starting points to disprove all varieties of the night view and the day view will be magnificently successful. All the essential ideas of the day view show up once again, if sometimes only vaguely, in psychophysics.

The following five examples demonstrate the day view within the context of practical, real, scientific work. The first case deals with Fechner's scientific realism, using the example of atomism. Advocating atomism, Fechner strives to "close off" the day view "at the bottom," as a pendant to the idea that the linkage of all phenomena in God's consciousness completes the day view "from the top." The second study depicts the destiny of Fechner's mind-body theory—his identity view, or psychophysical parallelism, as it was usually called. It perhaps best reveals how Fechner's intentions for the day view have survived in twentieth-century logical empiricism. The third case deals with psychophysics, the origins of modern empirical psychology. The fourth study looks at Fechner's cosmological theory of evolution and its universal "principle of the tendency toward equilibrium," including a thesis about the origin of life. And finally, we shall discuss Fechner's indeterminism, his notion that an undetermined element is inherent to all events; and how he tries to design a system of mathematical statistics that does justice to the fact of objective randomness.

4 SCIENTIFIC REALISM AND THE REALITY OF ATOMS

ONE MAJOR TOPIC in the philosophy of science of recent years has been the pros and cons of "scientific realism."[1] A realist, in this sense, is someone who contends that reasons leading us to consider a well-confirmed and scientifically tenable theory are themselves sound reasons for presupposing the factual existence of theoretical entities contained in that theory. Thus the realist does not dogmatically claim that theoretical—meaning unobservable—things, processes, events, and so on, are real. What he says is that all the sound reasons in favor of an empirical theory also provide good reason to accept the ontology supporting that theory. In contrast, the antirealist—and there are many varieties of this conviction (the instrumentalist, constructive empiricist, conventionalist, positivist . . .)—contends that theoretical entities included in a theory are fictions, employed merely for the purpose of ordering our factual observations and bringing them into context, to assist in the prediction of further observations. Even the greatest success will not alter the fictitious nature of theoretical constructs.[2]

A scientific realist regarding atoms, in other words, an atomist, therefore believes that atoms in fact exist, although they are not observable. The empirical success of theories about atoms, theories bringing them into context with visible phenomena, is sufficient to convince the atomist of the real existence of atoms.

The antirealist regarding atoms holds that experimental success, no matter how conclusive it may be, never provides sufficient reason for presupposing the existence of atoms, unless of course atoms could actually be observed in such experiments.

Fechner was among the first natural scientists and philosophers to deal theoretically with the question of scientific realism. He decided very early in favor of atomism and it became a leading concept for him. After leaving his university chair in physics, Fechner elaborated his hitherto practically applied atomism into a position of philosophy of science. This turn is interesting in itself. But what makes it fascinating is the fact that probably the most famous antirealist in the history of physics, the physicist Ernst Mach, arrived at his own position precisely by analyzing Fechner's. Mach's position, as we now know, was to have great influence on twentieth-century physical science (and philosophy of science). In fact, Mach initially agreed entirely with Fechner's atomism. His subsequent alienation from atomism had much to do with the (preliminary) waiving of metaphysics and ontology, as Fechner later propagated in psychophysics. Mach developed his own position—as it were—based on Fechner, and yet challenging Fechner.

I shall first portray Fechner's early atomism, followed by the philosophical justification he elaborated for atomism and realism, and then trace the very entangled path down which Mach ventured subsequent to Fechner.

4.1 Fechner's Early Writing on Atomism

Fechner reports how, as a young "professor engrossed in the study of atoms," he argued with his friend Hermann Lotze, who "though still a student, was intellectually quite mature" and who rejected atomism.[3] At the age of twenty-five Fechner wrote his first publication (to our knowledge) involving his position on the question of atoms. At that time in Germany intense disputes raged throughout the sciences, supporting two opposed realistic positions: one was dynamism, the other was atomism.[4] Dynamism was a collective term covering all those approaches following Kant's *Metaphysical Foundations of Science* (1786) in explaining matter as the interaction of forces of attraction and repulsion. For the dynamists, then, forces were the fundamental theoretical entities upon which all visible phenomena rest. At the time Fechner wrote his first essays, dynamism had been pressed to defend itself. Its fate was tied to the philosophy of nature.

In the end, physicists favored atomism as it was taught in French physics, preferring it to dynamism.

But the issues were not as simple for atomists as they had been for eighteenth century Newtonian physicists. In the nineteenth century, namely, the question of so-called "imponderables" became important and it was intimately linked to choosing between dynamism and atomism. The atomists presupposed not only the existence of weighable atoms but also—depending on one's particular position—the existence of one or more non-weighable fluids that could be incorporated into the Newtonian world picture in order to explain newly discovered phenomena of light, heat, electricity, and magnetism. Naturally, this threatened to dissolve Newtonian physics itself. If we allow individual imponderables commanding their own forces in all fields, we will soon have ubiquitous 'explanations' like that offered by the doctor in Molière's play *The Imaginary Invalid*, saying that the somnolent effect of opium is due to its "*vis dormitiva.*" Dynamism is frank on this issue: it renounces the existence of atoms and goes straight for basic forces for constructing all kinds of phenomena.

Fechner's early articles are marked by an effort to (re)create a uniform picture for an atomist theory that was becoming increasingly messy. He tries to prove that it is possible to "reduce all material phenomena, the ponderables as well as the imponderables, to attraction activity governed by laws of gravity."[5] This remark was intended for the dynamists, who presupposed the existence of a fundamental repulsive force, but it also shows an effort to simplify and unify atomism. Fechner thought this could be achieved by imagining a planetary scheme for atoms:

> While compiling all the phenomena, I have long deliberated the option . . . of making whole atoms, or rather making molecular theory, depend on the laws of gravity; if we did not view the bonding of imponderables as if these were attached to the ponderable molecules like resting atmospheres, but instead, as if they were determined for closed trajectories due to attraction governed by the law of gravity, such as planets move around a sun; then we could think of each body in a sense as of a system of solar, relatively fixed molecules (which comprise the ponderable mass of the bodies) and molecules moving therein in a planetary manner, which represent the bonded imponderables of the body.[6]

Here, probably for the first time in the history of physics, is the idea of using a planetary model to represent atoms.[7]

Fechner imagines that the imponderables circle around the ponderable atoms in every direction and at every possible velocity. So when two atoms encounter each other, the additional attractive force of an approaching ponderable atom throws imponderable particles of the second atom off course. If the supervening power of gravity is not strong enough to attract the imponderables, they will escape from both ponderable atoms. This creates the appearance of a repulsive force.

Fechner uses this model to explain a whole series of phenomena. He tries to show that his model is superior to the so-called atmospheric model, in which imponderables are presented as the resting atmospheres of ponderable atoms. The only flaw he detects in his own theory is the use of feeble mathematics "for a project that would demand and exhaust all the profound reflection of someone like Laplace, if it were to be dealt with in a secure and productive manner to some degree."[8]

In the second treatise of 1828 Fechner more carefully details his idea. He comes to the conclusion that ponderable atoms must exhibit a "density many billion times greater" than the bodies which they make up.[9] He also tries to explain cohesion, establishing it on the power of gravity, by proposing that ponderable atoms exercise a power of gravity in their direct surroundings that is greater than the earth's gravitational force. It is remarkable how much physical evidence Fechner brings into play in support of this model. He refutes Kant's claim that without the existence of a fundamental repulsive force all matter would melt down to one point, with the counterclaim that in an infinite universe no single point would be distinguishable, towards which all matter must gravitate. And besides this, individual bodies would have to attract each other and thus would constantly change their direction while moving towards a common center point, if there were such a point. Anticipating the cosmology he later developed in 1873 (and which we will discuss in chapter 7), Fechner assumes that originally matter was arranged in "all conceivable possible relations and ways." This idea well suits the notion that the "infinite world is the exhaustive or the laboring self-exhaustive manifestation of a concept or an idea, which itself and in itself is an infinite number of things." From this notion Fechner concludes that a general concept of matter would be that of "space with a talent for attraction" encompassing every degree of attractive force and all the different regions of space: "Thus the manifestation of this concept in the world covers the infinite diversity and unequal distribution of *perceptible* matter by itself."[10] In a footnote Fechner also mentions a hunch that the form, spontaneity, and vitality of organic bodies are perhaps explicable in terms of the indefiniteness attributable to any system having three or more atoms. So even here in this early article we discover that conceptual mixture so typical of Fechner, a blend of competence in natural science and philosophy of nature reminiscent of Schelling.

In his German rendition of Biot's work Fechner also inserted a chapter of his own on "the plausible original state of physical objects," written entirely in terms of atomism. He writes that although some time ago dynamism was applauded in Germany, it meanwhile had lost most of its supporters. All foreign physicists were atomists. It is likely that during his stay in Paris in 1827 (see chapter 1) Fechner found sustained confirmation among the "foreign physicists" for an atomism he himself already advocated.

In his most important essay in physics, which appeared in 1845, Fechner propounds a synthesis of his atomism (resulting from theoretical rumination) and his experimental research in electricity and magnetism.[11] He seems to have abandoned the attempts made in 1826 and 1828 to disregard fundamental forces of repulsion; at least he no longer mentions it.

He begins with André Marie Ampère's law for conductors of electric current. Ampère had split electric current up into mathematical entities, called electric elements, and conjectured that these electric elements have central forces and attract one another. Later he set up a hypothesis for molecular electricity providing for flowing electric fluids, and in which the forces depend on the relative velocities of the current. Adopting these notions, Fechner tries to sketch a fundamental physical model of electricity conduction for Ampère's purely phenomenological theory mentioned first above. For him electrical current has an atomic structure, meaning that it consists of "electric particles." Using two "fundamental axioms" he wants to link—with "at least partial necessity"—"Faraday's induction phenomena" to "Ampère's electrodynamic phenomena," which hitherto were related only on grounds of an "empirical rule" (defined by Lenz).

Fechner's first principle states that the effect possessed by an Ampère electric element occurs because of positive and negative electric particles passing through the same element of space in different directions. The second principle states that similar electrical particles moving through the same space in the same direction attract each other, and they repel particles moving in the opposite direction. Dissimilar electricity acts conversely. If no flow of electricity occurs, dissimilar electrical particles join to become neutral pairs. This is Fechner's model for explaining induction.

Fechner lists some novel empirical results that should ensue from his model, but for which experimental proof is difficult. He also surmises that his theory should proffer a way of measuring the velocity of electricity. Wilhelm Weber had already suggested how to achieve this in practice. In 1857 Fechner's guesswork finally led to Weber and Kohlrausch's discovery of the natural constant c; it coincided with the value that Maxwell reported in 1862.

In 1846 Wilhelm Weber published "the universal fundamental law of electrical effect" that Fechner had already announced in his article dating 1845. This law rests on Fechner's principles and brings together Coulomb's fundamental law of electrostatics, Ampère's fundamental law of electrodynamics, and the law of induction, and fashions them into one single law. The only disadvantage of Fechner's contribution was that it was purely qualitative. Weber formulated the law quantitatively. The main feature of Weber's summary is that it introduces a force that depends on the relative velocities and acceleration of reciprocally effective electric particles. Throughout his life, Weber shared Fechner's fundamental atomist view, not only in teaching electrodynamics. He too, assumed a planetary model for atoms.[12]

In Germany Weber's electrodynamics remained the dominant theory until Maxwell's electromagnetic theory gained acceptance towards the late 1880s. Weber's theory provided a framework in which scientists sought to couch not only electrodynamics, but also theories about matter, light, and heat, as well as all known electrical and magnetic phenomena.[13] H. A. Lorentz's theory of electrons presented at the end of the nineteenth century was in part a re-discovery of the Weber-Fechner scheme. In our concept of the electron, Fechner's notion of an atomic "electric particle" has survived to this very day.

4.2 The "Theory of Atoms"

During the first half of the nineteenth century, particularly between 1830 and 1860, most chemists felt that the theory of atoms was a comfortable working hypothesis and useful in the laboratory, but that otherwise it was to be treated with discretion. Many shared the opinion that chemical facts and laws could be equally well formulated without recourse to atoms, using only terms such as "compound weight," "mass part," or "ratio number." As a matter of personal temperament, one either decided that atoms are pure fiction lacking all reality, or one retreated to a skeptical position claiming that natural science will never succeed at proving or disproving their existence. Thus both reactions considered the atom not an object of experience, but "merely a hypothesis." Gradually, however, atomism underwent reevaluation, due to—among other things—Cannizaro's revival of Dalton's atomic theory (1860), the kinetic theory of gases, and the discoveries of radioactivity and x-rays. As the century waned, however, once again a wave of denial arose, in part propelled by Ernst Mach. Finally, Einstein's, Smoluchowski's, and Jean Perrin's works on Brownian motion (1905–08), studies that even "converted" Wilhelm Ostwald to atomism, broke that ban and the atom acquired widespread recognition.[14]

Nonetheless, let us return to the time around mid-century, a period when Mach still acknowledged atomism, and recall the general scientific situation prevalent then in German states. After the revolution in July 1830 and particularly after the unsuccessful revolution of 1848, a noticeable shift occurred in the relationship between philosophy and the natural sciences. Respect for philosophy had reached an all-time low, while science continued to expand its leadership function. Two suggestions were made for renewing philosophy: One was that philosophy should abandon its own metaphysical foundations and subject itself to science by making science's philosophical basis explicit; the other was that it should revert to Kant and revive epistemology from the start. Many scholars endorsed a combination of the two approaches as a viable solution.[15]

Against the backdrop of this general constellation, for a philosopher the scene in chemistry must have looked fairly peculiar and paradoxical. On the one hand, the concept of the atom was a useful instrument for the scientist, on the other hand scientists felt not at all responsible for deciding whether or not atoms truly exist. A number of German philosophers therefore hoped that by answering this and similar questions, they could put a halt to generally anti-philosophical sentiments and establish a new and independent role for philosophy. They saw philosophy as the provider of metaphysical foundations for hypotheses that go beyond experience. In this period, philosophy actually did try to productively fulfill that role.[16] Concerning the constitution of matter, most philosophers favored dynamism, finding it to be a superior and better-argued position. And this philosophical trend towards dynamism was also supported by a large number of scientists.

But that development led philosophers to a dead end. The chemists remained practical atomists and theoretical skeptics and (except for a few) were not convinced by dynamism. Philosophy's suggestions for science went unnoticed and turned increasingly into metaphysics of neither empirical nor practical relevance. Conversely, in science, particularly in chemistry, the demand arose for "abstinence from philosophy," a reluctance to engage in the elaboration of abstract theories. And this trend did not help diminish scientific prejudice about philosophy. Neither philosophy of science in the narrower meaning of the term nor Kantian epistemology can say more about reality than science itself can (and is willing to).

A book reacting seismographically to this state of things appeared in 1855, namely Gustav Theodor Fechner's *Concerning the Physical and the Philosophical Theory of Atoms*.[17] Fechner sought a new way out of the bungled circumstances by first requesting both parties to make compromises in their concepts. He demands that scientists take the making of their own concepts more seriously and acknowledge that the success of relevant scientific theories at least makes the existence of atoms *likely*. He demands that philosophers hand over to scientific method the competence on issues of reality. But otherwise, philosophers should free themselves from the scientists' skepticism, and instead conceptually anticipate the final goal, the "perfection" toward which science in the past seems to have striven. It is the philosophers' task to supplement science, to shape and refine it, even to use it for philosophical purposes. Fechner's program embraces a Solomonic revalorization of both philosophy and science. Science becomes philosophically relevant, but philosophy conceptually anticipates science.

As the title itself indicates, this work has two parts. The first part explains atomic theory as a "view . . . founded in nature"[18] and attempts, as far as possible "to establish it physically, i.e. by connecting facts."[19] In the second part Fechner discusses "what we may demand philosophically, based on what is materially established, and due to our desire to discover a purely conceptual end."[20] By a "demand" or a philosophical-conceptual "end" Fechner means what Peirce later came to call "abduction": The most plausible explanation that can be gleaned from known facts in a given field.

In the first part, the "physical theory of atoms," Fechner presents two types of arguments supporting the existence of atoms.[21] We can call the first type "inference to the best explanation": The success of a theory, which in comparison to other theories in its field offers the best explanation, provides sound reason for acknowledging the existence of unobservable entities upon which that theory rests. The second type of argument is "inference to the most comprehensive explanation" (the one with a "maximal range"): The success of a theory whose explanations thus far can interconnect the greatest number of phenomena, which otherwise would remain disparate and require several theories, provides good reason for acknowledging the existence of the unobservable entities upon which that theory rests.

Fechner expounds three reasons for the first type of argument supporting why he finds physics based on atomism more explanatory than physics established on dynamism:[22]

(1)   Wave theory can only explain the differing refraction of colored light rays, if it is based on the assumption that ether has an atomic structure.

(2)   Likewise, the explanation for the polarization of light through transversal vibrations can only be correct if it is based on the presupposition of atomically structured ether.

The question of whether or not we acknowledge atomism is vital for undulation theory, just as the question of whether or not we acknowledge undulation theory is vital for physics.[23]

(3)   Only by presupposing that matter has an atomic structure can we explain why heat rays are at maximum when perpendicular to the surface of a body.

But not only does atomic physics provide an explanation of phenomena superior to that given by dynamic physics, it also provides the more comprehensive explanation (demonstrating the second type of argument), for which Fechner offers two reasons:

(1)   The phenomena of heat radiation and heat conduction can only be combined in one single explanation if we follow Fourier in saying that heat conduction is the result of the heat radiation of individual atoms.

(2)   In contrast to dynamic physics, atomic physics can more uniformly and precisely link together the heterogeneous macroscopic properties of bodies, meaning that it "subjects them to the same principles of equilibrium and motion."[24] The examples Fechner gives of such properties are density, solidity, viscosity, heat expansion, proportions of compound weights, and aggregate state. He discusses isomerism and crystallization at some length. (In the book's second edition (1864) he also tries to show how the combination of magnetic and electric phenomena rests on the assumption that matter is structured atomically.) All in all, atomism is necessary for producing the link between macroscopic and microscopic physics: "Through the theory of atoms, chemistry, crystallography, etc., the doctrine of the smallest things whatsoever is brought under the rule of the same general principles of equilibrium and motion which science uses everywhere to achieve clarity and success, and becomes herewith for the first time a consistent coherent system. Without the theory of atoms, this context falls apart."[25]

Fechner well knows that both of his kinds of inferences cannot claim to be valid deductions. His arguments for atomism are—like inductive inferences—merely probabilistic inferences. But not having sufficient reasons for a hypothesis does not mean that we lack every reason for it: "We do not want to pretend that atomism itself is something absolutely certain, neither in the sense in which some philosophers speak of the absolute certainty of their systems, nor in the sense in which some immediate experience as such is immediately certain; even we retain a margin for belief . . . ; but in our endeavor to link material things, it is advisable to adhere to atomism as the most probable, accessible theory, until we have something more probable and more accessible."[26]

Fechner carefully analyzes the protest that even the best theory cannot make plausible the existence of the unobservable entities to which it refers, as long as what is unobservable does not become observable. Thus inferences to the best and to the most comprehensive explanation can never be compelling: "Now the philosopher has the habit of saying: as a mathematical fiction all of that may be fine for deducting empirical knowledge; but more sophisticated reasons prohibit us from assuming that it is real."[27]

In order to refute this objection, Fechner appeals to the essence of mathematics. Mathematics is a "purely formal science that employs no tricks to get more out of things than they contain."[28] Therefore if, by using a theory of atoms, we can "improve deductions about reality, that is, we can improve prediction and . . . production" when compared to using a theory incompatible with atomism, "this must be seen as evidence that the former itself is more empirical than the latter"; in other words: the hypothesis of atoms corresponds to reality. For we cannot infer reality from mathematics, we can infer it only from what is real:

> Wherever the limitations of our senses make something not directly perceptible, a decision about whether or not it is real must be based on the context of what is perceptible and the possibility of one thing being reduced to another thing that is perceptible. . . . We can hold a prerequisite all the more for proven conclusions; the totality of mathematical—and thus stern—conclusions prove themselves in reality. . . . The foundations, upon which mathematics calculates something real, themselves have the character of being real.[29]

Six years later Fechner summarizes this argument as follows: "The proof of the reality of atoms lies in the mathematical necessity of needing it."[30]

The second of Fechner's replies to objectors could have been taken directly from the arsenal of pragmatism. To acknowledge as real something that one has thus far not acknowledged as real, means to modify one's prevenient behavior. To believe that atoms are real means "to bestow upon the hypothesis of atoms the privilege of determining the direction of research, the form and method of presentation, links to other scientific areas, and reasons for general opinions."[31] If, like the chemist in his laboratory, we notice that by imagining atoms we "can better deal with what is visible and tangible, than if we did not imagine atoms"[32] then this means that we in fact do not view them as fictitious, but take them to be real. "Knowledge, belief, and action" are all connected in the end.[33] At any rate, there is no doubt about what the concept of atoms achieves for "apparent reality." If the scientist cannot do without reference to things incapable of being demonstrated by the senses, then he must bear the consequences, be consistent, and acknowledge their reality. Philosophy should not criticize this logical consistency immanent in physics; on the contrary, philosophy should demand it of physics "preventing it from remaining merely physics, and making it philosophical physics."

> "For the philosophy inherent in physics, like philosophy in every science, does not consist in dropping at some point the general categories and methods characteristic of that science and then switching over to categories and methods of philosophy; it means to consistently and thoroughly improve precisely those scientific categories and methods. So, entirely apart from the theory of atoms' philosophical verification(s), the physicist is formally, philosophically forced to presuppose it. . . . The theory of atoms is demanded by a philosophy that consists in the internal harmony of science itself."[34]

In the second, philosophical part of his theory of atoms, Fechner seeks to glean special ideas about the composition of atoms from the foundations laid out in the first part. He finds the most probable hypothesis in point atomism, which he adopts from Ampère, Cauchy, Seguin, and Moigno.[35] This notion, already elaborated in the eighteenth century by Rudjer Boscovic (Roger Boscovich) and by Kant (in *Monadologia physica*, 1756) holds that so-called "simple atoms" are centers of force that occupy a certain position, but are not extended. For Fechner, saying that atoms are like points means that "the implications that we deduct from the presuppositions of atoms regarding apparent reality . . . will be found infinitely precise, the smaller we conceive the atoms to be."[36] Fechner imagines the atoms of the chemist to be compounds of qualitatively non-distinguishable point atoms. He also agrees with the atmospheric model of atoms as presented by Poisson (1829) and Wilhelmy (1851), which he thought had found general acceptance. In this theory every ponderable atom is surrounded by an atmosphere of imponderable ether atoms. All ponderable atoms attract each other and ether atoms; all ether atoms are mutually repellent.[37]

Although even Helmholtz had adhered to point atomism in 1847, by the late 1860s apparently no one advocated it any longer, so Fechner's approach was soon outdated. F. A. Lange complained in 1875 that a "world structure built on central points of force" obliterates any distinction it may have from dynamism and therefore can itself be counted as a dynamic view. And in general, such speculation "dispels the essence of mass and the atom, making it a hypostatized concept" and contradicts the newest findings in chemistry, which, using the concept of valence can, for the first time, "predict the existence of new elements."[38]

Fechner begins the second part with an intriguing discussion on methodology. As mentioned earlier, he assigns to philosophy the duty of finding a "philosophical closure," or a "philosophical view of nature" built upon the foundation provided by the "exact sciences": "What is new about the metaphysics I have in mind—for it is metaphysics—is to abandon all the attempts at finding a foundation for metaphysics and make physical science the foundation we need . . . and thereby in fact to elevate the title of metaphysics, making it something that comes after physics, instead of something a priori or underlying physics. . . . We want to make the entirety of physical science the substructure for metaphysics, so that metaphysics represents only the highest peak, and this peak can be linked to other peaks."[39]

The task of this kind of metaphysics would be to deal with "the most general concepts and the border concepts of the given" and to conceptually evaluate the path that science has thus far taken, "thinking it through to the end." The methods coming into play for this are not as certain as those of material science, but they are at least better founded than the methods suggested by Schelling, Hegel, and Herbart—those philosophers most often listed by Fechner as representatives of metaphysics gone awry. If we understand metaphysical reality as "the conceptual scope of reality capable of being experienced," then we can apply this philosophical concept legitimately. This way we could "bestow upon atoms metaphysical reality instead of a physical one, in case we are unable to sharpen our senses sufficiently to see and feel atoms for what they really are."[40] Presupposing something metaphysically real in this sense is not done arbitrarily or a priori, it must be justified empirically by what it implies for physical reality. Point atoms, as "simple entities" are, and intend to be, "nothing other than the ultimate move made by philosophy, which someday physics itself must make in order to be perfect and thus advance into a state of metaphysics, the general doctrine of the scope of the given."[41] Thus Fechner advocates, as it later came to be called, inductive metaphysics.

A "metaphysical reflection on forces" concludes the book. Here Fechner tries to lay out a "general force law of nature" intended to bring together gravitation and various molecular phenomena by introducing short-distance atomic forces.[42] This set forth a theme he originally suggested in the 1820s. The principle of inertia describes a system consisting of only one atom. Newton's law of gravity expresses the relationship holding between two mass points. For each additional system with n>2 atoms, an additional law holds with a new simple force of the nth level, that combines itself with the simple forces of the lower level subsystems to equate the total force: "In every combination of any number of particles a force is at work, whose intensity and direction is determined at once by the relations of the combination of all those particles."[43] This means that in every system of more than one atom a force is at work that emerges from the system as a whole (it "depends jointly on the relationships of the combination of all parts of the system"[44]) and cannot be seen as the sum of partial forces. Binary and ternary forces are attracting, while both forces of the next higher order are repellent, and so on, a pair of attracting forces always following a pair of repellent forces. The forces are conversely proportional r^n, whereby n becomes larger with the order of a force. All forces having n>2 are thus only effective in terms of molecular dimensions and can be neglected in terms of macroscopic distances, the higher the order they belong to.[45]

Fechner hopes that his approach will result in an improved explanation for molecular phenomena, for fundamental qualities of the chemical elements, for crystallographic phenomena, viscosity, aggregate states, and above all, for organic systems. In its defense he appeals to Wilhelm Weber's doctrine of electricity; his friend had explicated an analogous law of force for electrical charges. According to Weber the reciprocal effect of two electrical charges ("electric masses") is also dependent on the acceleration they are subjected to by the influence of other charges.[46]

4.3 Realism Includes Phenomenalism

We acquainted ourselves with Fechner's scientific realism by looking at atomism. But this does not exhaust his notion of the character and function of physical theories, as expressed in the *Theory of Atoms*. Surprisingly enough, Fechner associates realism with radical phenomenalism: Theories in physics refer to appearances. The tension existing between these two concepts, being concepts we normally consider antithetical, lends Fechner's philosophy of science its original, albeit problematic distinction:

> For science, nature in general is merely a world of external material appearance.[47]
>
> If we ask generally, what the world consists of in the last instance, the answer is appearance ... Laws and definitions of appearance, links and relationships among appearances, including the possibility of future and novel appearances. This is all there is, and there is nothing behind it. We would have no way of talking about a world that did not appear to itself and others.[48]

Verificationism, the influential philosophical program propagated by logical empiricism in the twentieth century, can be traced historically at least back to this moment.

Fechner's phenomenalism begins with empiristic criticism of the concept of noumena, as noted above in chapter 2. As we saw there, he uses the term "*Ding an sich*" not only in Kant's meaning, but also subsumes under it the concept of substance in Spinoza's sense, the concept of the absolute taken from Schelling and Hegel, and the concept of the "real essence" found in the writings of Herbart. Noumena cannot be demonstrated in experience, nor do we have for them—in contrast to legitimate metaphysical border concepts—even the slightest bit of empirical evidence in support of their presupposed existence. Absolutely no consequences for explaining physical reality ensue from presupposing noumena. Arguments of scientific realism do not apply. So in using the term 'noumenon,' we are not designating anything real.

In the book *On the Soul*, published in 1861, Fechner shapes his ruminations into a radical critique of the traditional concept of substance and essence:

> There is no *essence* of things, other than the immutable immanent conditions under which those things prevail.
>
> What is immutably or lawfully connected in such a manner that we can only have one to the same extent to which we have or can have the other, is a thing of the same essence, and this link is the hard core, it is what is essential to essence.[49]

Why is it then that we speak of things at all, instead of restricting all our discourse to appearances from the onset? By nature we are endowed with a need to "conceptually hypostatize" the way in which appearances are related.[50] To speak of things in lieu of appearances implicitly expresses the belief that other phenomena, phenomena that would occur under different circumstances, are lawfully related to the appearances we now have before us. The concept of things, like the concept of natural laws, is a means by which we imply what is not yet given: we pass beyond what *is* given; we can infer what our experience tells us would happen, if certain other circumstantial conditions held. The concept of a thing therefore links appearances together to create a unit and permits conclusions about the unknown. In writing about the soul, Fechner says: A *thing* is "whatever it is that holds individual appearances together; it must be thought of as the source, the criterion, the end of an idea or the end of all those inferences that are used to link appearances in thought and to deduct the unknown from the known, although it never shows itself, indeed—it can only be understood as something abstract."[51]

Fechner's phenomenalism issues severe consequences not only for the concept of noumena, but also for physics' concept of force. All attempts are vain that try to reduce matter to forces or to somehow understand power as a "mythological being" that perhaps "originates in the realm of ideas but interferes in things physical."[52] Physics knows only mass in space and time, and laws describing the motion of mass. The concept of force must therefore be founded on the concept of law: "Nothing is given except what is visible and tangible, motion and laws of motion. Where is there room for force? It remains undiscoverable both in a common and in a philosophical sense. 'Force' in physics is nothing more than an auxiliary term used in describing the laws of equilibrium and motion; every clear notion of physical force can be reduced to that."[53] The concept of force favored by the dynamistic philosopher, in contrast, is a priori and not immediately demonstrable. (In the book's second edition Fechner reproaches Schopenhauer and the materialist Ludwig Büchner [author of *Kraft und Stoff*, 1855] for not clearly stating to which experiences their concepts of force refer.[54]) Confronted with the claim that he had inadmissibly eliminated the concept of 'cause' (being the condition of inertia and motion) in mechanics, Fechner replies: "Indeed, I do not eliminate it, I merely transpose it, I reduce it to its true meaning."[55] In a passage elsewhere Fechner also calls the concept of force a "relational concept, one that has meaning only regarding the combination of matter."[56] Trying to make matter out of force is like trying to create sounds out of the intervals between tones.

Together with Fechner, Lotze contributed a clear and carefully calibrated thesis on the concept of force. He distinguishes between fictions, hypotheses, and abridgments in scientific vernacular. Presupposing the reality of atoms in physics is not a fiction (like the assumption of a circle being an n-angle with infinitely many corners), it is a hypothesis. The concept of force is neither a fiction nor a hypothesis, it is "nothing but a linguistic abridgment" representing the experience of two bodies approaching or moving away from each other at a certain speed.[57]

Phenomenalistic reservations about the concept of force must also have consequences for methods of measuring force. The standard for measuring force is nothing other than the measure of acceleration of mass compared to another control mass: "The *measurement* of force consists in using as a unit of measurement a positive or negative change (acceleration, protraction, generation, or suspension) in the velocity of a given mass of matter and comparing this with another one."[58] This conception will be of importance for our discussion later on.

In another context, Fechner treats mass in a similar manner. While he does think of the mass of a body quite traditionally as *quantitas materiae*, meaning the "*number* of particles or atoms having the same (or having been reduced to the same) effect, which it contains," he thinks of the mass of an *atom* as being represented "by the intensity of the (attractive or repulsive) effect, that it has on a given particle at a given distance; if it imparts to the other a doubled acceleration, then we say that it has twice the mass; or by the reciprocal value of acceleration, which the particle has undergone by the same force of motion."[59] Once again, we will also be dealing with this definition later on.

For now we must take a look at how Fechner intends to combine scientific realism with phenomenalism. If the world consists of "appearances," then should it not be true that non-appearing entities employed in natural science—such as atoms—are necessarily fictions? If atoms cannot appear, they cannot be a part of the real world. This unwelcome conclusion (unwelcome for Fechner's position) however, would only ensue if the world were made up only of those appearances that *are given to all humans* (or could, in principle, be given to them all). But Fechner's phenomenalism does not demand that, and for good reasons. It is likely that appearances exist that are not given for any human being.

In defense of this position Fechner notes that we already always acknowledge the existence of appearances that are not accessible to us, for example, when we think of other persons as of beings with minds. To view others as beings with minds means to assume that they have appearances of a very special kind that are not accessible to me and cannot be accessible to me. That another person experiences inner phenomena similarly to the way that I do so myself cannot be proven with absolute certainty, but for interaction with other people we need that conviction in order to explain our social world and have perspective to our actions. We subsume all the inner appearances that the other person has under the concept of the soul. This context, this "thing" or "tie of appearances" or as Fechner calls it, is something that really exists *over and above* the reality of the appearances. The concept of a "thing" is a bundle of possible relationships of appearances. The result is the concept of the unity of consciousness, as well as the concept of an object in general: "A conceptual thing is true and sound, is heightened reality, as long as it correctly represents the relationship of appearances in common reality."[60]

Now it is important to realize that Fechner's critique of noumena cannot be widened to cover the "conceptual thing" (the unity of consciousness, atoms . . .). To assume that others have minds (or that atoms exist) is a different sort of assumption than presupposing noumena. For conceptual things we have cogent probabilistic inferences (inferences based on analogy) with premises based on our own experience and the behavior of others (in the case of atoms: the premises are based on observable material phenomena), while we have absolutely no experience of any kind from which we can make inferences about noumena. Thus if I legitimately acknowledge the existence of appearances in other minds, that is, self-phenomena that are not mine, then obviously I can also accept that there may be appearances that are not given (or cannot be given) for all mankind.[61] Furthermore, we *need* to presuppose unobservable things in explaining our world and guiding our actions. If we limit ourselves exclusively to "what is valid within finite and restricted realms of experience" we retreat—in Fechner's opinion—to austere, unproductive materialism. In everyday life as well as in science it's the "conceptual things" that provide perspective for our actions.[62]

To summarize: In *Theory of Atoms*, Fechner

(1) advocates *scientific realism:* The only sound reason we have for assuming that unobservable entities are real is a scientific theory containing those entities that is successful in explaining and predicting (observable) phenomena. (If two competing theories differ in their presuppositions about unobservable entities, then it is more likely that the unobservable things presupposed by the more successful theory are real.)

(2) He also claims that *atomic physics is superior* to the dynamism popular at the time.

These two points justify *atomism.*

(3) He also endorses *phenomenalism:* The world consists of appearances, and things are lawfully connected complexes of appearances. Under precisely specified circumstances there is a possibility of non-given appearances. It

follows from (3) that the concept of noumena is nonsense and that the concept of force is merely an auxiliary term, being merely a reference to mass and acceleration.

It is worth noting that Fechner's *Theory of Atoms* caused quite a stir among philosophers. His attack on the established philosophy of his time was considered particularly shocking. It was found outright reckless and caustic. Friedrich Albert Lange wrote that Fechner "explicitly used *The Theory of Atoms* to renounce philosophy, making Büchner's critique look comparatively mild."[63]

Among Fechner's critical commentators were his friends Christian Hermann Weisse, Hermann Lotze, and Moritz Drobisch, as well as the philosophers Hermann Ulrici and Julius Schaller.[64] With the exception of Lotze, all of Fechner's philosophical critics were advocates of dynamism, some consenting to more, others to fewer atomic elements in it. The number of intermediate theories was huge.

4.4 Mach Turns to Anti-Atomism

Sometime between 1863 and 1872 Ernst Mach (1838–1916) came to abandon atomism in Fechner's sense to become one of the late nineteenth century's most prominent anti-atomists. I see this change in Mach's thinking as the result of his analysis of Fechner's philosophy of science. Mach's philosophical and methodological views—particularly the denial of absolute space and the analysis of the concept of mass—came to play a decisive part in the physics and philosophy of the late nineteenth and early twentieth centuries. This provides all the more reason to wonder why he rejected atoms, particularly because he was so thoroughly wrong. In the following I would like to show that Mach's anti-atomism was not a one-time mistake, but that it relates to his most successful views. We will also see that Mach borrowed his famous definition of mass from Fechner.

Although much has been written about Mach's rejection of atomic theory, most of it fairly inconclusive because it neglects Fechner's influence on Mach.[65] I believe that we can better understand Mach's change of mind and what it relates to, if we view it consistently as a reaction to Fechner's (and in part to Johann Friedrich Herbart's) work. Let me elucidate this.

In 1855, just around the time when Fechner's *Theory of Atoms* was published, seventeen-year-old Ernst Mach "instinctively," as he says, discovered that the concept of noumena "is an idle illusion"—and this insight remained significant for the rest of his life.[66] Perhaps reading Fechner had nudged him in the right direction? Whatever else may have been the case, Mach's first lectures held in 1861–1862 as a young *Privatdozent* for ongoing physicians, were erected entirely on Fechner's physical atomism. These lectures were published in 1863 with the title *Compendium of Physics for Physicians*. Following a brief review of John Stuart Mill's "research methods," Mach, making explicit reference to Fechner, proceeds to deal in detail with "atomic theory . . . , which—as it were—constitutes the philosophical epitome of physics." He includes almost an entire page from Fechner's book and lists exactly the same reasons supporting atomism that Fechner had brought forth beforehand.[67] He diverges from Fechner's descriptions only by placing increased emphasis on the atmospheric model that Fechner had treated merely marginally in *Theory of Atoms*; Mach moves it up to the front of his reflections. Not only does Mach find atomic physics superior to dynamic physics, he also acknowledges Fechner's scientific realism. Yet on the whole, Mach's early enthusiasm for atomism is comparatively whimsical and lacks Fechner's elaborate distinctions. The single passage suggesting that Mach may have considered atomism a purely fictional and heuristic instrument for the study of appearances is found in the foreword to the *Compendium*:

> I let atomic theory occupy center stage, not because I think that it is the ulti-mate and best, requiring no further support, but because it orders appearances in a simple and clear context. If I may put it this way: We can think of atomic theory as a formula that has already achieved some success and will continue to do so. In fact, whatever metaphysical views we may have about matter in the future, all the findings we have gained through atomic theory will be translatable into those views, just as formulas stated in terms of polar coordinates are expressible in terms of parallel coordinates.[68]

But here Mach is merely reflecting the possibility that atomism may, at some later date, prove to be wrong, and suggests that even the effort will not have been in vain. At the time, Fechner probably would have thought this a case resulting from premature relativism regarding atomism, since Mach had no *concrete* evidence that questioned the superiority of the theory of atoms. But basically he could also have endorsed Mach's opinion. If we view the assumption of the atomic constitution of matter as an inductive probabilistic inference, we must be prepared that it might be wrong.

In 1864 Mach published a brief note, written in the same style as the *Com-pendium*, announcing his intention to translate the findings gained by atomism into "concepts of a different metaphysical view of matter."[69] From this we can conclude that Mach advocated realistic atomism at least until 1863 and in 1864 still accepted that physics needs metaphysics (in Fechner's sense) for maintaining consistent and clear order, even though he meanwhile believed to have found a real alternative to the metaphysics of atomism.[70]

This all changed radically with Mach's subsequent publication centered thematically on atomism, namely a booklet published in 1872 titled *History and Root of the Principle of the Conservation of Energy*. He writes:

> We want to remember one thing, namely that in researching nature what counts is knowledge about the relations among appearances. What we imagine to exist beyond appearances exists only in our mind. Its value lies only in assisting memory or being a formula that adapts itself easily to our cultural standpoint, because its devise is arbitrary and irrelevant.[71]

> The ultimate enigmas upon which science is established must be facts, or—if they are hypotheses—they must be capable of becoming facts. If hypotheses are designed such that their object can never be tested by the senses, as is the case for mechanic molecular theory, then the researcher has gone beyond what science demands of him. Science requires facts, and this "extra" information is inappropriate . . . In this case molecules are nothing but a worthless idea.[72]

Mach thus rejects scientific realism and the atomism associated therewith.

Seeking reasons for Mach's unexpected adversity to these notions, we find only his opinion stated in the passage quoted above, that hypotheses which refer to unobservable things are not testable and therefore exist only "in the mind." He does not explain *why* unobservable things exist only in the mind. Our only recourse for further interpreting this is to draw relevant conclusions from available texts.

In *Lectures on Psychophysics* (1863) Mach seems to have thought thus: If we, like the chemists, have sound reasons for acknowledging the existence of atoms, then these cannot be exhausted by assuming characteristic compound weights; instead they must also be transformable individually into equally sound reasons for assuming some specific *physical properties* of atoms. The degree of necessity we attach to presupposing atoms must also apply to our assumption of certain properties belonging to atoms, otherwise that assumption remains unjustified. Let us therefore examine which properties may belong to atoms: "How are we to imagine atoms? Colorful, bright, sounding, solid?"[73] Certainly not in any of these ways, for these secondary qualities are the products of atoms stimulating our senses and therefore cannot themselves be properties attributable to individual atoms. What about spatiality? After studying Herbart's treatment of space, Mach was convinced that it is not even necessary to think of atoms as spatial: "Indeed, we need not even imagine atoms as extended in space. For as we have seen [in Herbart's work] space is nothing original and results very probably from an indirect co-effect of several reals."[74]

According to Herbart the spatiality of our sensations is caused by the reminiscence and reproduction of previously experienced series of ideas stored in consciousness. Our sensations are three-dimensional because through vision we can follow-through perceptions in the directions of the three coordinates of space. However, as Mach discovered, Herbart could not explain why some series of perceptions are only one-dimensional—for example, why hearing is one-dimensional, even though our ears, like our eyes and our sense of touch, can follow through series of perceptions in all different directions.[75]

> Now if someone who could only hear would try to develop a worldview based on his perceived linear space, he would fall considerably short of his goal because this type of space cannot encompass the variety of real relations. But we have no *more* reason to believe that we can press the entire world, including things unobservable, into the space presented to our eyes. Yet this is the case for all molecular theories. We possess one sense that, in terms of the variety of relations it can grasp, is richer than any other. It is the mind. It is superior to the senses. It alone can establish a lasting and sufficient worldview.[76]

As long as we have no series of perceptions for atoms, there is no justification for assuming that the 'series of perceptions' that, for us, make up the object "atom" constitute diversity of a three-dimensional kind.

We are also not justified in attributing solidity to atoms, as Fechner taught: "Even physicists have noticed the difficulty of imagining atoms as something *material*, therefore some view them as merely centers of force. Yet a center of force in itself is actually nothing. And what is it supposed to mean to say that one center of force acts upon another?"[77]

Yet all these difficulties are not what make Mach into an anti-metaphysical anti-atomist. He first becomes a devotee of the metaphysical doctrine of monads. Since the success of physics has led us to believe that appearances are caused by atoms, and we know nothing to say about atoms, our last recourse is to attribute non-material properties to them: "Let's confess it straightaway! We cannot reasonably discover any external side to atoms, so if we are to think anything at all, we must attribute an internal side to them, an inwardness [subjectivity] *analogous* to our own souls. In fact, how could a soul originate as a combination of atoms in an organism, if its germ were not already contained in the atom?"[78]

This "physical-psychological monadology"[79] constituted the fundamental postulate of Mach's idealistic period, of which he writes in subsequent autobiographical remarks. He was very much influenced by Herbart's philosophy. According to Herbart, as we have seen in chapter 1, a substance is capable of no other form of modification except self-preservation in the face of disturbance caused by other substances. The substance called 'soul' modifies itself by having ideas, the substance called 'matter' changes inner states, which is expressed in terms of its spatiality.[80]

Compounding Herbart's sway over Mach, Fechner's *Elements of Psychophysics*, published in 1860, became the decisive factor during this phase of development.[81] Mach was so enthusiastic about Fechner's book that he immediately undertook experiments during the fall break of 1860 to see whether Weber's Law also holds for our sense of time.[82] In 1861 and 1862 he intended to perform in Vienna a "Study on the Acuteness of the Senses in the Mentally Ill," but abandoned the project due to a lack of support from his colleagues.[83] In the spring of 1863 and in the following winter terms 1863–1864 he lectured in Vienna on Fechner's psychophysics, followed by lectures in Graz in 1864–1865, where he had meanwhile been bestowed a chair.[84] His lectures were held and published almost simultaneously in installments in a journal for physicians. Mach believed that Fechner's psychophysics finally had "discovered for psychology methods of experiment and observation similar to those well established for physics."[85]

But Fechner's vehement opposition to Herbart's monadology soon dampened Mach's enthusiasm. In a letter written in 1864 Fechner tries to explain to Mach that Herbart's influence will distract him from the direction he had taken with psychophysics.[86] He would only accept Mach's offer of dedicating to him his book *Analysis of Sensations* if that dedication were meant to express that Mach had been *stimulated* by Fechner, but not if it would create the impression that Fechner *acknowledges* Mach's standpoint, because this—as a result of Herbart's influence—was not the case. In the future Mach himself, Fechner continued, would only regret having prematurely acknowledged Fechner's ideas. As a result, Mach postponed publishing the *Analysis of Sensations* for all of twenty-two years, but then he did unmistakably admit his debt to Fechner, even if it were only a mere "stimulation": "Twenty-five years ago my natural inclination for the issues dealt with here was most strongly *stimulated* by *Fechner's* 'Elements of Psychophysics.'"[87] Finally, Mach dedicated the *Analysis* to the British mathematician and philosopher Karl Pearson (1857–1936). But as late as 1911, at the age of 73, Mach mentioned that Fechner's reaction to his manuscript continued to haunt him in his sleep.[88]

Which faults did Fechner detect in Herbart's monadology that caused him to decline Mach's offer of dedication and thus refuse association with Herbart? At least two can be found. As mentioned above, in Fechner's mind a substance whose essential property is an inaccessible inner state is a new "noumenon." He also calls Herbart's "reals" (a term for simple qualitative units that make up reality) "schemes behind the given."[89] In Fechner's eyes, Herbart adheres to an inadmissible metaphysics. While it is true that Fechner's point atoms and Herbart's "reals" are both unobservables, Herbart's "reals" bring absolutely no advantage for science dealing with physical phenomena and can therefore not claim to be real. Point atoms, in contrast, can be empirically justified. The second reason for rejecting Herbart's doctrines is that for Fechner the soul, or consciousness, is not a property belonging to fundamental elements of the world itself, as Herbart taught, but instead, consciousness is a functional property of systems as such. "The mind is not in atoms, but in systems."[90] This conviction was antithetical not only to Mach's opinion, but also to that of other advocates of monadical doctrines; it opposed Hermann Lotze (who did not consider himself a Herbartian) and the astrophysicist Karl Friedrich Zöllner, Robert Grassmann, later also Ernst Haeckel (who spoke of "atom souls"), Eduard von Hartmann, and the anthropologist Georg Gerland.[91]

We know from lecture notes taken in 1872 that Mach was deeply hurt by the expostulation directed towards his monadology. He notes one of Zöllner's theses that had stirred a commotion: "Sensitivity is a general attribute of matter, more general than mobility"—and adds: "I myself have been reproached for saying that."[92]

Thus Mach himself was torn: On the one hand, consistent material atomism perfectly suited Herbart's doctrine of monads, which in turn was useful for explaining the psychical part of the world. In addition, Fechner's methods for measuring the psyche seemed to enhance and perfect Herbart's mathematical psychology such that these also went well together.[93] On the other hand, Herbart's monadology does not combine well with Fechner's atomism and in the long run—neither with psychophysics. The crux is the metaphysical assumptions on which the respective theories rest: both the metaphysics of atomism (which Fechner thinks are acceptable and necessary) and the metaphysics of Herbart's monad doctrine (which Fechner thinks are objectionable and superfluous). If one could dispense with all metaphysics—whether acceptable or objectionable —one could easily combine the useful elements of Fechner's methodology for physics with Herbart's psychology and with psychophysics.

It attests to the fact that Mach at the time was preoccupied with establishing a strong link between psychology and physics that he titled a public lecture given in Vienna the previous summer (1864): "Concerning the Relation of Fundamental Questions in Physics and Psychology." He himself later described this period as a time of torment and personal tension: "My idealistic mood [brought about by questioning noumena] disagreed with studies in physics. The tension increased through acquaintance with Herbart's mathematical psychology and Fechner's psychophysics, which seemed to provide both acceptable and objectionable elements."[94]

The conflict between science and metaphysics could not be solved by Herbart's monadology. The solution lay in an anti-metaphysical modification of Fechner's phenomenalism. In retrospect Mach wrote: "My only desire is to occupy a standpoint in physics that I must not abandon as soon as I peer into the domain of another science, for they should all constitute one single whole."[95] Thus the theme of the unity of science was already audible in his lectures on psychophysics. In the final passages of these lectures Mach expresses the hope to have shown that "physics, physiology, and psychology are indestructibly connected, such that each one of these sciences can only be successful in combination with the others, and yet each one of them is an auxiliary science to the other two."[96]

These lectures of psychophysics demonstrate how much Mach's concept of metaphysics had come under pressure and how ambivalent his reaction was. On the other hand, he desired to avoid "all metaphysical issues": "The question of what the soul is, is as unimportant for psychology as the question of the essence of matter is for physics."[97] But on the other hand, as we have seen, he does postulate the existence of atoms and sees that in every respect physics must assume "that there is a *finite number of beings and forces* active in phenomena." Objecting to Wundt's rejection of every metaphysical foundation for science he claims that "we would never have come to mechanics, even less to psychology" if we had proceeded entirely without metaphysics.[98]

And in eventually bridging the schism with anti-metaphysics Mach could also closely follow his role model Fechner. In *Elements of Psychophysics* Fechner had shown how to circumvent metaphysical disputes by sticking to consistent empiricism and phenomenalism. In the introduction to his work he states unequivocally that any "accurate doctrine of the functional or dependency relations existing between the body and the soul"[99] (i.e., the definition of psychophysics) is entirely independent of all metaphysical opinions about the body and soul and their connection:

> Now it is neither the task nor the intent of this book to embark upon profound or news-breaking discussions about the fundamental issues of the relation of body to soul. May each person seek to solve that puzzle (if he sees it as such) as he will. . . .
>
> In our case it makes no difference, whether one considers the body and soul two different ways in which one being can appear, or as two externally connected beings, or the soul as a point at the nexus of other points of a similar or dissimilar nature, or whether one wants to reject a basic view at all; all that matters is that one acknowledges *experienced* relationships existing between the body and soul and allows us to pursue the same in terms of experience, even if it means trying the most forced representations of the same. For in the following we rely solely on experienced relationships between the body and the soul.[100]

Precisely this freedom from basic metaphysical assumptions in psychophysics found Mach's undivided admiration. In lectures on psychophysics he writes: "*Fechner* executed all his experiments in psychophysics without hypothesizing about the essence of the psyche, without making a single assumption about those processes that may lie between stimulus and sensation. And this . . . is not at all necessary, if all we are seeking is facts of experience. Progress in science requires no precipitation in this respect."[101] Now, since psychophysics has attained success in a controversial scientific subject exactly by remaining free of metaphysics, does it not make all the more sense to follow this same strategy in other well-established sciences? Mach had such esteem for non-metaphysical psychophysics that he thought it worth trying to liberate science as a whole from metaphysics.

In his writing of 1872 Mach began testing anti-metaphysics. If atoms are "conceptual entities" and conceptual things enjoy no metaphysical reality, existing merely as fictions in the mind, then what should prevent us—for the purpose of improved order among appearances—from imagining them other than being three-dimensional, for instance, why not four-dimensional? Mach's suggestion for a new "conceptual thing" is "a molecule having five atoms in a space having four dimensions."[102] He sees the merits of such a model in improved explanation of chemical isomorphism. He also believes that by defining the compound heats for every pair of atoms in four-dimensional space, he can achieve a more rational way of abbreviating the names of chemical compounds.

Moreover, he imagines improving the theory of electricity. He wrote: "I believe to have shown that we can record, estimate, and use the findings of modern natural science without exactly being apostles of the mechanistic notion of nature; I have shown that the mechanistic view is not necessary for knowledge about phenomena and can be substituted well by another theory, and finally, that the mechanistic notion of phenomena can even be obstructive."[103] Thus Mach played Fechner, the psychophysicist and anti-metaphysician, against Fechner, the atomist and metaphysician.

Mach seems to have been particularly proud of the idea of a fourth dimension. He was probably prodded to at last publish his ideas, ideas that his colleagues had initially rejected, after learning of Zöllners view that the universe is four-dimensional and having read Bernhard Riemann's published habilitation paper. A remark made by Helmholtz in 1871, to the effect that we should not deduct "the principles of theoretical physics" from "purely hypothetical assumptions about the atomic design of natural bodies" may also have encouraged him to publish.[104]

We have yet to deal with the difficulty of how Fechner could view psychophysics as incompatible to Herbart's monadology, although he was of the opinion that psychophysics is the non-metaphysical crux of the metaphysical mind-body problem. To be liberated from metaphysics means for Fechner to lack metaphysical presuppositions. It does not mean that an anti-metaphysical foundation is of no consequence for metaphysical positions. Just as, in his opinion, physics can do without a metaphysical foundation and needs none until a much later stage of development, when it embraces the metaphysical assumption of the point atom as its crowning philosophical end, so also will non-metaphysical psychophysics someday renounce its neutrality and make certain metaphysical assumptions about the relation of body and mind and the role of the psychical in the world in general appear more probable than others. He felt that Herbart's assumptions had already proven themselves improbable at an early stage in psychophysics.

Fechner had set up psychophysics so entirely without metaphysics in order to irremovably anchor within empirical science the identity view of body and soul and the "day view"—after all the blows he had taken because of them. He thought that by establishing a science strictly without metaphysics it should be all the more possible to make certain assumptions about its metaphysical "crown." Just as atomism is the perfect philosophical epitome of physics, so, one day, will the identity view and the "day view" crown psychophysics: "This is not to say that the doctrine [of psychophysics] developed here entirely neglects how to understand the fundamental relationship of body and mind, nor that it has no influence on that understanding. On the contrary. But do not confuse the effects that we expect it to have, and that partially have already happened, with the fundament of the doctrine. This fundament is in fact entirely empirical and we reject every *presupposition* from the onset."[105]

It was tragic for Fechner that Mach was the one to take up psychophysics: Precisely those measures which Fechner had employed in order to safeguard as best as possible the metaphysics of the future—are the ones Mach will use to destroy metaphysics once and for all.

Mach, in turn, will have been particularly happy about finding an anti-metaphysical solution because this way he could get rid of Fechner's speculation on plant-souls and on the soul of the universe and forget about point atoms and Herbart's monads. It is very likely that the physicians to whom Mach lectured on psychophysics were not overly enthusiastic about Fechner's world-soul fantasies and Mach had endured his share of ridicule because of them. The late 1850s and the 1860s were the heyday of materialism, and the materialistic world-view as propagated by Büchner, Moleschott, and Vogt was widespread, particularly among physicians.[106]

Mach's thorough lack of metaphysics, also refusing the idea of a philosophical closure of science in Fechner's sense, is, however, not antithetical to Fechner's "day view"; on the contrary. Mach's doctrine stating that the ultimate components of physics are simple and indivisible sensory elements is nothing other than —as it were—a "secularized day view." The fact that Mach and Fechner shared common motives was well known up until Mach's death, as testified in young Viktor Kraft's memorial address for Mach: "As tempting as this image of a world of atoms [in the mechanistic worldview of, for instance, Du Bois-Reymond] was,—its unbridgeable antitheses of nature and consciousness, of true reality and sensory phenomena left it deeply unsatisfying. One felt urged to pass beyond this dichotomy. To overcome it, Fechner designed his system of the universal world soul. And for Mach this was also to become his motive for creating a different, new understanding of nature and the world."[107]

Mach borrowed from Fechner not only the abstract method of designing science without metaphysics and a way to bridge the gap between physics and psychology, he also found in Fechner's treatment of the concepts of force and mass a concrete model for reducing fundamental scientific concepts to factual phenomena. Mach's famous definition of mass stated in 1868 is a slight modification and precise formulation of an approach Fechner had already presented. Fechner, as we have seen, defined the measurement of force of a body of a certain mass via the comparison of its acceleration with the acceleration of a control body of the same mass. In analogy, Mach defined mass as the reciprocal ratio of speed, which a body endures via the influence of a control body with the mass of one. He then took Newton's second law as a definition of the concept of force. This way Mach thought he had depicted the laws of mechanics in terms of pure functional dependencies existing among phenomena, without any metaphysical obscurities in the concept of mass, and to have set up mechanics non-circularly.

In summary: Mach rejected Fechner's scientific realism and limited Fechner's phenomenalism to phenomena accessible solely *by humans* because:

(1) All the sciences can only be unified if they are not rivals regarding their metaphysical assumptions. If physics continues to be metaphysical, we must change our standpoint when moving from physics to psychology. But this impedes the discovery of connections between scientific fields. If we make a modified form of Fechner's phenomenalism the basis for physics, then we can easily link the sciences.

(2) By thinking of science as free of metaphysics we can elegantly dump all of those far-reaching metaphysical extrapolations that Fechner thought he could legitimize with science, without losing the merits of Fechner's phenomenalism.

(3) The case of psychophysics teaches us that scientific progress is possible for philosophically controversial topics if we ignore all metaphysical prerequisites and concentrate only on the phenomena (in Fechner's sense).

It is perhaps exaggerated and schematized, yet helpful to say: Fechner minus metaphysics plus a dash of Herbart equals Mach.

5 PSYCHOPHYSICAL PARALLELISM

The Mind-Body Problem

IT IS WIDELY HELD that the current debate on the mind-body problem in analytic philosophy began during the 1950s at two distinct sources: one in America, deriving from Herbert Feigl's writings, and the other in Australia, related to writings by U. T. Place and J. J. C. Smart.[1] Jaegwon Kim recently wrote that "it was the papers by Smart and Feigl that introduced the mind-body problem as a mainstream metaphysical Problematik of analytical philosophy, and launched the debate that has continued to this day."[2] Nonetheless, it is not at all obvious why these particular articles sparked a debate, nor why Feigl's work in particular came to play such a prominent part in it, nor how and to what extent Feigl's approach rests on the logical empiricism he endorsed.

Following the quotation cited, Kim offers an explanation backed by a widespread (mis)conception of logical empiricism. He claims that work concerning the mind-body relation done prior to Feigl and Smart dealt either with the logic of mental terms—as Wittgenstein's and Ryle's work had—and therefore missed the point, or lacked the sophistication of our modern approaches. One exception, C. D. Broad's laudable work, could not alter this, for it "unfortunately . . . failed to connect with the mind-body debate in the second half of this century, especially in its important early stages."[3] Kim seems to extend his verdict on

Ryle and Wittgenstein to include all authors writing on the mind-body problem throughout the decades preceding Feigl and Smart.

If we ask what distinguishes the young mind-body dispute of the late 1950s from older debates on the topic, we are told that Feigl and his friends and precursors of the Vienna Circle introduced new methods of logical analysis for solving or dissolving the mind-body problem. Feigl himself would probably have given that very answer. Others might say that the debate grew out of general frustration with Cartesian dualism and that it acquired its own specific character in dealing with the problems created by refuting that position.[4] The story goes that reflection on the mind-body relation was horribly wrapped in Cartesian obscurity and confusion until Feigl and the Australian materialists entered the scene. Their "brain state theory," writes Kim, "helped set basic parameters and constraints for the debates that were to come—a set of broadly physicalist assumptions and aspirations that still guide and constrain our thinking today."[5]

Interpreting the difference between the older and younger mind-body debate in the United States in this way may contain a grain of truth, but from the perspective of German-speaking scholars, it is entirely wrong. Seen against the backdrop of nineteenth-century German and Austrian philosophy, Feigl's approach was neither novel nor audacious; he merely revived a tradition that had once been a mainstream topic turned unfashionable; to be exact, he modified and spelled out one specific traditional position.

It is time to readjust our appraisal of Feigl. I intend to show that Feigl's treatment of the mind-body problem upheld an active anti-Cartesian tradition; it follows a pattern in philosophy that was widespread in German-speaking countries throughout the nineteenth century and well into the twentieth, even after World War I.[6] According to Thomas Kuhn's categories, not only Feigl, but almost all the scholars who discussed the mind-body problem within the Vienna Circle and similar movements, were doing "normal science," guided by one single paradigm, so there was nothing revolutionary about Feigl's endeavors. Clearly, Feigl's solution is characterized by the particular twist he gave to the dominant paradigm—an originally neo-Kantian attitude passed on to him by his mentor Moritz Schlick.

In order to understand Feigl's project, we need to first take a look at how the mind-body relation was discussed from mid-nineteenth century onward. So sections 1 and 2 of this chapter will deal with psychophysical parallelism, its popularity during the second half of the nineteenth century and how it was subsequently treated up to the late 1920s. My aim is to capture the setting in which young Herbert Feigl must have encountered the issue when he took up his university studies in Vienna in 1922. The third section deals with the special twist that Moritz Schlick and Rudolf Carnap gave the issue in their writings from that period. Finally, in the fourth section I analyze Feigl's fundamental essay "The 'Mental' and the 'Physical'" (1958) and discuss how it compares to positions he had advocated prior writing it.

5.1 Psychophysical Parallelism Dates Back to the 1850s

In order to properly understand the twentieth-century mind-body debate, we must turn our attention to the 1850s.[7] At that time German-speaking scientists were engaged in a quarrel over materialism that came to be known as the "materialism dispute." In reaction and opposition to German idealism's metaphysical and speculative post-Kantian philosophy, authors like Carl Vogt, Ludwig Büchner, and Jacob Moleschott propagated a very radical, albeit philosophically indigent, materialism identifying mental processes with physical processes. Vogt, for example, stated that any astute scientist must come to the conclusion "that all those capacities that we consider to be activities of the soul are merely functions of brain substance; or, put in simple terms, that thoughts issue from the brain just as gall is produced by the liver or urine by the kidneys."[8] Büchner refers to Rudolf Virchow, who wrote: "An expert on nature acknowledges only (material) objects and their properties; whatever goes beyond that is transcendental, and transcendence means intellectual confusion."[9]

As Büchner's reference suggests, at that time the materialistically motivated movement also aimed to weaken religious dominance; it contributed notably to political liberalism prevalent in 1848 and afterward. The outcome was that hardly any natural scientist or otherwise educated person dared risk seriously adhering to Christian or Cartesian dualism in solving the mind-body enigma. In one famous case the physiologist Rudolph Wagner gave a lecture at a congress for German natural scientists and physicians that started the whole materialism dispute. He insisted that for ethical reasons science must maintain belief in a personal God and in immortality, not only when scientific proof is lacking but even when science seems to disprove it. Needless to say, materialists scoffed.

As uncouth and simple as both the materialists' and their opponents' opinions were, combined with turbulent progress in physiology and gradual alienation from idealistic philosophy of nature *(Naturphilosophie),* the dispute over materialism aroused more scientific interest in the question of how mind is possible in a wholly physical world. Any solution offered for the mind-body puzzle that violated scientific conceptions or was reminiscent of substance dualism was strictly rejected. The latter hypothesis never had many devotees among scholars in Germany anyway.

The introduction of Darwinism pressed the relevance of finding the mind's place in physical nature and increased support for materialism. Slowly this movement became "monism"—led initially by Ernst Haeckel, Darwin's advocate in Germany, and then by the founder of physical chemistry, Wilhelm Ostwald. In fact, it is most likely that in their youth all our (German-speaking) heroes of logical positivism devoured the monists' books, as Carnap himself admitted.[10] It is also known that Moritz Schlick played a prominent role in a monistic organization.

In terms of providing a serious philosophical position, the second edition of Friedrich Albert Lange's history of materialism, published in 1873–75,[11] offered the most effective and sophisticated criticism of early popular materialism. While Lange defended Büchner against the claim that materialism terminates in a loss of morals, and also admitted that as a *method* materialism was not only feasible but also necessary for scientific work, he went to great lengths to analyze the difficulties, weaknesses, and contradictions inherent in materialism, if taken as a serious philosophical position. This critique in turn decidedly raised momentum for neo-Kantianism and contributed considerably to a revival of philosophy in general in Germany after 1860.

Lange ventured beyond offering a mere critique of materialism by affirming a doctrine that was to determine the course of the mind-body debate well into the twentieth century, namely the theory of "psychophysical parallelism." Along with many other scientists and philosophers of the period, Lange viewed psychophysical parallelism as compatible with science and science's materialistic inclination, without necessitating recourse to crude materialism of the type disseminated by Büchner and others. Simultaneously, psychophysical parallelism promised to provide a sophisticated program of empirical scientific research into the mind-body relation.

Psychophysical parallelism had been established and developed by the physicist, philosopher, and psychologist Gustav Theodor Fechner. First mention of his theory dates from the 1820s, but the contents became well known through his mature work, *Elements of Psychophysics*, in 1860.[12] This work marks a turning point in the history of experimental and quantitative psychology, and, I claim, also marks a crucial moment in the history of the mind-body debate and the history, or—if one prefers—the prehistory of scientific philosophy in general. Still in 1911, the Oxford philosopher Thomas Case noted that Fechner's work had attracted "both psychologists and metaphysicians by its novel way of putting the relations between the physical and the psychical" and that it "almost revolutionized metaphysics of body and soul." He even claimed that, as a consequence, Fechner's metaphysics "contains the master-key to the philosophy of the present [!] moment."[13]

Fechner himself did not use the term "psychophysical parallelism" to designate his standpoint. My guess is that this designation has been taken from Alexander Bain's book *Mind and Body* (1874), published in an authorized German translation fourteen years after Fechner's main work; but it may equally be credited to the indefatigable psychologist Wilhelm Wundt.[14]

A widespread misconception pervading pertinent English literature confuses this type of parallelism with forms of Cartesian doctrine of two noninteracting substances, such as doctrines of occasionalism or preestablished harmony.[15] Psychophysical parallelism means the exact opposite: It denies the Cartesian division of the world into extended substance (matter) and nonextended substance (mind).[16] While this conception is congruous with Leibniz's notion of noncausal "conformity of the soul and the organic body,"[17] at the same time it entirely rejects the theological and metaphysical explanation that Leibniz offered for it. Psychophysical parallelism has an entirely different explanation. It propounds a kind of *aspect* dualism that must be strictly distinguished from what should preferably be called "Cartesian parallelism."

In fact, it is best to distinguish three different kinds of psychophysical parallelism (not only regarding Fechner, but in general), each built upon the other.[18] The *primary* form of psychophysical parallelism is an *empirical postulate*—a methodical rule for researching the mind-body relation, claiming that there is a consistent correlation between mental and physical phenomena. In the living human body, mental events or processes are regularly and lawfully accompanied by physical events and processes in the brain; or, as Fechner put it, they are "functionally dependent" on them. A particular physical state corresponds to every mental state; for every mental event there is a correlated brain state.

It is important to emphasize that functional dependence between the mental and physical says nothing about the causal nature of the relationship; causal influence is neither claimed nor denied. This type of psychophysical parallelism refrains from all causal interpretation of the mind-body relation. Fechner said that it is neutral regarding every imaginable "metaphysical closure" compatible with it. This sort of parallelism constitutes the factual foundation for any and every ambitious explanation of the relation holding between the body and the mind, whether or not such explanations ultimately turn out to be causal and interactive.

As a maxim for research, psychophysical parallelism is not only neutral in terms of any causal interpretation that may later seem necessary, it is also neutral regarding the exact nature of the correlation holding among mental and bodily phenomena—namely, whether it is one to one or one to many—and also neutral in terms of precisely which mechanism physically manifests the mental. Understood this way, psychophysical parallelism presupposes nothing about the exact nature of the mental and the physical and how these relate. It is to be taken as a metaphysics-free description of phenomena on which any advanced and scientifically acknowledged mind-body theory must be founded. In his endeavor to clearly state—without any recourse to metaphysics—just how the mental depends on the physical, Fechner came quite close to what we today call "supervenience."[19]

Many scholars, who were skeptical in other respects, found this type of parallelism thoroughly agreeable. William James, for instance, confined himself—as he said—to "empirical parallelism," although he rejected all stronger forms of parallelism (see below). "By keeping to it," he wrote in *Principles of Psychology*, "our psychology will remain positivistic and non-metaphysical; and although this is certainly only a provisional halting-place, and things must some day be more thoroughly thought out, we shall abide there in this book."[20] The *second,* stronger, form of psychophysical parallelism is a *metaphysical theory* about the relationship between the body and the mind. It adds to the primary form of parallelism a certain interpretation, or enhances it, by providing a metaphysical explanation for the alleged correlation. Fechner called his own interpretation the "identity view" of the body and soul. It provides philosophical underpinnings for functional dependence, including the following theses: (1) A living human being is not to be considered a conglomeration of two substances—a human being is one single entity; (2) the properties of this entity are considered mental when they are perceived inwardly, meaning from the perspective of the entity itself; and (3) the entity is considered something physical, when it is viewed from the outside, meaning from a perspective that is not the perspective of the entity itself. The mental and the physical are therefore two different aspects of one and the same entity. This position is also sometimes called double aspect theory, or—more correctly—the "doctrine of two perspectives."

The theory suggests that each human being has double access to, or has two perspectives of, himself: When I am aware of myself in a way in which no one else can be aware of me, I am aware of mental processes. When I am aware of myself in a way in which other persons can also perceive me (for example, when I see myself in a mirror), then I see the same processes in a physical, objective form; I appear to myself as a physical, material being.

This second form of psychophysical parallelism abandons the neutrality implied by the primary form and takes a stand on the true nature of the mind-body relation. It is defined as noncausal and therefore noninteractionist. But this noncausal interpretation is not merely postulated *per fiat,* as is the case for Cartesian parallelism; instead, it results from the definition of the psychical and the physical in terms of the perspective in which something is given. Viewing the physical as something that causes the mental, or vice versa, results from scrambling differing perspectives. Wherever causality may be found in the world, it will not be within the mind-body relation. We can demonstrate that distinguishing perspectives is nothing mysterious by considering a bent coin. It would be ridiculous to say that a dent on the head's side causes a bulge on the tail's side. While both sides of the coin are intimately connected, their joint occurrence has nothing to do with causality; they are merely two sides of one underlying substrate—two aspects that appear parallel to each other when the coin is damaged.

Obviously, the metaphysical identity view is not the only logically possible improvement on empirical parallelism. Reductive materialism and Cartesian interactionism can also be seen as being auxiliary to it. Fechner finds all these theories that build upon empirical parallelism metaphysical, not because they lack empirical significance or because they are speculative, but because, ultimately, no finite experience can prove them. In Fechner's opinion, any meaningful interpretation of empirical parallelism must be conceivable as something that anticipates future experiences. The status of an improvement on empirical parallelism achieved by amending parallelism with metaphysical interpretation is—evaluated epistemologically—in principle no different than the status of a normal law of nature: Based on inductive generalization, both refer hypothetically to future experiences.

Some of the benefits of psychophysical parallelism presented as an empirical postulate can also be found in psychophysical parallelism presented as an identity view. First and foremost the identity view provides a nonarbitrary way of defining those claims of materialism that are reasonable, as well as imposing limits upon it. It allows for nonreductive materialism and dismisses crude reductive materialism, without reverting to antimaterialism. Materialism can thus be upheld as a research avenue while being dropped as a universal metaphysical doctrine. Another important benefit is that this stance confers upon psychology the autonomy it requires for explaining the mental and its phenomenal reality without colliding with the causality of physical reality. And, finally, the notion offers the additional benefit that it does not infringe on the autonomy of philosophy. Philosophy is not condemned to skepticism, but it can work on a reasonable explanation for the mind-body relation, one that goes beyond neutral scientific description.

It is noteworthy that Ernst Mach, one of the earliest and most enthusiastic devotees of Fechner's psychophysical parallelism, ultimately abandoned Fechner's own amendment to the empirical postulate and instead tried to do without any explanation whatsoever—not only in terms of the psychophysical relationship but also for all relations among phenomena in the whole of science.[21] Mach wanted to restrict natural science exclusively to those neutral functional dependencies among phenomena, which Fechner had only meant to be a provisional stage of psychophysics. In doing so, Mach desired to banish causal claims not only from psychophysics but also from physics and psychology. This indicates that Mach's prime motive for rejecting causal explanation and scientific realism originated in his preoccupation with mind-body theory, rather than in his work on physics or from some basal animosity to atoms. It also shows that Fechner actually (if perhaps unintentionally) headed an antimetaphysical movement skeptical of causation that Mach picked up and furthered, and that ultimately led to logical empiricism and beyond.[22]

Basically, the identity view form of psychophysical parallelism was supported by four arguments: First, none of our experiences compels us to acknowledge the reality of a thinking substance independent of a material bearer of mental properties. Second, the realm of physical phenomena and processes is causally closed; this means that each event is caused by another physical event and in physics there are no "gaps" in which the mental could "intervene" with the physical. The same holds for phenomena in the psychical realm: they, in turn, can only be explained in mental terms. Third, the law of the conservation of energy shows that physical energy can only be transformed into or derived from other physical energy. Therefore, the physical can neither affect the mental nor vice versa. And the fourth argument for the identity view—and Fechner considered this one the most important—is that it is simple and frugal. All other amendments to the basic empirical fact of the psychophysical relation are metaphysically stronger than the identity view because, for the purpose of explanation, they involve more causality than the identity version does.

In its *third* form, psychophysical parallelism is a cosmological thesis stretching beyond the range of human life. It claims that even inorganic processes have a psychical side to them. Fechner was convinced that we can, by reasoning from analogy, plausibly assume in a scientifically respectable way that there exists a psychical dimension other than the realm of inner human experience. He believed that his identity view applies not only to humans and perhaps also to animals, but also to plants, the earth, planets, and the whole universe. His argument rested on the premise that the mental must not necessarily correlate to a nervous system; it could also be realized in other material systems. This notion became popular in our times under the banner of functionalism. Fechner elaborated the idea several times beginning around 1848, but it met with resistance and ridicule—even as late as 1925, brought forth by Moritz Schlick.[23]

The way that Fechner heightened psychophysical parallelism in this third type of parallelism (to become full-blown panpsychism) led many of his contemporaries to also dismiss his identity view—I feel, unjustifiably—as entirely speculative and inappropriate. But even extending the view into cosmology was not simple nonsense; it actually represents the origin of what later came to be called "inductive metaphysics," as opposed to dogmatic metaphysics. In order to avoid being mistaken for panpsychists and to explicitly limit psychophysical parallelism to living human beings, many authors preferred the term "psychophysiological" to "psychophysical" parallelism.[24]

At first glance one would think that a cosmological type of parallelism could be interpreted as pure Spinozism. But Spinoza saw the difference between mental and material attributes as something ontological and objective, something that refers to real intrinsic properties; whereas Fechner and many of his followers viewed the distinction as epistemological, based on the perspective from which the substance is investigated. This difference between Fechner and Spinoza demonstrates that Spinozism is more strongly tied to Cartesian dualism than is Fechner's parallelism.

Another difference is how Fechner treats teleology. While Spinoza rejected all teleological assumptions, Fechner chose to do the exact opposite and used psychophysical parallelism to argue *for* a teleological view of nature. According to this interpretation, the purposiveness of the mental inner side as seen from the outer perspective is completely compatible with mechanistic, nonteleological natural necessity, including the Darwinian version of it. Leibniz would have agreed to a similar type of reconciliation, one stating that causal laws to which bodies are subjected are compatible with the laws of final causes that hold for activity of the soul.[25]

Another difference from Spinoza is Fechner's treatment of the concept of substance. Very early on, Fechner noticed that letting psychophysical parallelism depend unquestioningly on the concept of substance is very problematic, both for the identity view version and for the cosmological version. That sort of a substance would be a strange metaphysical entity, neither purely mental nor purely material and thus even worse than the notion of noumenon, a concept he opposed energetically. In order to dispense of this undesirable entity, he suggested a phenomenalistic conception of substance: a substance is nothing but a bundle of lawfully connected appearances. And since physical appearances are nomologically connected to other physical appearances as well as often connected to psychical appearances, we end up with it being entirely admissible to speak of material substances that also possess mental properties. Readers may recognize that this is precisely the source of Mach's view that substance is nothing but a "complex of (sensory) elements."

But most scholars failed to notice Fechner's early phenomenalistic modification of the identity view, so that for a long time it was regarded as faulty and obscure metaphysics. This was particularly the case when, in the last years of his life, Fechner came to consider all appearances, whether mental or material, to be appearances in the mind of God, thus landing in "objective idealism" of a sort similar to that of Charles Sanders Peirce.

5.2 Psychophysical Parallelism from Fechner to Feigl

From a philosophical point of view the most pressing problem for psychophysical parallelism was the question of the precise role attributable to causality. It was one thing to dismiss causal interaction between the body and the soul, but to determine which role legitimately remains for the causality of nature and the causality of the mind—without forgoing psychophysical parallelism—was quite another matter. It seems that Fechner favored various options at different times: When he was young he tended to think that there are two different sorts of causality and that these are neither exclusive of one another nor intolerant of one another: physical causality in the realm of physical phenomena and psychical causality in the realm of inner experience. (Thus, on this issue Fechner also tended toward Leibniz's interpretation, which says that "bodies are active as if souls did not exist . . . and souls likewise, as if there were no bodies, and yet both move as if one had influenced the other."[26]) But throughout the phase represented by his major works, Fechner limited causal efficacy to that realm of reality that underlies all appearances of both types of aspect. In his old age, as mentioned previously, he adhered to objective idealism, which says that the correct place for causality is within the sphere of the mental. The distinctions separating these three views are subtle and tend to vanish if we take Fechner's phenomenalist dissolution of the concept of substance seriously. "Neutral monism" as it was later propagated by William James and taken up by Ernst Mach and Bertrand Russell became the logical outcome of these ruminations.[27]

As the controversy continued, several other variations were suggested and many new distinctions were introduced, raising the complexity of the issue.[28] The discourse centered mainly around the role played by causality in the second type of psychophysical parallelism. Some scholars limited causal efficacy—and thus also reality—to the realm of the physical, ending up with "materialistic" parallelism implying epiphenomenalism for the mental. For many, this seemed to be the price to be paid for psychophysiological parallelism absent of panpsychism. The result implies a discontinuity in causality for the realm of the mental.

Others assumed the opposite—namely, that the realm of the mental is primary—and this led to causal inefficacy on the material side. Besides materialistic and idealistic parallelism, a third type was suggested, occasionally called "realistic monism" or "monistic parallelism." It held both the psychical and the physical sides for equally causally inefficacious epiphenomena of an underlying and causally efficacious actual reality. So, in the end, parallelism itself encompassed all those philosophical positions that it had originally intended to conquer! We have already noticed Ernst Mach's reaction to this confusing situation. He cut the Gordian knot by entirely forgoing causality and permitting solely functional dependence.

But the most significant form of psychophysical parallelism of interest here is not Ernst Mach's, but rather what is called "critical realism." The main advocate of this interpretation during Fechner's and Mach's time was the Austrian philosopher Alois Riehl. He wrote on the mind-body problem in 1872 and then extensively again in 1887.[29] Riehl defended the second type of the monistic form of psychophysical parallelism, which assumed that the reality underlying physical and psychical aspects of our perception is identical with Kant's noumenon. Since he shared this and other concepts with Kant, he is usually considered a neo-

Kantian. But contrary to the other—for the most part Marburger—scholars he interpreted noumena as objective and causally effective reality independent of human consciousness, and he defended, in contrast to Kant, the notion that noumena are to a certain degree recognizable. Riehl labeled this mind-body conception "identity theory" and "realistic monism," thereby idiosyncratically constricting the traditional meanings of those terms (This contradicts Place's claim that in 1933 the American psychologist E. G. Boring may well have been the first to use the term "identity theory."[30])

Let us now briefly discuss Wilhelm Wundt's opinion. Wundt, the principal representative of "new psychology" in Germany advocated an interesting form of partial parallelism. On the one hand he was an outspoken opponent of the "theory of reciprocal effect," and therefore there are many passages in his writing where he unrestrictedly endorses psychophysical parallelism at least as a research maxim. But on the other hand he wants parallelism confined to those physical and mental events for which we have actual proof that they are parallel.[31] In his opinion parallelism applies "*only to those elementary* psychical processes (sensations), *to which alone* certain limited movements run parallel." Parallelism is "merely the parallel running of elementary physical and mental events, never parallel movements amongst complex performances on both sides."[32] Wundt does not claim that there is thought without brain activity. It would seem, rather, that he struggles with a distinction familiar to present-day mind-body study—namely, the distinction between type identity and token identity. Elementary mental processes, Wundt maintained, are type identical with corresponding physiological processes (each occurrence of a specific sensation always corresponds to a specific physical event), while one and the same complex or higher mental event can, at different times, also be accompanied by differing physiological processes. While it is possible on the elementary level to know the psychical meaning of physical events, this is no longer possible on the sophisticated level.[33] It is true that the perceptual contents of our mental life are linked to physiological events, but the "mental configuration" of these contents, "being what links them according to logical and ethical standards," can no longer be bound to physiological events.[34] The outcome of Wundt's partial parallelism is a very complex theory of volition.

In spite of the complications and modifications, psychophysical parallelism —at least in its empirical, methodological version—was endorsed by the vast majority of both psychologists and physiologists well into the twentieth century. To them it seemed to be a scientific and philosophically respectable doctrine that honored the autonomy of psychology, permitting it to peacefully coexist along-side physiology and science in general. The experimental psychologist Georg Elias Müller gave perhaps the most elaborate formulation of psychophysical parallelism for this group when he split it up into five different "psychophysical axioms" as components presupposed by psychophysics.[35]

It also gave philosophers enough room to exercise sagacity in criticizing deviating positions and to discover new ways to fill in the outline provided. Toward the end of the nineteenth century idealistic notions gained more signi-ficance. Charles Sanders Peirce said: "The new invention of Monism enables a man to be perfectly materialist in substance, and as idealistic as he likes in words."[36] Of course, an author not at home in the ongoing philosophical dis-cussion had less interest in and awareness of the distinctions separating the vari-ous forms of psychophysical parallelism. But the result was a widespread diluted type of psychophysical parallelism that obscured many of the important dis-tinctions that Fechner and subsequent thinkers had introduced into the debate.

Fechner's adherents included many prominent names. For instance, a letter from 1922 addressed to a Swiss journal and dealing with the theory of relativity shows that Albert Einstein adhered to Fechner's ideas: "To guard against the collision of the various sorts of 'realities' with which physics and psychology deal, Spinoza and Fechner invented the doctrine of psychophysical parallelism, which, to be frank, satisfies me entirely."[37] Although Niels Bohr apparently never mentioned it in print, he adopted Fechner's psychophysical parallelism as taught by his philosophy mentor, his father's close friend Harald Höffding.[38] In his successful book *Outlines of Psychology*, Höffding had discussed the identity theory at length and praised Fechner as the first "to construct the theory of the relation between the mind and body based on the consequences ensuing from the axiom of the conservation of energy."[39] It is no coincidence that as late as 1932 John von Neumann phrased the distinction made in quantum mechanics between the observer and the system under observation in terms taken from the "principle of psychophysical parallelism." He made reference to Bohr, who was the first to have pointed out that "the, in formal respects unavoidable, du-plicity in describing nature in quantum mechanics" is related to psychophysical parallelism as a fundamental principle of the scientific worldview.[40]

Of course, psychophysical parallelism also had opponents. Many natural sci-entists were not particularly interested in philosophical controversies and did not wish to be involved in anything resembling philosophy of nature in the post-Kantian tradition, with which Fechner initially had been closely associated.[41]

Simply using the expression "identity theory" or "identity view" suggested proximity to F. W. J. Schelling's "identity philosophy," or at least it seemed so for
scholars like Hermann von Helmholtz. This criticism could be easily refuted by
noting that the first form of psychophysical parallelism was limited to being a
mere empirical postulate. Helmholtz was not even willing to concede that and
opposed even the more or less harmless form of psychophysical parallelism. He
argued the incompatibility of free will and determinism. In his opinion, the
realm of the mental, with all its voluntary and spontaneous activity, should not
be mixed with nomological and necessary processes of nature, as psychophysical
parallelism mixes them, and that even in natural science, for the time being, one
must tolerate interactionism.[42] His student Heinrich Hertz advocated a similar
opinion in the introduction to *Principles of Mechanics* in 1894.

It was entirely natural for other critics to reject parallelism's pan-psychical
implications, arising in its generalized third form.[43] But it is surprising to discover that a willingness to adopt such an idiosyncratic and highly speculative
consequence of Fechner's doctrine was much greater then than it would be today.
For example, in a private letter, the physicist H. A. Lorentz admitted in 1915 that
he believed in Fechner's psychophysical parallelism and came to the conclusion
that "the mental and the material are inviolably tied to one another, they are
two sides of the same thing. The material world is a way in which the *Weltgeist*
appears, since the smallest particle of matter has a soul, or whatever one chooses
to call it. This is all closely tied to Fechner's views . . . and I think that we have to
assume something similar."[44]

By way of reviving a bit of the atmosphere in which parallelism enjoyed such
widespread recognition over such a long period of time and in order to also
demonstrate the significance that the law of the conservation of energy had for
this controversy—even for philosophers—I would like to quote a passage taken
from a letter of 1875 from the philosopher Hans Vaihinger to Friedrich Albert
Lange, mentioned above, written forty years prior to Lorentz's letter. In this letter Vaihinger deals with some of the motives that made the identity theory—at
least as an empirical postulate—so attractive for the scientifically enlightened
public. (When he writes about "moderate occasionalism" he means something
like psychophysical parallelism of the first type.[45] This letter was intended to
deny rumors—that Lange had heard—claiming that Vaihinger had converted
to occasionalism.)

> I made the following distinction: a scientist has two options: either "moderate
> *occasionalism*," or *Spinozism* as rectified by Kant. For a scientist may only *either*

say: *occasionally* certain brain activity occurs *simultaneously* with certain psychological events; but he may not permit them an inner connection at all; he makes no hypothesis about how they are connected, he states only the *fact* that the totally inclusive cycle of mechanical causality in the brain is *accompanied* in some mysterious way by psychical phenomena. If he were not satisfied by this provisionary and insufficient notion, a notion that serves only those who are anxious and overly careful, then the scientist would have to proceed towards the wider *Spinozian* hypothesis, which says that whatever appears to us to be an *external material event*, is—for us—inwardly a *sensation*; and I added that this latter opinion, which after Kant has been advocated by *Fechner, Zöllner, Wundt, Bain*, and others, and which is also your view, seems to me to be the only possible consequence of the *Law of the Conservation of Energy*. So you see, my dear professor, that "occasionalism" is hardly perilous. It is merely a provisionary stopover for those unwilling to address the other conclusion; and for those persons advocating an intermediate position it is at least better than either opposite position, viz., *materialism* or *spiritualism*, both of which violate the law of the conservation of energy by allowing physical things to "become" psychical and psychical things to "effect" physical things and be involved in the "mechanic series of causes."[46]

During the late 1870s the arguments and methods supporting Fechner's psychophysics were more frequently attacked by neo-Kantians.[47] But that hardly damaged the peaceful and fruitful rule of psychophysical parallelism within German-speaking culture. What abruptly ended that rule was a new chapter that the philosopher Christoph Sigwart added to the second edition of his *Logic* in 1893 in an attempt to refute psychophysical parallelism and demonstrate its intolerable conclusions.[48] (We must remember that opposing psychophysical parallelism and subsequently adopting a form of psychophysical interactionism did not necessarily mean that one embraced Cartesian substance dualism.) Sigwart tried to show that neither the concept of causality nor the principle of the conservation of energy encompass parallelism and that only the doctrine of reciprocal effect between the mental and the physical is philosophically permissible and valid.

As if Sigwart had opened the locks, a flood of refutations against parallelism broke through. The author of a dissertation in Vienna in 1928 noted dryly that the ensuing dispute over parallelism was surpassed only by the Trojan War.[49] The most influential critiques of parallelism after Sigwart were written by Wilhelm Dilthey, Carl Stumpf, and Heinrich Rickert, but there were also many other authors who spoke out against parallelism, who were of lesser importance for academic philosophy or less interested in the relationship between philosophy and natural science. Most critics were bothered by parallelism's proximity to materialism, which robbed the human soul of causal efficacy and subjected the mind to determinism.

Dilthey was one of the founders of an antinaturalistic movement that came to be called *Lebensphilosophie* and sought an autonomous fundament for the sciences of the spirit—that is, the humanities. In the 1880s he had already come to the conclusion that a "correlation" between the mind and body that was understood as being noncausal was "the worst of all metaphysical hypotheses" and that the various attempts of his coevals to establish empirical psychology were nothing more than "poor metaphysics."[50] In 1894 Dilthey read two essays to the Prussian Academy of the Sciences, to which he had belonged since 1887, contrasting two types of psychology: One was descriptive and analytic, a type to which he himself subscribed.[51] This type of psychology strove to describe and analyze real psychological experience. The other was explanatory and constructive psychology, a method used in contemporary scientific psychology, going beyond actual experience and postulating an abstract psychological reality in an entirely hypothetical and deductive manner.

In this context Dilthey branded psychophysical parallelism an essential but unfounded and hypothetical construct for the new psychology. He reproached its advocates for their "refined materialism" that reduces the "most powerful mental facts" to "mere accompaniments of our bodily life." Its deterministic consequences, he thought, had already begun to disintegrate "political economics, criminal law, and constitutional law."[52] The experimental psychologist Hermann Ebbinghaus, who adhered to psychophysical parallelism, replied skillfully to these heavy attacks and defended the new psychology against Dilthey's accusations.[53] Debates over the Dilthey-Ebbinghaus controversy lasted well into the Weimar Republic period and left traces that are more or less noticeable to this very day.

Although William James, as we saw earlier, provisionally advocated "empirical parallelism," he criticized identity theory as early as 1879 in a way similar to Dilthey. Simply by calling it *automaton-theory* he made it clear that in his opinion parallelism degrades man to a mere automaton, such that "whatever mind went with it would be there only as an 'epiphenomenon,' an inert spectator . . . whose opposition or whose furtherance would be alike powerless over the occurrences themselves." But mind, according to James, must have some effect on the body, otherwise it would not have been able to outlive the "struggle for survival." James came to the conclusion that "to urge the automaton theory upon us, as it is now urged, on purely *a priori* and *quasi*-metaphysical grounds, is an *unwarrantable impertinence in the present state of psychology*."[54] Thirty-two years after publication, James's critique was echoed by Edmund Husserl, who protested that parallelism treats the psyche as a "merely dependent modification of the physical, at best as a secondary parallel accompaniment" and that it interprets all beings as having "a psychophysical nature unequivocally determined by fixed laws."[55]

Dilthey arranged Carl Stumpf's appointment to a chair for psychology in the department of philosophy in Berlin in order to prevent parallelists like Ebbinghaus, Wundt, or Benno Erdmann from attaining this position, to which they claimed rights. (What he could not prevent, however, was the call for Friedrich Paulsen for another chair in Berlin. Paulsen's interpretation of psychophysical parallelism, however, did not tend toward materialism, but in the exact opposite direction, toward panpsychism.) Although Stumpf was a leading psychologist at the time, who emphasized experimental and scientific methods, he defended interactionism vehemently in a well-received opening speech at the third congress for psychology in 1896 in Munich.[56] It is possible that he was then still influenced by his teachers Franz Brentano and Rudolph Hermann Lotze, who belonged to a small group of older scientists cum philosophers of the nineteenth century refusing to follow the fashion of psychophysical parallelism and advocating an interactionist position instead.[57] Stumpf found parallelism obscure and ambiguous, a theory that, if examined carefully, actually represents a concealed form of dualism, since it assumes two different realities. He also claimed that since according to Darwin's theory of evolution all reality must be causally efficacious—then causal efficacy must also be attributable to the mental.

The third prominent antiparallelist was Heinrich Rickert, a leading advocate of decidedly antinaturalistic neo-Kantianism in southwest Germany. In a contribution to a *Festschrift* for Sigwart in 1900, Rickert claimed with astute elegance that any concession made to parallelism that weakens the relation of psychophysical causality inevitably leads to intolerable panpsychism.[58] He tried to show that the mind-body problem is a pseudo problem, originating in unqualified attempts to reunite the sciences of physics and psychology after such great effort had been made to distinguish and divorce them—the former as the science of quantity and the latter as the science of quality. Rickert believed in a special type of causality holding for the realm of qualities that is different from "mechanical causality" as found in the realm of quantity and therefore not subject to the law of the conservation of energy. He emphasized that both determinism and parallelism are useless categories for historians and that the discipline of history, dealing as it does with real human activity, must assume psychophysical causality and interactionism. If we start with this concept, instead of some kind of parallelism, human action appears to be an exception to determinism. In history, individual actions of civilized humans have nothing in common with mechanical causality of the kind found in natural science.

Since this kind of antinaturalistic neo-Kantianism surged mainly in Germany, throughout the 1890s resistance to psychophysical parallelism was greater in Germany than in Austria. There were frank interactionists in Austria too, however, and among their most eminent advocates were Franz Brentano, Wilhelm Jerusalem and—less obviously—Alois Höfler, although their motives for resisting parallelism differed from those of Germany's neo-Kantians.[59] The Viennese academics and Hapsburgian culture in general seem to have been more favorable for identity theory, a fact attested by the work of Ernst Mach, Friedrich Jodl, Ewald Hering (who later taught in Prague), as well as Josef Breuer and—at some distance—Sigmund Freud.[60] There were even parallelists among the followers of Brentano, the dualist. (A late member of this group was Gustav Bergmann, who subsequently was to become a member of the Vienna Circle. After Feigl helped him get a job in Iowa in 1939 he tried to enhance psychophysical parallelism by combining it with Brentano's concept of intentionality and with methodological behaviorism.[61] Like Herbert Feigl, he was involved in bringing psychophysical parallelism to the United States, although to a lesser degree.)

As early as 1896 Friedrich Jodl had phrased identity theory in terms of a two-language theory, a development for which both Feigl and Schlick later claimed the credit (and which, incidentally, came up again with Donald Davidson's "anomalous monism"). Jodl thought that physiological and psychological descriptions for a state or process in a living organism are identical and refer to the same event, although they take on different forms.[62] We can probably trace this early two-language theory back to Hippolyte Taine, who in 1870 had already compared the relation of descriptions for the mental and the physical with the relation of two languages that mutually augment and elucidate one another.[63] Höffding also advocated a two-language theory when writing that brain processes and processes of consciousness refer to one another "as if one and the same fact were expressed in two different languages."[64]

It is only a small step from Riehl's and Jodl's identity-theoretic interpretations of parallelism to Moritz Schlick and his Viennese colleague Robert Reininger. In 1916 the latter dedicated an entire book to The Psychophysical Problem[65] and taught a course on Gustav Theodor Fechner during the summer term at the

University of Vienna, almost two years after the young Herbert Feigl (1902–88) had come to Vienna—it was perhaps the one single course ever to be given dealing solely with Fechner.[66]

5.3 Schlick and Carnap Enter the Scene

Considering all that has been said, it is not surprising that philosophers well educated in natural science, as Moritz Schlick and Rudolf Carnap were, stood squarely within the tradition of psychophysical parallelism when it came to dealing with the mind-body problem. In *General Theory of Knowledge,* published in 1925, Schlick referred to himself explicitly as an advocate of that doctrine.[67] He stressed, however, that his own position is more radical than that of common parallelism and surpasses it in two respects: First, his position includes the "reduction of psychology to brain physiology" in the sense that there is an "identity" of reality such that "two different systems of concepts"—psychological concepts and concepts of physics—refer to it, and, second, his parallelism is not of a metaphysical but of a purely epistemological nature.[68] In a letter to Ernst Cassirer in 1927, Schlick wrote the following: "The psychophysical parallelism in which I firmly believe is not a parallelism of two 'sides' or indeed 'ways of appearing' of what is real, rather, it is a harmless parallelism of two differently generated concepts. Many oral discussions on this point have convinced me (and others) that this way we can really get rid of the psychophysical problem once and for all."[69]

Schlick's solution for the mind-body problem reflects two components of his philosophy, which originate from diverging traditions and therefore appear at first contradictory. On the one hand we have Schlick's critical realism, which (besides naive realism) rejects positivism and every other form of "immanence philosophy" while simultaneously accepting a reality that transcends the given. On the other, we have Schlick's positivistic inheritance that views reality as consisting of qualities, whether or not they are actually given for consciousness. Schlick explicitly used a positivistic strategy adopted from Richard Avenarius claiming that the riddles of the mind-body connection (and other challenges) can be seen as an inappropriate use of "introjection" (more on this later).

Schlick's realism rests on a threefold distinction: First, there is a realm of noumena, consisting of complexes of qualities, which must not necessarily be given to any consciousness. Second, there exists reality characterized by the quantitative concepts of natural science; it results from eliminating (secondary) qualities in the course of scientific progress. And, third, there exist our intuitive perceptual events with which reality (in the second sense) is represented in consciousness—namely, experience. In understanding reality we must learn to distinguish "knowing" *(kennen)* from "recognizing" *(erkennen)*. In this sense, noumena can never be directly known—they are never given to consciousness —but we can at least partially recognize them by their causal effects and thus determine their place in the network of objective relationships by characterizing them with scientific and quantitative concepts. Recognition consists of assigning systems of symbols to circumstances. However, objectively recognized reality, which to a certain extent also encompasses what is not given, is represented through our acquaintance with our perceptual subjective experience, for only in this way can we have access to the realm of noumena. But since these qualities and complexes of qualities themselves are part of reality, they can in turn be described using scientific concepts.

Now, these distinctions imply a very specific meaning for "psychical" and "physical." For Schlick, the concept of the psychical refers to what is at all given (The Given), meaning what is identical to "content of consciousness." Reality is called physical, "inasmuch as it is described by the spatio-temporal quantitative system of concepts provided by natural science."[70] It is important to note the special role of spatial extension in this distinction. Schlick insists that space appears in two ways that must be kept strictly separate: one is perceptually imaginable space as we know it in sight, from touch, and from our kinesthetic sensations and so on; the other is physical space as conceptually construed by natural science. If we do not make this distinction and use introjection instead, meaning that we locate mental properties inside the brain or that we attribute experiential extension to the physical, we suffer from the fundamental confusion that Schlick considered to be the source of the mind-body problem. Schlick remarked with amazement that—in spite of all their differences—Avenarius and Kant nonetheless were both able to avoid that kind of unfounded introjection.

So we must now pose the crucial question regarding how Schlick intends to avoid introjection in the relationship of the psychical to the physical. This can best be done using Schlick's example of person *A* looking at a red flower and that person's brain processes ("with *A* having an open skull exposing the brain") being observed by another person, person *B*.[71] *B* is interested in those cerebral processes that are necessary and sufficient for *A* to see a flower. *A* is not acquainted with the noumena, the flower, at all, but she can comprehend it using scientific terms; she can employ botanical and physical terms of classification, she can describe its molecules, and so forth. Thus *A* can recognize the flower in a scientific way. But *A* also undergoes a perceptual event; she experiences the flower in a way that can be described as "red," "which in the same sense is something very real in itself, just as the transcendental object 'flower' is real." Person *B* does this: His experience shows him that the same reality that *A* describes as "red" can—using a physical term—be described as a brain process of such and such a kind. But *B* cannot only know about *A*'s brain and her mental world, he can also have a perceptual experience of *A*'s brain.

Schlick felt that this example shows well just how the mind-body problem arises by confusing characterized reality with the terms used in that very characterization, or confusing it with its perceptual representative. The first mistake is to think of the actual brain process in *A* as a physical concept of the brain process. The result is an unwarranted reduplication of reality: Instead of assuming just one reality, which is either described as physical or mental, a distinction is made between the reality of *A*'s brain and that of her consciousness. It is this sort of confusion that encourages the question of how both realities are related.

Another mistake is to confuse the concept that a physicist might have of *A*'s brain for the real intuitive experience that *B* has of that brain. All three realities, says Schlick, the flower itself and the contents of *A*'s and *B*'s consciousness, are equally valid and must each be understood for itself. And for all three of these realities it is clear from the start that they are causally tied to one another: The first causes the second and this causes the third. For what *A* and *B* know, we have a "parallelism of ways of description: both psychological and physical concepts can be applied to them."[72]

If we compare Schlick's discussion with other, earlier attempts made at psychophysical parallelism, we inevitably fall back on Alois Riehl. He seems to be the true representative of "Spinozism rectified by Kant," mentioned by Vaihinger in the letter quoted above. As I already noted, Riehl considered noumena to be at least partially recognizable *(erkennbar)*, even if they are not direct objects of experience. In a way similar to Schlick's, he was also convinced that scientific progress consists in increasingly freeing scientific objects from secondary qualities and reducing those objects to primary qualities. He also thought that spatial extension was a property capable of being experienced, therefore deserving a status similar to that of color and taste.[73] In dealing with the psychophysical problem, he stressed, just as Schlick did, the "definite identity of that process which underlies at the same time physical and psychical phenomena." He rejected "the hypothesis so popular today that physical and psychical correspond" because it involves "some hidden dualism."[74] And, finally, Riehl and Schlick entirely agree on disposing of metaphysics. Neither wants to turn the identity of the physical and mental into a "theory of the universe" but to confine it to those "points at which the objective and the subjective world actually touch"—as Riehl put it.[75]

In commemorating Schlick, Feigl claimed that Schlick's solution to the mind-body problem differed entirely from all traditional metaphysical solutions. "Neither materialism," he wrote, "nor spiritualism is being maintained here, neither monism nor dualism, neither parallelism, the double-aspect theory, nor interactionism in the usual sense." But Feigl did admit that Schlick came closest to identity theory "as found, say, in the 'philosophical monism' of Alois Riehl's." Nonetheless, Feigl hurried to make clear, "even this must first be divested of its metaphysical character." He concluded that "Schlick's solution is best described, no doubt, as a two-language theory."[76] Still, I find Feigl's attempt to detach Schlick from the tradition of psychophysical parallelism exaggerated. The obvious conformity of Schlick's and Riehl's views demonstrates Schlick's involvement in that traditional debate. If we compare Schlick to his predecessors, we see that he did not express a more radical, effective antimetaphysical or materialistic attitude, nor did he make a progressive "semantic ascent" (Quine), meaning a linguistic analysis of the problem in the manner of twentieth-century analytic philosophy.

Let us now consider how Carnap dealt with the problem prior to turning toward physicalism. There is not much to say, since his treatment of the problem was relatively brief. But, naturally, that does not mean that he found it insignificant. In *The Logical Structure of the World* he even called the psychophysical relation the central problem of metaphysics.[77] He said that the "essence problem" of the psychophysical relation lies in the difficulty of understanding and explaining the surprising parallelism of such heterogeneous phenomena as that of the mental and the body.[78] In his opinion, only three different metaphysical solutions need be considered seriously: the hypothesis of mutual effect, the identity thesis, and the thesis of parallelism without identity. However, none of the three hypotheses is better than any of the others, for strong arguments refute all three. Carnap's most important argument was the standard objection to identity theory, stating that identity is an empty term, as long as it is not entirely clear what it means to "underlie an inner and outer side."[79]

Carnap's radical solution to the "essence problem" of the psychophysical relationship is well known and follows the pattern provided by his general critique of metaphysics. The fact that the given can be ordered in two parallel series should be accepted without reserve. If the issue of "interpreting" or "explaining" parallelism persists, this can only be seen as an unqualified inclination toward metaphysics. Within the means provided by the system of constitution such issues can no longer be stated seriously or meaningfully. "The question of how to interpret the finding [that the Given can be ordered in parallel series] lies beyond the scope of science. This can be seen in the very fact that it cannot be expressed in constitutive terms. . . . The question of interpreting that parallelness belongs to metaphysics."[80] Science investigates functional dependencies, not "essence relationships." Carnap mentions Ernst Mach as the chief advocate of this interpretation.

Compared to the previous development discussed thus far in these investigations, neither Carnap's nor Schlick's interpretation of the mind-body problem appears to be particularly revolutionary. We can certainly say that Schlick's and Carnap's solutions convey and focus a tension that was already prevalent in Fechner's treatment of the problem and that ruled the whole ensuing discourse. It is a tension and a dilemma, if you will, between the antimetaphysical and empiristic tendency of psychophysical parallelism of the first type and the realism of the second type of parallelism's identity theory. The problem with which the Fechner tradition struggled was the following: If we want solely to deal with the facts, then parallelism can only be understood as research heuristics. But this would mean that we dismiss any explanation for a very strange regularity that obviously seems to suggest some underlying causal mechanism that would make it understandable. But accepting the simplest imaginable explanation for this regularity—namely, the identity theory—means to transcend the direct realm of facts and invite panpsychism or similar metaphysics.

Confronted with this dilemma, Schlick—properly following the tradition set forth by Riehl and other "critical realists"—opted for realism and tried to modify the concept of the physical object as much as necessary in order to make undesirable metaphysical consequences vanish. The best elucidation of the parallelism between mind and body is seen as the two-language theory. The unobservable realm underlying the different conceptual constructs is—as it were—tamed by realism. Carnap, though, was more willing to follow up Fechner's original solution, which was later radicalized by Ernst Mach, William James, and Bertrand Russell, thus dealing with the other side of the dilemma. Like many others in logical empiricism, he strove to demonstrate that natural science can describe the world without losing something in the process and without recourse to explanations that transcend the given. If we shun every reference to realms and objects that are inaccessible to experience and restrict natural science to the description of what is observable, we can retain meaningful science without unwelcome metaphysics.

5.4 Psychophysical Parallelism in the United States: Herbert Feigl

For a long time Herbert Feigl was a devotee of Schlick's critical realism and its related realistic solution for the mind-body enigma. Since he set up both the subject and the name index for the second edition of the *General Theory of Knowledge* and helped Schlick with corrections, he must have been well acquainted with his teacher's views.[81] Thus it is not surprising that—as he himself reports—he "opposed Carnap's phenomenalism from the start in Vienna" and that he was involved "in a standing dispute with Carnap . . . over the 'realism,' 'subjective idealism' or the 'phenomenalism' issues."[82] His Viennese friends must have made life quite difficult for him. He later recalled being scorned for advocating realism: "You metaphysician! they told me in Vienna. Imagine! This was the worst thing that could happen to a philosopher at that time."[83] To his "great chagrin" he watched his teacher and friend finally give in and under Wittgenstein's influence become "a positivist in terms of the problem of reality." That was too much for young Herbert, for—as he later recalled—he himself was "temporarily overwhelmed" by Carnap, a thinker "tremendously resourceful in discussion . . . who has thought through everything a hundred times more fully than is evident from his publications."[84]

In the period between the late 1920s and 1958, when Feigl's essay was published, the Vienna Circle turned to embrace physicalism. We are uncertain whether Feigl went along with this change in every respect. In retrospect he described the first phase of physicalism as an error: In his opinion it soon became obvious that mental states could be neither identified with overt behavior nor reduced to neurophysiological states. But in Carnap's retreat from the principle of verification and his concern after 1956 with bilateral reduction laws as a method for introducing mental terms Feigl saw a revision of the original view of the *Logical Structure of the World* leading back to critical realism and a two-language theory for mind and body. He considered two factors responsible for reinstating "clarified critical realism": One was Tarski's "pure semantics," which Carnap further developed, and the other, the "pure pragmatics" demonstrated in Wilfried Sellars's work. These developments encouraged Feigl to return to his own previous interpretations and those of his mentor, Schlick.

In his first publication on the mind-body problem, in 1934, after the general turn to physicalism, Feigl held the relationship between the physical and the mental to be a *logical* identity between two descriptions of the given, a description in psychological and a description in physical vernacular.[85] It was a more radical version of Schlick's notion interpreting the identity spoken of in identity theory as a relation holding between realities.[86] This focus can be understood as a concession made to logical behaviorism, which at that time was quite popular within the Vienna Circle. But by 1958 at the latest, Feigl returned to Schlick's views and those he himself had held prior to 1934. (Incidentally, in the foreword to the essay he remarks that he was initially introduced to "philosophical monism" by reading Alois Riehl and that he found "essentially the same position again in Moritz Schlick."[87] Growing criticism regarding behaviorism at that time might have encouraged Feigl.[88] The most significant change was that now Feigl no longer saw the identity of the mental and the physical as a necessary, but as an empirical identity.

In several passages of the essay Feigl asks what distinguishes his own identity theory from parallelism and concedes that the distinction is not of an empirical nature. "The step from parallelism to the identity view is essentially a question of philosophical interpretation." Thus, deciding for one of the two positions is similar to making a choice between phenomenalism and realism or between the regularity theory and the modality theory of causality—things that cannot be decided empirically. The principle of frugality, or "inductive simplicity," demands that we forgo the doubling of realities, such as parallelism assumes, in favor of identity theory.[89] The advantage of this theory is that it "removes the duality of two sets of correlated events and replaces it with . . . two ways of knowing the *same* event—one direct, the other indirect." If we admit a synthetic element in the psychophysical relation, then "there is something which purely physical theory does not and cannot account for."[90] The translation of the mental into terms of the physical still assumed in 1934 has totally disappeared here. In lieu of a two-language theory we now have Feigl's "double-knowledge, double-designation view,"[91] which is nothing other than a revival of the second form of psychophysical parallelism.

The preceding discussion shows once again that even twentieth-century logical empiricism is more firmly rooted in tradition than has often been assumed and more so than the logical empiricists themselves conceded. And we have also seen the wealth of options, distinctions, and arguments that lay and still lie waiting in the tradition of identity theory, and which, unfortunately, have frequently been lost in contemporary discourse. In the present tiny renaissance of the identity theory it is time to recall forgotten history. In this exposition I limited myself to the tradition leading up to Herbert Feigl and neglected the Australian version of identity theory.[92] An excellent comparison of both schools of thought, which elaborates the differences separating them, is available, however, and it confirms my view of Feigl's proximity to parallelism.[93]

But the story told here also shows (and I intend to suggest it to my readers repeatedly) that the mind-body problem is not simply one problem amongst many, nor one that logical empiricism could have skipped elaborating. The twists and turns in discussing this very problem has formed logical empiricism in essential ways. Do not forget that the prehistory of logical empiricism roots not only in logic and physics but also particularly in psychology and physiology. At times, these two latter disciplines were of greater interest in the nineteenth century than the former two. I mentioned earlier that a tendency of many empiricists in the tradition of logical positivism to replace causality with functional dependency can be explained by reviewing the discussion of the mind-body problem. Their antimetaphysical inclination also originated not (only) in physics, but was due to the efforts made to find a serviceable scientific basis for emerging empirical psychology. Even the preference for "description" over "explanation" resulted from psychophysics' neutrality regarding causality. Something similar can be said for the origins of phenomenalistic critique of the concept of substance in the discussion on parallelism. And the early logical empiricists' antirealism, as expressed at the dawn of the twentieth century, has roots not only in physics, but also in the endeavors of psychophysical parallelism to prevent from the start any sort of conclusion about a panpsychical side of the world. It seems an irony of history that it was precisely Feigl's theory, a realistic variation of parallelism, that "survived." In light of what I have reported it seems even thoroughly possible that Carnap's philosophical neutrality in epistemological matters, as he elaborated it in the system of the *Logical Structure of the World,* and that pervades all of his work, in the end is obligated to Fechner's demand for neutrality in psychophysical parallelism as a maxim of research and to the two-language theory related to it.

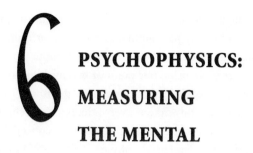

6

PSYCHOPHYSICS:
MEASURING
THE MENTAL

6.1 Basic Concepts

IN HIS "NOTES ON SCIENCE," Charles Peirce wrote of: "Fechner's psycho-physical law (at which the utter ignorance of German philosophical professors of the mathematical theory of metrics, their fondness of expressing opinions about matters of which they are ignorant, and the awe with which they are copied by Americans, has made it the fashion to sneer)."[1] It would take an elaborate study, if not several of them, to properly and adequately present Fechner's *Psychophysics* and the controversy it sparked. Many of Fechner's contemporaries discussed psychophysics, and for the most diverse of reasons. So it is somewhat peculiar that to this day very little interest has been shown for the history of the subject.[2] Perhaps one reason for dwindling interest in psychophysics after Fechner's death was that the "pre-paradigmatic" period of psychophysics (as Thomas Kuhn might have put it, meaning the phase when incomplete approaches vie for acceptance) was fairly long, so that a historian of science would have felt obliged to take sides. Also, although it is a vivacious academic discipline (particularly in the U.S.), in terms of subject matter psychophysics today is regarded more a methodical auxiliary science or a part of mathematical psychology than belonging to psychology proper. Naturally, the complexity and difficulty of the material itself is also inhibiting: When sorting through the debates in early psychophysics, one is never quite sure of reward for struggling through thicket and clearing some ground.

Let us focus on one point of major philosophical interest. We can elucidate Fechner's basic idea of measurement without catering to all the other notions in its tow, and without over-indulging in Fechner-scholasticism. Based on the day view, Fechner developed a special concept of measurement applicable to *both* physics and psychophysics. This concept in turn crucially influenced Ernst Mach's theory of measurement, thus becoming highly significant for modern physics.

We will concentrate on the group of persons directly involved with Fechner's concept of measurement. The most intriguing among them were the scientists and philosophers who favored a combination of mechanistic and neo-Kantian standpoints. A look at those standpoints creates a contrast that reveals the true contour of Fechner's theory of measurement. The material presented here should be sufficient for supporting the thesis that, among other things, Fechner's alternative concept of measurement contributed to the downfall of the mechanistic worldview.

Fechner defines "psychophysics" as "the exact science of the functional relations of dependence among body and soul, more generally, between the corporeal and the mental, the physical and the psychological, world."[3] Fechner distinguishes outer from inner psychophysics. *Outer psychophysics* deals with how sensation depends on (external) stimuli; *inner psychophysics* deals with how psychophysical excitation relates to sensation. The term "psychophysical excitation" designates the activity of that part of a physical system that normally bears the psychical side. In the case of human beings these are "certain activities in the brain," or neural activity.

As we saw in chapter 2, according to Fechner, all science is concerned with *appearances*. Any further rumination about things existing "behind" the appearances has no part in natural science. Fechner's *Psychophysics* is strictly neutral on this matter; it is free of metaphysics (except for chapter 45 in volume 2, where Fechner shows how well psychophysics applies to the day view). Fechner wrote: "All expositions and investigations in psychophysics relate only to the appearance side of the physical and mental world; they relate to whatever is directly accessible via either inward or outward perception, or can be inferred from appearances ...; in other words: they relate to what is physical in terms of physics and chemistry, and to what is mental in terms of psychology."[4] Today we would say that psychophysics explains psychical data as a function of the physical body, but does not provide a basic explanatory model in support of that function.

The objects with which outer psychophysics is concerned are given directly as both inward and outward appearances, while inner psychophysics must study anatomy, physiology, and psychology, in order to approach its subject, namely that aspect of the physical entity that bears psychological phenomena. This is because when someone thinks, that person cannot herself simultaneously investigate the brain activity supporting those thoughts. Awareness of brain activity is not the same as thinking the thought expressed by that activity.[5]

The reverse is also true: A living brain cannot scientifically investigate itself and simultaneously think. Some kind of 'psychophysical uncertainty relation' (analogous to Heisenberg's relation) holds: Measuring psychophysical activity excludes measuring psychical activity, and vice versa. We can study each individually—the body and the mind—but not both at once when taken together.[6] Fechner mentions this as one reason for why few attempts had been made at psychophysics prior to his addressing the problem.

Marshall has said, and I believe she is right, that Fechner would have been pleased to learn of the Copenhagen interpretation of quantum theory. Perhaps when developing the concept of complementarity, Niels Bohr actually made use of Fechner's idea without real awareness of its authorship, having received it from his esteemed teacher Harald Höffding. Höffding, at least, was familiar with Fechner's work and was certainly convinced that Fechner's identity view was correct.[7]

6.2 The General Principle of Measurement and Measuring Sensation

Fechner takes two steps in approaching the problem of measuring sensation. In the first step his effort is to determine the general criteria that are both necessary and sufficient for claiming that something is measurable at all—for both physical and psychological quantities. After 1858 Fechner called these criteria "the general principle of measurement" and considered it one of the lasting accomplishments of psychophysics.[8] In the second step he makes use of the principle of measurement to gauge magnitudes of sensation, i.e., to find a way to "apply" the principle to psychological phenomena "empirically."

In order for a magnitude to be measurable the principle of measurement requires that the following criteria be fulfilled:

(1) Magnitudes of a quantity must appear in various ways and exhibit constant quality; they must be conceivable as phenomena that increase and diminish in a continuous manner.[9]

(2) There must be a degree of difference, such that we can reliably discern whether or not it is equal to the degree of difference given between any two other magnitudes.[10] The difference must be reproducible or found in many replications.

(3) It must be discernable under which circumstances a magnitude is zero.

How can these conditions define the measurability of a magnitude? They enable us to "count equalities,"[11] which is—by definition—the art of mensuration. When the three conditions are fulfilled, we obtain a margin value (a difference) that is a unit capable of being compared to other margins (differences). We can then conceive of every other value as consisting of n number of units, beginning with zero. We count how often we divide the object to be measured into equal units, i.e., how often we can generate those units by summating equal differences of value.

Formalized, Fechner's argumentation runs thus:[12]

Find the unit of measurement appropriate for measuring a certain magnitude. The unit of measurement for x is defined as a quantity of difference in magnitude (x_0, x), whereby x_0 is a magnitude having the quantity of zero.

Notational key: "o" is the symbol for concatenation, i.e., 'linking' the value differences 'in a series'; "<" means "is weaker than"; "~" means "is equal in strength." (x_0, x_1) is the difference in magnitude serving as a unit according to condition 2, whereby $(x_0, x_1) << (x_0, x)$.

Now assume that an infinite number of replicates of the unit magnitude are producible or discoverable in sequences; meaning that

$(x_0, x_1) \sim (x_0, x_2) \sim ... \sim (x_{i-1}, x_1) \sim ...$

whereby $x_0 < x_1 < x_2 ... < x_i < ...$

Thus it is true by definition[13] that

$(x_0, x_i) \sim [(x_0, x_1) \circ (x_1, x_2) \circ ... \circ (x_{i-1}, x_i)]$.

But this means that $(x_0, x) \sim i \circ (x_0, x_1)$.

Therefore there is an $n \in N$ such that

$(x_0, x_n) \leq (x_0, x) < (x_0, x_{n+1})$.

n is "how many times the sought equality occurs"; it is thus the magnitude given the value x.

Fechner justifies this procedure by alluding to how infinitesimal calculus deals with spatial dimensions: "A curve or a surface is considered given; but instead of comprehending it as a given whole, infinitesimal calculus has it grow in increments."[14] This is exactly how we must also find a way to represent other quantities as sums of infinitesimally or finitely sized increments that are equal in size. Fechner, therefore, considers mathematical quantities to be the result of a perceptual constructive process occurring in space and time. This is an interpretation of mathematical quantities that was common at the time and was not abandoned until after the formalistic standpoint in mathematics was introduced through Weierstrass's arithmetical treatment of analysis (1872).

Fechner's principle of measurement is purely theoretical and does not tell us how we are to reproduce or discover the unit magnitude; nor does it tell us which means and procedures are necessary to achieve the required comparisons; nor how to increase the reliability of comparisons. Depending on the type of magnitude in question, very different challenges arise.

Now, how does the principle apply to mental phenomena? The first condition is easily fulfilled, for we know by experience that: "Vividness of consciousness waxes and wanes, attention varies in degree; sensations, feelings, instincts, volitions are weak or strong."[15] Therefore, we can think of sensations in terms of magnitudes that are continuously generated by gradually emanating from nonexistence (the zero state, or from the subliminal to supraliminal state).

Applied to sensations, the second condition requires that we can discern that certain increments of sensation—differences in sensation—are equal to others. This condition is not easily fulfilled, for we are seldom able to reliably make such comparisons. It seems as if we could—at the most—say that two sensations are equally strong when they are simultaneously supraliminal. And in this case we can only be certain that one sensation is stronger than another if the two differ immensely. We are normally not certain in evaluating the equality of sensations. And in many cases an effort to compare our own sensations is frustrated from the onset by the fact that we cannot treat those sensations as if they were material objects. When comparing the lengths of two material objects, for instance, we can place them side by side. That is impossible for sensations: "It will never be possible to lay one sensation next to another such that one can be used to gauge the other."[16] Yet these complications are not insurmountable for sensation measurement. When measuring physical objects we make use of "standards," that is, we use instruments to increase our acuity of discernment and to enable comparison in circumstances where we cannot directly compare two quantities. If, for instance, we cannot lay two lengths aside one another, we can first place a stick next to one, and then next to the other, and thus arrive at a comparison.

By using a steelyard (although it has no divisions) we can determine the equality (or inequality) of two weights.

What kind of gauge could we use to measure sensory magnitudes? It seems as if there is no appropriate instrument for this purpose because it would have to provide some way of placing sensations side by side, placing, as it were, one "standard unit of sensation" next to the sensation to be measured.

Fechner's solution for this difficulty rests on two insights about *material* standards; these are fundamental and momentous not only for Fechner's argumentation, but also for his entire theoretical standpoint. Continuing the numeration used above for the criteria of the principle of measurement, these are:

(4) A definition for a standard unit for measuring physical magnitudes can be obtained by relating that unit to other dimensions with which it is linked functionally.

(5) In order to obtain standards for measuring physical magnitudes we must rely on the mental impressions that are produced in us by material quantities.

The task, then, is to find the standard for measuring psychical objects in analogy to facts (4) and (5).

Regarding (4): When we measure physical objects we find that dimensions are not "pure" or "abstract" and read-off as such, but that they are manifested via other, "concrete" standards, against which they are read-off:

> When we measure time, do we compare sections of time directly with time? When we measure space, do we compare pieces of space directly with space? Actually, we set these objects up against an *external* standard, using a standard for time that does not consist of time and a standard for space that does not consist of space, and a standard for matter that is not material. In order to measure one of these things we need both of the other two. For every ell used to measure extension we need not only the length of the ell itself, but also the matter of which it is made and then the placing of the ell next to the length to be measured, which occurs in time; for every movement of a clock's hand with which we measure time, we need not only the duration of the hand's movement, but also the matter of which it is made and the movement of the hand through space. Even weights cannot be measured abstractly against other weights; using scales involves regard to space and time.[17]

Fechner had already advocated this view in *Theory of Atoms*, independently of discussing the issues of psychophysics.[18] And in his last article, published in 1887, he writes that we cannot "measure units of time . . . abstractly, disregarding the units of space with which they are connected, whether we do so by observing the movements of heavenly bodies or by reading a clock."[19] The same holds for measuring force: "Instead of measuring force itself (which is impossible) we measure the velocities related to and dependent on it, which are attributed to equal masses, or we measure masses exhibiting the same velocities."[20]

Note that Fechner advocates a "correlative" interpretation of measurement. The prerequisite for measuring any dimension Q is a directly observable dimension R and an instrument of measurement A, that represents different values of R in correlation to values of Q. Correlation is such that if values of A are ordered in terms of the dimension Q, then they are also ordered in terms of R. Varying values of R are *defined* by an inter-subjective, specified and repeatable calibration of the measuring instrument A. They themselves need not be re-measured. The function that describes the correlation between Q and R relative to A (and on which measurement of Q using R and A rests) is what Fechner calls a "standard formula." Normally we seek (or devise) a measurement apparatus that renders an unequivocal correlation between the values of Q and those of R, such that the value of R can be considered a direct representation of the value Q.

Using the example for measuring time, Fechner points out that we measure the dimension "time" by using the dimension "distance," as that distance is manifested by the movement of clocks or the sun. Time values cannot be gauged "abstractly," viz., without any relation whatsoever to spatial values. Normally clocks are built in such a way that we have an unequivocal representation of time quantities in spatial quantities. If we know the length of a distance (the distance to be covered by the clock's hand), then we immediately also know the amount of time assigned to it. For measuring time, the "measurement formula" on which the construction of the clock relies is a linear function $d = c\,t$, for constant c.

What does this tell us about measuring sensations? In order to get a standard unit for the measurement of sensations we must look for a dimension that is linked to sensations in a way analogous to how the standard for time is linked to spatial values, or the length of a rod is linked to matter: "Perhaps we can find a standard for sensations by examining something else to which sensations are well-linked, just as the length of an ell is measured with the material of an ell."[21] That "something else" is, in Fechner's opinion—as we all know—the external stimulus. The (abstract) standard unit for measuring the intensity of sensations must be found by studying the (concrete) stimulus to which the sensation is related. "We will use the stimulus, the means provoking the sensation, as it were, as an ell for sensation."[22]

But we also need a gauging apparatus for which we know the "measurement formula." The only "apparatus" that is suitable, because it combines the variable values of sensation intensity E with values of the physical stimulus, is the living human body. We know from experience that sensation increases as a function of stimulus intensity, such that (for a fixed range) E correlates isotonically to R (i.e., for all pairs E_m, E_n there exist correlated stimulus values R_m, R_n, such that: if E_m is greater than E_n, then R_m is greater than R_n).

Obviously, the measurement formula realized by a human body is much more complex than that of a manmade clock. The primary challenge regarding sensation is to find the "measurement formula" correlating the external stimulus with inner sensation.

Of course, this interpretation only makes sense if we think of the relation of stimulus to sensation as a non-causal context. Just as the material of a yardstick is not the *cause* of its length and the distance covered by the hand of a clock is not the *cause* of the time represented, likewise, a stimulus does not cause a sensation. While the stimulus is the cause for psychophysical excitation occurring in the brain or in the central nervous system in general, it is merely functionally (that is, non-causally) linked to the sensation. This result suits Fechner's identity view perfectly, for which the physical and the mental are two different ways of interpreting one and the same substance and therefore no causal relationship is necessary between them. It looks as if Fechner developed this notion of measurement with his identity view in mind.

Regarding (5): Before we see how Fechner intends to set up a *psychical* standard based on stimuli, it is important to understand how, in finding *physical* dimensions, he construes the relationship between the values of "concrete" dimensions and the value of the "abstract" dimension to be gauged. If we measure the length of an object with a yardstick, then we say that the length consists of *n* number of sub-lengths following one another in a series in one direction, in which each has the length of the yardstick. But this means to *infer from sense impressions* that the yardstick is the same length as that part of the object next to which it was laid. We count how often we have had the *impression* of equality while placing the yardstick next to the object again and again.

Or, to use the example of measuring time: Looking at a clock calibrated in minutes, we infer that a minute has passed from our *mental impression* that the distance covered by the clock's hand equals that between two lines (or, that from *n* impressions of that kind, that *n* minutes have passed). It is true, therefore, that: "Physical measurement is most generally and ultimately based on the fact that equal amounts of equally sized *psychical* impressions are produced by equal amounts of equally sized *physical* causes; the number of times they occur is determined by the number of psychical impressions, whereby the size of the cause producing one single impression is used as a unit of measurement."[23]

Thus we derive the standards for measuring abstract physical dimensions from a premise that expresses an awareness of the equality of a standard to an object being measured, and this awareness rests on our subjective mental impressions of concrete physical quantities. In other words: To measure means to discover that a standard (or that standard multiplied, or only a part of it) is equal to what is being measured. Discovering that equality rests on the *subjective* condition that the standard and whatever is being measured *seem* equal to the observer. We can never get past that subjective 'hurdle.' In order to prevent infinite regress, we must, at some point, simply concede that what *appears* to be equal *is* equal! In other words: Any discovery that two objects are equal in terms of their primary qualities rests on knowledge of their secondary qualities. The observer's subjective impressions play a fundamental part in observations of physical measurement: the observer cannot be eliminated.

This would mean that all mensuration rests ultimately on subjective impressions, which may vary from person to person. Is it then at all possible to ascertain objectivity in mensuration, as physics claims? Fechner discusses this issue in connection with his Day View. He begins with the question: "In any given case, what do we actually do to convince ourselves that this is a case of same material states or circumstances?"[24] Obviously, in this case we may not "view the equality or inequality of subjective individual perceptions as decisive for objective equality or inequality."

But this does not mean that subjective conditions play no part in, or can be eliminated from science concerned with measurement: "Most importantly, we must consider that equality and inequality in the objective conditions of phenomena, only when taken together with the subjective conditions, determine the consequences of the phenomena in terms of causal law; therefore, in terms of that law we should not draw our conclusions solely from the former [i.e., the objective conditions]."[25] We know that things appear differently to us when we take different stances towards them or when the state of our sense organs alters. But no stance-less, no 'non-individual' view of the world is possible. Therefore, in order to achieve objective comparability we must distinguish *one* perspective that everyone can use. And we must devise a method for keeping this perspective as constant as possible.

The same holds for measuring. We can make the subjective component of measurement objective by limiting its role to the function of comparing coincidences between the measuring instrument and the thing measured; this would be something like comparing the extension of a yardstick to the extension of the thing being measured. For when we compare lengths with one another, there is not much variation in the "subjective conditions" of different people.

> At different places on the earth, different observers will see the moon in a different position and movement in relation to the sun, so that, for instance, while there may be a total eclipse of the sun at one place, it is not so everywhere; here again, the objective conditions of the appearance are the same, but not the subjective ones, such that the moon's path appears differently. Now, for the scientist, naturally everything depends—while maintaining an equality of subjective conditions for appearances—on making temporal and spatial constraints of given phenomena congruent with the constraints of immutable* standards for space and time, or with standards permitting controllable changes, or with parts of those standards, and, by always using these standards in the same way, moving from the smallest to the largest, the scientist extends and pursues the laws according to which phenomena themselves change when there are changes in the dimensions within that range of phenomena. From the changes in the phenomena he then also draws conclusions about the standards and is able to convince himself—with more or less certainty—of the equality or inequality of objectively material circumstances.
>
> * This immutability must, in turn, also be determined for the equality of subjective conditions of appearance.[26]

Fechner's view of scientific measurement is very important.

6.3 Applying the Principle of Measurement to Gauging Sensations

What does all this mean for finding units of measurement appropriate for mental phenomena? As we have seen, when we measure things materially we discover the size of the (material) thing measured by concluding from the (mental) impression that the standard and the thing measured are equal. In order to measure psychical dimensions we must—as it were—reverse the order of inference and conclude information about the psychical magnitudes from information about the physical magnitudes: "Now, just as we . . . can only find the physical size by relating the physical to the psychical, conversely, we can only find the psychical magnitude by following up the relation in the reverse order."[27] This reverse inference is permissible after we have set up and calibrated the physical scale such that equal psychical magnitudes result for equally-dimensioned intervals on the physical scale: "Instead of precise equal divisions of a material standard, what we need here [i.e., when devising a psychological standard] is the precise determination and statement of a law permitting us to relate the increments of stimulation and sensation to one another and then draw conclusions from the intensity of the stimulus about the intensity of the sensation."[28]

This would be simple, if we could conclude equal differences in sensation from equal differences in stimuli (using the gauging instrument "human being"), just as we did when measuring time, namely, concluding equal portions of time from equal distances covered by the clock's hand. But even the simplest experience teaches us that such inference is invalid. We need to find "a function grounded in reality" that holds between the intensity of stimuli and the intensity of sensations, "in order to make this all applicable to reality."[29]

But that task appears impossible. If we wish to represent a sensation as a function of a stimulus, the intensity of the sensation must be measurable independent of the stimulus. It seems circular: In order to measure sensations, we must already have measured them.

Fechner finds release from this circle in the fact that in particular cases, under certain circumstances, we can actually register barely noticeable differences of intensity among sensations as being equal.[30] Fechner believes that this is the case for "just noticeable differences" ΔE. A just noticeable difference (abbreviated as j.n.d.; the modern term is 'limit') is that increase in intensity of a stimulus that first becomes noticeable when the stimulus slowly increases. Subjective aberrations occurring throughout this discernment of equality must be reduced by using methods of "error theory" (the modern term is 'methods of adjustment').[31]

Now, for external stimuli of varying strength we can check which increase in the stimulus ΔR would be necessary in each case in order to attain the same constant increase in sensation. As we have seen in chapter 1, section 1.1, Ernst Heinrich Weber discovered in 1834 and 1846 that a difference between two stimuli is always perceived as being of the same intensity when the *ratio* holding between those stimuli remains the same. Therefore:

$\Delta R/R$ = constant (Weber's Law),

whereby ΔR is the increment of stimulus that must be added to any initial stimulus R in order to get the just noticeable difference. ΔR is the relative threshold for the stimulus, or the limit at which difference is discernible. Armed with

Weber's Law and the fact we know by our own experience, viz., that the barely noticeable difference in sensation ΔE is constant for us (meaning that it is always perceived as being of the same intensity, no matter which initial stimulus we chose to minimally increase) we can find, within the various sections of the stimulus scale, a functional relationship between the constant increase in sensation and the increase in stimulus necessary for it. If c is a positive constant, then:

$\Delta E = c \cdot \Delta R/R.$

Fechner calls this the *basic formula*.[32]

This basic formula holds only if, as we said, ΔE actually remains the same. This prerequisite is the "hypothesis of difference." Many have doubted its validity and replaced it by the so-called "hypothesis of relation." According to this, $\Delta E/E$ is constant. The blind Belgian physicist Plateau,[33] who was once a student of Quetelet, was the first to set up this hypothesis (1872), but he later discarded it. Brentano, Grotenfelt, and Merkel also thought it was right. After a deduction as simple as Fechner's, it leads to a different formula for measuring:

$E = c \cdot R^p.$

Stevens and his school also used this formula paradigmatically.

However, this does not yet meet the requirements of the general principle of measurement. All we know at this point is which intensity of stimulation corresponds to a unit of sensation. But the principle of measurement demands that we must be in a position to specify the number of units ΔE that make up *each* difference of sensation, whether or not we can discern that difference directly (as we can for ΔE).

Fechner therefore is challenged to find a psychological scale for E, such that all equally well distinguishable pairs of stimuli (not only the relative differences, i.e., the differential thresholds of sensation) can be represented on a continuum by scale differences of equal size. If, in addition, we can determine the *stimulus threshold*, in other words: the magnitude of the stimulus that corresponds to a sensation's 'state of zero,' then, according to the principle of measurement, we should be able to interpret every sensation as the sum of increments in sensation ΔE, beginning with zero. Fechner says that "By viewing a sensation in its entirety as something that, beginning at zero, grows in constant increments dγ, which are defined as the function of appertaining stimuli increments dß within various sections of the scale of stimuli, we get the magnitude of the entire sensation γ by the accretion of its increments from zero up to γ, which corresponds to a given stimulus ß."[34]

Fechner executes this as follows:[35] In the basic formula let ΔE and ΔR be infinitesimally small. It then follows, that

$dE = c \cdot dR/R$, therefore

$E = \int_{R_0}^{R} c \cdot dR/R$, where R_0 is the intensity of the stimulus at which the sensation is zero.

(Fechner also assumes that sensation does not diminish when the stimulus decreases, but diminishes at a certain intensity of the stimulus. He calls this the *threshold principle*.[36])

It follows that: (*) $E = c \cdot \ln R + C$.

The constant for integration C is determined by using the threshold value R_0 for R. We know that sensation disappears at R_0.

$E_0 = 0 = c \cdot \ln R_0 + C$, therefore

$C = -c \cdot \ln R_0$.

Entered into (*) the result is:

$E = c \cdot (\ln R - \ln R_0)$

$= z \cdot (\log R - \log R_0)$

$= z \cdot \log R/R_0$.

If we assume that R can be measured by its threshold value, i.e., if we set $R_0 = 1$, we finally get the *measurement formula* as follows:

$E = z \cdot \log R$. (Fechner's Law)

(Negative values of E are interpreted as subconscious psychological events.) This meets the conditions of the measurement principle, giving us a method for measuring sensations by using the stimulus on which they depend: "With this formula for measurement we have defined a general function that is not only valid for cases of similar sensations, but a function holding between the degree of the basic value of a stimulation and the degree of the corresponding sensation, such that starting with the ratios of quantities of the former we can calculate the quantity of the latter, which gives us the measurement of the sensation."[37]

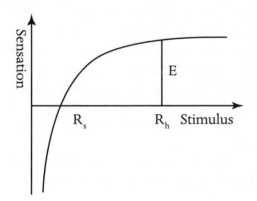

We can summarize Fechner's idea of measuring mental phenomena as follows: Look for differences in sensations (meaning: increments of sensation) that are actually perceived as having the same intensity. Then make this subjective judgment objective by using adjustment methods to reduce individual aberrations found in judging the equality of sensations. Now test sensations of varying strength to find which increase in stimulus is necessary for attaining the increase in sensation that is subjectively felt to be the same as others. Express these experiences as a monotone function between the increment of sensation found to be constant and the increment of stimulus required for it. Use this function of stimulus to construct a standard for measuring sensation.

The validity of almost every step of this procedure has been doubted. Particularly controversial were: the assumption that the just noticeable difference is actually always the same, the notion that R_0 is a unit of R, the idea that it is permissible to move from ΔE to dE and to integrate them,[38] the concept of negative sensations and the claim that sensation is zero at the absolute threshold. Most principal objections to measuring sensations emerged within Fechner's lifetime and he replied to them in detail. Actually, the controversy remains unsolved to this very day.[39]

Fechner's Law differs from the formula for measuring time not only by being more complex, but also because it states a rule for an *indirect* relation. Time 'depends' directly on the distance covered by the hand of the clock, but sensation depends only indirectly on external stimulation. It is directly related to psychophysical excitation, which in turn is caused by external stimulation. By using Weber's Law, then, for measuring sensations, Fechner has inserted an additional causal relation between the concrete and the abstract standard, one that (usually) is not present in physical data. If it were possible to directly measure the *vis viva* of the psychophysical excitation of the sensation in question, we could exploit that as concrete data for directly gauging the sensation.

I believe that Fechner means *precisely* this fact, and no other, when he writes that the physical world is immediately mensurable, while the phenomena of the psychological world are not.[40] He does not mean that inner perception is inferior to outer perception in any respect, simply because its data can only be compared in a less rigorous manner. On the contrary, Fechner always stresses that his principle of measurement is in principle the same for both cases, for physical and for psychological data.[41]

To see whether and how Fechner's ideas have survived in today's psychophysics we must distinguish three elements: Fechner's logarithmic formula,

Fechner's theory of measurement (disregarding the question of whether or not the law is correct as stated), and Fechner's methods for constructing subjective scales. In terms of the first, for years Fechner's work was overshadowed by S. S. Stevens's "new psychophysics" which paradigmatically stresses the power law $E = c \cdot R^n$. More recently, ways have been found to reconcile Fechner's and Steven's differences on a higher level. One essay suggests "harmonizing" them.[42] Regarding the theory of measurement, the distinction of four types of scales (nominal scales, ordinal scales, interval scales, and ratio scales) that Stevens made in 1946 had an immense paradigmatic effect (on both psychophysics and the philosophy of science). In his memoirs one prominent psychophysicist wrote that Stevens's distinction of scale types "stood like the Decalogue."[43] Stevens's definition did incorporate much of the Fechner-Mach tradition. But it remains impartial in terms of distinguishing between a Fechner-Machian kind of philosophy of science and one of mechanism/neo-Kantianism. We cannot deal with the issue here, but it is worth noting that in terms of the philosophy of science and philosophy in general, Stevens contributed anew to obscuring the facts. In 1975 Prytulak wrote a valuable critique pro Fechner and Mach. From Fechner's standpoint, one must first work out a great amount of preparatory theory before setting up a scale—one should not simply adopt "readymade" scales (such as those of Stevens). Generally, though, it does seem that Fechner's philosophy of measurement has been widely accepted within psychophysics (at least within a certain sector).

From a contemporary mathematical perspective we can, assuming that Weber's Law is valid, view Fechner's method of deriving the formula for measurement as an approximation.[44] Fechner thought he could use it for all kinds of Weber functions. A central task of psychophysics is to develop a generalized form (the measurement of structures of differences; particularly formulated in terms of probability) of Fechner's challenge: 'Starting with subjective evaluations, how can we measure sensation?' This challenge enjoyed a renaissance, as Mausfeld put it, at the end of the twentieth century.[45]

The notion that, in order to find standards for measuring mental phenomena, we must revert the order of inference used for measuring things in the physical world, had led Fechner to find his "formula for measurement." I believe that this notion also persuaded him to view mental data as statistical data. When we measure physical things, the use of adjustment methods amounts to beginning with varying (mental) estimates concerning equalities, estimates that have been made by various persons (or one and the same person at different times) based on their (mental) perceptual impressions, and then inferring the most probable value. In order to measure mental phenomena we must—in analogy—proceed conversely: Beginning with purely physical differences attributable to the most different kinds of relative stimulus thresholds that are felt to be similar, we must infer the most probable psychical magnitude. The greater part of volume 1 of *Psychophysics* is dedicated to developing various methods for this purpose.

Interpreting Fechner in this way would also explain why the scales that Fechner set up are called "confusion scales." In contemporary psychophysics these are understood as scales that result from judgments made by a test person asked to distinguish and compare stimuli that physically slightly differ from one another.[46]

For the philosophy of science it is important to determine the exact standing of Fechner's logarithmic formula. It has a double status.

(1) First, it is not a law at all, but a formula for a standard, meaning that it is a conventional *definition* of the functional relationship holding between the standard and whatever is being measured. It is a definition, however, that does not extend beyond experience. It *stipulates* the standard for measuring sensation conventionally. The formula rests on three empirical principles: Weber's Law, the basic formula, and the existence of the absolute threshold. (The analogy to measuring time would be: Define which distance covered by the clock's hand corresponds to a particular unit of time.)

(2) Fechner holds his logarithmical formula also to be an empirical law, namely a theoretical *law of inner psychophysics*. That law remains theoretical because the internal stimulus (the psychophysical excitation that directly carries psychological phenomena) cannot be measured directly, it must be inferred from directly measurable physical data. For Fechner the fundamental status of the formula is more significant as a law of inner psychophysics than as a law of outer psychophysics. He saw it as a law that is as important for psychology as the law of gravity is for astronomy; for inner psychophysics it is a *basic,* or a *fundamental law.*

To Fechner, deriving the formula and finding a standard for measuring sensations useful in outer psychophysics was an important, yet merely a preparatory sub-step in establishing psychophysics itself. Psychophysics cannot pass the provisionary phase and use the standards developed for factual gauging in outer psychophysics until it has been shown that the logarithmical relation is strictly valid for the relation between neural excitation and sensation. In that case, psychophysics could become an autonomous natural science with its own progressive research program investigating how higher psychological functions, such as memory, attention, and self-consciousness are represented psychophysically.

If the relation between neural excitation and sensation were linear, meaning that the logarithmic relation is not a law of *inner* psychophysics, then this would reduce psychophysics to an insignificant appendage of physiology.[47] Fechner was not as concerned with it being a logarithmical formula as he was interested in it being interpreted as a strict empirical law for inner psychophysics. He was always aware of whether a critic belittled the importance of the formula as a law for inner psychophysics and he composed his replies accordingly.

> While Weber's Law, when applied to the relationship of stimulus and sensation, has only limited validity in terms of outer psychophysics, it is, when it is applied to how sensation is related to the *vis viva,* or any other particular function of the underlying psychophysical motion, unrestrictedly valid for the world of inner psychophysics. . . . Thus we can predict that this law. . . , once we exactly understand its application to psychophysical events, will be just as important and fundamental for understanding the mind-body relation as the law of gravity is for understanding heavenly movements. It is just as simple as the fundamental laws of reality.
>
> Therefore, while for outer psychophysics psychological standards can only to a certain degree rest on Weber's Law, inner psychophysics will find its unconditional underpinnings precisely therein. Yet, presently these are merely thoughts and prognoses, which the future must confirm.[48]
>
> Our practical interest in outer psychophysics . . . diminishes in comparison to the profound scientific value that inner psychophysics has for psychology, physiology, and philosophy by naming the fundamental criteria for the relation of mind and body; experiments in outer psychophysics can only increase this value by providing a foundation for the probabilistic inference required in stating these criteria.[49]

As late as 1882 Fechner wrote that "psychophysics would only be a poor appendage to physiology" (meaning it could not promise to provide general conclusions and confirmation for the Day View) if we were to drop the principle of the psychophysical threshold for inner psychophysics and to ignore the synechism principle.[50]

6.4 Objections to Quantifying Psychical Phenomena

Jules Tannery's Critique

Debate on Fechner's psychophysics initially centered on two issues: 1) whether the fundamental formula and the measurement formula could correctly be derived from observation and whether the two are technically and mathematically inviolable, and 2) whether Fechner's Law is by nature physiological, psychological, or psychophysical, i.e., whether it sets up a relation between the external stimulus and inner psychophysical excitation, or between sensitivity and awareness of sensation, or is—as Fechner thought—a relation between the sensation and psychophysical activity (i.e., neural excitation).[51]

A third set of problems soon surfaced, dealing with the mensuration of sensations and psychological magnitudes in general. We can date the commencement of debate on facets of the theory of mensuration fairly accurately to the year 1875. When and whether this debate ever ceased is a matter of controversy. At any rate, it continued well into the twentieth century.

In December 1874, the Parisian journal *Revue Scientifique* began running a series of articles by Théodule Ribot (1839–1916) dealing with contemporary psychology in Germany.[52] After sketching Fechner's psychophysics, Ribot dealt comprehensively with Wilhelm Wundt's work. A year later, the *Revue* printed an anonymous letter to the editor responding to Ribot's first article. Replies came not only from Ribot himself, but also, somewhat later, from the Belgian philosopher and psychophysicist Joseph Delboeuf (1831–1896), as well as from Wilhelm Wundt, at the time living in Zurich, who later in Leipzig was to establish the first laboratory for psychology in the world. These replies were also printed in the *Revue*. The two latter reactions were obviously also presented to the author of the anonymous letter prior to publication, because his answer to them was also printed in the same edition. To this Delboeuf again replied.[53]

Both of these astute and bitterly ironic anonymous letters created a stir. They concisely anticipate all the objections that would be brought forth against Fechner's notion of measuring psychical dimensions in the years to come. The anonymous critic writing these letters was the young mathematician Jules Tannery (1848–1910), who had just obtained a doctorate for his work on Fuchs's theory of linear differential equations (Fuchsian functions).[54] Tannery, who presumably was acquainted with Fechner's work only by reading Ribot's article on it, assailed Fechner's theory on three points:

(1) Summation (additivity) and equality make no sense in terms of sensations. There exists no definition for what the sum or the difference of two sensations is supposed to mean: "The only dimensions that can be measured directly are those for which we can define equality and summation."[55]

Using a drastic example, Tannery illustrates how summation (additivity) and equality do not apply to sensations: We believe that Venus of Milo is beautiful.

Which spatial dimensions could be used to measure her beauty? Assuming that we were to find her arms, would the beauty of her arms be of the same kind of beauty as that of the entire statue? Assuming that we re-attached the arms to the statue, would we have added one aesthetic sensation to another? Would the final sensation be of the same kind that we had for the original torso?

(2) Sensations are not homogeneous. That is the underlying reason for why summation (additivity) and equality do not apply to them: "The essential characteristic of directly mensurable dimensions is *homogeneity:* whatever is added, such that something increases, is of the same exact kind as that which was already there: length, surface, and time are dimensions of this kind. If we add one length to another, both of them are of the same kind and essence and their sums are also of the same kind. Directly measurable dimensions necessarily have this quality, because measurement itself requires that dimensions of the same kind be comparable."[56]

Tannery illustrates the lack of homogeneity in sensation with the example of the sensation of heat: If you hold an object in your hand and the heat of that object increases, at some point the threshold of pain is reached. The original sensation (heat) is of an entirely different kind than the final sensation (pain). Entirely different nerves are involved in those sensations. And what is true for these two extremely different sensations also holds for all those in between, albeit to a lesser degree.

(3) Fechner's logarithmic equation only makes sense if we view it not as an empirical law, but simply as a conventional stipulation. The lack of homogeneity in sensation and the ensuing lack of applicability of summation (additivity) and equality can be compensated by *defining* sensation differentials (the just noticeable differences) as being equal. Equating them with the relative increase in stimulus is also purely a matter of convention. And everything we derive there from can also be merely matters of convention. Such a procedure is legitimate, but it is arbitrary and leads us nowhere.

What in the world, Tannery asks, is a sensation differential? We clarify nothing simply by making it smaller! "Every kind of definition is permissible; but I find this introducing of transcendental mathematics into such an obscure and unknown realm rather rash. We should be skeptical, monsieur, of pompous expressions and of a mélange of sensations, logarithms, and stimuli. *Je vous avoue que j'ai ce logarithme sur le coeur.*"[57]

Tannery thinks that Fechner opts—from among all logically possible definitions—for exactly this one because he made an improper inference from the nature of the stimulus to the sensation. He was misled from the onset by researching the stimulus and confusing the sensation with the stimulus. It is nonsense to say that the definition has been derived from sensation itself. "Someone concerned solely with stimuli would never have thought of it." In other words: The notion that sensations are measurable was derived from the homogeneity of the stimulus provoking sensation, and that homogeneity being projected onto the sensation. (This came to be known as the "stimulus fallacy."[58])

It is surprising that in spite of his crushing criticism, Tannery did not consider Fechner's formula entirely worthless. Even an arbitrary definition is good for something. The definition can be used in the same way that physics uses the thermometer to define temperature. For contingent reasons and convenience we have come to view the height of a column of mercury as a measure of heat. Dimensions for which summation and equality are meaningful terms "seem to be given only in the realm of the abstract, i.e., in pure mathematics. In reality we do not measure the things themselves, we cannot observe them, we know nothing about them. We merely measure lengths; the variations of those lengths are related in a complex and unknown way to the variations of the things themselves."[59]

As long as we do not understand that complex relation, we cannot really speak of measurement. Compared to Tannery's arguments, the replies of his adversaries published in the *Revue* are thoroughly unsatisfactory. Some do not really deal with the criticism brought forth, some refer to numerous authorities instead of offering counter-arguments (Wundt makes use of Laplace's concepts of *fortune morale* and *fortune physique*—which was a clever move within the French context), some openly confess their own lack of mathematical competence (Ribot's tactic), some paraphrase Fechner's arguments in a diluted version and allude to the newest, as yet unpublished research findings which will eventually clear up all the difficulties (Delboeuf).

The debate that Tannery set off was ardently continued in Ribot's *Revue philosophique*. It reached a peak with Henri Bergson's influential *Essai sur les données immédiates de la conscience* written in 1889, putting a temporary end to it. A great number of Bergson's arguments contra "mixing succession and simultaneity, duration and extension, quality and quantity" rest on conclusions he reached by thoroughly criticizing Fechner's and Delboeuf's psychophysics. He mentions "Fechner's keenest critic," Jules Tannery, and Tannery's contribution to the *Revue philosophique* of 1875.[60]

Tannery's critique provided an argumentative scheme that buttressed the interests of three different groups. Each group had deeper reasons not openly expressed in the critique (or suggested only in an encoded manner). One such group was that of the neo-Kantian university professors yearning to gain acceptance for their own methods of epistemology as distinguished from empirical, scientific psychology.[61] A second group included physicists and physiologists, members and associates of the Helmholtz School; these scientists resisted a naturalistic theory in psychology that would be independent of, yet equal to, physiology. And finally there was a group of psychologists succeeding Fechner who wanted to liberate psychophysics of the last remnants of nature-philosophical speculation and transform it into a theory that could withstand neo-Kantians and physiologists.

If we follow up the interaction of these groups carefully, we can examine quantification in psychology more thoroughly and appropriately than E. B. Titchener and E. G. Boring have done.[62] Neither of those authors investigated the context surrounding the so-called *quantity objection* (a name given to Tannery's kind of objection to measuring sensation in American literature since Boring), and this resulted in irrelevant opinions about the motives of all the critics involved. For the physicists and the physiologists, Tannery's objection was a corollary of the mechanistic worldview governing science. For neo-Kantian philosophers fighting for predominance, Tannery's objection was also a consequence of their (historically rather questionable) interpretation of Kant, one that enabled them to go along with the physiologists, without sacrificing their own independence. For psychologists, the objection provided an opportunity to either take sides with the physiologists at the risk of losing their own scientific identity, or to pursue an independent third way between neo-Kantian philosophy and established natural science.

In contrast, ever since Titchener's and Boring's opinions on the matter were voiced, the *quantity objection* has weirdly been related to *introspection*: "Introspection, the [quantity] objection runs, does not show that a sensation of great magnitude ever contains other sensations of lesser magnitude in the way that a heavy weight may (supposedly) be made up of a number of smaller weights. 'Our feeling of pink,' said James, 'is surely not a portion of our feeling of scarlet; nor does the light of an electric arc seem to contain that of a tallow-candle in itself.'"[63]

Recently it has even been claimed that the *quantity objection* was not about whether it is appropriate to measure sensations, but about whether introspection is an appropriate method for psychology and physiology. It has been argued that Fechner's critics used their objections to stress the importance of introspection for science, thereby playing off traditional philosophy for psychology. Traditional philosophy had always viewed introspection as a legitimate source of knowledge.[64] But in truth the question of whether introspection is acceptable in psychology was either insignificant, or it took the opposite form of Hornstein's claim. It is an irony of history that precisely those philosophical authors most conspicuously opposing Fechner's work were neo-Kantians, who— as such—were outright objectors to introspective methods in psychology.

G. E. Müller, A. Stadler, and F. A. Müller

We can fairly accurately pinpoint exactly when Tannery's objection initiated a crisis in psychophysics. It was around 1878, a significant period for our study. Around this time the participants in the debate were remarkably active. Each of them made an effort to open a new chapter in the science of psychology by settling an important and as yet unsolved controversial issue. And each was stunned that the other was pursuing a similar objective, but obtaining different results.

Fechner's formulas and methods had unleashed enough 'paradigmatic energy' to start-off a new, normal, scientific tradition. But it was unclear whether his approach was compatible with the mechanistic tendency preferred in all other natural sciences. In 1878 Georg Elias Müller intended to set things straight with *On the Foundations of Psychophysics* [Zur Grundlegung der Psychophysik]. This book suggested how to make use of Fechner's fundamental law in physiology. (In 1868 Helmholtz's follower J. Bernstein had expounded Fechner's psychophysical law as a physiological law describing how stimuli are related to brain activity.[65]) Müller suggested that the Fechner measurement formula be generalized, thus making it compatible with aberrations from Weber's Law. In his opinion sensation is not proportional to the logarithm of the stimulus, but to the logarithm of a certain function of the stimulus. His writing met with great approval and after Lotze resigned from his university chair in Göttingen to follow a call to Berlin, Müller was appointed to that chair.

The publication of Müller's book crossed with that of Fechner's *The Case for Psychophysics* [*In Sachen der Psychophysik*] (1877) in which Fechner replied to his critics and defended his fundamental psychophysical view against being interpreted physiologically. Müller, unaware of Fechner's book at the time he completed his own, reacted quickly and published a review in the *Göttingischen gelehrten Anzeigen* [Göttingen Scholarly Reviews] stating that Fechner's book altered nothing of his findings and that the psychophysical view was "at a disadvantage" compared to the physiological interpretation. And, as if that were not enough, Müller associates the philosophical core of Fechner's psychophysical idea, namely the identity view, with the unscientific machinations of earlier nature philosophy: "I fancy we still find in this author a rudiment of a manner of thought popular decades ago, whose obscurity in terms of the philosophy of nature the author himself had once understood to clarify so well."[66] Whether intentionally or not, Müller did distort Fechner's mind-body theory. Fechner felt the blow and wrote later that being reproached for obscurity was "quite exaggerated."[67]

Unfamiliar with Müller's work, Otto Liebmann, a neo-Kantian who had also briefly studied in Leipzig, published a friendly review of Fechner's book in defense of the case of psychophysics. He wrote that "after Lotze's and Helmholtz's studies" Fechner's *Elements of Psychophysics* was widely accepted as "a revolutionary subject within the field of scientific psychology." He continued to say that all the objections raised against Fechner "almost without exception dealt with individual passages and did not question Fechner's general principle."[68] Yet such positive evaluation from a neo-Kantian side was to disappear radically within the year.

In 1878 the Belgian Joseph Delboeuf, a professor in Liège, published a review of Fechner's book almost sixty pages in length, but did not mention Müller's *Foundations*. He regretted Fechner's lack of response to Tannery's highly consequential critique. Delboeuf admits only seemingly having defended Fechner's Law in his own reply to Tannery in the *Revue scientifique*. In truth he abandoned it after reading Tannery's appraisal. At the time he already felt that Tannery's arguments were "absolutely correct."[69]

The article in question made Tannery's critique known in Germany and while there may have been some previous awareness of it, that criticism could now no longer be ignored. Impressed by Delboeuf's review, the neo-Kantian August Stadler (1850–1910) wrote an article within a year in which he compares himself to Delboeuf as "striking a blow at the root of the law."[70] (Stadler was a close friend and among the first followers of Hermann Cohen [1842–1918] from the period in Berlin around 1872. We shall discuss Cohen, the founder of the Marburg School of neo-Kantianism, later.) While Stadler did defend Fechner against some of Delboeuf's complaints, he strongly assailed the psychophysical law.

Stadler argued as follows: Weber's Law says that the intensity of a sensation increases by the increment ΔE when the increase of the stimulus c ΔR is reached, for which c is the constant in Weber's law and ΔR denotes the relative stimulus threshold (R for German *Reizschwelle*). But sensation does not increase when the increase in stimulus is smaller than ΔR. Put in modern terms, it follows from this that sensation is a discontinuous non-monotone function of the stimulus. "The essence of the relation of size between ΔE and ΔR is discontinuity."[71] Fechner's fundamental formula, however, employs a continuous and monotone function. Therefore, it is incompatible with Weber's Law: "Weber's Law involves . . . the fact of the differential threshold, and any mathematical formula that [like Fechner's Law] ignores this fact is not an adequate form of that law."[72] He concludes: "The logarithmic curves which have become favorite descriptions of the psychophysical law lack all empirical truth."[73] In line with G. E. Müller he proposes in the end that Weber's Law should be interpreted "physically." In this case we should investigate "not the alteration of sensations, but the change in the central molecular motion that accompanies that alteration."[74] Two years later Stadler heightens his criticism by denying—as Tannery had done—that any fundamental formula is applicable at all: "The intensity of a sensation cannot be understood as the sum of this or that many simple degrees of sensation."[75]

As if Stadler and Delboeuf had broken the dam, a flood of criticism burst through. The main concern was not so much the details of the theory, but the general notion of measuring sensations. Stadler's metaphor was used frequently; many felt called to "chop down the tree at the roots."

The period around 1878–1879 is interesting not only because of this crisis in psychophysics. It also marks a decisive political turning point, which Hans Rosenberg has called "The Beginning of the End of the Liberal Age" in Germany. In reaction to two assaults on the emperor, Bismarck enforced a domestic change in political policy in favor of conservatism, using laws adverse to the social democrats. Köhnke has convincingly shown how this change influenced university philosophy. Neo-Kantianism, which had originally developed to censor idealism, was itself going through an idealistic phase at that time: evading guidance by science, distancing itself from all genetic, naturalistic, psychological and social scientific forms of explanation, and instead becoming a "science of generally valid values" that are to be justified independently of nature, striving for "objectification," "strict necessity," and "unconditional generalization"—in Hermann Cohen's words.

For the first time the Marburg School of neo-Kantians were perceived as a unified movement; they derived their understanding of Kant from Cohen's interpretation. Their criticism of psychophysics identified them as the "Cohenian School" [*Cohensche Schule*].[76] For them the problem of determining a functional relationship between a stimulus and a sensation became "the most important challenge in the borderline field between epistemology and psychology."[77] Cohen announced the philosophical essay prize theme for 1880–1881 to be "evaluating" Kant's second mathematical principle "in light of the psychophysical challenge."[78]

In 1882 Ferdinand August Müller (1858–1888) was the first to write a dissertation under Cohen's supervision setting himself the task of examining Stadler's objection to Fechner in detail and systematically and comprehensively evaluating Fechner's problem "from the perspective and according to the method of Kantian philosophy." In the foreword he states clearly from whence he will approach the issue and how radical his attack will be: "Since Hermann Cohen has rediscovered the basic intention of Kantian philosophy in his major works '*Kant's Theory of Experience*' [1871] and '*Kant's Foundation of Ethics* [1877] and, particularly in the former, acknowledges the stimulus to be the object of sensation, the proper framework for condemning all attempts at measuring the psychological is available."[79] Müller called Fechner's thesis, proposing that there is a functional connection between the sensation expressed numerically and the stimulus expressed numerically, the "axiom of psychophysics."

Müller attacks Fechner's approach from two angles. For one, he tries to prove that if psychophysics were true in Fechner's meaning of the word, then "anyone would have the right to devise his own psychophysics."[80] On the other side, he attacks from the position of transcendental critique coming from neo-Kantian philosophy.

As Fechner himself had shown,[81] we can think of the general problem that Weber's Law is meant to solve, as a function having the following form:

$$f(R) \, \Delta R = \text{const.},$$

where R stands for stimulus [*Reiz*] and ΔR corresponds to the just noticeable difference (j.n.d.) of sensation. If $f(R) = c/R$ results in an empirically determined function, we get Weber's Law by inserting it. If experience should prove that Weber's Law is invalid, we can always find a new function adapted to experience and that can take the place of Weber's Law by "using one of the well-known formulas for interpolation." Weber's Law can be substituted in this way, but the fundamental formula, and therefore psychophysics, cannot. For if we set

$$f'(E) \cdot \Delta E = f(R) \, \Delta R,$$

which would be the general case of the standards formula, then f' is an entirely arbitrarily selected function that cannot be determined empirically. It is a hypothesis that cannot be tested empirically. If we make f' a constant, as Fechner does, such that DE is constant (hypothesis of difference), then it is true that we can derive Weber's law. But if we make $\Delta E/E$ constant, as Plateau and Brentano do, (hypothesis of relation), then experience expressed in Weber's Law is equally deducible. But only one of the two hypotheses can be correct. As a matter of fact, "infinitely many formulas of measurement, i.e., relations holding between sensation and stimulus, can be devised, which contain this experimentally given relation [i.e., Weber's Law] between the stimulus and the difference of stimuli."[82] Müller makes reference to work done by Ewald Hering, in which this was allegedly demonstrated for the first time.[83] But in his opinion the deeper reason for the failure of Fechner's procedure is because it uses "numerical symbols" to represent sensation.

This leads up to the second attack: Müller considers the "axiom" of psychophysics to be just as a priori as the axioms of geometry. And just as we can only prove axioms of geometry by "insight into the foundations upon which the possibility of that knowledge rests," we would also have to prove for psychophysics that it is possible "by the rules of human intuition."[84] The question of the possibility of psychophysics is, therefore, "transcendental."

However, we can derive the characteristics of scientific experience only from mathematics and pure natural science (i.e., mechanics), since these are what is "really given." The objective reality found in these sciences is made possible as pure intuition in space and time, by the manifold impressions of the senses being related to the object through the synthesis of the power of imagination.

Psychophysics strives to achieve "objective knowledge about the magnitudes of psychological phenomena." So if, for psychophysics, a concept exists that provides the required unity, it can only be the concept of magnitude. The diversity given in intuition first becomes a synthetic unity in the material *object*. In terms of the concept of magnitudes, this means that only objects can be attributed with size.

After a detailed report on Kant's distinction between intensive and extensive magnitudes and in making heavy reference to Cohen's writings, Müller comes to the conclusion that: "Sensation is not a function of the stimulus, but instead, the stimulus is the object of the sensation, therefore, according to the findings of our transcendental exposition, sensation cannot be represented in the form of numbers at all, because we can only have knowledge of objects."[85]

In the second part of the dissertation Müller tries to give a purely psychological interpretation of Weber's efforts, which in his opinion "really cannot serve as the foundation of psychophysics." And finally, in the third part, he settles accounts with his namesake Georg Elias Müller, who lacked respect for "the simplest principles of epistemology," although the title page of his book (1878) proclaimed him a "docent for philosophy."[86] Here we find the seeds of the conflict that would later end in institutionally separating psychologists from philosophers.[87]

Hermann Cohen's Principle of the Infinitesimal Method

F. A. Müller's writing skillfully treated both sets of problems—the philosophical as well as the psychophysical, but ultimately rested on an irremediable tension between technically informed and constructive debate on psychophysics and the destructive neo-Kantian criticism of the same. Müller's dissertation supervisor, Hermann Cohen (1842–1918) subsequently concentrated on the epistemological aspect of the problem. He also had received the impulse for his critique from his student and friend August Stadler.[88] In *Das Princip der Infinitesimalmethode und seine Geschichte* [The Principle of Infinitesimal Method and Its History] (1883) he deals with this criticism in a larger context. It marks the beginning of "epistemological idealism" for the Marburg School of neo-Kantianism. Together with other neo-Kantian movements, by the turn of the century this idealism had become the "leading philosophy in Germany."[89]

In order to explain Cohen's judgment of Fechner and empirical psychology in general, we must widen the scope of our study. Cohen gave himself the task of discovering which levels must be distinguished in creating a real object and how these are particularly constituted. First of all, whatever can appear as an object of experience must be representable in space and time. It must "correspond to the fundamental laws and categories of *pure mathematical understanding*" [*Anschauung*].[90] We must not be misled into thinking that objects could be given for us independently of consciousness and without the aid of reason. But rather, something first becomes a Given when it has become a content of consciousness in understanding.[91]

It is important to note that for Cohen sensation is absent at this level of creating an object. On the contrary, he would like to keep the capacity to be represented in space and time free of any reference to sensation:

> In order to define pure sensibility, we must first have achieved *pure intuition*, and avoid all comparison with sensation. Space that is intuited in *pure terms* is *not* sensed. This epistemologically critical meaning of the psychological "chimera" of

space puts an end to all discussion about the *image* [*Vorstellung*] of space, which as such naturally contains the element of sensation, yes, is even founded on it. Pure intuition of space in contrast is that very element of epistemologically critical abstraction that in connection with thought is expanded to be the first mathematical principle. This justifies *avoiding sensation* when reevaluating intuition and making it a priori.[92]

Of course, this interpretation made untenable all empirical theories of space that were popular during Cohen's lifetime, including those of Herbart, Lotze, Helmholtz, Stumpf, and Mach, and it discredited the notion of "extensive sensations" among which Fechner counted the "notion of spatial extension by sight or touch."[93]

According to Cohen's scheme, we cannot *sense* objects in intuition, i.e., through the representation of space and time, but instead, we can only grasp them in pure intuition, because they are magnitudes of intuition. If, in thought, we grasp these objects mathematically, then we are in fact dealing—in Cohen's terms—with extensive magnitudes as pure mathematical objects. But these mathematical objects do not guarantee the reality of the object, because "as such, they are founded on analogies," they "depend" entirely "on unstable comparisons," or, in other words, they are relative. But they do guarantee the "objectivity of things" and make up the first "*instance of constitution* for the synthetic knowledge of *objects*."[94] But to believe that this alone constitutes the entire object is to confuse "material knowledge with the formal condition of knowledge."[95]

So how do we get from formal to material knowledge? In order to achieve that, we need an intensive magnitude instead of an extensive magnitude. Cohen uses the term "fulfill" for this purpose: Reality strives to "stabilize and fulfill" the "suspension" of extensive magnitudes, it wants to "form the geometric bodies into physical bodies."[96] This fulfillment happens by a "cognitive means of reality" that accompanies intuition. This cognitive means is given in the concept of the infinitesimal. The infinitesimal value becomes the "idealistic lever for all knowledge of nature." Without incalculable minuteness, all mathematics would be "irrelevant," "nothing would be founded," everything would dissolve into "relativity": "What is infinitesimally small is, as an intensive quantity, reality in this specific and clear sense: it provides science with *that which is real*, i.e., with that which all natural science presupposes and pursues; and it does this by being and constituting what is real."[97] The standards used in representing objects in intuition are based on arbitrarily selected units. But representation in the calculus employs a qualitative unit, namely the unit of what is endlessly small, or, the unit of what brings forth reality. It is defined as an intensive magnitude, not an extensive magnitude. A real object is then a sum of "infinitesimal intensive realities."[98]

Here Cohen is referring to Fechner's *Atomic Theory*. Knowing this helps us understand Cohen's peculiar and enigmatic metaphysics of the infinitesimal:

> Obviously, all these considerations compel us to the following answer: Only by distinguishing *theoretical* elements from *sensible* elements or *elements of intuition* are we able to determine the course which Fechner and other idealistic atomists consciously pursue. The *units of reality are the "points of definition"* for *matter*. And these units of reality are *theoretical units*, which, as such, lack extension, and can therefore be combined with it. This combination is necessary for solving the problems of natural research. The solution demands that forms and elements of thought and understanding are isolated. This interpretation of elements defines epistemological idealism.[99]

In light of what we know about Fechner's scientific realism, as discussed in chapter 4, Cohen's approach appears to be an idealistic distortion of it. Expressed in modern terms we would have to say: Not only a few non-formal concepts (i.e., non-mathematical, non-intuitive concepts) in science, such as the concept of the "atom," are of a theoretical nature, but *all* concepts are theoretical.[100] But for a concept to be theoretical means that it is not given; it is a concept that we have created. From Cohen's special thesis it would follow that one single concept would be sufficient as a foundation for natural science, namely the concept of being infinitesimal.

From Fechner's viewpoint (which in these terms is certainly largely in line with our own contemporary view) Cohen's conclusion would be nonsense: Fechner would say that introducing theoretical concepts into a theory is only meaningful if that theory also includes concepts of observation (or concepts that are theoretical in terms of a theory other than the one in which they occur). And it is also not permissible to interpret a mathematical idealization, such as the concept of being infinitesimal, in a scientifically realistic way. It looks as if all other objects of science are produced from it magically.

Now, what is important for our attempt to understand the contemporary objections to Fechner's method for measuring the mental is to know how Cohen dealt with sensations. If our interpretation is correct, then in Cohen's theory sensation must have an extremely subordinate and deficient status. In fact, Cohen sees sensations as "merely subjective," as "hypothetical" because they are elements of consciousness; describing them merely brings "subjective facts" to light, "because *sensation* by definition characterizes the *indescribable* part of consciousness."[101] Therefore, Cohen must find any phenomenalistic-empiristic foundation for the empirical sciences—such as Fechner's—doomed from the onset: "Whoever considers sense perception to have a natural claim to reality, or to provide direct certainty, or be an absolute source of reality, finds himself beyond instruction in critical epistemology. If sense perception contained a sufficient *criterion for objectivity*, it would require no further critical equipment: we would know immediately how to comprehend infinite nature."[102] Instead, perception and its reality only characterize a relation that must first be "made a priori" and "objectified" using special "cognitive skills." So, in these terms, Cohen considers it the main task of all scientific knowledge to "invalidate subjectivity" in sensation.[103] The objectification process takes place within the two-phase object-constitution process described above. On one hand, sensation is reduced to understanding, in other words, reduced to the spatio-temporal definition of the content of sensation. On the other, the sensation content is produced as reality by what is infinitesimal. At the same time, this also makes whatever is infinitesimal the unifying element of consciousness: "Unity is a form to which the *consciousness of thought* aspires, in contrast to that of sensation, and even that of intuition.... Quality in terms of *mathematical natural science* rests on the definition of that kind of reality, for which infinitesimal calculation provides the unit of measurement."[104]

If we did not continually measure reality using the unit of infinitesimal calculation (as a dimension of reality, in contrast to space and time as dimensions of intuition), all existence would be "only capable of sensitive qualification."

One of the few examples Cohen provides to illustrate the process of objectifying a sensation is that of objectifying heat sensation when measuring temperature: "A diagnosis of fever does not rest on the subjective sensation of temperature as expressed by the patient, but on the facts discovered by objectively measuring the temperature. This measuring itself presupposes an intensive reality, on which the concept of heat motion is founded. Hence, *science's* ubiquitous claim: it provides sensation with the objectivity it otherwise lacks."[105]

Cohen's critical epistemology results in a peculiar split in human experience. On one hand we do *experience* sensations, but on the other, when taken for what they are, those sensations have—as it were—absolutely no value in terms of knowledge. Once the sensation has been "purified and made a priori,"[106] the outcome is knowledge, but not because we have experienced a sensation, but instead—as it were—in spite of that sensation. This can perhaps be best illustrated using an example that Cohen borrows from Descartes: In the third meditation Descartes wrote that he finds within himself two different notions of the sun: "One is, as it were, created by the senses, and it could most easily be counted among those which I think are acquired; it makes the sun seem very small to me. The other, in contrast, results from calculations in astronomy, meaning that it is either gained from concepts that are innate within me or which I have produced in some other way; it presents the sun to me as something several times larger than the earth."[107]

Cohen interprets this as saying that basically only in the science of mathematical physics "are all things *given* and available for philosophical inquiry; stars are not *given* in the heavens, instead, in the science of *astronomy* we call all those things given which we distinguish as serious products and improvements of *thought*, as being based on perception. Perception is not achieved by the eye; it is done by the raisons de l'astronomie. This is what Descartes's classical sun example tells us."[108] Cohen concludes that critical epistemology cannot be bothered with the psychological process of obtaining knowledge, its concern is to harvest knowledge: its business is scientific findings. This approach systematically obscures and conceals how human beings move from sensations to science, the way Descartes had not only an idea of the sun derived from astronomy but also an idea of the sun gained through sense perception. All our quotidian life, where we communicate through and about our sensations without any type of objective science, becomes a totally puzzling world. It is a mystery how people can manage at all without understanding infinitesimal calculation and objectification processes.

No wonder Fechner's psychophysics diametrically opposes this type of critical epistemology. Other than disapproving of Cohen's sharp divorce of our everyday relation from our scientific relation to the world, in aiming to measure sensation, Fechner made sensation itself an object for science. Cohen found that ridiculous.

Yet, prior to discussing Fechner's work in detail, Cohen does acknowledge the merits of psychophysics. Its worth lies in the fact that it "empirically confirms that sensation is an intensive magnitude and solely an intensive magnitude."[109] This is proven, according to Cohen, by the concept of something being "just barely noticeable," a concept he defends against all objections. The quality of being just barely noticeable and the related law of threshold had contributed to dismissing consciousness as a "receptacle in which psychological events occur and take shape."[110] He also found it to Fechner's credit that it was now no longer proper to speak of the proportions of matter and consciousness "in the manner taken for granted by Herbart." It dispensed with dualistic prejudice. Neither does Cohen object to "defining the *natural procedure of sensation* as a magnitude." But, and herein lies his attack, this type of definition is only meaningful if we view sensation as an *intensive magnitude* of reality, that is, as a unit of production, and not as a unit of measurement. One must not mistake the intensive magnitude of a sensation for its intensity. The intensity of a sensation is caused by the stimulus. The reality of a sensation understood as an intensive magnitude rests, in contrast, on the relation of the stimulus, such as that expressed in the concept of the just noticeable difference.

But it is a "fundamental mistake," an "error in principle and method" to want to produce a ratio of magnitudes between the stimulus and the sensation: "As gratefully as we acknowledge the importance of the concept of the just noticeable difference, we must decisively reject the *notion of the fundamental formula.*"[111] According to Cohen, the fundamental formula equates the intensive magnitude of sensation with its extensive magnitude, namely the magnitude of the external stimulus. It is possible to estimate the similarity of one sensation to another, thus equating one intensive magnitude with another. But it is never permissible to equate a sensation with an extensive magnitude, as it is done in the standard formula "to which sensation *as such* has no access." As we have seen above, Cohen thinks that we can only make the *content* of sensation, i.e., the stimulus, extensive, but not the sensation itself. Sensation cannot be expressed in extensive units of stimulus. It would mean a reversal of the order in which an object is constituted. Instead, the extensive magnitude of a stimulus has the intensive magnitude of sensation "as its indispensable *prerequisite.*"[112]

Cohen considers it an "erroneous method" to represent sensation as a function of the stimulus. According to the doctrine of the constitution of objects we can describe the stimulus as an extensive magnitude x, which can be thought of as resulting from the differential d x. But sensation is only imaginable as d y, an intensive unit, which produces no y. This is because: "Whatever objectivity can be gained in and through consciousness is provided by infinitesimal calculation, wherever it is applied to the physical world. *If it is applied to the psyche, then the psychological becomes eo ipso physical.*"[113]

This means that sensation has been confused with its own content, i.e., with the stimulus as the objectified sensation. We can easily detect Tannery's objection in this reproach; Cohen was probably acquainted with it through Stadler.

In any case, Cohen remarks that Stadler's essay of 1878 instigated "epistemological criticism of the psychophysical problem." And another objection of Tannery's, namely that one sensation cannot be a part of another, also surfaces in Cohen's writing.[114]

But this does not mean that Fechner's basic law was worthless for Cohen. It can be significant, if interpreted physically. It is not the sensation that is a logarithmic function of the stimulus, but the "molecular central motion." Thus two physical values are brought together, just as Stadler had previously proposed.

Cohen also reflects on the reasons for Fechner's fallacy. In his opinion Fechner was misled by his desire to find a solution for the mind-body problem that reduces the adverseness of consciousness and matter to a minimum: "The trace of the psyche . . . was to be discovered in material motion, that trace that remains hidden in the "night view." And in sensation, the daylight of consciousness, matter and the body were to be fixed, those aspects which only the night view chooses to consider heterogeneous."[115] Like G. E. Müller before him, Cohen denies that Fechner's view is scientific and demotes it to a nature-philosophical myth: "This is the difference between mythos and science: Science deals with matter, whereas myth presumes consciousness."

We could mistake this phrase for a plea for the strongest type of materialism. But Cohen states explicitly that *his* day view, the "day view of our epistemological idealism" considers matter to be the objective and objectifying content of *consciousness*: "For reality is—consciousness; this does not mean the material content of consciousness, but the lawful basic gestalt of scientific consciousness, the kind of unity in consciousness, the principle of knowledge. The psychophysical problem can be eradicated by a critical epistemological definition of reality."[116] These are the last lines of Cohen's critique.

Cohen's critical epistemology remained influential for philosophy in Germany at least until World War I and beyond in two respects: It set the rules governing the relationship between philosophy and psychology: Between pure physiology and pure critical epistemology there was no room left for an empirical objective psychology gleaning independent knowledge of the psyche via methods of measuring. At the most, as Kant had once said, there was room for "comparative description of psychological events." At the same time, an idea surpassing those of Kant, namely the idea that besides reason and sensation there is another "cognitive skill" that must be applied in order to constitute a scientific object—that idea remained virulent for a long time. It was soon discovered that thinking of infinitesimal calculation as that means was much too narrow. But the notion remained that there are forms of knowledge, symbolic forms, with which reason itself synthesizes images to create a scientific object.[117]

The other main representative of the Marburg School of neo-Kantianism, Paul Natorp (1854–1924) later propagated a concept of the psyche bringing forth a research method unlike Cohen's, namely the concept of "reconstructing" the psychological out of objectification. But this psychology also diametrically opposed empirical psychology. In Natorp's opinion, physiology is the sole objective, empirical psychology: "The fact remains that the scientific *explanation* of psychical appearances can only mean *objectifying it to become a natural process* and is only possible using the method of natural science. . . . The explanation of the psychical is necessarily physiological."[118] But a psychology aimed at dealing with the subjective presence of appearances prior to all definition must proceed with an entirely different method. That method differs completely from objectifying knowledge in natural science and may not be confused with it.

Helmholtz's Follower Johannes von Kries

Cohen provided an important illustration for the opinion that followed Tannery, namely that measuring sensation confused sensation with its intentional object, the stimulus. Sensations themselves cannot be objectified, meaning for Cohen that they cannot be described in terms of space and time; this can only be done for the stimulus. Sensations cannot be made up of weaker sensations, because stronger sensations are qualitatively different from weaker sensations. If we ignore Cohen's infinitesimal metaphysics, which was strictly rejected by mathematicians,[119] we see that the neo-Kantian critique of psychophysics was quite attractive for many natural scientists. All of the physiologists and physicists who critically discussed the issue of measuring sensation (except for Ewald Hering) were heavily influenced by neo-Kantianism. They saw this epistemology as a way to bestow—as it were—higher philosophical honor on the mechanistic understanding of nature. Simultaneously, it enables us to elegantly circumvent the problem of consciousness and sensation in a mechanistic universe.

From among the many relevant scholars, Johannes von Kries (1853–1923) is a typical representative who unearthed far-reaching aspects of the problem, when viewed philosophically. Von Kries had studied physiology and medicine, in Halle under the supervision of Volkmann, in Leipzig with Karl Ludwig, and finally in Zurich. After completing his studies he worked for a year in Berlin with Hermann Helmholtz and then accepted an assistant position under Ludwig in 1877. In 1880 he was given a chair for physiology in Freiburg. Von Kries's special fields were the mechanics of muscle movement, particularly that of the heart muscle, and physiology of the senses, particularly color theory. In the new edition of Helmholtz's *Physiological Optics*, which he coedited, he developed a theory of space with which he wanted to demonstrate that Helmholtz's own empiristic theory of space "did not contradict" Kantian doctrine of space, but "on the contrary is an indispensable foundation for it, without which it more or less would float freely."[120] Von Kries was highly respected during his lifetime. After Ludwig's death in 1895 Von Kries received a call to Leipzig and after Du Bois-Reymond's death a call to Berlin, but in both cases he chose to stay in Freiburg.

From a very early age, Von Kries was captivated by philosophy, especially Kant's philosophy, and it remained a lifelong interest. He was so fascinated by it, that he was torn about which field should be his real occupation. His basic idea in epistemology is a fairly orthodox neo-Kantian interpretation of mathematics and its relation to experience. He was intrigued by the question of why it is that mathematics has such a binding validity for the scientific world view: "In our attempt to gain scientific understanding of the world, a natural development . . . led to the practice of completely ignoring those descriptions which refer to our sensations in the narrow sense (sweet, cold, red, and so on), and to selectively seize those which are founded on our fixed notions of time and space."[121] External events are understood as the movements of material points, i.e., as a process defined by space and time. As long as we continue in this way to "found scientific thinking about reality" on notions of space and time, we will also be bound in psychology to the particular nature of these notions. Although the specific contents of mathematics "possess an evidence (an a priori validity) that is independent of experience," that content is nevertheless binding for the scientific worldview. This means that definitions which are part of the worldview must remain within the framework provided by mathematics. Mathematics thus "compellingly determines the inner relations and contexts" for empirical science, particularly the concept of equality.[122]

This is one of the sources of Von Kries's critique on Fechner. The other source is the doctrine of physical dimensions that was systematically elaborated in the 1870s by setting up a 'centimeter-gram-second system.' Looking back over his life, Von Kries claims that the influence of this development on his thought was "hardly less profound" than that of Kant's philosophy.

When I became acquainted with these things, I experienced it as an insight of the greatest importance. The system that theoretical physics had set up, or was in the process of establishing, obtained a most impressive elegance and transparency. But above all it became clear what physical measurement means. It became evident that the *concept of equality*, with its full mathematical stringency, is a part of the quantitative statements about all kinds of empirical circumstances. If only values of the same dimension can be considered equal, then all such statements can be understood as statements about relations of spatial and temporal distances and sizes of mass or as statements about unnamed numbers."[123]

Von Kries elaborated his critique in an essay dated 1882.[124] First he emphasizes that every form of measurement, whether it is theoretic or practical, rests on "equating things that are not identical," meaning, for instance, a portion of a measuring stick with whatever is being measured. If we want to make a claim about theoretical measurability, we must prove it both permissible and meaningful to equate two non-identical "pieces," or "elements" as Kries puts it. We have practical measurability if we are actually in the position to confirm the equality of real given pieces. This is the case when certain pieces (objects or processes) are actually available and can be used as standards that remain unchanged in the relevant respect when comparisons are being made with other pieces. If a criterion of equality of this sort is available, each value can be compared with a standard value and we can count how often the unit is contained in the value. When measuring both space and time it is understandable and meaningful, without further methodological preparation, to say of two nonidentical things that in terms of the value in question they are equal. If this were not possible, then all of geometry would be a "figment of our imagination." Space and time therefore are also theoretically measurable.

But they can both also be measured practically. For measuring distances and space we have standards that "cover the same amount of distance at various places within the area to be measured," for measuring time we use events within the world that "always take up the *same amount of time* although they happen at different times."[125]

But things look different for measuring mass. We can only speak of confirming the equality of two masses if we are measuring the same stuff; only under this condition can we speak of a comparison of different parts. Mass is therefore only theoretically measurable for things made of the same stuff. It only makes sense to compare the masses of things made of different stuff if we have first arbitrarily *stipulated*, "we shall use as the unit of mass for each substance that quantum which has the same *weight* as a certain quantum of a certain sub-stance."[126] We say of two objects that they are equal in weight if they hold the balance on a steelyard. The way we establish this depends on simplicity and expediency regarding certain empirical laws. For measuring mass we rely on the experience that the additivity of mass remains when we add masses of differing substances. Additivity means that equality of weight exists not only for those quantities whose masses have been defined as equal, but also for equal multiples of them. Although we make use of this experience, the definition is arbitrary. We could just as well establish the definition that two bodies are equal in mass if they can be heated from 0° to 1° Celsius by equal quantities of heat. Once we have established the unit, determining the mass of a body heavier than one unit means to count the units needed to balance the weight of that body.

Now, if we look at other quantities in physics, quantities that are neither mass, space, nor temporal magnitudes (these being the intensive magnitudes), we notice that these other quantities are all—without exception—somehow combined with space, time, and mass. We establish what equality is supposed to mean for these quantities by defining them as a combination of spatial, temporal and mass magnitudes. Using several examples (force, quantity of electricity, intensity of current, temperature . . .) Von Kries illustrates that in physics every claim of equality rests ultimately on a claim of equality in terms of space and/or time and/or mass. Thus, it is true for physics that "in the last instance always only values of length, time, and mass are compared one to another, and the reduction of all other values to these is mediated by expedient definition that takes the factual relations into consideration."[127] It is this act of defining that guarantees objectivity in physics.

Von Kries has herewith prepared everything needed for criticizing the measurement of sensation as an intensive value. According to the analysis of measurement up to this point we can only speak meaningfully of measuring sensation if we can discover equality existing among sensations (or increments of sensation). But it is immediately obvious that sensations lack that "kind of sameness that elements have, an equality that is characteristic of our notions of space and time."[128] Tannery would say that sensations are not homogeneous. Values are of the same kind only if we can conceive of them as being made up of sub-values of the same kind. And this is not the case for sensations. We cannot think of a sensation as being divided up in such a way that it consists of smaller sensations that are its parts: "A loud sound is not a multiple of quieter sounds the way that a foot is twelve times the inch it contains and a minute is sixty times the second in it."[129] This passage shows once again clearly that Von Kries is not appealing to introspection or inner perception as a source of knowledge, but instead he is referring to conceivability, or intuition [*Anschauung*] as it is meant in Kant's philosophy. if the former were the case, Von Kries would have had to also claim that space is not measurable, because we cannot see that an inch is contained twelve times in each foot.

But if, in analogy to measuring mass, for example, we wanted to *define* equality for the purpose of measuring sensation, we would be at liberty to do so. But we could not justify doing so in any way. We have no experience that proves one definition to be more expedient than the other. Any definition would be entirely arbitrary and worthless. Of course, we could establish that the just noticeable increment of a sensation is always the same, whether the sensation being just barely increased is strong or weak. But none of our experience compels us to accept this. We would be equally justified in saying that those increments in sensation are equal which correspond to the same increments of the stimulus. "One is just as correct as the other."[130] It would be equally nonsense to want to represent sensations analogous to intensive values in physics as a combination of spatial, temporal, and mass values. It is just as meaningless and useless to equate one sensation with another as it would be to say that a centimeter is equal to a second.

Von Kries generalizes his findings: Intensive magnitudes are not measurable, unless we *define* them as magnitudes of space, time and mass.

With this claim Von Kries sees himself in "basic opposition" to Fechner. Fechner's fault was to search for a standard for measuring sensation that is *useful in terms of our experience*, instead of developing an expedient *definition* of what equality of sensation allegedly means. But we cannot speak of a correct standard, or distinguish an expedient from a useless standard, prior to establishing what it allegedly means to say that two values are equal. For instance, we cannot search for an appropriate standard for measuring space—i.e., a solid body—if we are not sure of what the equality of spatial dimensions means *at all*. Von Kries concludes that: "In my opinion the whole attempt to measure the intensive values of inner life is nothing more than a careless and unjustified transfer of what is permitted for intensive values in physics."[131]

In evaluating where Fechner went wrong, Von Kries suggests the same diagnosis that Tannery had made previously: Fechner had confused sensation with "objective values," meaning the stimulus. When we try to estimate objective space, time, or mass values by relying merely on our sensations, we do not make a statement about the value of the sensation, but merely about the value of the stimulus. But if we ignore the stimulus when the sensation is present we must admit that it is senseless to claim that one pain is ten times stronger than another.

On a topic to be discussed later, in chapter 8, it is interesting to note at this point that a few years later Von Kries developed a logical interpretation of probability based on the very arguments that he used to reject the measurement of sensation. If we think of probability as a subjectively felt degree of certainty or expectation, then it is just as impossible to find a standard for measuring that degree as it is senseless to find one for measuring the intensity of a sensation. Probability, if it is to mean anything at all, must be understood in terms of logic. Ludwig Wittgenstein later adopted this idea from Von Kries. In the hands of Friedrich Waismann and Rudolf Carnap it became the logical interpretation of probability as we know it today.

Adolf Elsas, Physicist in Marburg

Adolf Elsas (1855–1895)[132] was a physicist in Marburg and a friend of Cohen's. He won the prize in the 1881 essay competition mentioned above, for which Cohen had proposed the theme requesting the treatment of how Kant's second principle of mathematics as stated in the *Critique of Pure Reason* relates to the problem of psychophysics.[133] Later, in 1886, Elsas wrote a longer treatise *On Psychophysics* [Über die Psychophysik]. After attempting to show that Fechner's deduction of the measurement formula is not only not compelling, but even contradictory, he discusses the following 'cardinal' question: "Is psychophysics as Fechner proposes it possible at all? I say: No; any application of mathematics is restricted to what is physical, which corresponds to sensation, if we let it be psychical."[134] Elsas's epistemological critique remains within the argumentative outline sketched by Cohen. The way he naively translates Cohen's terminology into the vernacular of physics shows perhaps most clearly how the Marburg School, in its stand on Fechner and nascent naturalistic psychology in general, took sides with mechanistic physics and physiology. Elsas expressed that tendency even more candidly than Cohen had done.

Elsas's independent contribution to the debate that took the issue further than Cohen and Von Kries deals with an argument about the role that causality plays in measurement. Elsas argues that two values are equal if we are aware of the identity of their production.[135] He associates this with Cohen's theory of infinitesimals. One mathematical function claims to generate a value from one (or more) other. To be functionally related therefore implies not only that there is a relation holding among two values, but also that there is some 'causal nexus' between them:

> The link, which is governed by natural law, serves as it were as a basis and fundament for the mathematical connection.
> [We will] always only discover values where we simultaneously find a causal relationship, which links the physical values to mathematical values and thereby first makes them be values.[136]

Even if a functional relationship does not always imply that there is a causal relationship, for example, in the case where the number of swings of a pendulum is a function of the pendulum's length, if we look more closely we find "that nonetheless, some causal condition is the basis of the functional connection."[137]

Elsas's critique of Fechner follows from this. If, as Fechner claims, there is no causal relationship between the body and the mind, then the connection between stimulus and sensation also cannot be expressed mathematically. If Weber's Law were really an expression of the causal connection between stimulus and sensation, then Fechner's psychophysics would be "real psychophysics of the mind." But this option is dismissed because sensation is neither an extensive magnitude, nor—as Kant and Gauss have shown—can it be made extensive. If Weber's Law has any meaning at all, all that remains is to see it as a connection between stimulus and psychophysical excitation, and this demotes psychophysics to common physiology and physics.

Elsas finds that psychical magnitudes cannot be 'made extensive,' as, for example, temperature can, because only mechanical causes can be formulated mathematically: "All cause of experience is force and energy, and general mechanics makes these concepts magnitudes by reducing them to movement."[138] Naturally, this does not work for sensation, because it is not the cause of anything, it is caused by something else. What can be measured are the intensity of the external stimulus and the energy of the psychophysical movement.

Ultimately Elsas dismisses all endeavors in naturalistic psychology: "Mathematical psychology, psychophysics and physiological psychology—three absurd names! Mathematics cannot be applied any more than the concepts of movement and force can be applied; physics ends where causality no longer rules; and physiology has no further purpose, once it has finished measuring an organism."[139]

In a review commending Elsas's book, Kurd Lasswitz said there can be "no such thing as psychophysics."[140]

Physics on the one side, and neo-Kantian epistemology on the other had staked out their scope of validity such that there was no room left for psychology.

But this makes it seem as if within science there is no place at all for sensation: "And sensation? It is not an object of scientific knowledge; it is not a part of nature; it has no reality for the mathematical physicist; it cannot be treated mathematically as a quantum. We claim this with all severity."[141] Thus Elsas advocates the same position as Emil Du Bois-Reymond, whom we encountered in chapter 3. A few years later he was supported by Natorp, writing: "In reality contemporary science knows no other qualities than those that can be objectified in a quantitative expression. It considers those that cannot be objectified, the so-called qualities of perception, as merely subjectively present and as of no further use for science, whose task is to objectify what is physical."[142] It seems as if every neo-Kantian and mechanist felt pressed to claim that sensation has no place within a mechanistic universe.

There is a direct connection between Natorp's view and Rudolf Carnap's *Logical Structure of the World* published in 1928. We can say that generally early twentieth-century logical empiricism was not merely a furtherance of Mach's theory of elements, but was also intimately related to neo-Kantianism. For Carnap, too, as a foundation of science, sensations are out of the question. "Structural properties" are what correspond to the flow of experience and guarantee scientific objectivity: "Certain *structural properties* concur with all flow of experience. Science must restrict itself to statements about such structural properties, because it should be objective. And it *can* restrict itself to statements about structure, . . . because all objects of knowledge are not contents, but form, and these can be presented as structures."[143] Moritz Schlick also favored this solution.[144]

Hermann Helmholtz (1821–1894) produced some work in 1887 on "*Counting and Measurement in Epistemological Terms.*" He did not explicitly mention the dispute on psychophysics. But it is clear that this work evolved contextually from the debate on psychophysics, since it mentions Elsas and also appeared in a *Festschrift* honoring Eduard Zeller, who had been involved in the debate in 1881–1882.[145] Like Von Kries before him, Helmholtz stresses the importance of equality for measuring. We need a comparative method for discovering equality. Helmholtz probably viewed the issue mainly in terms of the theory of space, since he was working on that at the time. The problem solved in the paper "Counting and Measurement" was how to formally represent the conditions governing extensive magnitudes; these being the very criteria that Von Kries and Tannery had demanded for measurement. Helmholtz's description of these conditions remains the standard notion of extensive magnitudes to this day. He added to Von Kries's criteria the requirement of an empirical operation for combining two objects exhibiting the property to be measured, such that their concatenation also exhibits that property. Length, for instance, can be measured, because it is possible to link two or more lengths in such a way that a total length results, which, in turn, is made up of the subsections.[146]

Let us see how Von Kries, Elsas, and Helmholtz could have described measuring temperature in a way to which a Marburg philosopher could have consented (putting aside for the moment all further reasoning that Cohen's infinitesimal theory would have additionally implied). According to this concept, temperature would be an intensive magnitude. But we have no operation available for it like those for extensive magnitudes, which can be represented formally as addition. Placing two sticks of equal length one behind the other results in a doubled length, but pouring two equally warm and equal volumes of water together does not double the temperature.

One solution is to revert to another magnitude as an extensive representative, for example, the extension of the column of mercury. This representative must have very specific properties:

(1) It must depend causally on the temperature. This is the case for the extension of the column of mercury. From experience we know that it is caused by the temperature of the body with which it comes into contact. (This could be called "the Elsas condition.")

(2) The representative must allow the distinction of similar elements. This, too, is the case for the extension of the column of mercury. We can subdivide it into equal sections and count these sections. (This would be "the Von Kries and Tannery condition.")

(3) The countable parts of the representative must be capable of additive linkage, such that the relation of two temperatures can be reduced to a process of addition. (Helmholtz's condition.)

However, in this procedure the measurability of (intensive) temperature is bought with the fact that the numbers gained are dependent on the (extensive) representative. Therefore, Helmholtz sees the measurement of temperature not as obtaining information about the temperature of the object but as information about how the extension of the column of mercury reacts to temperature.

Part of the measurement is thus a special property of the thermal substance. What we superficially designate 'measuring temperature' is actually not measuring temperature; it is a column of mercury's reaction to temperature. Therefore, measuring temperature, as well as measuring any kind of intensive magnitude is "coefficient measurement," in other words: determining the numbers of ratio,[147] but not the discovery of an objective property of a body itself. "We must not forget," writes Helmholtz, "that the scale of temperature fixed in this way is derived with a degree of arbitrariness from special properties of especially amenable bodies, but not from the essence of heat."[148]

Now, how did empirical psychologists react to the various attacks made by physiologists, physicists, and philosophers, and directed towards Fechner's concept of measuring sensation? I cannot fully depict their reaction here. But Wilhelm Wundt took refuge in interpreting the psychophysical law psychologically, not refuting the attacks, but simply making them irrelevant. This entirely abandons Fechner's revolutionary notion of the principles of measurement (the "measurement formula") and concedes to the adversaries that (particularly psychical) magnitudes can always only be measured for magnitudes of the same kind. This concession went so far that Wundt ultimately betrayed his own roots and denied outright what he had previously attested in 1887 and 1901 in his commemorative writings in Fechner's honor, namely that experimental psychology began with Fechner's psychophysics.[149] His denial also liberated him from any suspicion of association with philosophy of nature, a stigma perhaps still attached to Fechner's original psychophysical opus (or, which his contemporaries thought to be the case):

> But the psychological insight that sensations can only be directly compared to other sensations, or put more generally, that psychical magnitudes can only be measured using other psychical magnitudes of the same kind, and that therefore we can never use physical stimuli as standards of measurement, but only as aids with which we relate sensations of a certain intensity to one another under precisely defined conditions—just this psychological insight freed psychophysics from its initial metaphysical prejudices, transforming it into an integrated area of psychology proper. In this respect, the so-called 'psychophysical methods of measurement' have become 'psychological methods of measurement.'"[150]

It is a historical error to think that psychophysics—as is generally believed today —first introduced experimental methods into psychology.[151]

Wundt claims that Fechner's methods do not belong to experimental psychology, but to the *physiology of the senses*, just like the work of Johannes Müller and E. H. Weber—the origins of Fechner's methods. Experimental *psychology*, in contrast, began when interest in and understanding of the mental increased and it became clear that "reliable self-observation" could be achieved experimentally. The sixth edition of Wundt's *Physiological Psychology* [Physiologischer Psychologie] clearly bears this attitude.[152]

6.5 Ernst Mach's Theory of Measurement

Now that we have seen how the Helmholtz School and neo-Kantianism in Germany reacted to Fechner's principle of measurement, let us look at the repercussions it had within Austria-Hungary. Ernst Mach's (1838–1916) reaction was crucial; it headed in a direction entirely different from that of the mechanists. His basic idea was that it is not psychophysics that needs re-interpretation in physiological terms—so that it suits normal mechanistic philosophy of science, instead: philosophy of science must change so that it harmonizes with psychophysics! Instead of evaluating the measurement of the mental against principles of material measurement, we should interpret the measurement of material dimensions in terms of Fechner's principle of measurement. This way we need not eliminate sensations from the scientific worldview, as Elsas had most candidly demanded, instead, this theory returns to sensations their primary right. Sensations thus remain a part of nature and constitute the starting point for natural science! Mach finds it fundamentally wrong to believe that physics can do without sensations. "Color, sound, heat, pressure, space, time, and so on" constitute "the basic elements of the physical (and also psychological) world."[153] Thus, for Mach, space, time and mass lose all the fundamental privileges that they enjoy within the mechanistic worldview and become "sensory elements" among many others. The resulting task for science is to examine the actual dependencies relating those elements: Physics examines the dependencies of things outside of our bodies, psychology examines those within us, and psychophysics deals with those crossing the boundary between inside and out.

What is Mach's underlying notion of measurement for physical dimensions? In short, it is that I measure a phenomenon that I experience, meaning that I have a sensation of it as one of its features, by numerically representing the behavior of an external observational element serving the purpose of being a feature of my sensation, and this happens in such a way that the order inherent in the external feature correlates isotonically with the order within the sensation: If the sensation becomes stronger, the external feature also increases. Thus, the discovery of objective equality rests on the comparison of two sensations.

Measuring Temperature

By way of illustration let us return to the procedure of measuring temperature, now as Mach developed it in 1896 in *Principles of the Theory of Heat* [Principien der Wärmelehre]:[154] "We know [by experience] that the *volume* of a body can function as a *feature* or an indicator of its *state of temperature*, and that a change in volume can be understood as a change in its temperature."[155] Expansion indicates higher temperature, contraction indicates lower temperature. We are at liberty to select the characteristic to be used as a feature inasmuch as we may basically select any characteristic which experience shows to vary with variations in temperature, such as electrical conductor resistance, dielectric constants, refraction exponents, and so forth. Nothing compels us to rely on space, time, or mass. A certain "arbitrariness" is inherent to selecting any particular feature and a certain "agreement" confirms that choice. Which considerations shape that choice? They are of a purely expedient nature. We notice that we can better predict our sensations, meaning that we can deal with life more sovereignly, if we substitute the feature called "sensation of heat" with a feature of a physical thing. In a sense, we let that thing "experience" the temperature for us.[156] But doing this does not mean that we have somehow 'objectified' *one* sensation by transforming it into another, as perhaps a student of Cohen might have said; what we have done instead is to produce a relation between *two* elements, namely between the sensation of heat and an observed volume: "The sensation of heat and volume are two *distinct* elements of observation. We know by experience that they are generally connected; how, and to which extent they are connected, is something which—once again—only experience can teach us."[157] It is well known that disregarding the affirmative experience we have had using thermometers, there nonetheless are cases in which the expansion of the column of mercury is *not* characteristic of heat sensation. We sometimes feel cold, although it is 'actually' warm, and sometimes we perspire when it is 'actually' cold. 'Actually' here means: what the thermometer indicates. We should not conclude that the thermometer measures the real temperature and we experience merely a subjective, unreliable sensation. Such cases show that whatever

we decided upon when selecting the principle and when devising thermometers was guided by purposes entirely foreign to that one purpose of finding a
feature that properly serves as a characteristic of our sensation in *every* situation. If this were our sole objective we should be prepared for a long search, and
one that may be in vain. To complicate matters further, everyone has their own
life history and reacts differently. For much of our daily experience the thermometer is quite useful as an inter-subjective "substitute for sensation."

But if it was not predominantly its help in mastering daily life that led to selecting the volume of mercury as an indicator of heat, why did we choose it?
Well, according to the definition given above, physics examines the relations of
things outside our bodies. This means that physics strives to discover a characteristic of our sensation of heat such that science can make productive, simple,
and ordered claims about the *relations* existing between *this* characteristic and
other phenomena in the physical world. We know by experience that hot things
glow and melt, cold things stiffen and freeze, we know that we can sear a steak
on a hot stove, and so on. Therefore, if we are clever in selecting the feature representing heat sensation, this will enable us to better understand the relations
between things in the external world and to make predictions superior to those
we would get sticking with pure heat sensation. Another purpose served by selecting a particular feature is the advantage that physical measurement brings
for reducing those intra- and inter-subjective deviations associated with our
sensations, thereby creating inter-subjective science. Mach writes: "Instead of
using the individual hands and feet that each person always has with him without perceiving a noticeable spatial change in them, we quickly select a generally
more accessible standard, one that fulfills the conditions of immutability to a
greater degree, and this introduced an era of greater precision."[158] This clearly
demonstrates that Mach took measuring instruments to be—as it were—
extended or improved sense organs, while neo-Kantians viewed measuring instruments as a means for eliminating and overcoming sensations by objectifying them.

Another idea of Mach's is to make the selection of a characteristic depend on
the sensitivity with which it measures. A thermometer often indicates distinctions that go unnoticed by us, meaning that the thermometer differentiates to
a degree that does not correspond to distinctions we make when sensing heat.
Selecting the thermometer then, to indicate heat, only makes sense if we wish
not only to expediently predict our own sensations as they depend on conditions
in our environment, but if we also are predominantly interested in representing
the laws for the behavior of material things amongst themselves. Physics thus

alters our naturally given scale of sensation, enhances it, and increases sensitivity by including the causes and effects of heat and coldness.

Up to this point we have spoken as if selecting expansion as the characteristic feature of heat sensation were the only choice required in order to accomplish the measurement of heat. But this is not the case. Three further decisions are needed:

(1) Determining a "thermoscopic substance." We must select a standard substance (for instance, mercury) whose expansion is to be related to the state of heat of all other matter.

(2) Defining equality of heat. We must determine under which circumstances a defined quantum of any material is said to have the same temperature as a standard quantum of the standard substance (such as the standard expansion of the column of mercury).

(3) "Agreement regarding the principle of correlation." Finally, we must define the functional relation for assigning the (normally real) series of numbers, whose elements are to function as the temperature of the substance, to the various volumes of the selected standard substance.

All of these definitions must also reflect practical experience and simplicity requirements. The fact that we order the expansion of the thermoscopic substance numerically is due to the fact that numbers are the best system of ordering symbols at our disposal that can be "augmented and refined indefinitely."[159] By allotting numbers as names for special states of the thermoscopic substance we obtain a highly convenient system for achieving order in our experience of the links of properties of physical things, we maintain a synoptic view, and we can quickly accommodate new experiences.[160]

What follows from all of this? It follows that measuring a magnitude is not to discover a state of a thing, but to discover a relation, namely, the relation existing between the thing measured and the standard of measurement.

> Designating a state of heat by using a *number* [is founded] on an *agreement* . . . There are states of heat in nature, but the concept of temperature relies on our arbitrary *definition*, which could have been entirely different. But up to the most recent times, scholars laboring in this field seem to more or less consciously be searching for a natural standard for temperature, a real temperature, a kind of Platonic idea of temperature, of which the temperatures read off the thermometer are merely an imperfect, imprecise expression. . . .
>
> Once again: We are always only dealing with a temperature scale that can be safely and precisely *produced*, and, in general, *compared*, but we are never dealing with a temperature scale that is "real" or "natural."[161]

According to Mach we must arrive at the same conclusion for dimensions of space and time. Our space and time sensations, too, are *substituted* (in this case it occurs in practices acquired before we had physics) by the features of clocks and rulers. The choice of these standards for measurement ensues from our goal of achieving optimal inter-subjective comparability.

Mach attempted a psychological explanation for why we think that we must find a "true" temperature. His explanation is that we have replaced the order of our sensations of heat with the order of a physical series of states in such a way that the latter and the former are "not entirely parallel." "This is precisely why we secretly and subconsciously harbor a notion of our original sense of heat, a notion that was replaced by these poorly coordinated characteristics, and that notion is at the *core* of our ideas" [about heat].[162] This "shadowy core" acquires the imagined role of being the 'real' temperature, which is at most approximated by indications of the thermometer, but never exactly portrayed. And when we measure space and time, our notions of absolute space and absolute time are merely relics of the pre-physical perception of space and time before we knew of physics, prior to the institution of clocks and other standards:

> Newton's notion of "absolute time," "absolute space," and so forth, . . . origi-nated analogously. *Sensations of duration* play the same part for variously meas-ured portions of time as the sensation of heat does [for measured temperatures] in the case mentioned. The situation is similar for space.[163]

This would appear to be a harmless, modest remark. Einstein found it ex-plosive. Defining physical time and physical space in terms of the behavior of clocks and standards comes fairly close to the fundamental insights of the Spe-cial Theory of Relativity. In a letter dated January 6, 1948, addressed to Michele Besso, Einstein emphasized twice that besides Mach's *Mechanics*, he had also been influenced by Mach's *Theory of Heat.*[164]

Now, obviously in his analyses of measuring temperature Mach borrowed and improved the five conditions that Fechner had laid down for the principle of measurement. The congruence in criteria numbers four and five is particu-larly important: Fechner says that we need a "concrete standard" for measuring, one that is "functionally related" to the object being measured. Mach expresses this in terms of "features" [*Merkmale*] that proxy for the sensations.

Incidentally, in 1887 Max Planck made a statement similar to Mach's:

> Physics . . . deals with the description of *all* forms of appearance, not only with those given us by our sense of motion, but also with those given by our muscle

senses, our sense of temperature, sense of color, and so on. The fundamental concepts are to be derived directly from these specific sensations accordingly. While it is not possible to precisely measure a temperature using our sense of temperature, just as we cannot measure force by our muscle sense, or nuances of color by our sense for color, because the acuity of our sensations is insufficient, we are forced to look for other phenomena to accomplish this purpose, phenomena which—according to our experience—are somehow related to those sensations mentioned, and have the advantage of quantitative measurement . . . for temperature this is expansion, for force it is acceleration, for color it is wave length, and so on. . . . We acknowledge that the only sturdy and unassailable starting points are the phenomena as they are provided by sensations.[165]

In a famous speech at the University of Leiden on December 12, 1908, Planck made a radical about-face, severely dismissing Mach's expositions.[166]

We are now in a position to understand Mach's (and thus also, Fechner's) judgment on neo-Kantians and mechanists. Homogeneity and additivity can no longer be viewed as necessary criteria for measuring magnitudes, as Tannery and Von Kries had claimed. While it may be the case that physicists always found it expedient to use length in measuring dimensions, nothing necessitates this. Elsas mistakes measurement for the discovery of the real cause of a phenomenon. What Mach calls a convention, Elsas thinks is an empirical hypothesis about a causal relation. For Von Kries measurement also implies agreement in terms of equality, but, like Tannery, he thinks that this type of agreement only makes sense if it ultimately refers to the 'objective' equality of spaces, times, and/or masses (or, ultimately: equal length). To this Mach would also reply: It is very practical to *substitute* the equality of magnitudes with equality of lengths (recall the thermometer example), but it is not necessary for objective measurement. The series of our temperature sensations ranging from cold to hot also includes a concept of equality, and it is factually applicable, it is actually used by us, and serves its purpose, without requiring prior knowledge of physics or reducing heat sensations to a certain length (for instance, the length of the column of mercury), thereby making the sensation 'extensive.'

Once Again: Measuring Sensation

Now, how should we view sensation measurement from Mach's standpoint? It is entirely analogous to that of measuring temperature. It rests on an *agreement* to use the intensity of stimulus as the characteristic *substituted* for sensation. We decide that all increments of stimulus that create the just noticeable difference should—*when seen as a feature of the sensation*—be equal, although they may differ physically by having different intensities. And finally, we present the number associated with the sensation as a logarithmic function of the intensity of the stimulus, because doing so is practical, but also for lack of a practical alternative.

Fechner saw things similarly, but unfortunately he did not elaborate his ideas in sufficient detail. He hinted subtly in less familiar passages of his work, such as a reply to a critique by F. A. Müller:

> The author proves, using Kantian principles, that 'size is attributable only to objects,' such that naturally there can be no psychical sizes, and even less a standard for measuring them. But the author will certainly agree that some change is involved when a sensation of light or sound becomes weaker or stronger, and that we can speak of differing *magnitudes* of that change; if he does not agree with this, then he uses language differently from the rest of us. I believe to have shown that we can treat the magnitudes of such changes as changes of value and starting with premises from experience we can arrive at conclusions in line with experience; what should I care about Kantian definitions.[167]

The second-to-last sentence significantly summarizes the essential point with incomparable accuracy. Someone like Fechner who finds the measurement of sensation meaningful does not claim that it leads to improved or more precise knowledge of the 'true' or 'real essence' of sensation or to have identified objectivity in subjectivity, and notions along this line. He claims only that *in one particular respect* he has found a way to understand the relation of man to his world, better than he would without this form of measurement. Therefore, we can only answer the question of whether Fechner's ideas on measuring the mental make sense, by *testing* whether we can 'use' the findings for anything, i.e., whether we arrive at "conclusions that conform with experience," at which we would not, or not easily, arrive without his research.

Following Mach, it is important to stress the fundamental distinction separating psychophysics from physical measurement. Psychophysics strives to produce those agreements about measuring sensation that *most precisely* reflect the judgments that people make based on their sensations, when they are confronted with certain stimuli. This is different for measuring temperature, where it is possible that one and the same object sometimes feels cool, and sometimes warm, although the thermometer indicates the same temperature in both cases. Since the objective of physics is to know about the relations among physical things and their properties, it sometimes happens that new characteristics selected to describe the behavior of things (such as readings on the thermometer) are "not exactly parallel"—as Mach says—to the original features (heat sensations).[168] Yet the objective of psychophysics is precisely to discover a feature of the sensation *itself* (in contrast to what it sensed), such that this feature is as parallel to the original feature as possible. Here sensation is—as it were—its own characteristic.

Psychophysics fulfils primarily two purposes. One, as a contemporary advocate explains it, is "to explore how the physical world is transformed in human perception."[169] But, in analogy to physics, it also aims to investigate the relationships existing between various psychological standard magnitudes, for example, how subjective brightness of color relates to the sensation of its degree of satiation. Combined with methodical considerations, a great number of factors must be taken into consideration. This makes contemporary psychophysics a quite exciting, albeit somewhat disorderly science. The field is also particularly interesting because, following Fechner's lead, it investigates subjective dimensions primarily in terms of statistical values. Much has been accomplished, but we cannot dwell on it here.

There is yet another difficulty in dealing with the historical dispute on psychophysics. We are tempted to locate all authors denying the measurability of sensations in groups of mechanists or neo-Kantians. But that does not always reflect the truth. Prior to Fechner and Mach, "measurement" *meant* nothing other than providing an extensive magnitude or reducing something to such a value. According to that definition it is trivially true that Fechner's sensation measurement is not 'correct' measurement at all. One criterion for determining whether any particular scholar involved in the historical dispute on psychophysics opposed Fechner's measurement of sensations ('measurement' understood here in terms of present-day philosophy of science) is to examine whether that author rejected any assignment of numbers to sensations, even going *beyond the historical usage of the term 'measurement.'* If the author does so, then he is probably a mechanist or neo-Kantian, if not, then he accepts 'measurement' of sensation as we understand it today.

Ernst Mach's writings also contain the view that sensations cannot really be measured.[170] This is (as stressed by Thomas Kuhn) an instance of how word usage changes greatly throughout the course of any scientific revolution; previous designations are given new meanings, or new designations are used with old meanings. Here Mach uses the term "measurement" to mean "measurement on a rational scale," as was common practice at the time. Measurement employing a nominal or ordinal scale was not considered measurement at all. What we today consider measurement on an interval scale was then considered, as we have seen above, a special case of rational measurement. So when Mach says that sensations cannot be measured he means only that they cannot be measured using any rational scale. In the same context, however, he emphasizes that sensations can be "characterized and inventoried numerically, using psychophysical methods."[171] Fechner would have considered this in itself a measurement.

In a letter to Fechner, Wilhelm Weber writes ideas similar to those of Mach when commenting on Fechner's first outline of the measurement formula. We cannot say, for instance, that light intensities are magnitudes, but we could phrase the measurement formula in such a way that "the relations of *vires vivae* [are] proportional to the numerical variation of the numbers of degrees of intensity that are distinguished in sensation."[172] Weber thereby distinguishes measuring a magnitude from assigning ordinal numbers to a (weakly reflexive) ordered series of magnitudes. The same is true for many psychologists, like Hermann Ebbinghaus, who, filling pages, tries to prove that measuring sensations is impossible, yet wonderfully demonstrates that sensations, as we would put it today, can, as a structure of difference, be represented as intensive magnitudes on a differential scale; which is thoroughly in line with Fechner's thought.[173]

Even Moritz Schlick displayed this attitude: While we cannot, as he wrote in 1910, scientifically speak of measuring the intensity of sensations, since it is not something to which addition or subdivision apply, we can, however, say that we "assign certain numbers to specific intensities of sensation, such that the former are clearly designated by the latter in this way." Nevertheless, "mental events" will never be subject to "exact scientific concepts," because mental phenomena cannot be interpreted mechanically, like temperature, and the numerical assignment used in measuring sensation is arbitrary.[174]

Although Ernst Mach defended Fechner's principle of standards this does not imply that he would have approved of Fechner's interpretation of the measurement formula. In fact, Mach denied the validity of Fechner's Law for *inner psychophysics* (but not for outer psychophysics). Instead, he presumed that sensation depends linearly on psychophysical activity (neural excitation directly underlying a sensation): "The final nerve excitation and the sensation, which are unalterably intertwined, cannot be anything other than proportional to one another.... The law of psychophysics is true ... for the relation of the first stimulus to the last nerve excitation to which conscious sensation is connected."[175] Here Mach propagates a physiological (he calls it "organic") interpretation of the law of psychophysics.

Ewald Hering (1834–1918), an influential physiologist and important opponent to Helmholtz and the Helmholtz School agreed with this interpretation at least as of 1875.[176] Hering had studied in Leipzig under, among others, E. H. Weber and had also heard some lectures given by Fechner. From 1865–1870 he had been at the Imperial Academy for Military Medicine in Vienna, and then received a call to the University of Prague, where he remained until 1895. Mach taught in Prague from 1867–1895, so he and Hering were colleagues for a period of all of twenty-five years. In 1895 Hering returned to Leipzig as successor to Karl Ludwig. But he did not approve of the physiological interpretation of the psychophysical law because he wanted to soften Fechner's Law and make it compatible with the mechanistic worldview; this had been the intention of the neo-Kantians and the physiologists of the Helmholtz School. Instead, Hering, along with Mach, was convinced that the assumption of the proportionality of body and mind "better harmonizes with Fechner's philosophy than his own psychophysical law."[177] Hering thought he could show that we would be unable to adequately perceive our environment and find our way about in it if the basic law of inner psychophysics were logarithmic.[178] A linear relation would much better guarantee strict parallelism between psychological and physical events, so that each mental property corresponds to exactly one physical property. Very early in his career, Mach had addressed this as a "heuristic principle of psychophysical research": "All mental details correspond to physical details."[179]

Fechner dealt at length with Mach's and Hering's objections. Complying with proportionality would have meant giving up the notion of an inner threshold. But Fechner felt that such compliance would lead to some kind of eliminative materialism or the reductive identification of mental and physical. This would omit the need for natural science to accept the mental as an autonomous field. It follows from Fechner's view that one and the same mental state can be represented by differing physical states—an entirely natural view for a functionalist. In contrast, Hering and Mach advocate isomorphism between the two realms. One specific mental state or process can be realized by only one specific physical state.[180]

Here we cannot pursue in detail just how Fechner, Mach, and Hering discussed this problem. At present, Fechner's theory of measurement is of greater interest.

Our investigation of how Fechner was received in Austria-Hungary would not be complete without mentioning the criticism voiced by the philosopher Franz Brentano (1838–1917). In 1874, the year he received a call to Vienna, Brentano published the first volume of *Psychology from an Empirical Standpoint*

[Psychologie vom empirischen Standpunkt], in which he also criticized Fechner's attempt to measure intensities of sensation. He basically reproaches Fechner for confusing the measurement of psychological phenomena with physical measurement. "Color is not sight, sounds are not hearing, heat is not the sensation of heat."[181] He also raised the objection that while it is true that we can claim that just noticeable differences of sensation are also equally noticeable, it does not necessarily follow from this that they are equal in size. Fechner's Law, therefore, does not deal with the link between stimulus and sensation, but between the awareness of sensations and their intensities. Brentano's reaction, then, is, similar to Wundt's, a plea for interpreting Fechner's Law psychologically. In 1877 Fechner replied to Brentano's objections and showed how the philosopher's views terminate in the hypothesis of relation that Plateau had already suggested in 1873.[182]

6.6 Measurement Theory and the Day View

At the beginning of this chapter we suggested that Fechner's day view is profoundly involved in his theory of measurement. We can now elucidate that claim. Fechner's (and Mach's) theory of measurement is based on the conviction that it is not arbitrary and that it does make good scientific sense to discern the equality of sensations, whether we evaluate just noticeable differences of sensation, sensation of heat, or the sensation that the length of something to be measured equals the length of a certain standard. Evaluating the equality of subjective magnitudes is therefore not only scientifically meaningful, it is also necessary. Even the strongest means of neo-Kantian "objectification" cannot abolish sensation.[183]

While it does make good sense in science to substitute mechanical characteristics for sensations, this good sense is relative to the overall goal of natural science. It is important for physical measurement that we reduce inter- and intra-individual variations in judgment in a way that enhances communication and improves our ability to cope with the world. But even if we achieve the goal of reducing all physical dimensions to values of length, physics would still depend on sensation because: 1) nothing can be substituted that 'was not already there from the start.' We can only replace those sensations which we actually have; and 2) at some point we must read a scale (or, since today this is done by instruments and passed on to computers for evaluation: we must devise and correctly adjust a reading instrument). Mach put this nicely (and simultaneously gave a psychological explanation for Von Kries's reasoning):

> The introduction of generally comparable, so-called absolute standards of measurement in physics, the reduction of all physical measurement to the units of centimeters, grams, seconds (length, mass, time) has a peculiar consequence. We already have a tendency to think that what is physically comprehensible and measurable, what can be collectively established . . . is "objective" and "real" compared to subjective sensations. This notion is apparently motivated psychologically (but not logically) by the concept of absolute standards of measurement. It would seem that what we commonly call sensations are entirely superfluous in physics. But if we look closer, our system of units of measurement can be simplified further yet. For the number expressing the value of mass is a ratio of acceleration, and measuring time relies on measuring angles and arc lengths. Thus, the *measurement of length* is the foundation for *all* measurement. Yet, we do not measure *pure* space, we need a physical standard, and this re-introduces the whole system of diverse sensations. Only intuitions provided by the senses can lead to the development of equations in physics, and those equations are the interpretation of the ideas. And although an equation contains only numbers of spatial measurement, these represent merely a *principle for ordering*, one that instructs us as to which parts of the series of *sensory elements* we must use to piece together our *image of the world*.[184]

This shows that natural science is not only compatible with the day view; it actually presupposes and necessitates the day view.

The last difference between the Mach-Fechner concept of measurement and the mechanist/neo-Kantian concept is of a fundamental nature, having more to do with the *history* of natural science than with science itself. Somewhat simplified, mechanists and neo-Kantians saw the historical development of the natural sciences as comprised of two periods: First, the prehistory of scientific measurement, when objectification of physical dimensions was still unknown (or was known only for a few spatial, temporal, or weight values). And second, the period of insight into the art of objectification, which has been gradually and continually perfected.[185]

Ernst Mach contrasted this with an entirely different picture, one that allowed an interpretation of the history of science to incorporate varying human purposes and goals in the course of human history and not to pretend, as the neo-Kantians did, that "the history of philosophical ideas represents strictly logical continuity and strictly logical progress for pertinent notions."[186] In Ernst Mach's approach, man is not 'compelled' by causality, as it were from outside, to define physical dimensions, as, for instance, Elsas had thought, but instead, we are relatively free to choose how we devise our concepts. We select those definitions which we believe to be the most favorable for attaining our particular theoretical or practical objectives in terms of the means and experience at our disposal. Since not only our objectives, but also our experience changes over time, so also, will the views of natural science regarding the measurement of dimensions, change. Mach also tried to name anthropological and biological reasons for why space and time are so important for the foundations of physics: The greatest factor was that in order to survive, the human species required good practice with tools. It was important to know how to use lengths, times, and weights. Had our olfactory sense been crucial for survival, we would perhaps today have a physics of odors, from which mechanics would have been derived at a much higher level. Mach thus offers an evolutionary naturalistic explanation of the mechanist's emphasis on mechanics. From the mechanist's perspective, however, it remains a riddle why humans ever began to measure and weigh natural objects, instead of relying exclusively on their sense perceptions.

The following scheme compares both concepts of measurement.

Mechanistic Picture of the World

1. Physical measurement is the objectification of sensation: The stimulus causing a sensation is represented extensively.

2. Intensive magnitudes are measurable by reducing them to their causal extensive magnitudes (if such exist).

3. Sensations cannot be measured.

4. All of physics necessarily rests on the extensive values: Space, time, and mass.

Fechner-Mach

1. Physical measurement is an ordering activity: We discover where an object belongs in terms of a certain order of physical characteristics.

2. Intensive magnitudes are measurable, once we have determined which physical reaction is to be taken as a substitute for our own direct perceptual reaction.

3. Sensations are measurable.

4. It is expedient to reduce physical dimensions to spatial, temporal, and mass magnitudes, for which, however, any establishment of equality rests on a purely subjective evaluation. If, unlike physics, one's objective is not inter-subjectivity, one is at liberty to select magnitudes other than mechanical magnitudes as a foundation for physical magnitudes.

5. Measuring instruments objectify the qualities of sensation and thereby eliminate them from the scientific picture of the world.

5. Measuring instruments 'refine' and 'extend' sense organs. They 'sense' for us. The observer and his sensations cannot be eliminated. Instead, sensations constitute the foundations from which physics proceeds.

6. The world of physics begins with measurement and over the course of time changes only in terms of the amount accomplished.

6. Measurement depends on human objectives. Since our goals (biological, cultural, economic, and so forth) change, the structure of physics changes, also.

In light of the severe controversy and the profound change in the meaning of the concept of measurement that engulfed the discussion of psychophysics, we may say that Mach and Fechner revolutionized the concept of measurement. In 1967 Brian Ellis came to a similar evaluation, writing: "Ernst Mach was the first to make a decisive break with this outlook [in traditional measurement theory]. His analyses of quantitative physical concepts, particularly mass and temperature, were the first important advances in the logic of measurement since pre-Socratic times. They were important because they succeeded in changing men's views about the significance of the numerical assignments that are made in the course of measurement, a change in viewpoint that was necessary for the two great twentieth-century revolutions in physical theory—relativity and quantum theory."[187] If we agree with Ellis, and we have little reason not to, then we must apply his statement on Mach also to Fechner. Vindicating Fechner today, of course, no longer helps him. But it prods us to realize that significant scientific progress (not only, but also) springs from convictions in *Naturphilosophie*.

7 SELF-ORGANIZATION AND IRREVERSIBILITY

Order Originating from Chaos

THE NINETEENTH CENTURY witnessed basically two explanations for organic life. One tenet held that laws of physics and chemistry provide sufficient explanation. The other was that any adequate explanation must take recourse to some special vital force (or similar properties and entities) common to organisms. After 1850 we find fewer proponents for this latter belief, because the mechanistic worldview governed most of science. Vitalism did not resurge until the 1890s. But alongside these two major views, the nineteenth century also knew of a third option for explaining life: This tradition held that while living organisms do strictly obey laws of physics and chemistry, and while there is neither a vital force nor are there other ad hoc postulated entities that explain life and its development, nonetheless, as necessary as physics and chemistry are for this purpose, they are hardly sufficient. For the world of organisms is also subject to physical regularities of an entirely novel category: the laws of self-organization. These alone distinguish living organic systems from purely mechanical ones.

Fechner was a prominent advocate of this tradition. It has gone almost unnoticed, but Fechner also elaborated a theory of the constitution and the development of living beings founded on the phenomenon of self-organization.[1] His idea influenced scholars of the most diverse fields, like Ewald Hering in physiology, Sigmund Freud in psychoanalysis, Charles Sanders Peirce in philosophy, Ernst Mach in physics, Georg Gerland in anthropology, Friedrich Ratzel in geography and biology, and Ludwig Von Bertalanffy in systems theory.

I would like to demonstrate the significance of Fechner's theory of development for a tradition stretching from *Naturphilosophie* around 1800 right down to present-day theories of self-organization. Chapter 7 has three parts: The first briefly reviews Fechner's basic thoughts on the topic. The second deals with the *Naturphilosophie* tradition and scientific underpinnings of Fechner's idea. The third sketches the historical effects of the notion.

7.1 Life and Organic Development

Fechner elaborated his basic ideas for explaining life in *Ideas on the History of Organism Creation and Development* [Einige Ideen zur Schöpfungs und Entwickelungsgeschichte der Organismen], published in 1873. He proposed five main ideas, four of them of a scientific nature, and one nature-philosophical idea.

His *first idea* is a novel definition of life: The difference between living matter and inanimate matter reflects a difference in the mechanical states of motion in material particles.[2] In contrast to inorganic molecules, neither chemical constitution nor the aggregate state is essential for organic molecules; instead, their outstanding feature is the way the particles move. It is characteristic of organic particles that their movements are more complex and irregular than those of inorganic particles. In an inorganic molecule, individual particles move only in thermic oscillations within strict limits and fixed stable positions, without changing their positions to one another. In contrast, in viable organic molecules, inner forces cause the particles to continually and irregularly alter their position and relation to one another.

The type of motion within a system of particles is controlled by either external or internal forces, or by both. Fechner calls subjection to inner forces "spontaneity" and dependence on external forces "receptivity." In inorganic molecules, only the thermic oscillation, i.e., the oscillation around the fixed stable positions, is spontaneous; otherwise, inorganic molecules are only receptive in terms of location change. In contrast, organic molecules can spontaneously alter the arrangements of their particles.[3] Yet there is no clear demarcation between organic and inorganic types of motion. Higher-level organisms are mixed systems consisting of combinations of organic and inorganic molecules.

For Fechner, this first idea does not contradict a chemical theory of life. Special chemical combinations are mere *sufficient* conditions for the presence of motion in an organic system. Since we cannot deny that the type of motion required by life may be realized by more than one single chemical compound, we have no reason to believe that any specific chemical alliance is a *necessary* condition of life.

The *second idea* links organic and inorganic types of motion: A general principle, the "Principle of Tendency toward Stability," governs both organic and inorganic matter, encompassing and connecting "all laws of evolution [thus also Darwin's Laws]."[4]

Fechner distinguishes several forms of stability: A closed system is *absolutely stable*, if the state of the system no longer changes. *Full stability* is attributable to a system whose states reoccur periodically. And finally, *approximate stability* means that the reproduction of states is achieved only approximately.

Take, for instance, the case of a system of material particles occupying finite space that is closed or subject to constant external influence. If we add the prerequisite that particle motion is clearly determined, meaning that under the same conditions, the same effects occur, then this system, once it has reached full stability, cannot spontaneously revoke it. For if it reverts to a previously achieved state (or reverts to approximately that state), then that state will occur repeatedly, and in the same way, and the intermediate states of the system will also repeat themselves.[5]

Fechner puts it this way: Any closed system, or system subject to external influences, and occupying finite space, is governed by the *principle of the tendency toward stability*, in other words: "due to its inner forces, it gradually approximates a so-called stable state, without relapsing, meaning that it approximates a state in which its parts periodically, i.e., at equal intervals, return to the same positions and the same relations of movement regarding one another."[6]

A system's degree of stability thus increases continually; the system strives for full stability. But the principle is only true for the system as a whole. It is quite possible that the stability of the system's subparts temporarily decreases. But in the long run a closed system does, on the whole, strive for ever-increasing stability.[7]

Can reasons be found for this principle? Fechner suggests both theoretical and empirical grounds. A closed system containing two bodies governed by Newton's Laws and to which any initial velocities in any directions are attributable (except infinite movement), is stable from the onset. Movement cannot be theoretically determined for systems containing more than two bodies. But for open systems as they exist in our world, the more constant the external forces are to which such systems are subjected (or, the greater the system's degree of closedness), the more these systems tend to approximate a stable state. This general experience gives us reason to assume that closed multi-populated systems will, in the short or long run, also come close to full stability.

Fechner illustrates this using the solar system: Over the course of time the solar system has come near to a state of full stability and that stability does not decrease, "at least every astronomer *believes* [as Fechner is careful to note] in the stability of the relations within the solar system in this sense, inasmuch as the calculations that have been made to date provide no reason to doubt it."[8] Taken literally, the principle of the tendency toward stability is true only for the world as a whole, since all systems within the world are more or less subject to external influences. Since the "principle of the conservation of energy" is also true, the world as an entirety can only achieve asymptotic approximation to full stability, never absolute stability.[9]

Yet the principle of stability also has approximate validity for open systems within the world. In an effort to achieve stability, each system strives to compensate the external effects as much as possible, it struggles to maintain its approximate stability vis à vis the external world to the greatest possible extent. Life processes in organisms are ordered in certain periods and cycles and acquire a certain rhythm: "Sleeping and waking, the circulatory system, peristaltic contractions in the alimentary canal, respiratory rhythm, the more or less periodic intake of food and sexual performance are among these. . . . Organisms are made for periodic functions, they are made for stable circumstances."

Fechner knows, of course, that stability in organisms differs from that of mechanical systems; he uses the concept of dynamic equilibrium, as the previous citation continues: "Indeed, in the case of metabolism, to which organisms are subject, not always the *same* particles, but *equivalent* particles return periodically to the same position; yet nothing prohibits generalizing the concept of stability to also include this case."[10] Under normal circumstances, after completing its period of growth, an organism maintains a fairly stable state "within itself and in relation to the external world." But as an organism ages, its organic states of movement become increasingly inorganic, its parts become more fixed and rigid, movement slows down, "until finally the entire organism collapses into an inorganic state, and this would terminate all organic life, if it had not, during its lifetime, been able to split off parts that repeat the life process."[11] This does not mean that the stability of an organism waxes monotonically over time. It need only increase in the long run; it can also sometimes wane, as long as the system to which the organism belongs increases stability on the whole.

Fechner's *third idea* is another principle, the "principle of relative differentiation": The conditions under which organic creatures exist depend one on another and are complimentary. The origin and development of new kinds occurs for the greater part in order to enhance the mutual stability of organisms.[12]

By introducing this principle, Fechner, on the one hand, wants to do justice to the fact that organisms could not survive as closed systems because they can only maintain stability through a constant supply of energy from the outside and by relating to other living beings. On the other, he would like to add a new principle to Darwinian Theory.

Fechner distinguishes two ways in which systems partition themselves: One is to *split up*, the other is to *differentiate*. A system splits itself by distinguishing original parts of itself in terms of size and shape, but not in terms of the state of its inner motion. It differentiates by distinguishing parts in terms of their inner constitution. Differentiation is done in one of two ways: *Fortuitous* differentiation is caused by "the changes in organisms resulting from indeterminable and purposeless efficacious forces in nature,"[13] i.e., variation in the Darwinian sense; *relative* differentiation maintains and increases the original system's stability. Natural selection, finally, being the "principle of the struggle for survival," is for us, according to Fechner, "a corrective to relative and random differentiation, but it is demoted to a rather secondary and subordinate position."[14] Thus Fechner sees his theory as consistent with Darwin's theory of evolution and a supplement to it.

We should, according to Fechner, imagine the origin and development of organisms as follows: A parent organism can only maintain or increase stability by splitting itself into two organisms which, in terms of their state of stability, mutually complement each other and also create a more stable relationship with the rest of their environment. Obviously, according to Fechner, the development of an organism must be understood in terms of the biotope and ecology system to which it belongs. The world as a whole is on its way to becoming a balanced ecological system in which "Everything fits together as well as possible." The "struggle for survival" is important in this process because it "awards superiority to the best mutually adjusted complementary members."[15] Fechner felt that this explains the phenomenon of co-adaptation much better than Darwinism had.

Fechner's *fourth idea* inverts the usual notion of the origins of life: According to it, organisms did not develop from the movements of nonliving matter. Instead, in its early epoch the universe was in a state comparable to an organic state. Over the course of time, gradually organic and finally inorganic states developed from this original—as Fechner called it—"cosmorganic" state.[16]

Fechner concluded that from his first and second idea: If the state of motion in the inorganic world is more stable than that of the organic world, then life cannot have developed from nonliving matter, for that would have meant that the origination of life weakened the stability of the universe, instead of increasing it. We must assume then, that before life originated, the world was in a state even less stable than that which we know today as living organic matter. Fechner writes of a "chaotic, totally unstable state" of the world, which supposedly "stretches backwards indefinitely." The "particles of the cosmorganic system [were] initially disordered."[17] The primeval universe was enormously extended and highly spontaneous.

Following Kant-Laplace cosmology, Fechner sketches a scenario for the development of the universe from its original cosmorganic state. Inherent gravitation caused all particles to fall toward a common center of gravity in the universe. The closer they got to each other, the more the forces of gravity of neighboring particles distracted each one from the direct path toward the center of gravity. In the ensuing anomalous chaos, at some point in space and time, a situation arose similar to that of a former state. Governed by the principle of stability, this state was then attained repeatedly. Pockets of approximate stability developed, but these were still sufficiently capable of spontaneous changes that they could be described as an organic state of motion. Over time, however, these organic parts also increased in stability, becoming the source of inorganic matter. In its terminal phase the universe will have lost almost all spontaneity and will asymptotically approximate an "inert state," in which the only motion remaining will be that of thermal oscillation within inorganic particles.

As a corollary, Fechner concludes from the first four ideas that on the whole the world's states' capacity for "variability" has continually decreased over time, meaning that in earlier times new biological kinds developed quicker and more frequently.

Summarizing up to this point, we can say that for Fechner mechanics should be supplemented by three further hypotheses, if it is to explain the origin and development of organisms:

Postulate 1: The world contains objective randomness.

Postulate 2: The universe and all organisms in it are subject to an irreversible self-organizing process headed for approximate stability, meaning a process that gives order to a system not from external forces, but from within.

Postulate 3: The growing stability of the universe is achieved by an increase in the differentiation of its subsystems, i.e., by an increase in complexity and the degree of organization.

In the final chapter of his book, Fechner adds to the aforementioned four ideas a nature-philosophical idea for the purpose of explaining the relation between physical and psychical development in the world, or how matter relates to consciousness. The resulting "psychophysical worldview" ties in closely with his psychophysical identity view of body and soul, in which "consciousness in general" is interpreted as "the inner phenomenon of what appears externally as a physical process." As in all of his other works, Fechner here also finds that this more adequately represents the "fundamental facts of the relation between body and soul" than any other interpretation.[18]

Expressed more accurately, the *fifth idea* construes the connection between the stages of development of the physical universe and the degree of development of its psychical side. True to his psychophysical worldview, Fechner sees the psychical development of the world as strictly correlated to its physical development: "The physical tendency toward stability [is] the carrier of a psychical tendency to create and maintain just those states, from which the physical originates."[19]

By thus enabling the transfer of the law of stability onto the realm of the psychical, Fechner believes to have found a "cardinal fundament for psychology," as important for the psychical as the law of gravity is for the physical world.[20]

Toward which states does the universe strive psychologically? In terms of everything we know about consciousness through experience (of our own, and that of others), "all motives and purposes of action are by nature essentially" related to "pleasure and pain."[21] We therefore have good reason to believe that, seen psychologically, each organism and the world as a whole pursue the aim of achieving the greatest possible state of pleasure for itself. Every psychological activity of an organism that crosses the threshold of consciousness will be experienced as all the more pleasurable, the nearer it comes to stability after reaching a certain limit, and it is experienced as all the more painful, the further it removes itself therefrom. This is true for individuals as well as for parts of individuals, and for the entire world. Living, conscious beings are therefore, when seen from the outside, mechanical systems subject to the principle of the tendency toward stability and relative differentiation, while they inwardly, on their psychical side, tend toward pleasure. The physical tendency toward stability is identical to the psychical tendency toward pleasure. Over time, previous conscious activities become subconscious and leave traces within the organism, which to a certain degree can be hereditary. What is now subconscious in both the organism and in the world as a whole is "only the residuum or inheritance of previously conscious creativity and effort."[22]

If the fifth idea is correct, then we can describe the development of the world, seen from the inside, meaning from the perspective of consciousness, as a teleological process. The tendency toward stability is, to use Ernst Mayr's terminology, a "teleomatic" process, i.e., a natural process that quasi 'automatically' accomplishes an ultimate goal.[23] By attributing an inner meaning to this teleomatic process, namely the goal of attaining a certain state of consciousness, Fechner can combine "mechanical causality," as it was then called, and teleology to become a *causa finaliter efficiens*, without disturbing or reversing the causal order of the development of the world: "Just as the quivering of nerves is not a sensation itself, but the sensation that we feel belongs to the externally visible nerve-quivering as a form of self-phenomenon, so also are the physical tendencies of nature themselves not purposeful tendencies, being merely in consciousness or true for consciousness, but they can also have such [tendencies] as a way of appearing to themselves, and the law governing effects of the mind, namely something appearing to itself, can correspond to the law governing the outcome of those physical tendencies."[24] With this trick Fechner tried to liberate teleology from its unscientific odium and reconcile it with the natural science of his time.

7.2 The Philosophical and the Scientific Context

To which tradition do Fechner's five ideas belong? Almost all of them spring from Lorenz Oken's (1779–1951) writings. In *Ideas*, Fechner does not mention Oken (or any similar influence), and for good reason. Perhaps he was also no longer aware of being formed by Oken, since all of fifty years had passed since he had perused Oken's book as a young man.[25]

Oken said the world has a material (real) side and an immaterial (ideal) side. The ideal and the real are identical, differing merely in form. The shape of the ideal is pure unity, while the shape of the real is a ruinous state, diversity, disunion. God is the Whole, "the whole world lies enclosed within Him, and there is nothing but God." The development of the world is an evolving of a vague idea into a real certainty. "The creation of the world is nothing other than an act of self-awareness, the self-appearing of God." Thus, *Naturphilosophie* is about the Absolute becoming real, it is about "God eternally transubstantiating himself into the world."[26]

Oken calls God's act of becoming aware of himself the primordial act, the entelechy of God, the source of activity of all things, of all the forces in the world, and of all motion. As the Absolute, God depends on Himself only; He is free. Oken calls the movement that originates in God's primordial act 'dynamic' in contrast to mechanical motion: "Mechanical motion, lasting infinitely and caused by mere mechanical movements, by impulses, is nonsense."[27] At some point every mechanical movement must have a dynamic source.

Oken sees God's activity, i.e., the way in which the universe as a whole unfolds, as the very properties of a living organism. Life, then, did not develop later, it is an immanent property of the universe: "Life is nothing new, it is not something that entered the world after creation, it is original, an idea, a thought of God's, entelechy itself with all its consequences, with motion."[28] In the next step, Oken tries to show that the development of life and the presence and development of consciousness *within* the world can be explained by the vitality of the universe as a whole and its higher form of consciousness, or—stated less metaphysically, from the fact that the world is a self-organizing system.

Of course, for this we need a concept of organism that fits both the universe as well as the living beings in the world. According to Oken a living organism is

an individual, complete, closed, body stimulated and moved by itself . . . Self-stimulation of organic elements means life . . .

After all, a self-moving mass is organic. Whatever is inorganic cannot move itself, because each inorganic thing is only a piece of the whole. . . . The organic is destroyed as soon as it loses its ability to move; the inorganic is destroyed if movement enters it. Hence, motion is the soul, through which life rises above nonliving matter.[29]

Oken's definition is not a pseudo-explanation of life, as is, for example, hylozoism. Hylozoism explains life by deriving it from a vitality already inherent in matter. For Oken, in contrast, it is the *kind of movement* that makes a body a living organism, not a *qualitas occulta* of matter itself.[30]

Now, how can we link the self-organization of the universe to that of organisms? By viewing organisms as the outcome of subdividing the world-organism into differentiated entities. Since self-stimulation and self-movement of all organisms originated in divine activity of the universe, continued divine action manifests itself in the living beings of the world, i.e., in the plants, in animals, and in human beings. Hence, according to Oken, biology, or pneumatology, is the doctrine of "the whole within the part."

A consequence of this is that organic beings are subject to a polar process, or dual developmental tendencies pulling in opposite directions: They represent an ideal Whole and are thus self-moving, but they are also real entities as parts of a larger Whole and are thus moved by that Whole: "Every living being is a double. It is something that exists for itself and something immersed in the Absolute. Each one therefore undergoes two processes, a process of individuation that animates it, and a process of universalizing that annihilates. In the process of annihilation the finite thing seeks to become the Absolute itself, in the process of animation it seeks to manifest the diversity of the universe, and yet remain an individual."[31] The absoluteness contained within man himself also establishes man's freedom. Man is an image of the Absolute and therefore free in his action.

Oken also vividly paints his version of the development of the organic world: The sea is the primeval organism from which all organic beings originate. It is one single blubbery slimy organism. This primeval slime consists of bubbles, called slime points or "*infusoria*," which constitute all organisms. Plants and animals developed from the differentiation of this primeval organism. The development of the embryo recapitulates phylogenesis. The human being is the perfect living being and animals are only lasting phases of human ontogenesis.

Now, how does Fechner translate Oken's notions into his own? He interprets the dynamic, voluntary auto-mobility of organisms as the spontaneity and anomalousness of motion in the cosmorganic state; the "annihilation process" to which the organism is subject becomes the "tendency for stability," the "primeval slime" becomes the "cosmorganic state" of the universe in its early phase and the development of the species originating in the mutual form of the primordial organism becomes "relative differentiation." After all, for both Oken and Fechner, the development of the universe's material, physical side correlates to its (conscious or subconscious) ideal, psychical side.

However, Fechner modeled his *Ideas* on more than mere nature-philosophical notions. Obviously, he tried to re-construe Oken's thought to make it compatible with the prevailing contemporary scientific worldview. Laplace's mechanistic worldview was of central importance: Fechner reduced the difference between organic and inorganic things to the type of *movement* exhibited by the thing's individual parts, and he favored, as a cosmogony, the Kant-Laplace nebular hypothesis.[32] His concept of the development of the earth is compatible with Lyell's uniformitarianism. And last, but not least, his concept can, albeit modestly, accommodate Darwin's theory of descent.

In establishing the principle of stability Fechner abandoned abstract reasoning to make use of an explanatory approach employed by his friend, the astrophysicist Karl Friedrich Zöllner, in explaining the periodicity and distribution of sunspots.[33] Zöllner thought that sunspots are caused by local cooling of the sun's surface, occurring when the atmosphere is still and clear, when additional heat is lost to the surroundings. This cooling, in turn, disturbs the atmospheric balance and creates turbidity, which in turn causes the previously cooled area to reheat. This re-warming returns balance to the atmosphere and it becomes still and clear once more. Consequently, some areas cool off again and the whole process begins anew. Since there are no further conditions, gradually the intermittent oscillatory occurrence of sunspots occurs at constant intervals.

The wide distribution of grouped sunspots can be explained by the fact that the conditions creating a sunspot in one place also hold for an area much larger than that of a single spot. Thus it is more probable that within a certain range, other spots will occur simultaneously, more so than outside of that range. But this also means that within certain limits, sunspots mutually promote one another. This reflects "a tendency toward the coexistence of similar states"; sunspots tend to occur in groups.

Zöllner attributed broader validity to this type of explanation, making it applicable to all phenomena of the kind "that are caused by the cooperation of a larger number of individual phenomena. If the latter are of an oscillatory or intermittent nature, and if they are linked by the tendency toward coexistence of similar states, as, for instance, several clocks attached to the same board—then, in general, over time the resulting aggregate phenomenon will acquire a periodic nature."[34] Fechner also mentions Huyghen's demonstration that two pendulum clocks will assimilate their swings if there is a rigid link between them,[35] viewing this as confirmation for the notion that two independent, themselves approximately stable systems will, if combined, in turn result in one approximately stable system.

A year before Fechner published his *Ideas*, Zöllner had already expressed a thought similar to Fechner's fifth idea, namely that the physical tendency toward stability correlates to a psychical tendency toward pleasure. Zöllner imagines a law governing all motion of matter, whether "it occurs in inorganic or organic natural bodies":

> All performances of natural beings are determined by the feeling of pleasure or pain, in such a way, that the movements within a closed realm of phenomena behave as if they pursued a subconscious purpose, namely that of reducing the number of pain sensations to a minimum. . . .
>
> Obviously, in terms of psychical significance, this principle is *optimistic*, because it expresses the continuous approximation of all changes in the world toward a state in which, in accordance with the circumstances given, the universe exhibits *minimal pain*, in other words, *maximal pleasure.*[36]

That this point summarily dismisses Schopenhauer's pessimism is difficult to ignore. Zöllner's law was much discussed in philosophical circles of his time, until about 1878, when the "idealistic turn" of neo-Kantianism propagated a novel mood in treating nature-philosophical issues, bearing a sentiment opposed to naturalistic approaches.[37]

As I mentioned earlier, Fechner viewed organisms as systems in equilibrium. At that time the concept [*Fließgleichgewicht*] was already familiar under another term. Emil Du Bois-Reymond (1818–1896), for instance, viewed the difference between organisms and crystals as a difference in the equilibrium contained within each system: "What distinguishes the living from the dead, the plant and the animal in terms of its bodily functions, from crystals, is ultimately this: crystal matter is stably balanced, while living beings are subject to a flow of matter, and the matter therein is more or less in perfect dynamic balance."[38]

In using the term "dynamic balance" [*dynamischers Gleichgewicht*], Du Bois-Reymond mentioned work by Willem Smaasens on the theory of electricity. In 1846 Smaasen (1820–1850) called the state of electricity in a conductor, through which stationary current passes, "dynamic balance."[39] At one end the conductor receives only as much current as it can pass on at the other. By applying this notion to living beings, Du Bois-Reymond concluded that the forces in crystals are basically no different from those in living beings, but that both are "together incommensurable," just as a factory that at one end receives raw materials and energy and at the other puts out products cannot be compared with a simple building: "It is a mistake to view the occurrence of living beings on earth, or on any other planet, as something supernatural, as anything other than a very complex mechanical problem."[40]

Except for the last thought, Fechner could agree with this. After all, his theory rests on the insight that dynamic balance alone cannot explain organisms. Since for living beings, balance is not something determined from outside of themselves, as is the case for electrical conductors or waterfalls, we must assume that there is also a non-mechanical law (that is nevertheless not supernatural) that independently guides and maintains a dynamic equilibrium *within* the system, without violating any other mechanical law.

7.3 From Fechner to Freud and Peirce

Let us now examine some of the effects that Fechner's ideas had. Because of Fechner's influence (not only of his *Ideas* of 1873, but also of other writings), many conceptions of *Naturphilosophie* remained, as it were, in a "secularized" form, adjusted to natural science and made anonymous, in part well into the twentieth century.

Throughout Fechner's remaining lifetime, biological science took little or no note of his *Ideas*. At that time in Germany, Darwinism was approaching its peak and Ernst Haeckel (1834–1919) dominated the scene. Haeckel's work promised more tangible perspectives for empirical research in biology than could Fechner's abstract concepts. It was not obvious how Fechner's notions could blend with biological and physiological research; Fechner had not suggested any new problems to be solved in practice. Simultaneously, Haeckel's penning satisfied some of the public's thirst for a tangible and clear worldview, which, for whatever reasons, favored a naturalistic explanation of man and nature. Haeckel's notions of the animation of atoms, the "soul of the cell," spontaneous generation ("autogenesis") of the organic world from nonliving matter, the development of protoplasm—although all these ideas were no less fantastic than Fechner's— they met with much approval even outside the scientific community.

Fechner calls his mind-body theory a "synechological" theory because it explains the role of consciousness in the world as a *functional* property of certain systems *as wholes* and not, as monadical theories had done, as an irreducible property of certain smallest units. Such theories about animated monads were popular after about 1870. It was Zöllner, by the way, who reintroduced the idea of "animated atoms" prior to Haeckel.[41]

In spite of Fechner's strong opposition to any kind of monadology, several attempts were made to combine his theory with Haeckel's notion that the atom has a soul. If we need a theory of souls, then it should at least be one that closely corresponds to the atomist-mechanistic worldview and which does not, as Fechner's had done, relate to holistic system properties foreign to the mechanistic interpretation of the world. One such mediating approach was provided as early as 1875 by the geographer Georg Gerland, to whom we shall return shortly. Similar monadical modifications of Fechner's ideas, mixed with thoughts adopted from Schopenhauer, surfaced frequently around the turn of the century, for instance in the writings of the neurologist Paul Julius Möbius (1853–1907), writings of the philosopher Friedrich Paulsen (1846–1908) and even in those of one of Fechner's (and Haeckel's) most ardent followers, the author Wilhelm Bölsche (1861–1939).[42]

Fechner's intuitions were in sharp contrast to Haeckel's fundamental opinions, they were more detailed and sophisticated, and more ambiguous. His avowal of Darwinism was respected, but it was soon thought that his new theory, while it was well meant, was actually a burden for Darwinism, and ultimately directly opposed Darwin and Haeckel. The Darwinist Oscar Schmidt (1823–1886) from Strasbourg conceded in a review of Fechner's books "that in fact it cannot be proven, that at some point inorganic states became organic," but an assumption of this kind is in itself logically necessary. The cardinal point of Fechner's belief, namely, that a transition "of inorganic into non-previously existing organic matter" is impossible, is, in comparison, much less plausible.[43]

One of Fechner's anonymous followers commended him for showing that Kant-Laplace cosmogony and the doctrine of descent could be combined "to become a monistic world-interpretation even without the abiogenetic hypothesis." But unfortunately, the necessary additional hypothesis, namely that inorganic matter comes from organic matter, is not tenable.[44] The reviewer here is most likely alluding (like O. Schmidt, see above) to Haeckel's claim that to deny the *generatio aequivoca* is to question the universal validity of the law of causality, in order to make a monistic worldview possible, based only on the hypothesis of spontaneous generation.

A flurry of less pleasing reviews was also published. Alois Riehl (the realistic neo-Kantian in Graz, 1844–1924) found Fechner's "redesign" of Darwinism, to which he himself adhered, "unsuccessful," and criticized Fechner's "very whimsical metaphysics," as part of his "dream aspect of philosophy." Maximilian Perty (a zoologist in Bern, 1804–1884, influenced by *Naturphilosophie*) claimed to have expressed the same views as Fechner, but forty years earlier. Carl Snell (a mathematician and physicist in Jena, 1806–1886, and a non-Darwinian like Perty) said that in combining mechanical causality with teleology, Fechner was unnecessarily embracing the dogma of the exclusive rule of atom mechanics. And finally, Hermann Ulrici welcomed Fechner's theory of the spontaneous movements of organic molecules, claiming that Fechner thereby reinstated the "rights of the exiled vital force." He saw Fechner's writing as adverse to Darwinism.[45]

The only contemporary natural scientist to openly express agreement with Fechner's cosmorganic theory of the origin of life was one professor for physiology in Jena, William Thierry Preyer (1841–1897). In a letter dated January 2, 1874 he wrote to Fechner that he "was entirely convinced of the enormous significance [of Fechner's *Ideas*] upon first reading them and in spite of many differences in the details, [he] would welcome this writing enthusiastically as great progress, if the great thoughts were not quite so embryonic, but had been born more ordered and detailed."[46] In a lecture given in 1875 he claimed to "heartily agree" with Fechner "on one point, namely, that inorganic nature was not prior to organic nature,"[47] in other words, on the very point that was untenable for Haeckel (also in Jena at the time) and his followers.

F. A. Lange commended Fechner's book as "valuable work on issues instigated by Darwin," but he criticized Fechner's (and Zöllner's) ideas as "presently simply audacious hasty metaphysical thoughts, entirely lacking proof and explication." These are additionally superfluous, because the principle of stability can be "derived directly from the principle of the struggle for survival."[48]

Of all those who dealt with Fechner's thought, Ewald Hering, whom I mentioned in chapter 6, perhaps most seriously tried to combine it with the empirical gritty work of the physiologist, which caused intense further development and differentiation of Fechner's ideas. While Hering rarely explicitly mentions Fechner, their basic ideas are so strikingly similar that some sort of direct influence is quite likely.[49]

Fechner saw the question of life as an issue in physics, particularly even an issue in mechanics: the characteristic trait of life is spontaneous movement of organic molecules. Hering viewed the problem as a genuinely chemical issue, metabolism being what distinguishes living substance from nonliving matter. Each type of nerve substance has its own specific metabolism, its own specific energy in the sense elaborated by Johannes Müller.[50] In metabolism, energy is added by assimilation and simultaneously released by dissimilation. These processes are so interconnected that there is constant movement within the organism. Hering distinguishes allonomous from autonomous assimilation and dissimilation, depending on whether or not the living substance is influenced by external stimuli. A living substance is in "autonomous balance" if its autonomous assimilation and dissimilation are equal and neither of the two processes is greater.[51] If a stimulus affects the substance, this disturbs its autonomous balance. As a consequence of such disturbance, the substance strives "with increased energy to return to its previous state."[52]

But there are also states of forced allonomous balance, when a stimulus continuously affects a substance, compelling that substance to adapt to it. All these cases involve a "metabolic inner self-regulation within the living substance."[53]

Although Hering's ideas on the autonomy of life met with great approval in Austria-Hungary, they were defeated in the long run. The purely mechanistic interpretation of life ultimately remained dominant, as it was taught by the school of Helmholtz and Emil Du Bois-Reymond. That school denied any special metabolism for nerve substance; it saw the activity of nerves as nothing other than physical conductivity, occurring similarly in all species.

Fechner's concept of stability also has a certain affinity to the concept of *milieu intérieur* used by the French physiologist Claude Bernard (1813–1878), with whom, incidentally, Preyer had studied. Bernard distinguished the milieu surrounding an organism, a milieu that can greatly vary, from the inner milieu, which must remain constant if life is to be maintained at all. He studied the importance of constancy for sugar, water, oxygen, heat and the principles of regulation. We find the concept of inner milieu in Bernard's writings as early as 1854; but it did not become a central concept of his theory until the late 1870s, shortly before his death. In the twentieth century Bernard's theory was further developed by Walter B. Cannon (1871–1945) and L. J. Henderson (1878–1942). It was Cannon, who, in this connection, coined the term "homeostasis" in 1926.[54]

Ernst Mach said that he came upon the fundamental idea for his principle of economy during the time around 1870–1871.[55] While further developing this principle into a kind of evolutionary epistemology, Mach became acquainted with Hering's theory of balance and Fechner's principle of stability. Mach sees the essential trait of organic substance in the fact that, like a waterfall, it constantly changes its contents, and yet it seeks to uphold its "dynamic state of equilibrium" against external influences, meaning that it strives for the greatest possible stability. This interpretation solves an apparent conflict in Darwin's theory, namely the contradiction between the ability of the organism to pass on inheritable traits and its ability to adapt. Also, Boltzmann's phrasing of the second law of thermo-dynamics in terms of the theory of probability actually says nothing other than that every system left to itself returns to the most stable state.

In the *Theory of Heat*, Mach names other authors occupied with the "tendency toward stability": first Fechner, then Hering, Avenarius and Petzoldt.[56] However, Mach is wary of the terms "striving" or "tendency." That vocabulary relapses into Aristotelian physics, which unscientifically postulated bodies striving towards their natural places, and similar notions.

For Mach the principle of stability itself was important enough as a characteristic of organic substance. But what he found more important was the elucidation it enabled for the nature of "thought as an expression of organic life."[57] "I seek a similarity of form, or relationship of form between the psychical and the corresponding physical, or vice versa."[58] Following Fechner's psychophysics, as of 1863 Mach advocated the view that every stable psychical form, meaning thoughts and sensations, corresponds to physical forms, such that we can conclude one from the other.[59]

Here we probably have the key to Mach's most important application of Fechner's principle of stability: the conceptually economic biological interpretation of the historical development of empirical knowledge. According to this theory of the development of theories, the history of empirical science is part of a biological process of survival and adaptation. The more economical a law is, meaning the more it expresses the mere connection of phenomena and nothing further, the more favorable it becomes biologically, and the longer it lasts, the greater is its stability.[60]

Joseph Petzoldt (1862–1929),[61] a student of Avenarius, and enthusiastic follower and friend of Mach, was the most ardent advocate of Fechner's principle of stability. Beginning in 1890 he emphasized its importance repeatedly and tried to make it productive for all kinds of issues, particularly in philosophy and psychology.[62] He engaged in a discussion on the law of stability with another of Mach's followers, the physicist Anton Lampa from Prague. Lampa objected that the principle of stability is also valid for entropy reductions (which contradicts the second law) and it presupposes that the transition to greater stability is the rule, although according to statistical mechanics it is highly improbable. Petzoldt replied that the second law is merely a special case of the more general principle of stability and that the directedness of all processes is due to the principle of stability, being the "most general law of events and development."[63]

As we have seen above, Oken thought that two processes occur in an organism: one is individualizing and animating, the other is universalizing and annihilating. Fechner interprets one of these processes as that spontaneous, voluntary movement of living beings, which appears from the outside to be chaotic irregularity, the other is the essence of all those processes subject to the tendency toward stability, i.e., those that repeatedly strive to reinstate previous states. We find this duality of processes again, only slightly modified, in the meta-psychology of Sigmund Freud (1856–1939), particularly in *Beyond the Principle of Pleasure* [Jenseits des Lustprinzips] (1920),[64] where he introduces the distinction between life instinct and death instinct. At the onset of this writing, life instinct is still identified with the libido and death instinct is identical to ego drive. But gradually Freud clears up the point that because of its primary narcissism, the ego also follows instincts of a more libidinal nature, and for this reason we must think of the ego's instincts as more heterogeneous than we originally imagined them to be. Therefore, it is unsatisfactory to classify drives solely in terms of ego instincts or sexual instincts; instead, we must make the distinction between life instincts and death instincts, *both* of which can be at work in the ego.

Just as Fechner's psychophysics teaches us, Freud also attributed a certain quantity of endogenous excitation to the psychical apparatus. (For Fechner that quantity of excitation determined the "degree of its *vis viva*.") Variation in tension is regulated by the instincts. An instinct is "a drive within the living organism to return to a previous state."[65] Drives, therefore, do not reflect "pressure to change and develop," they are "an expression of the *conservative* nature of life."

The death instinct acts through the principle of pleasure, whose function is to "make the psychical apparatus lack excitation, or to keep the excitation constant, or minimize it as much as possible."[66] The reduction of inner stimulus tension is experienced as pleasure; an increase in excitation is experienced as pain. "As a special case," the principle of pleasure is, as Freud writes, subordinate to "*Fechner's* principle of the *tendency toward stability*," or what Freud calls the principle of constancy. Since the intention of the psychical apparatus is to eliminate excitation, Freud also calls it the nirvana principle. He also maintained that Fechner's interpretation of pleasure and pain is basically the same as that discovered in psychoanalytical study.[67]

But we should not ignore one difference between Fechner's and Freud's views: In Fechner's opinion it is not the *reduction* of excitation that leads to pleasure, as it is for Freud, it is the greater *stability* of the movements of excitation that accomplishes this. It is possible that Freud assumed that every increase in the stability of a movement is accompanied by a reduction of the quantity of excitation, so that it would not be a change of issue to consider Fechner's principle of stability also a principle of the tendency to reduce excitation. But this is no thought of Fechner's.

While Freud views it as the objective of the death instinct to return organic matter to a nonliving state, returning namely to that inorganic state from which all organic life originated, he sees the objective of the life instinct as "uniting organic life to make increasingly larger units," i.e., combining two different germ cells in a specific way. "What is lifeless, is prior to life," an organism returns to an inorganic state.[68]

Freud also notes a similarity between his classification of instincts and Hering's work, although he does not dare identify the assimilatory process in Hering's sense with the life instinct and the dissimilatory process with the death instinct.[69]

It would have been consistent if Freud had come to the same conclusion that Fechner and Oken had already argued, namely that life instincts are directed at nothing other than reproducing a previous state, the state of universal vivacity. Although he himself sees this thought as a logical conclusion of his theory, Freud won't really risk embracing it. He finds the hypothesis too fantastic, more a Platonic mythos than a scientific explanation: "Should we, following the suggestion of the poet philosopher [Plato], risk presupposing that living substance, at the time it was animated, was torn into tiny particles which, ever since, have been driven by sexual instinct to reunite? . . . Or that this is how these dispersed particles of living substance achieve multiple cellularity and ultimately transfer a dense instinct for reunion to the germ cells? This is, I believe, the point at which we must break off."[70] How strong nature-philosophical thought still was, and simultaneously, how great was the fear of being accused of *Naturphilosophie*!

Freud's cooperation with Josef Breuer (1842–1925) between 1882–1895 was decisive for the formulation of psychoanalysis. Breuer, too, had been strongly influenced by Fechner, perhaps via his teacher Hering.[71] In a letter to the author Marie von Ebner-Eschenbach he called Fechner's pantheism "the religion of the twentieth century."[72] In a lecture given in 1902, discussing the relation of teleology and Darwinism, he concludes, "the teleological argument, in its religious and deistic wording, takes the true cause of organic purposes outside of the causal relation, and thereby deprives it of all comprehension."[73] But if we take psychical events and creative reason as being causally determined, as Fechner's pantheism does, then perhaps we could find a way to solve the conflict between causality and teleology. "This [statement by Fechner that the world's creative activity is determined] and other pantheistic statements may, perhaps, be compatible with scientific assumptions."[74] In a letter to Brentano, Breuer explains that like Fechner's *Ideas*, his own lecture did not doubt Darwin's theory, but merely questioned the claim that it is a "sufficient reason" for the evolution of the species. Sharing Fechner's "belief" that there must be more to it, does not imply that one is making a scientific statement in favor of belief, but simply that one is more like a physician, who, when he cannot decide among two plausible diagnoses, hopes that the less serious illness is the case.[75] Breuer found Fechner's belief acceptable because, in his opinion, it "ignores no essential fact of our knowledge, it requires no *sacrificium intellectus*."[76]

Let us further pursue our portrayal of Fechner's effect on his contemporaries. In anthropology an attempt had been made, using the principle of stability, to explain why it is that similar forms of tools and other inventions occurred in different cultures without contact. Georg Gerland (1833–1919) taught that such artifacts were due to common accomplishments of mankind during its earliest phase, and that these had been reintroduced in later times according to Fechner's law of stability. Everything achieved after that became known via diffusion. The geographer, cultural anthropologist, and founder of biogeography Friedrich Ratzel (1844–1904) also took up this idea.[77]

We even easily find Fechner's principles in twentieth-century Gestalt psychology. In 1920 Wolfgang Köhler (1887–1967) wrote about "structure's tendency to pregnancy" and (following Max Wertheimer) "simple design's tendency to occur," "on the pregnancy of Gestalt": "In all processes that terminate in time-independent final states [stationary states], the type of distribution shifts towards a minimum of structural energy."[78] And this also holds for psychophysical gestalts. In 1925 Köhler expresses it more clearly: The dynamics of systems that are sufficiently functionally connected is subject to "the principle of stationary distribution that arises spontaneously from its inner dynamics." This principle states that due to self-regulation, a suitable system, as a whole, will "tend toward a stable state in comparison to its surroundings," "it will approximate a stationary state in light of prevailing conditions." This principle "provides the sole general conceptual fundament used by the theory of physical gestalt." Köhler also quotes Petzoldt, Avenarius, and Mach on this matter. In contrast to mechanistic distribution, stationary distribution is characterized by the fact that the states of a system mutually "determine" each other in various areas.[79]

This thought is also echoed by Wolfgang Metzger (1899–1979), when writing that "—among others—there also exist kinds of events which, when left to themselves, are capable of their own appropriate order. . . . There are—besides circumstances of externally determined order, which no one denies—also natural, inner, technical orders, that are not coerced, but voluntary."[80] In 1873 Fechner himself had surmised "that the tendency of every material system left to itself to group its particles in an orderly way and take on an ordered external shape [*Gestalt*] has to do with the principle of a tendency toward stability. Even the realm of the mental seems to be governed by this principle."[81] Alexander Herzberg, who in the 1920s was one of Berlin's leading physicians and psychologists and a zealous member of Petzoldt's *Gesellschaft für empirische Philosophie*, expressed the opinion in 1929 that "some schools in psychology that disagree on fundamental points or are even quite distant from one another" such as Freud's psychoanalysis, Alfred Adler's individual psychology, and Max Wertheimer, Kurt Lewin, and Kurt Koffkas's gestalt psychology, "all converge on a general law of psychical events, or any kind of events, namely the principle of stability" as stated by Fechner and Petzoldt.[82] He saw "the future of unified psychology" in this conformity among the various schools in psychology.

Elements of the evolutionary metaphysics of the American philosopher Charles Sanders Peirce (1839–1914) saliently harmonize with Fechner's ideas, in part right down to the details. While Peirce never straightforwardly cites Fechner's *Ideas*, we are quite certain that he (as well as his friend William James) was familiar with Fechner's writing. He was one of the first to analyze Fechner's psychophysics and, in cooperation with Joseph Jastrow, he introduced it in the U.S.[83]

Peirce elaborated his kind of evolutionary metaphysics during the 1890s in a series of articles printed in the journal *The Monist*. He supplements and modifies Fechner's conceptions: He finds new, handy, catchwords for Fechner's ideas, although at times his fairly odd neologisms also miss the mark. He also alters the emphasis to become the evaluation of three different possible theories of evolution (Darwinism, Lamarck's theory, and the theory of catastrophe). He gives Fechner's theory, which he favors, an obvious Lamarckian hue. And finally, Peirce generalizes the range of application for the principle of stability to the extent that he explains the evolution of regularity from chance by the tendency to build habits, in other words, the tendency toward stability, and sees in this even the possibility of "a natural history of laws and nature."[84] The fact that the universe will not attain stability until far in the future indicates, according to Peirce, the gradual evolution of the laws of nature themselves.

Peirce modifies Fechner's view decisively by radicalizing the psychophysical identity view to become objective idealism: "All matter is really mind."[85] This notion is totally foreign to Fechner. Peirce's turn towards extreme idealism pervades his entire manner of exposition: While Fechner strives to first lay down a well-established empirical basis for natural science, prior to erecting nature-philosophical suppositions upon it, Peirce is more inclined to allow his idealistic opinions to permeate his scientific theories themselves.

So, where do Peirce and Fechner agree? What happens to Fechner's ideas when formulated by Peirce? Just like Fechner, Peirce also thinks that spontaneous, unstable motion is the characteristic trait of life. Life-slime (or protoplasm), the focus of biological study, is highly unstable. Its typical activity can be described as "molecules flying asunder" or as "molecular disturbance." Life is pure spontaneity.[86] For Peirce contingency and spontaneity are the marks of "feeling": "Wherever chance-spontaneity is found, there, in the same proportion, feeling exists. In fact, chance is but the outward aspect of that which within itself is feeling."[87] For the principle of stability we find, in Peirce's writings, the expression of a "tendency toward a final state." But normally Fechner's concept is translated as "tendency toward habit taking" or "principle of habit." For Peirce this principle is primarily a mental law, he even calls it the psychical law *par excellence*. Processes, to which final causation is attributable, are irreversible and they "tend asymptotically toward bringing about an ultimate state of things. If teleological is too strong a word to apply to them, we might invent the word *finious*, to express their tendency toward a final state."[88] Peirce deals extensively with what he calls "Cosmogonic Philosophy"; this would be what I call Fechner's fourth idea. Like Fechner, Peirce assumes that the universe developed from a state of pure spontaneity and objectively existing, absolute chance via the gradual development of habits of ever-greater regularity, higher order, and stricter lawfulness. Wherever we find uniformity in the world, it can be explained as a tendency toward habit:

> a Cosmogonic Philosophy . . . would suppose that in the beginning—infinitely remote—there was a chaos of unpersonalized feeling, which being without connection or regularity would properly be without existence. This feeling, sporting here and there in pure arbitrariness, would have started the germ of a generalizing tendency. Its other sportings would be evanescent, but this would have a growing virtue. Thus, the tendency to habit would be started; and from this with the other principles of evolution all the regularities of the universe would be evolved. At any time, however, an element of pure chance survives and will remain until the world becomes an absolutely perfect, rational and symmetrical system, in which mind is at last crystallized in the infinitely distant future.[89]

Compare this to Fechner's notion of the Unconscious, set out in the *Zend-Avesta*:

> Innumerable things in nature, probably everything that we are aware of as constant unconscious contrivances and works of nature, can be considered the residuum of a once conscious process, which, as it were, has become fixed, or crystallized, similar to how natural science actually assumes that all that is firm was once fluid and agile and gradually stiffened. . . . In the end, it is wrong to seek the primeval source of consciousness in the unconscious. The contrary is true. Instead

of consciousness originating in the unconscious, the unconscious evolved from consciousness; first, because every initial creation of something new happens in clear consciousness, while every repetition, inasmuch as it only repeats something else, enters the unconscious, or semi-consciousness.[90]

Fechner's third idea, relative differentiation, corresponds to Peirce's doctrine of "agapasticism" (from *agapé*, love) or the "agapastic theory of evolution,"[91] sketched in an article appropriately titled "Evolutionary Love." For Peirce this type of evolution makes up the core of Lamarckism, which he plays off against Darwin's theory of "evolution by sporting" (the doctrine of "tychasticism"; from *tyché*: chance) and the theory of "evolution by mechanical necessity" (the doctrine of "anacasticism"; from *ananke*: necessity).[92]

Peirce tries to idealistically reinterpret Fechner's fifth idea, namely, that the physical evolution of the world correlates to its psychical evolution. He does this by attempting to demonstrate that the physical side of the world evolved from the psychical side. Fechner would have been able to assent to Peirce's description of the *outcome* of such evolution, but he would not have agreed with such a conception of the evolution process itself. Phrased in an identity-view manner (still conforming with Fechner), Peirce's version of the psychophysical world-view reads, for starters, as follows: "Viewing a thing from the outside, considering its relations of action and reaction with other things, it appears as matter. Viewing it from the inside, looking at its immediate character as feeling, it appears as consciousness." And he continues, even surpassing Fechner's identity view: "These two views [from within and from without] are combined when we remember that mechanical laws are nothing but acquired habits, like all the regularities of mind, including the tendency to take habits, itself . . ."[93] Thus, for Peirce even physical laws are the *result* of psychical habit. He writes explicitly that he views "physical law as derived and special, the psychical law [of habit] alone as primordial."[94] He calls this position "idealism" and proclaims it the only comprehensible theory of the universe. He contrasts it with "materialism," which reverses the relation of physical and psychical laws, and "monism" (which he also calls "neutralism"), which views physical laws as "independent" of the psychical law. From the dependence of physical laws on psychical law Peirce radically concludes that external appearances themselves are nothing other than a very particular class of inner appearances. "Matter is effete mind."[95]

On Fechner's interpretation, every psychical law can also, from an external perspective, be described as a physical law. The reason for this is, as we have seen in chapter 2, that the psychical and the physical are appearances of one and the same substance. Therefore, as Spinoza also says, a succession of psychical phenomena must correspond to a succession of physical phenomena. So the laws of change applying to inner phenomena must have the same origin and have equal bearing as the laws of external phenomena; they reflect one another. In Fechner's opinion it would be nonsense to say that physical appearances are the result of psychical appearances, as Peirce does. For Fechner there exist no psychical phenomena irrespective of any material carrier possessing them. So we cannot say that the physical evolved in dependence on the psychical.

Peirce also gives a logical reason for rejecting Fechner's "neutralism." If we can explain all the uniformity in the world with the general principles of evolution, "chance" and "habit," then we must be consistent and view the physical side of the world also as a product of evolution: "Neutralism is sufficiently condemned by the logical maxim known as Ockham's razor, i.e., that not more independent elements are to be supposed than necessary. By placing the inward and outward aspects of substance on a par, it seems to render both primordial."[96] While on Peirce's view the sameness of origin was merely apparent, Fechner considered it to be real.

Fechner provided essentially two reasons for why the organic world can only be adequately explained by assuming laws of tendency that satisfy the principle of stability. First, it offers the only plausible explanation for the fact that the speed of phylogenetic development, and with it the variability of organisms, decreased over time and continues to decrease. And second, only the principle of stability renders it plausible that species adapt at all. Lacking a principle of stability it would be very unlikely that the "contrivances" "capable of sustenance and reproduction" are the ones to survive.[97] An argument like this was not brought forth again until 1962, then by Herbert Simon. Simon showed that complex systems evolve only if intermediate forms of development remain stable to a certain extent and do not disintegrate into their composite parts after every unsuccessful trial at adaptation. It follows that hierarchically organized complex systems exhibiting stable intermediate forms will develop much more quickly than nonhierarchical, non-modularly designed systems of the same size.[98]

7.4 Self-Organization Today

In a study on the genesis and development of concepts of self-organization, Krohn, Küppers, and Paslack write that two characteristics distinguish modern concepts of self-organization from older concepts: on the one hand, systems are open for a flow of material and energy, and on the other, they are operationally closed, or structurally determined.[99] These authors date the rise of new concepts from the 1940s to 1960s. Compared to the tradition I have portrayed above, we can say that, naturally, the vocabulary used was different and the particularly interesting applications of the theory were others, but actually Fechner and his successors had, in principle, already fully developed the basic modern conception of self-organization (in the sense meant by Krohn et al). As we have seen, it was already well known that organisms are open systems in dynamic equilibrium. And what today is called the operational closedness of systems is clearly already present in Fechner's concept of stability. The system creates its own structure from within, without external coercion. It is crucial that the development of that structure not be explained in terms of a mysterious efficacious vital force, supposedly present only in organisms, but instead, that it is explained in terms of a naturally given and empirically testable tendency. In addition, Fechner and some of his successors condensed their theoretical concepts to the point of being structured models enabling the formulation of general and nomological statements that could have served as the heart of new programs of research. (This fact severs the Fechner tradition from that of Oken, which partially has the concepts, but no clearly phrased principles at its disposal.)

Contemporary advocates of the theory of self-organization have not been blind to the fact that their notions are related to Fechner's ideas and to *Naturphilosophie*. Ilya Prigogine and P. Glansdorff see their own ideas and those of Manfred Eigen anticipated in Fechner's writing of 1873:

> A ce point de vue, il est intéressant de signaler qu'en 1873, Fechner apercevait déjà dans la notion de stabilité le principe fondamental responsable du passage du 'non-organique' à l' 'organique.' Bien que la notion de stabilité utilisée par l'auteur comportait plusieurs équivoques, les idées qu'il a défendues restent intéressantes á rappeler, aujourd'hui. On peut leur reconnaître en effet, un caractère d'anticipation à l'histoire d'un règne naturel considéré comme une séquence d'instabilités, établissant une succession de niveaux différents d'organisation. La même conception, présentée sous une forme plus correcte et plus moderne, conduit directement à notre interprétation actuelle du sujet, ainsi d'ailleurs qu'à celle qu'on retrouve à la base des travaux contemporains de Eigen.[100]

In short: The modern history of the concept of self-organization began in 1873.

INDETERMINISM

From Freedom to
the Laws of Chance

8.1 Fechner's Indeterminism

WE SAW IN the previous chapter that according to Fechner the universe evolved from primordial, irregular, spontaneous chaos and continues to progress toward the greatest order. We also saw that science, if it is to explain any kind of evolutionary process in the world, must work with the fact that there is objective anomalousness. We also saw, a few chapters back, that Fechner views animated beings as something that can cause novel, non-predictable effects that are not determined by laws of nature. Now, how does this fit together: Fechner, the philosopher, advocates objective indeterminacy, while Fechner, the physicist, argues that the physical world is causally closed? His first try at addressing this dilemma can be found in two lectures given in 1849 at a public meeting of the then three year-old Royal Saxon Society of the Sciences in Leipzig, where he was vice secretary. The lectures were titled: "On the Mathematical Treatment of Organic Beings and Processes" and "On the Law of Causality."[1]

Fechner's ambition in these lectures was to show that "accurate scientific research" is compatible with an indeterministic worldview and that there are even arguments in support of the notion that the world is actually not predetermined, although we have no empirical method by which we can decide the matter. This is the first serious attempt known to explain that a certain degree of "non-determinedness" or "indetermination" is inherent in the world—claiming that the world "really depends . . . on freedom," as Fechner puts it—and that this indetermination is not simply derived "from our lack of knowledge about the conditions."[2] In elaborating this thought, Fechner stretches the concepts of "determinism" and "indeterminism" to apply to all natural events. Before Fechner, those concepts had been applied only to the doctrines of the predetermination, or freedom, of will.

In these two lectures Fechner implicitly lays out various different possible types of indetermination. Before examining the lectures more closely, let us paraphrase these four types:

Type 1: Indetermination due to a lack of determinedness (of the empirical attributes) of the scientific object itself;

Type 2: Indetermination due to violation of the principle of causality;

Type 3: Indetermination due to the continuous occurrence of novel initial conditions throughout the course of the evolution of the world (indeterminism through emergence); and

Type 4: Indetermination due to an objective, insurmountable limitation in predictability.

Fechner believes that we can dismiss the second type of indetermination from the onset on the premise that it is unscientific. Yet science can work with indetermination of the types 1, 3, and 4 because these are compatible with the principle of causality's validity, i.e., they are compatible with a presupposition of "thorough," or "inviolable lawfulness." Although Fechner dismisses type 2 as implausible, he believes that it cannot be refuted by (human) experience. Type 4 is more than just compatible with natural science. It can be shown (though not experimentally) that this kind of indetermination actually manifests itself in the world. If this were true, it would, according to Fechner, make it plausible to assume that indetermination of types 1 and 3 also rule in the world. Accepting indeterminism of type 4 would also depend on the assumption that types 1 and 3 are true.

The first lecture examines the claim that "in comparison to nonliving matter, organic circumstances can only to a lesser degree, or not at all, be subjected to mathematical treatment."[3] Fechner says that the arguments usually brought forth in support of this claim run as follows: 1) Organisms, by virtue of the efficacious mind within them, acquire "the nature of being free and undetermined." Mathematics, however, is applicable only to things governed by necessity. Therefore, what is organic cannot be treated mathematically. 2) In comparison to inorganic things, organic form, development, and motion are much too complex for accurate treatment. They behave "by nature in a manner incommensurable with mathematical definition." (This second argument can be upheld even if one rejects the nature of non-determinedness as stated in the first argument.)

In response to the first argument, Fechner does not outright deny the presence of a spiritual element in organic things, as we might expect from a physicist. He stays neutral and shows that the premises lack sufficient reason. Even assuming an element of freedom in the organic, it would not follow that mathematics are inapplicable. No one claims that organic things are *entirely* undetermined. So no matter how tiny we may suppose the part to be which *is* determined, that part would suffice for applying mathematics, it would vouchsafe a "sure attack."

If we consider the organic world indeterminate, it must always, somewhere, still exhibit an "expression of a basically regular ideal effort," because otherwise we would not call that indeterminacy freedom, we would, if anything, call it arbitrariness. Even an "ideal order" must have a bit of order "in the material or physical." Thus any presupposed spiritual principle cannot "disturb the processes of the body entirely irregularly."[4] We must expect that even an ideal order assigns "general norms" for events, giving them a "general character." However, within these fixed norms, events cannot be prevented from "freely taking on their particular shape, developing, expressing themselves, emancipating themselves from the laws of a necessarily causal process." Fechner says that the law of general norms governing all this is "as it were, boundless, also encompassing indetermination."[5]

We frequently encounter just this case when studying nonliving things. For this case, mathematics has its defined methods for dealing with non-determination. Thus, nothing refutes the notion that the laws of the organic world can be formulated mathematically, on the contrary, a great amount of evidence appears to support it. Fechner says: "regarding the application of mathematics to the organic [world] we are at no less disadvantage than when we apply it to the nonliving [world] in all those numerous cases, where we have only limited knowledge of the conditions of those events. In those cases we also encounter indeterminacy; but that indeterminacy does not overrule the use of mathematics, it merely lends it a special form, one which, in a certain way, includes the measuring and the kind of that indeterminacy itself."[6] Fechner mentions the oscillation of a string and the flapping of a banner in the wind as examples for indeterminate phenomena in the non-living realm. In both cases the size, the means of attachment, the material, and so on, sets a limit, a norm, within which a variety of motions are possible. The string produces only one pitch of a note, but its timbre is variable and changes with the slightest alteration in plucking; the pennant is tied to a ship's mast, but its individual movements in the wind are even less predictable than the movements of a person; likewise, the general equation of a cone cross-section is valid for many different models.

Fechner also does not deny the kind of indeterminacy mentioned in the second argument. He concedes, "no matter which of the more simple organic forms or movements we chose to study, we will not discover a finite and complete formula for a single one, a formula for expressing it accurately. We may, if we so desire, actually call this the incommensurability of the organic with mathematical determination."

But in Fechner's view this does not prove that the organic world cannot be dealt with mathematically at all. For "when we take a closer look, we see that this incommensurability applies equally well to nonliving forms and motion in nature. *It applies generally to all forms and motion in nature, following the same principle.*"[7]

Nevertheless, mathematical approximations still apply to all sorts of cases. The drawback is merely that an approximate description of forms and movements can never be fully accurate. Yet it is possible "to progress in approximation up to any desired degree,"[8] whether or not we think of the details of the processes as being determined. By way of example Fechner mentions the approximation of the egg-form using the curve of the fourth degree, the description of snail shells, of human skull shapes, determining human retina and lens curvature, and the spiral and quincunx position of leaves around the stems of certain plants. Even the physiognomy of the human face could be approximated to any desired degree by using formulas. But not only organic form, in principle organic development could also be described mathematically: "Thus how the chick in the egg develops its form could be represented using a special formula, and by altering the constants in the formula we could represent the development of every chick, nay, every bird. Just the *practical* execution of it is difficult."[9] But the practical difficulty is not limited to cases of organic things; we also encounter it when describing nonliving things, for instance, when describing meteorological phenomena.

The organic can therefore be mathematically defined as well or as poorly as the nonliving. The ideal of absolute precision is not achievable in inorganic science either. Still, mathematics does remain applicable, "whether or not indeterministic freedom is true, and whether or not the organic is subject to entirely different or the same laws as the nonliving."[10] The "exact scientist" can thus continue to use mathematics, "unperturbed and disregarding all philosophical qualms about the issues" of freedom and how the organic is related to the inorganic. Applying mathematics on the assumption that the object of investigation ultimately may remain undetermined is no different from applying mathematics based on the opposite assumption, namely that all objects are fully determined and we only arrive at statements about indeterminacy due to ignorance.

But this conviction has its price. We can no longer categorically distinguish a law of nature that *explains* phenomena (or that discovers the unknown "ultimate and inalterable causes of events . . . in their visible effects," as Helmholtz put it in 1847) from a purely mathematical description (or a "general rule" as Helmholtz would call it).[11] In the first argument the opponent is characterized as someone who believes that mathematics is only applicable to domains whose components are related by necessity. Does it follow, then, from the appropriateness of applying mathematics to living phenomena, that these phenomena are also governed by necessity? According to Kant's understanding of natural science, mathematical treatment of nature includes the necessity and objectivity of its laws. Successfully applying mathematics to things of nature therefore demonstrates that the link between cause and effect is one of necessity.

Fechner's conclusion is different: If we can subject the realm of organic things to mathematical calculation, even on the assumption that no particular case is governed by necessity, and if the same holds for the realm of nonliving things, then we cannot conclude from the mathematical nature of science that it discovers objectively necessary relations among events in the world. "The purely mathematical is a product of our imagination."[12] The idea of the absolute determinedness of objects follows from unduly projecting upon nature itself a notion of convergence of increasingly more precise values of measurement.

In the second lecture, "On the Law of Causality" Fechner wants to remove a more stubborn barrier to indeterministic thinking, namely the notion that the law of causality solely permits determinism: He starts with the question "of the supreme law governing events in nature and the mind, being the supreme judge of our conclusions inside the realm of all experience."[13] One law is more general than another if it either summarizes more special laws or if it lawfully connects more circumstances than the other law. A general law would therefore cover not only all other laws, but also all possible conditions and effects (Fechner speaks of "circumstances" yielding "results"). The most general law imaginable that fulfills this criterion states that "ubiquitously and at all times, inasmuch as the same circumstances reoccur, the same effect reoccurs; inasmuch as dissimilar circumstances occur, the same effect will not reoccur."[14] Alternatively, we might be tempted to think of a law that covers all events, but that leaves a choice between either different effects following one and the same cause, or different causes leading to the same effect. Such a law would be less generally applicable than the one suggested, because it could state nothing definite for some single cases, while the first law does provide a definition for all individual cases. Fechner therefore considers certain phenomena to be "lawfully determined" if the phenomena and their causes unequivocally reflect each other, i.e., if the causal chains do not branch over time, neither forwards, nor backwards.

Fechner advocates an almost 'chemical' concept of causality: The coincidence of two (or more) conditions leads to effects in the world. If hydrogen and chloride are brought together by an electric spark then hydrogen chloride is created, and its qualities differ from those of the original elements. Or, take a biological example: The combination of ovum and sperm causes a new creature to become.[15]

The "mere conceivability" of a supreme general law, however, does not, for Fechner, imply that it is "real or actually valid."[16] It must first be confirmed by experience. We might object that in order to test this we would have to consider the totality of all previous events as conditions. Since that totality never reoccurs, we do not have a law. Law only makes sense where application is repeatable. Fechner basically endorses this. But experience shows that when making use of laws of nature in explanation, we view the conditions more abstractly, the further removed in space and time the event to be explained lies in relation to ourselves. This can also be applied to causal law. Since neither the same conditions nor the same effects reoccur with the same precision, we can confirm laws only approximately. Experience teaches us that they do reoccur approximately. And within the framework of that approximation we can confirm a law. So the supreme law is valid "inasmuch as experience permits inference."[17]

Proof that this supreme law is valid for both the organic and the inorganic, for matter and for the psyche, is extremely important to Fechner: "Inasmuch as the physical circumstances are the same or dissimilar in the organic as they are in the inorganic, the physical results are the same or dissimilar and rules are transferable from one domain to the other, or they are not."[18]

Even if we were to assume that something like an "ideal principle—whether that be a soul, a vital principle, or a principle of purpose" is at work within organisms, this would not alter the validity of the law. The law permits effects in organisms that are not present in the nonliving—but only to the extent that the organic "exhibits special circumstances or means, which make this a condition." Therefore, the organic realm does not require a new kind of law, but different laws are valid, because different conditions hold, to which the laws apply. If we view "forces as the mediators of results," then every force must be capable of "characterization by its law." Forces are, additionally, not "independently existent, really isolated beings."

> The development of crystals in brine and the growth of the chick inside the egg happen governed by very different forces, but this does not mean that there is no law determining how—in line with the different material circumstances given in the brine and within the incubated egg—the physical results of development in both must be different; such a general law can characterize the more general force of development, of which the organic and the inorganic are only special cases.[19]

Up to this point we may view Fechner's second approach as a variation on the theory of causality typical for the nineteenth century. We may even view it as a plea for Laplacian determinism.[20] Fechner himself says that we can assume "perfectly inviolable regularity" existing throughout "the whole world of nature and the mind," as long as the supreme law holds.[21] We may take his idea, that the law of causality can be confirmed empirically, for inductivism à la John Stuart Mill; and we may view his theory of the relation of the organic to nonliving nature as—at least—an original variation on the dispute on contemporary vitalism.

After Fechner has—as it were—made a tribute to the law of causality, he engages all his effort to demonstrate that "every freedom or indetermination" is "compatible" with the law of causality, "indeed, even the greatest that we can imagine."[22] The sort of indetermination he envisions could be called 'indetermination via novelty.' In the course of ongoing evolution new initial conditions arise continuously, of a kind that had not previously existed. According to the most general law, then, these conditions also lead to results that have never yet been. The law "says that inasmuch as the same circumstances reoccur, the same results must also reoccur; but it says nothing about the kind of first result, anywhere and for any circumstances, nor about how the initial conditions themselves are determined. Here everything is left open."[23]

Each coincidence of conditions has some similarity to some former conditions, such that there is never an entirely new, disconnected effect. Yet, each co-incidence of conditions does also contain something new, something that has not previously happened, and that lends every event a facet of freedom:

> In fact, our law states that even future results can only be preconditioned and predefined in terms of former circumstances to the extent that those circumstances themselves are repetitions of yet earlier circumstances. But they are never complete repetitions. Each new space and time is accompanied afresh by novel circumstances or novel alterations of previous circumstances, which in turn demand something new, or demand altered results, which would not have been predetermined nor predeterminable by laws founded on what existed at some other time and place, or which would be so only to the extent that what is new still contains what is old; but what is old never remains totally unchanged. The world changes from place to place and evolves continuously over time. Observe, for instance, the evolution of the earth from the first phase onwards, through the various creations of organic kingdoms, right down to the creation of mankind and its continuous development. An indeterminist may link his whole doctrine of freedom to it, without at all violating the most complete lawfulness of nature and the mind.[24]

The origin of the first human was the result of certain conditions that were entirely new and would not have been explicable based on former conditions. The law that became effective during that event was not given before humans originated, but thereafter it is valid indefinitely. A similar explanation holds for the development of individual persons: A person is free to the extent that he has the capacity to produce novel and thus undetermined effects from new inner and external circumstances. But since former conditions "lay down the rules" in the course of a life, human character increasingly "determines itself," such that one reacts similarly under similar circumstances. But since circumstances are never repeated exactly,

> the liberty to define oneself as such and such in the future always remains. Continuously new circumstances always contain new reasons for indeterministic freedom.[25]
>
> Everything that is free, and thus also the most free Will that exists, therefore has its justification for ensuing from what has previously existed and for relating to it; but the direction it will take as a consequence of that justification remains undetermined and undeterminable, *to the extent that it is free*.[26]

Now, the determinist may rightly object that whatever appears to be new merely seems to be new and that in reality only an "alteration of former circumstances or a new combination of them" occurs, such that the future is merely a function of the past. The determinist could point to motion in the planetary system, for instance, which never entirely returns to its initial state, "but nevertheless . . . all motion within it is determined in all eternity by rules founded on what has previously been."[27] All would depend on the masses, distances, velocities, and directions involved. If this is true for the planetary system, then nothing prevents us from assuming that it holds for all other cases. What looks like a new circumstance merely appears to be one and is only a variation on a former circumstance.

To this Fechner replies that the determinist may be right, but that in fact he has "by far not succeeded at reducing what is new to older laws."[28] Using Imre Lakatos's jargon from the philosophy of science we could say that determinism is—at most—a 'research program,' but it is neither proven nor is it a priori necessary. As long as the research program is not finished and its options have not all been exhausted (which is humanly impossible anyway, even if determinism were true), even an indeterministic strategy of research remains an option and its underlying plausibility is at least as great as that of the opposite choice. The consequence of all this is that it is impossible to "settle" the "dispute" between determinism and indeterminism within finite time and "by experience."[29]

It is clear which side Fechner favors. He sees determinism as constituting a "degenerative problem-shift," as Lakatos put it. Fechner points to his friend's, Wilhelm Weber's, findings in physics to the effect that the force which two electric particles exert on one another does not depend only on their electrical mass (charge), their distance from one another, and their relative speed, but also on "the relative acceleration . . . , which is attributable to them in part as a result of the duration of the movement already present in them, in part as a result of the forces of other bodies effecting them."[30] But from this it follows, "The most general law governing the outcome of the effects of two particles can be altered by the presence of a third, a fourth, and so on, in a manner which becomes increasingly difficult to describe, the more the complexity increases; the context of the whole has an influence which cannot be calculated from the combination of details. [Therefore, we cannot uphold the deterministic view that] everything in nature can be reduced to elementary forces between one and another particle, and that the laws of these forces and their union provide the principle for calculating everything that occurs in nature."[31] Here Fechner proposes that we acknowledge emergent properties.

What did Fechner's argument for 'indeterminism via novelty' accomplish?

For one, it shows how natural science is (at least) compatible with the assumption that objectively undetermined, i.e., causeless, events happen. Since a law does not exist prior to the first occurrence of the conditions it states, we cannot say that undetermined events violate the law of causality. A law must exist prior to its violation! Second, Fechner showed that human freedom can be interpreted as a naturalistic property already manifested in the world of phenomena. We need not, as Kant does, take refuge in noumena, in order to harmonize the necessity of causality with the spontaneity of freedom. And third, Fechner resists the temptation indulged in by his contemporaries and their successors, namely of making a categorical distinction between nature and history. By allowing unique, undetermined events that set up rules for what follows them, the evolution of nature, too, acquires a historical character; and the physical manifestation of human freedom also 'naturalizes' history.

That would be a fine ending for the essay, but Fechner continues expounding his arguments. Even if the determinist were right, he could not avoid admitting a tiny residuum of indeterminism, because even in his world there must be a natural limit for the predictability of events. On the other hand, the indeterminist must also concede to the determinist that in his world there also remains the possibility of, in principle, unrestricted retro-calculation (retro-diction into the past) of all previous events. If we make these reciprocal concessions, we gain a "mediation, linking the deterministic and the indeterministic point of view" thereby negotiating the dispute for both sides.

Fechner begins this mediation by noting that the capacity to pre-calculate circumstances of a given degree of complexity depends on the complexity of the thing to be calculated itself. The more the world's complexity increases by the addition of more novel events, the more difficult it becomes to calculate the effects yielded by complicated circumstances, and the more this in turn demands that "the mind develops greater capacity." Since our brain, as the bearer of psychical events, is also the prerequisite for our mathematical abilities, the complexity of the brain would have to increase adequately to 'keep abreast' of the complexity of the world when predicting the future. This means that prognoses will always come up against limits; "in fact" there will always remain "indetermination regarding knowledge of the future for all time." If the evolution of the world is not finite, there will always be a future state of the world that is more complex than the previous system, such that that future state could not be predicted based upon that past system. "A worm will never be in a position to predict the behavior of an ape, an ape will never be in a position to predict the behavior of man, and man will never be able to predict God's behavior."[32] Even humans can only foresee their own development as persons to the extent that they have a "level of skill" that allows prediction. The calculation of "what will happen on a higher level of development" would require a being with higher mental capacities. Laplace's spirit (whom Fechner does not mention explicitly, but the connotation is intended) differs from man not only in that it must know all the initial conditions and the laws of world processes exactly, he must also exhibit a degree of mental development greater than that which the development of the world could ever make possible. It is obvious how the objective delimitation of predictability corresponds to infinite retro-calculation into the past. Just as it is possible to more deeply understand the motives of one's own actions in hindsight, so also can a backwards-calculating system of sufficient complexity in principle recalculate all previous facts of the world's development that are less complex than the calculating system itself. Now, if we complete Fechner's thought to include that the world gains its increasing degree of complexity only *through indetermination* (i.e., that the occurrence of novel events is a necessary and sufficient condition for an increase in complexity), then, in such a world, unlimited retro-calculation would remain, in principle, possible for such events that were novel when they first occurred.[33] If there is asymmetry between prediction and retro-calculation, as Fechner argues, then it is obvious that laws occurring at a later stage of the world's evolution cannot be reduced to preceding laws.

Fechner immediately addresses the obvious deterministic objection, that in this case at least a being of infinite complexity could foresee the course of the world for any chosen points in time. It would follow that our human constraints in prediction do not provide proof of real indetermination in things, but merely reflect our own finiteness and ignorance. For the determinist, therefore, the "residuum" of indeterminism is only apparently undetermined.

Fechner replies with a metaphor intended to convince us that even God cannot foresee the course of the world. Could a poet foresee the poem he will write tomorrow? Of course not, for when predicting it he would already be creating it, thus it would no longer be a prediction. And in like manner, God, as appearing Himself in the world (as the self-appearance of the world), cannot foresee a new phase of world development, "because at the very moment when he sees it for the first time, it would also exist for the first time. But what God cannot foresee, or predict, can hardly be foreseen or predicted by one of his finite creatures."

But if the course of the world is the external appearance of a divine thought, as Fechner believes according to the identity view, then the indeterminism inherent in every prediction is not illusory, but objectively "founded in the nature of things and knowledge": "Whatever cannot be predicted using all the rules of knowledge, whatever at the time of its origin cannot be adequately derived from what was prior to it, should always be viewed as something which in reality is undeterminable, and if the occurrence of these things is accompanied by a feeling of volition in man or God, then Free Will is something not determined by any prior thing."[34]

The indeterminist may also object to Fechner's "mediation" between determinism and indeterminism: If it is really possible to retrospectively calculate any chosen event from our knowledge about the present and the laws of nature, then human free will would no longer be free, it would be completely determined. If we can trace the reasons for free will "back to what is undetermined," then these reasons must have existed, making the will predetermined after all.

In this respect, Fechner doubts whether it makes sense to speak of "complete determination or predetermination," if it is actually realizable for neither finite nor infinite beings. And even if we accepted this indeterministic objection to the mediating point of view, the ethically disadvantageous consequences of determinism—being a thorn in the side of indeterminism—would be resolved.

If, in addition, we assume that "the law of divine world order . . . ubiquitously has the effect that the consequences of evil ultimately, with necessity, produce Good in a determined way," then undesirable aspects of both determinism and indeterminism vanish. Neither the mediating standpoint (which at least assumes objective predictability) nor the strict indeterministic interpretation (which holds indetermination via novelty as proven) is at conflict with the law of causality and thus also not at odds with natural science. Both views have the advantage that the natural scientist must not surrender his "belief in something that is factually indeterminable in what he calls his freedom," if he prefers to simultaneously acknowledge both causal law and the fact that natural causality is closed: "Even if, then, the assumption of incomplete effective lawfulness within the realm of events [meaning that the mind infringes on causality] . . . can never be fully disproved by experience, because experience itself has limits, it will appear improbable in terms of what we experience and unnecessary in terms of what we must demand."[35] Now, if we pause for a moment and take a closer look at the three kinds of indetermination that Fechner finds tenable (types 1, 3, and 4 on my list), we notice that he says little about how these types are related. It seems most likely that indetermination due to lack of determination and indetermination via novelty are in fact identical, because they refer to the same kind of freedom. So, how is indetermination due to limited predictability related to the other kinds of indetermination? At one point one might think that Fechner wanted to explain the growing complexity of the world by the occurrence of new initial conditions.[36] Since it is precisely the increase of complexity in the world that delimits its predictability, one could, perhaps, also use indetermination via novelty to explain limited predictability.

8.2 Excursus I: Freedom and Physiology

To which tradition do Fechner's views on freedom and indeterminism belong? His thought moved within two main contexts. One was the context given by the particular turn in late idealism introduced by post-Hegelian philosophy; the other was contemporary physiology's attitude toward late idealism. The second context involves a unique mixture of epigenetic thought and ideas in the philosophy of history prevailing at the dawn of the nineteenth century.

In the very first chapter of this book we saw that late idealism assumed a certain degree of non-pre-determinedness in events for the purpose of explaining vitality, freedom, and individuality in the world. It started with critique of Hegel's panlogicism. Fechner's indeterministic worldview looks like an attempt to make Christian Hermann Weisse's doctrine of freedom compatible with natural science, without sacrificing the order and precision of scientific methods, indeed, on the contrary: by supplementing them. In the following I would like to portray Weisse's doctrine of freedom in more detail than delivered in chapter 1. An excursion dealing with Lotze, Helmholtz, and Emil Du Bois-Reymond will show the intellectual context of Fechner's treatment of the issue of freedom in a world of scientific lawfulness and also show how late idealism influenced the physiology of the time.

Weisse's relation to Hegel is complicated. On the one hand, he acknowledges (at least early on) Hegel's logic and his method of philosophizing. But on the other he denies that Hegel's logic is exclusively valid and relevant for reality. While logic does contain all of being's principles of form, it does not contain being itself; it underlies being, but does not exhaust it. Weisse found Hegel's philosophical system "at once strongly attractive and severely repulsive. The formal truth of Hegel's philosophy, the superior excellence of its method and the desolate barrenness of its findings were all equally evident to me, and spurred me to seek a solution with all the strength at my disposal."[37] The mixture of assent and dissent in his relation to Hegel also led to Weisse's being considered in part a Hegelian, in part a decided opponent to Hegel, or a "half"- or "pseudo"-Hegelian.

On which points did he disagree with Hegel? Weisse thinks that Hegel has succumbed to a fundamental misunderstanding by confusing content and form. Concrete reality is not the yield of a formal, logically necessary evolution of an idea, as Hegel says, instead, it contains an element that transcends any formal necessity. Weisse sees the world as a "voluntary creation" and not as a kingdom of pure necessity. One could say of Hegel's philosophy "that it does probe freedom deeper than every past philosophy and represents it more completely; but nevertheless, according to it there is no freedom in reality—in the individual person, as well as in nature and in the world spirit as a whole, only beings acting with machine-like necessity."[38] Thus, in Hegel's philosophy the world seems to lie before us in "dull motionlessness," that at most permits a "sluggish, uniform gear and pendulum movement, as if the irresistible power of the abstract idea, whose definitions and categories wander about like heavily armored but tired warriors on a barren battlefield, had sucked all the life out of them and stripped them of all their fragrant flowers."[39] In an effort to revive and embellish reality, Weisse desires to return to freedom its rights "as a general attribute of natural being, in contrast to logical necessity."[40]

Thus the concept of freedom becomes pivotal for Weisse's philosophy, revolving around the question "of how freedom is related to necessity in nature and the world of the mental." The answer lies in the "central issue or point of departure for all philosophical research at the level to which Hegel raised it," although "Hegelian philosophy itself entirely lacks awareness of this central issue."[41] Freedom, Weisse says, manifests itself through the individual character of reality. "Individuality or uniqueness" cannot be entirely understood using general concepts, it is not "*preserved* in the generality of a pure thought"[42] as Hegel claims. Nature contains "a surplus of reality," an "incommensurability in terms of purely metaphysical concepts . . . , which Hegel denies it."[43]

As a consequence, Weisse greatly revalorizes individual variations and deviations: "Concrete reality" is distinguishable "from the concept of [concrete reality]" only via recognition that "the very moment of contingency common to all individual beings subsumed under a species is not external, *pejorative* negation, but is, instead, the explicit presence of that *immanent, dialectic negativity* by which natural creatures and intelligent individuals break loose of the generality of the concept and its rigid, lifeless, soulless necessity, and rework the concept to become their own reality—thereby attaining freedom. Actually, only on this view is freedom here postulated as being real and recognized for what it is, namely as the alternative, equally essential and true side of logical necessity."[44] From this Weisse also draws some conclusions about the concept of causality. While he does acknowledge the existence of laws of nature, he does not think that the individual exists out of necessity, the individual is not clearly determined: "Nature also ubiquitously displays a real, not a merely apparent *origination* of individuals occurring for the first time, as well as a *voluntary* coming forth from the lawfulness of physical events: entirely in line with our definition of freedom, namely, such that the laws of creation never include metaphysical or mathematical necessity of what really happens."[45]

Weisse clearly saw that supplementing and correcting Hegel's system by acknowledging freedom involves difficulties for the concept of knowledge. In Hegel's system knowledge occurs by grasping the aspect of the individual present in the concept of the general. But in Weisse's "system of freedom" there exists "a realm of nature, namely *volitional creation*, which as such cannot be recognized dialectically, only historically. Historical insight is related to dialectical insight as freedom is to necessity: it presupposes the same thing, just as freedom presupposes necessity, and yet it is more and it is superior."[46] If that which is individual can be understood exhaustively neither by using concepts from Hegel's metaphysics nor by using the laws of natural science, then we must ask whether historical insight is not the same as pure postulation of individuality. Weisse knew no answer to the question of what else might be required for knowledge, beyond simply registering individuality.

Hermann Lotze (1817–1881) seems to have experienced some sort of conversion to Weisse's concept of freedom.[47] As a youth he felt that free will is a delusion and we are at fault for allowing this madness to enter our thought. But by 1842 he apparently took freedom of will for granted.[48] During the same year, Lotze wrote his famous article "Life. Vitality" for Rudolph Wagner's *Handbook of Physiology* [Handbuch der Physiologie], rejecting all theories based on an irreducible vital force and developing his view of a specific kind of efficacy for freedom in the realm of empirical phenomena, attributing a central *physiological* function to it. He lays out an indeterministic model as a paradigm for physiology and pathology. By proving that this model is productive and empirically adequate he hopes, in turn, to empirically support the postulation of human freedom.

Lotze's line of thought is elaborate and requires several steps to explain. First he tries to show that ideas are at work in nature and in the behavior of organ-isms. Using scientific knowledge we can show, for instance, how within the solar system one state follows another, but we cannot use the same laws of nature to explain why the system commenced with precisely these and no other conditions, why exactly *this* particular process was realized and not any other possible process. Only the efficacy of an idea can explain this fact: "The specific first arrangement of mechanical masses and activities, from which, once it is given, everything else follows purely mechanically, is ubiquitous . . . not through the mechanism itself, but determined by the rule of an idea, which in this phenom-enon could express itself only in this one particular arrangement, and in no other."[49] But it is not science's task to busy itself with nature's ideas; here lies the border. In scientific terms, the efficacy of ideas is not an object of experience: "It is not unimportant to remember that each natural science can always only teach what can be necessarily derived from general laws under given conditions, but also that it cannot make up the existence of those conditions. Its tasks consist of making prognoses from the present to the future, and to discover the past by analyzing backwards . . . This is how astronomy calculates the positions of exis-tent stars, how each natural science calculates a portion of existent nature." The beginning of the astro-system, the origin of movements of heavenly bodies "is no longer a part of the science of *nature*, but of the science of creation, i.e., it is an object of religious and mythical, but not *exact* contemplation."[50]

Yet there is one field in science in which it is not only permissible, but un-avoidably mandatory that we contemplate the idea, namely, in Lotze's opinion, the science of animated bodies. When investigating living organisms we must not omit the "historical origin of the entire chain" of cause and effect: "Here mental intervention in the physical [world] is, continuously, an object of expe-rience and observation, and science cannot, like the others, become complete, if it fails to include this fact in its theory."[51] The living body, the organism, is a mechanism that is distinct from other mechanisms because it "not only pursues the aims predestined by the combination of its masses, it can also set itself new goals, and it is in a position to produce the means necessary for attaining those goals by making absolutely new mechanical moves." As early as in his first sci-entific work, his doctoral dissertation of 1838 in medicine, Lotze says explicitly, that this initial movement should be thought of as being fully contingent. While a mechanical system does exhibit strict necessity at all points within its limits, it has its origin in a contingent and non-necessary initial state: "a fortuito et contingente initio."[52]

According to Lotze, the soul is the means by which the organism, in conscious acts of will, can cause the non-determined new start of a mechanical movement. In having a soul, the body possesses "a principle of change ..., whose free, physically perfect lawless, irregular, contingent influence on the body itself belongs to the lawful affections of the same."[53] But the irregularity is itself no accidental property of an organism; for the organism it is a necessary function: "The living body, viewed as a mechanism, differs from all other mechanisms by the fact that it encompasses a principle of immanent disturbances, whose intensity and re-occurrence do not at all follow mathematical laws. Its irregularity is not accidental, it is part of the body's essence."[54] Now, it is obvious that a physical organism does not merely—as it were—suffer its own self-induced random disturbances, it must also react to them, in order to compensate for them. "The system must have auxiliary means for protecting itself against these disturbances. This is the main point of all general physiology."[55] By contending this, Lotze, on the one hand, in part conforms to Herbart's metaphysics, claiming that only "simple real things" (i.e., monads) exist and that their only activity consists in the self-preservation of their single and simple quality.[56] But on the other hand we notice his effort to gain a scientifically practical notion of human freedom. All of this is ultimately combined with elements from past humoral pathology: Confronted with various diseases and externally induced disturbances the body always reacts with the same type of compensation (Lotze says it reacts with a "crisis") and strives to re-attain the balance that has been disrupted.

He views bodily functions primarily in terms of their contribution to compensating repeated inner disturbances. Lotze asks "what a mechanism must be like in order to maintain balance under such conditions [of disruption]. It would take the mind of *Laplace* to answer this question broadly, to determine the total of mathematical possibilities."[57] The simplest compensating mechanism conceivable for this purpose is, according to Lotze, metabolism. Experience shows that it is also the most important one. Without disturbance we have no metabolism, without metabolism we have no life. But metabolism is not a mechanism that compensates each specific disturbance in a special way. A special compensation for each single disturbance would require a mechanism much too elaborate and complex for living bodies. Instead, metabolism is a general function of the body that works steadily in the same way, with which the body "dodges" all disturbances "in one and the same way," such that "the most specific idiosyncrasies of those disturbances vanish."[58] Metabolism is thus one of "nature's devices" for "preparing resistance to future disruption by steady progressive activity." Lotze also calls this activity "steady spontaneous change of efficacious masses."[59]

By interpreting metabolism this way, Lotze believes to have discovered the most fundamental process of life, the point of departure for easily explaining all other bodily functions: "Since we ... normally find that in the bodies of animals metabolism is used to regulate disruption, we may believe to have discovered the center of organic mechanisms, to which all other processes of animal economy are connected."[60] Thus an organism can be defined as a mechanical system that compensates, via metabolism, those irregular impulses forced upon it by the soul.

In an article titled "The Soul and Inner Life" [*Seele und Seelenleben*] in the handbook of 1846, Lotze weakened his position. He writes about three groups of phenomena that point to a soul: First, imagination, feeling, and desire, in which the phenomenon of consciousness appears; second, the unity of consciousness, and third, "the non-observable fact, nonetheless derived from observation" that whatever we call animated "voluntarily produces, as an active subject, movements, changes, deeds in general." "The peculiar free vitality" attributed thus to the soul is "not a fact of experience, but an assumption."[61] The concept of free will harbors too many difficulties to be the basis for any psychology. It is not a task of "physiological psychology" to determine a positive concept of the freedom of will. "But we must resist the presuppositions that make it [freedom] impossible, no matter what they include."[62]

Lotze emphasizes that he is unwilling to screen-off "aesthetic and ethical needs" from natural science: "I demand that the investigative mind activate itself entirely in studying even the most insignificant of objects and that it assert all its rights, not just those of theory, so that scientific findings, even if they cannot satisfy all of man's needs, at least do not hinder further satisfaction."[63] While Lotze wrote this, the so-called "School of 1847" was arising in Berlin, a biophysical school guided by Helmholtz,[64] a movement headed in exactly the direction that Lotze feared: in the direction of the total separation of science and ethics, the direction of excluding freedom from physiology, ignoring the inevitable tension that this creates between quotidian ethics and science.

In his epochal writing of 1847 "On the Conservation of Energy" Hermann Helmholtz wrote that "all effects in nature can be reduced to attracting and repulsing forces, the intensity of which depends only on the distance between the points acting upon one another. . . . The ultimate goal of theoretical natural science is therefore to discover the ultimate unchanging causes of events in nature."⁶⁵ These ultimate causes function according to "an unchanging law," "under the same external circumstances they always bring forth the same effect."

In further discourse on the subject, Helmholtz clearly demonstrates that he does find the issue of freedom, with which Fechner and Lotze are concerned, thoroughly legitimate. But he recommends excluding that issue from physiology. Physiology, namely, makes use of a concept of knowledge quite different from the concept adequate for the issue of freedom, assuming that any epistemological term is appropriate for freedom: "Now, whether all events can be reduced to such [ultimate unchanging causes], meaning whether nature must be entirely comprehensible, or whether it encompasses changes not subject to the law of necessary causality, changes that therefore belong to the realm of spontaneity, or freedom, this is not the place to make a decision on it; it is clear, anyway, that science, whose purpose is to understand nature, must proceed from the assumption that it is understandable, and it must draw conclusions and investigate according to this assumption, until, perhaps, irrefutable facts force it to recognize its constraints."⁶⁶ Physiology must therefore exclude freedom until hard evidence is available.

Helmholtz speaks similarly in a lecture "On Human Sight" (1855) which belongs to the founding documents of German neo-Kantianism, and in the first edition of *Handbook of Physiological Optics* (1867): We think of "conscious acts of volition and thought" as free; we claim decidedly that the "principle of the free will" "is independent of strict causal law." But in spite of the fact that we are confronted with "an exception" to causal law in the "case of action, which we know the best and most exactly," we must nevertheless acknowledge its validity, for this law concerns not only "real experience . . . , but also its comprehension."⁶⁷ In short, while we are free, this freedom is per definition not scientifically comprehendible. It can be postulated, but it cannot be grasped using scientific concepts. As late as 1847 Helmholtz conceded the possibility that experiences are conceivable which might lead to revising the concept of understanding; in 1855 and 1867, however, he sees acknowledgement of causality as an a priori category in Kant's sense as the irrevocable prerequisite for scientific thinking and knowledge whatsoever. This denies any validity at all—not to mention revision—to all empirical criticism.⁶⁸ Later editions of *Physiological Optics* no longer mention freedom of will. An awareness that freedom might indicate something yet incomprehensible has entirely vanished.

Another representative of the School of 1847, Emil Du Bois-Reymond was initially much more careful and thoughtful of what he said than Helmholtz had been. In the prologue to *Investigations on Animal Electricity* of 1848, which came to acquire a programmatic nature for the new trend in physiology, he no longer mentions acknowledging freedom *alongside* science. Du Bois-Reymond states unmistakably that "if our methods are sufficient, we should be able to achieve analytical mechanics for all life processes." This conviction rests on the insight "that there are no other changes in the physical world than movements." Thus "the events in organic beings must ultimately be reducible [to simple movements]. This reduction would yield analytic mechanics for those events. We see, then, that the challenge of analysis does not transcend our capacity, we see that in principle analytic mechanics would be sufficient, right down to the issue of personal freedom, which must be resolved by the talent of each individual to think abstractly."[69] In remarks made in 1885 on this passage Du Bois-Reymond writes that in his lecture "On the Limits of Understanding Nature" (1872) he had shown that this previously expressed opinion had been wrong. "Not only does the origin of movement [of the atoms] remain in the dark, but consciousness, even at the most fundamental level, cannot be explained mechanically."[70] This means that for us humans, consciousness remains a perennial riddle.

How does Fechner's position fit into this dispute on the role of freedom in physiology? Fechner's thought looks like an attempt to empirically establish Lotze's and Weisse's basic ideas as far as possible and rescue them for physical-physiological thinking, without questioning principles of physics, particularly not causality. On the one hand, along with Weisse and Lotze (and opposing Emil Du Bois-Reymond, Hegel, Kant and the mature Helmholtz), Fechner advocates acknowledging and respecting indeterminism as an empirical phenomenon, meaning that freedom and individuality can and must be seen naturalistically. On the other hand he goes along with Du Bois-Reymond and opposes Lotze in rejecting the notion that "the soul disturbs" the business of matter, and in claiming that the materialism of scientific research is legitimate and necessary.

It is true that in denying any specific life force, Lotze had destroyed the notion of a major qualitative difference between organic and inorganic nature. But he continues to embrace a distinction by suggesting that within the organic realm the soul is the point where indeterminism is at work. But now Fechner's identity view of the physical and the psychical allows destroying even this distinction. So Fechner is of one opinion with Du Bois-Reymond and Helmholtz, that there is no principal difference between the laws applying to living and dead matter. But this, in turn, paves the way for indeterminism that is not, as Lotze had suspected, limited to the realm of the organic, but is, instead, universally effective. Of course, this forces Fechner to argue that the whole world is animated.

8.3 Excursus II: Epigenesis and Philosophy of History

The second context in which we can place Fechner's indeterminism is made up of reciprocal effects existing between epigenetic thought in biology and a particular notion of the philosophy of history. The early nineteenth century witnessed many attempts at combining these two ideas. Owsei Temkin has shown that particularly the idea of ontogenetic development rendered it feasible to link those two approaches.[71]

In embryology a gradual transformation had been set off by Caspar Friedrich Wolff's *Theoria generationis* (1759), which changed the face of biology in general, moving it from the theory of pre-formation to the notion of epigenesis. Besides Ignaz Döllinger and his students Christian Pander (1794–1865) and Karl Ernst von Baer (1792–1876), Karl Friedrich Burdach (1796–1846) was also involved in this movement.

Epigenesis (the term was coined by William Harvey in 1651) defines life as the growing complexity of an initially homogeneous germ. This theory states that throughout the development of an organic being something new is created, such that the initial germ and the final product have no similarity. The doctrine of pre-formation, in contrast, taught that the egg contains all the pre-formed parts of a being, which, in the course of development, are merely unfolded or enlarged.[72]

In general, epigenesis suited contemporary German philosophy of history very well. The predominant names connected with it are Herder, Schelling, and Hegel. Combining the notion of epigenesis with philosophy of history led, as Temkin put it, to a "threefold parallelism between 1) ontogeny and the ages of man, 2) successive creation of species, and 3) the history of mankind through successive civilizations."[73] In the following let us more closely characterize Burdach's, the physiologist's, thoughts on this topic and compare them with Fechner's. Burdach, by the way, had been professionally active in Leipzig from 1793–1798 and 1799–1811. Von Baer, before he went to Döllinger, had been Burdach's student and later became his colleague and coworker.[74]

For a philosophically highly educated person like Burdach it must have seemed natural to at least consider applying the notion of epigenesis to nature as a whole and to ask whether or not the development and differentiation of the universe can be interpreted as an epigenetic process, thereby doing justice to diversity in nature and the development of numerous novel individual peculiarities of it. He must have found this type of generalization even more convincing, since he was strongly influenced by Schelling and thought of himself as an empirical Schellingian. According to Schelling, nature is not just a product (*natura naturata*) but also a subject; it is, itself, also productivity (*natura naturans*). What seems more natural than to compare the idea of biological epigenesis to the philosophical notion of productivity?

Burdach attempted this comparison in 1842. His discussion of the issue is so similar to the arguments that Fechner's uses in his "general law" that we wonder whether Fechner's thoughts were inspired by Burdach's approach.[75] In full length, the introduction to Burdach's reflections demonstrates the likeness to Fechner's scheme:

PERSONAL INDIVIDUALITY

King Salomon says that nothing new happens under the sun; in contrast Leibnitz, a royal in philosophy, claims that no leaf grows that is entirely the same as a previous leaf. In spite of this contradiction, each of them is right to a certain extent: The former, if we view the world in terms of generality and eternal laws; the latter, when we peer deeper into individual phenomena of infinite diversity. Nature, with its laws and forces, is forever the same with itself; the elements are always and everywhere the same; based on the same ground, similar phenomena come forth in every age and every region; and what excited the first human being who saw the sun will also move the last person to see the sun shine on earth. Yet what reoccurs is always only similar, never the same.

The majority of the elements allows a diversity of combinations and the coincidence of those combinations produces an infinite array of circumstances. Because the creative world force is essentially infinite, it is also infinite in being expressed: inexhaustible in its combinations, nature brings forth new things for all eternity; it tolerates no uniformity, never lets one law alone reign in blank obstinacy, but softens its rigidity by the intervening of another law.[76]

If we combine this idea of infinite diversity in nature, which never leads to the reoccurrence of exactly the same phenomena, with a slight modification of Fechner's most general law of causality, we get the basic idea of Fechner's indeterminism: New causes produce new effects; new causes cannot themselves be understood as the effects of other (past) causes. "The laws of the natural scientist are valid for what is new only inasmuch as what is new contains what is old."[77]

Burdach never tires of appealing to "the immeasurable abundance," the "unforeseeable," and the "innumerable diversity" of life. The notion that life "always eventuates in individualization"[78] certainly echoes Schelling's doctrine of freedom. Burdach takes the idea of individualization so far as to deny that the individual should be seen as the merely accidental specific difference of a general, cross-individual type: "The greatest individualization naturally belongs to mankind. Mankind is not an edition of stereotypes, reprinted in thousands and thousands of copies, instead it is the essence of mobile letters, whose various combination permits a diversity inexhaustible in eons."[79] The cause of that variation ("definitional reason for individuality") is, according to Burdach, not only the external conditions, but also the inheritance of acquired characteristics. And there is a third, additional, factor: "the power of the species": "Only the power of the *species* imparts upon an individual its whole individuality, just as it imprints its general character onto the individual. It is an infinite law that can only partially be fulfilled by single individuals: only an infinite diversity of individuals could fully manifest the [proto-] type of a species."[80] The laws in which the power of the species are manifested and which determine the "frequency," for instance, of mutations, can be discovered by collecting a large amount of data.[81] Burdach frequently mentions Adolphe Quetelet's studies, for example, when discussing the ratio of male to female newborns.[82] It is the "power of the whole," which is superior to the individual and determines the differences of sex; the species produces "proportions among the sexes that are adequate for its own concept."[83] Burdach finds the power of the species empirically provable: "[Within the realm of outer phenomena] we find above all relations that serve the same purpose in terms of the species as those which in nature the power of healing serves for the individual: Return from what is unusual and abnormal to what is normal, return from deviating tendencies toward one side or the other to the middle-of-the-road type."[84]

While Burdach did clearly advocate an interpretation of biological species shaped by the idea of populations, in this passage he also endorses essentialism. Essentialism (theory of types) views the essence of a biological species as something (an idea or substance) attributable to all members of the species. Population theory, in contrast, denies the existence of any objective, ideal type and defines the boundaries of a species as the constraints given the species by the limits of reproduction within natural populations.[85] According to Burdach, it is part of the essence of a species that all its members deviate from one another and no two are the same. But on the other hand, there is something like a species type, which apparently, as Quetelet supposes, is represented by the average of all variations. A similar discrepancy is apparent in the discussion on deformation. In Burdach's opinion no specimen can "be monstrous in itself," because there exists no ideal type by which to measure it. "Instead, all that occurs are forms that are appropriate for organic life, which are considered monstrous when they are inadequate in terms of the rest of the individual's design and belong to a different species or an inappropriate level of development."[86]

In closing this section, let us ask how Fechner's thought relates to Burdach's ideas. Burdach's strategy of beginning with the epigenesis of the world as a whole, as well as the anti-essentialist undercurrent, can both be found in Fechner's writing: Starting with the idea of the epigenesis of the world, Fechner questions the principle of the stability of substance, as it had been taught in philosophical tradition (especially in Kant's first analogy of experience). His criticism makes room for indeterminism, which does better justice to the uniqueness of the world's history and the uniqueness of individual events and agents than determinism can, because determinism presupposes a stability of substances throughout the course of the world and interprets change merely as the rearrangement of what is unchangeable.

I may note, that Lotze, too, questioned the stability of substance. He goes even as far as to agree with Heraclitus that "every idea of rigid substantiality is idle and error."[87] The reality of substances is mere appearance. Whoever thinks of substances as real "is lost in a world constructed of existing beings that are estranged from every relation and happening"—a sharp attack on Herbart's monadology. In Lotze's opinion 'substance realists' cannot explain how one substance affects another. He proposes, alternatively, "All that exists and appears to be substantial . . . is only the intersection of the averages of differing relations." Substances, then, would be nothing other than relations among accidentals!

8.4 "Collective Objects"

Fechner's contribution would still have been significant, if all that he had presented was his philosophical definition of indeterminism. We have no knowledge of any earlier attempt to advocate objective indeterminism so comprehensively, in quietude with natural science. Fechner was thus the first indeterminist. But he aimed to enhance indeterminism with mathematical statistics, and this made his ideas highly controversial. I would like to show how this notion heavily contributed to the downfall of Laplace's theory of probability and led to the rise of frequency theory in the twentieth century.[88] Nineteenth-century determinism considered indeterminism merely a result of human ignorance. Fechner's theory led to conceptualizing the theory of probability as an empirical science of (objective) phenomena of randomness in nature.

He developed a simple mathematical formalization, which, on the one hand, expresses the undefined character of individual phenomena, but on the other expresses the law-like character of its general norm, in other words: its distributions. His work in this field is epitomized in *Measuring Collectives*, which appeared ten years after his death. Fechner relies on two different traditions in mathematics: the moral statistics of the Belgian statistician and astronomer Adolphe Quetelet (1796–1874) and the error theory of the mathematical astronomers Gauss, Encke, Bessel, and Hauber.[89] Besides mathematical traditions, he also made use of statistics (at that time usually understood as political science), as it was taught by Gustav Rümelin (see below).

In his treatise *Sur l'homme* of 1835 (later re-titled *Physique sociale*), Quetelet tried to set up an analogy between Laplacian dynamics of heavenly bodies and a physics of human society.[90] Just as gravitational force, a constant cause, keeps a planet in orbit, so also is society governed by constant causes. The constancy and regularity of "moral phenomena" (crime, suicide, marriage, births, deaths . . .) demonstrates that continuous causes are also efficacious in society. And just as there are small deviations in the orbits of planets, caused by small, local disturbances (also called secondary, accidental, variable, or disturbance causes), without altering the center of gravity, so, likewise, are individual deviations in society, deviations in predispositions and behavior, reducible to small, local disturbances that disappear in the global view and balance out to become a stable average. In both cases the distribution of these disturbances can be represented by Gauss's law of error. Just as in astronomy the most probable correct value of a planet location can be found by averaging huge amounts of observational data, so, too, can we discover the constant causes of social development by averaging great amounts of data from social statistics. The individual idiosyncrasy of a single human being is entirely negligible.

We do not know when Fechner heard of Quetelet for the first time. He does not mention him in his papers from 1849, but we can well imagine that he had Quetelet's theory in mind while developing the idea of a "general" or "law-like character" for his lecture on "organic forms," an idea that can be expressed mathematically, but which includes no necessity regarding any individual case. If it wasn't Burdach who drew Quetelet to Fechner's attention, then it was probably at the latest his colleague in Leipzig, the mathematician Moritz Drobisch, who prior to Fechner had held the office of vice secretary at the Royal Saxon Society of the Sciences and, following Herbart's lead, introduced more mathematics into psychology.

Drobisch mentions Quetelet as early as 1834 in a book on Herbart's philosophy. In 1849 he wrote a review on a social statistics *mémoire*, written by Quetelet, that fit very well into the context of the debate on freedom carried on by Weisse, Fechner, Lotze, and late idealism in general. In this review Drobisch spurns the view, frequently expressed at the time, that the regularity of social phenomena proves the nonexistence of human freedom. This review was well noticed, encouraging Drobisch to expand the basic idea into a book, published in 1867.[91]

The earliest date of proof that Fechner was preoccupied with the mathematical aspect of statistics lies shortly after 1852. In *Measuring Collectives* Fechner reports that in the 1850s he had "procured from the relevant authorities the lists of ten Saxon lotteries from 1843 to and including 1852, each having from 32,000 to 34,000 numbers; these lists containing the winning numbers in the random sequence in which they were drawn."[92] Fechner used these lists as a "substitute for a probability urn . . . containing an unlimited number of equal amounts of black and white balls" (a method originally from Poisson), for the purpose of subjecting his deductions to "empirical verification."[93] In the second part of his book of 1856, where he defends himself against refutation from Schleiden, Fechner tires to answer questions such as: "Does the moon influence the weather? Does the moon influence vital processes, or the illness of humans and other earthly creatures?"[94] In this context he discusses great amounts of meteorological data gleaned from the most scattered scientific treatises. He concludes that all the scientists who had investigated the issue of the moon's influence on the weather had answered in the affirmative, without being able to state the cause of that influence. This whole study reads like an attempt to refute the opinion that small unpretentious effects, such as those brought forth by variable causes, are always negligibly small and balance out in the long run.

Fechner then used probability methods in his psychophysical publications after 1859 and later particularly in his investigations into aesthetics. His most important employment of probability theory, and one that was also known to Francis Galton,[95] is found in *Elements of Psychophysics*.[96] Fechner, in developing so-called "methods of the right and wrong cases," applies Gauss's law of error to the variations to which human judgment is subject when it is repeatedly confronted with one and the same stimulus and then evaluates that stimulus. Fechner uses the measure of precision h, that occurs in older phrasings of Gauss's law, as an expression of the sensation of difference. [Modern formalization is $h = 1/(\sigma\sqrt{2})$.]

Nowhere in *Psychophysics* does Fechner suggest that it is human freedom that makes a statistical law of judgment-variations for equal consistency of stimuli necessary. If this were his reason for using probability methods in psychophysics, then he would have to view variations in judgment not as erroneous deviations from one true value, but as voluntary individual reactions to stimuli. This conjecture is not improbable, because, after all, Fechner's *Psychophysics* went forth from his philosophical reflections. In the previous chapter we saw how Fechner viewed the organic world as a residuum of undetermined processes. In chapter 45 of *Measuring Collectives* he mentions in general that psychophysics provides "starting points" for views expressed in detail in *Zend-Avesta* and *On the Soul*: "The beginning and the first appendix to this work [*Psychophysics*] themselves evolved for the first time in context with contemplation made therein; thus it may seem natural that in the course of things we take recourse to the views expressed therein."[97] Between *Psychophysics* and *Measuring Collectives* Fechner wrote three articles dealing with the theory of error. In one of them he elaborated the concept of "central value," meaning the median, and presented a complete explanation of its properties and how to calculate it. In *Introduction to Aesthetics* of 1876 he used modern-type methods of ranking subjective judgments.[98] In spite of these publications, nothing was known of Fechner's comprehensive plan for new statistics. Wilhelm Wundt was amazed to find the manuscript for *Measuring Collectives* among the papers Fechner left behind at death: "No one knew that this work existed, neither Mrs. Fechner nor any of his friends and colleagues, although he must have borne the plan in his mind for twenty or more years and taken at least a decade to elaborate it."[99]

Let us take a closer look at *Measuring Collectives*.[100] Fechner sees the "main fruit" and the "main root" of his investigations in "the mathematical establishment and the empirical verification—reciprocally checking each other—of a generalization of Gauss's law of random deviations . . . , in which I relax the same from being restricted to symmetrical probability and the relational insignificance of bilateral deviations from the arithmetical mean, such that hitherto unknown lawful relations happen."[101] The objects that are subject to the generalized form of Gauss's law and the newly found relations are so-called collective objects. Fechner defines "collective object" as "an object that consists of an unknown number of specimens that vary randomly, held together by a concept of kind or species."[102] This is almost the very concept of 'collective' that the statistician and chancellor of the University of Tübingen, Gustav Rümelin (1815–1889) had introduced, except that Fechner added the decisive condition of objective random variation, which was foreign to Rümelin. In 1863 and 1874 Rümelin made a distinction between individual specimens that are typical for their kind and those that allow no direct conclusion about the nature of the kind as a whole. The former are compiled to make up the concept of a species, while the heterogeneous individuals, if they exhibit a common attribute, are subsumed under a "collective term." With collective terms we indicate groups of various members that only have one common characteristic (for instance, the inhabitants of a particular city), while the concept of a species only covers the individuals typical for that species (for example, when we speak of *the* human being, or *the* lion).[103] Rümelin emphasizes that to be "individual" does not imply being "undetermined" or beyond the law of causality. It is always possible to explain and reduce individual cases to constant causes. Statistics, the science of empirical characterization of human collectives by counting, is interested in the properties of the collective as a whole, not in its individuals.

The members of a collective object can, on Fechner's view, differ spatially or temporally, such that spatial and temporal collective objects can be differentiated. Since his theory is a theory of measurement, he is concerned solely with measurable attributes. Fechner deals only with properties that have continuous arguments; it apparently did not occur to him that his theory for measuring collectives must also apply to discrete arguments. Yet he did distinguish between one-dimensional and several-dimensional values (i.e., correlation among values).[104]

As it was for Rümelin, for Fechner the human being is also the most important collective object: humans of a particular gender, age, race, and so on, constitute collective objects in the narrower sense: "Anthropology, zoology, and botany deal, in general, essentially with collective objects, since they are not concerned with characteristics of individual specimens, but with what is attributable to a society of the same."[105] As an instance of a temporal collective object Fechner names the average temperature on the January 1 at the same place over a period of several years; meteorology provides many other examples. Other collective objects he considers are numbers of recruits (Quetelet's classical object of investigation), skull measurements, the weight of human organs, the size of rye stalks and seeds. We can even talk of collective objects in aesthetics, such as the size of exhibition paintings, books, and calling cards.

As we have seen above, Fechner's definition of "collective objects" has to do with "randomly varying specimens." Fechner makes it very clear that he understands "chance" as an external phenomenon existing in the real world. We sense that he was aware of the philosophical difficulties involved in such a view, but also that he tries to circumvent any justification:

> Since our concept of collective objects includes the concept of *random* variation in individual specimens, one might desire from the onset a definition of chance and an explanation of its nature. But an attempt to provide this from a philosophical point of view would hardly enhance the following investigation. We must be content to name the *factual* point of more negative than positive a nature that underlies what is to follow. By *random* variation of specimens I mean a variation that is just as independent of any *arbitrariness* in determining dimensions as it is from a natural law that governs the relations holding between dimensions. Although the one or the other may partake in defining the objects, the only real random changes are those that are independent of them."[106]

In other words: Take a collective object, measure the values of a particular attribute for individual specimens, and correct them, on the one hand in terms of error made when measuring, and on the other in terms of special external influence by laws of nature. The variations that remain are objectively of a random nature and objects of the doctrine of collective measurement.

Fechner knows that we can always still assume hidden, hitherto unknown causes of the laws-of-nature type for explaining any remaining variations. The insight he had as early as 1849, that it is impossible to decide between indeterminism and determinism on empirical grounds, now becomes a carefully stated salutary clause, leaving the option that there may be no objective chance, but without Fechner necessarily admitting that he believes in this possibility: "This does not mean to deny that from a *most general* point of view there is no chance, because under the existing laws of nature and existing conditions the size of each individual specimen can be considered as defined by necessity. But we speak of chance as long as we are unable to derive individual definition from such general regularities nor conclude them from available facts. Inasmuch as this is the case, that is where chance ends, and the application of the laws which I will present ends also, or is interrupted by it."[107] The assumption of chance may be metaphysical, but an assumption of its nonexistence is not less so. The empiricist in Fechner endorses acknowledging chance, if no empirical evidence for deterministic explanation is available.

In the obituary Wundt remarks that toward the end of his life Fechner avoided discussing philosophical issues with his colleagues, although he continued to engage himself passionately in lively debate. After having harvested so much conflict with his colleagues, scientists and philosophers alike, he changed his "tactics" without sacrificing his ambition of gaining success for his Day View. He devoted himself mainly to the task of establishing the Day View—as it were —'from the other side,' by multiplying and strengthening the empirical 'links' of natural science to the Day View by elaborating and applying scientific methods as exactly as possible.[108] The same is true for the *Theory of Measuring Collectives*, which can be read as another, albeit encoded, confirmation of Fechner's worldview.

What kinds of laws are involved in the doctrine of measuring collectives? They are different from 'the usual' laws of nature in various ways. First, they are laws of chance. The doctrine of the measurement of collectives attains a validity "for the subjection of chance under more general laws here in one domain and in one manner . . . , which hitherto had not underlain the view."[109] Fechner sees this new view as bearing the philosophically interesting part of the doctrine of measuring collectives. But the practically interesting part, its being a method for science, is much larger.

As the second special characteristic of these laws we should remember that they are about distributions: They show "how the specimens of a collective object are distributed in terms of size and number. By *distribution* we mean how the number of specimens of a given collective object changes with its size."[110] The theory of measuring collectives thus states the frequency with which the values of individual specimens of a collective object are distributed.

And finally, the laws of the theory of measuring collectives differ from other laws in that they cannot be used to explain individual cases: Since the values of members of a collective object are not subject to laws of nature, but depend on chance, "no law of chance can determine how large this or that *individual* specimen is, although [it can be determined] within which size limits a given number of the same with this or that degree of probability will remain."[111] This property of these new laws is particularly relevant for anthropology, zoology, and botany. Similar to Rümelin, Fechner also thinks that these fields do not deal with "a characteristics of individual specimens," but instead, with "what is attributable to all of them."[112]

From this it follows that the laws of the theory of measuring collectives can only be laws of probability. But it also makes a substantial difference, whether one views the laws as descriptions of nature or as a methodical means for avoiding error. Since a collective object "consists of specimens *varying randomly*, in any case the general laws of random probability—and every mathematician knows that there are such laws—apply here. In fact, the distribution ratios of collective objects are *governed* by them, while when determining dimensions in astronomy the same laws of probability are only used *secondarily* for defining the certainty of the calculated average measurements, and thus play a different and much less significant role than in the theory of measuring collective objects."[113] In astronomy and physics laws of probability are used for concluding the average as the true value from the repeated measurement of a single object, while "the specimens of a collective object, no matter how much they deviate from an arithmetical average or any other main value, . . . are equally real and true, and preferring one above another for a reason that is insignificant to them all naturally makes no sense."[114] And precisely this new application for the law of probability, namely, using it as a law of distribution for collective objects, requires that the doctrine of measuring collectives be set up as a new, independent discipline.

The purpose of the doctrine of measuring collectives should be to differentiate individual collective objects "by characteristic constants derived from their distribution ratios"[115] and to describe them as exhaustively yet frugally as possible. The first important step consists of examining whether or not a given compilation of data is actually a collective object. It is important to gather as much data as possible for an object of investigation, "so much, that the ideal laws of randomness, which strictly speaking apply only to an infinite amount [of data], can still be confirmed with an approximation sufficient for the desired degree of precision."[116] The object must also be "complete," meaning that it should not be "mutilated" by simply ignoring the specimens that lie above or below a certain size limit (today called 'outliers'). A collective object may not consist of a mélange of various objects ("disparate components"). And the combination of two collective objects does not in itself make another collective object.[117] If a set of given data exhibits more than one maximum, meaning that it has more than one "densest value," this indicates that we are dealing with a mixture of different objects.

The most important condition, however, that a data set must fulfill, in order to be considered a collective object, is its random variation. If we want to describe a group of data as a collective object, we must ascertain that individual specimens "are not related to one another by natural law dependency, resulting from laws of chance." Periodical meteorological values, for instance, are therefore not collective objects, "but the *non*-periodical [ones are], inasmuch as they can be considered random."[118] For this reason Fechner developed various simple ways for testing the homogeneity of data, for discovering whether a set of data is governed by chance (meaning whether it constitutes a collective object), or whether (and to which extent) the laws of randomness given in the distribution of values is "disturbed" by laws of nature of the more common kind. The test intends to demonstrate whether the specimens of a collective object are related by laws of nature or whether they differ due to pure randomness.

The basic idea of the test is as follows: Take a series of data for which we know that they "vary by chance," as Fechner puts it, meaning that it is irregular (or at random). Using this random series, then compare the group of data to be investigated and measure the degree of deviation and thus also the degree of dependency or independence among individual specimens. The ideal random series would actually be an urn containing innumerous balls exhibiting different numbers. Since no such urn is available to us, Fechner uses, as his own standard random series, the above-mentioned lists of drawings from Saxon lotteries, in which the numbers drawn stand in the sequence in which they were drawn. "Here, if anywhere" says Fechner, "chance plays its pure part."[119] He uses the lottery as the prototype of a collective object, whose components are successively independent of one another.

Then Fechner counts the sequences of even and odd numbers in the lottery lists and determines how often the successor of a number is similar to its predecessor by also being even or odd and how often it changes in this respect. For the Saxon lottery these numbers are approximately equal. If we determine the mean, for instance, from a large number of temporally ordered temperature values taken on one day, and count the sequences of values that lie above or below that value, and the changes, "then the number of series crucially outweighs the number of changes, proof of a dependency of successive meteorological day values over and above the laws of randomness."[120] When counting increasing or diminishing series and the change between sequences we also find a marked difference between the lottery list and the weather and thus also a second "proof, that the ascent and descent of meteorological values from day to day does not follow laws of pure randomness." In chapter 15 Fechner elaborates this idea to create a test for determining whether differences between two collective objects belonging to the same species,[121] are of a purely random nature or whether (and if affirmative, then to which extent) we are dealing with special external influences (season, gender, location, and so on).

We must always remember that the finiteness of every set of data is a source of error, if we wish to conclude properties of the corresponding collective by examining it. For this reason Fechner, following the lead of Poisson and Bernoulli, tries to answer the question of how much one should bet, in a given finite case, that the fluctuation of data is dependent on external causes or on pure chance.[122] For this purpose he divides the lottery numbers into groups of three, ten, fifty, and one hundred and studies their random behavior. He determines the absolute value of the difference between the sum of the even and the sum of the odd numbers and finds that empirically and theoretically the selections of a series exhibit the same random behavior. Later Fechner uses this basic idea when setting up a very simple test for estimating the reciprocal dependence of two statistical series, in other words, a kind of correlation index.[123]

The basis for investigating collective objects is the so-called "distribution table" stating how many specimens of a collective object one has found exhibiting the various values. Values are ordered according to size and differ from their predecessors by a constant interval corresponding to the precision of observation. Frequencies discovered in this way must be corrected, however, prior to more narrowly defining their mutual relations. The collective object must first be liberated of its "nonessential deviations." These exist because there is no distribution table available consisting of an infinite number of values and one must settle for the finite case. The nonessential deviations can be found by successively increasing the number of specimens and determining which deviations decrease with increasing size.[124] An additional correction is required because frequencies can always only be given for discrete values, due to limited precision of observation.

Now we are in the position to "derive *definitional items* or *components* of the collective object which guarantee one of the characteristics of the object and the possibility of comparing it with other objects in terms of quantity."[125] Prior to Fechner scholars had considered values to be the total number of specimens m in a distribution table, the corresponding arithmetical mean A, the deviations O, and perhaps also the extreme value E. Fechner found this list unsatisfactory and added a whole series of new elements, which cannot be traced back to those developed before his time. The most important are the median C (or the "central value," see above), for which the number of positive and negative deviations is equal, and the value of the "highest density" D, where the majority of the specimens of a collective object are to be found, thus being the most probable value of a collective object. A, C, and D are the "major values" to which all other values are related. Additional values include the "divide value" R which has an equal value sum above and below it; the "heaviest value" T, for which the product of the number of specimens of a value and the value itself is a maximum, and the "deviation heavy value" F, for which the product of the frequency of a specimen and the deviations from a major value is a maximum.

Fechner looks, as always, at the "mean deviation $\varepsilon = \Sigma \, \delta/m$, where δ designates the deviation from one of the three major values. Fechner used the jargon common in Germany, which originated with the astronomer Encke.[126] He also considers the "probable deviation," i.e., the median of standard deviation, the "square mean error" $q = \sqrt{\varepsilon} = \sqrt{\Sigma \, \delta^2/m}$, which since Pearson (1894) has been called standard deviation.[127]

The necessity of involving other major values besides A results from Fechner's conviction that asymmetrical distribution is the normal case. For symmetrical distribution all one needs is the deviations from the arithmetical mean, in order to calculate the most probable value. Besides this, in classical examples of symmetrical distributions, being error distributions, neither C nor D is really different from A.

Fechner's concept was entirely revolutionary, because it contradicted the whole statistical tradition of the nineteenth century. Both Quetelet and Francis Galton viewed normal (symmetrical) distribution as the key to the science of man. It was the work of Karl Pearson in the 1890s that robbed normal distribution of its predominant position.[128] According to Quetelet, asymmetrical distribution in the data indicated that insufficient data had been collected, so that secondary causes could not be balanced out and still hid the true value.[129] Naturally, this well suits Quetelet's essentialist attitude.

But for Fechner the asymmetry of collective objects is not part of a temporary external influence that creates misleading numbers, it is an objective natural phenomenon. It is an essential characteristic trait that distinguishes the law of randomness from all else. It is "nothing less than normal" writes Fechner, to apply Gauss's law of observation errors to collective objects: "In fact, it is, from the onset, something entirely different to deal with deviations from the arithmetical mean of the measurement caused by imprecise gauging instruments or senses and random external disturbances for a repeated measurement of one *single* object, and deviations from the arithmetical mean exhibited by *many* specimens of a collective object for reasons which lie in the nature of the object itself and its external influential factors. By no means could we say a priori, that in these deviations from the mean nature follows the law of error in observation, instead we would have to directly examine the same in terms of the collective object itself."[130] The chest sizes of recruits that comprised the first examined collective object did initially seem to confirm Gauss's law of distribution. But later it turned out that many other distributions are doubtless of an asymmetrical nature.[131]

As a general law of distribution for collective objects, Fechner develops a "two-sided" or "split Gaussian law" that treats the symmetrical case as simply a borderline case. According to this law, deviations are not calculated starting with A, as in normal distribution, but starting with D. The idea is then to deal with the left and right branch of the distribution curve separately, as if they originated in two different normal distribution curves that converge in D, but that differ in their degree of precision h. Different probabilities may result for values deviating equally far above and below D. Fechner discovers a series of lawful relations between the differing values that characterize two-sided distribution.[132] Incidentally, three years prior to the publication of *Measuring Collectives*, the biologist Hugo de Vries had the same idea. He represented "discontinuous variations" using what he called "half Galton curves."[133]

The last part of *Measuring Collectives* deals with special examples of concrete empirical distribution: Variations in numbers of recruits and students at different places and at different times, data which Fechner in part collected from bureaus in Leipzig, measurements of rye seeds and stems (he notes exactly just where and when [1863] he picked the rye in Leipzig's surroundings, the dimensions of paintings in galleries (he collected 10,558 measurements of painting sizes from twenty-two museum catalogs) and meteorological data (precipitation data from Geneva from the year 1845). He also appears to have intended to prove in an additional chapter that even classical astronomical error distributions can be asymmetrical. However, when supplementing that chapter the editor Lipps came to the conclusion that no asymmetry occurs in such series of error.

Now, if we pause and ask what Fechner's *Theory of Measuring Collectives* has to do with his philosophical ideas on indeterminism, the most probable answer seems to be that in developing the theory of measuring collectives Fechner realized the mathematics that he had already discussed as an option as early as 1849 and that suits his notion of the function of irregularity in the development of the universe. As we have seen, in "The Mathematical Treatment of Organic Designs and Processes" he envisioned mathematics that allow a general way of studying nature without implying necessity or strict determination of individual cases, such that objective indeterminacy of the individual can occur. The *Theory of Measuring Collectives* seems to be the detailed elaboration of this idea. Although the individual specimens of a collective object randomly vary, it is possible to formulate laws for the collective object as a whole.

Of course, Fechner never suggests a connection between his statistical theory and his philosophy. But this can be explained by his general hesitation to reveal his real philosophical motives in scientific contexts, and to prevent prolepsis of the empirical starting points that establish the Day View. This coyness even explains the switch in his terminology. In 1849 he spoke of "indetermination" and "indeterminism," as a philosopher does. Later he spoke of "the laws of randomness," as if that only meant 'probability theory,' like French scholars (Laplace, Poisson . . .) used the phrase *les lois du hasard* when they worked on probability theory.

The thesis that there is an underlying connection between mathematics in the *Theory of Measuring Collectives* and lectures given in 1849 is authenticated by three of Fechner's coevals, Wilhelm Wundt, the psychologist-philosopher Oswald Külpe, and the astronomer Heinrich Bruns from Leipzig. At least Wundt and Bruns had direct access to the handwritten papers that Fechner left unpublished.[134]

But it is also difficult to connect the ideas presented in the lecture "On the Law of Causality" with the *Theory of Measuring Collectives*. It would be particularly interesting to know how Fechner imagined the link between both types of indetermination (by novelty, type 3, and by objective limits to prediction, type 4) and collective objects. The most probable link conceivable seems to be this: The constant occurrence of new initial conditions leads to random variation in the attributes of a collective object, meaning that intrinsic novelty *explains* random variation. Random variation, in turn, explains the fact that there actually is an objective limit to predictability in nature. Seen empirically, this explanation is much more satisfying than the determinist argument that takes refuge to limited human knowledge in order to explain an incapacity for prediction. It is easily imaginable that Fechner preferred elaborating arguments provided by "mathematical treatment of the laws of causality," because these promised to be more productive and to have empirically relevant results. Novelty and non-predictability (indetermination types 3 and 4) cannot be proven, but we could be successful in seeking useful criteria for random indeterminacy (indetermination type 1) and finding conclusive criteria.

8.5 From Fechner to Von Mises

Fechner's contemporaries reacted to his ideas divergently. His lectures given in 1849 appear to have been totally ignored, with one significant exception. Ernst Mach seems to have been quite familiar with Fechner's writing. In a lecture given in 1872 on *The History and Root of the Axiom of Conservation of Effort*, Mach mentions Fechner's lecture "On the Law of Causality":

> Therein Fechner . . . formulated the law of causality very pointedly, stating 'that everywhere and at all times, inasmuch as the same circumstances reoccur, the same outcome reoccurs; to the extent that the same circumstances do not occur, the same outcome will not occur.' This, as Fechner says in a latter passage, 'relates whatever happens everywhere and at all times.'
>
> Now, I believe that I must add, and I have already done this elsewhere, that explicitly involving space and time in the law of causality is at least superfluous. Since we recognize what we call time and space by using other particular phenomena, spatial and temporal definitions are merely definitions based on other phenomena. When, for instance, we express the positions of heavenly bodies as functions of time, i.e., functions of the rotational angle of the earth, we have done nothing other than to determine how the positions of the heavenly bodies depend on *one another.*

Since physics then strove to represent every phenomenon as a function of another phenomenon, of space and time, the law of causality could be sufficiently characterized "if we say it presupposes the dependence of phenomena upon each other."[135]

Mach continues to reject the idea advocated by William Thomson and Clausius of the universe's death by heat, for this does not follow from the law of causality. "As soon as a certain number of phenomena is given, the rest are then also co-determined, but where the whole universe, the totality of all phenomena, is headed—if we may say that—is not predicted by the law of causality, and it cannot be discovered by any research, it is not a scientific question. This follows from the nature of the matter."[136] This whole context reads like an enhancement of Fechner's suggestions in "The Case for Atoms" of 1857. Here we find Ernst Mach's first criticism of causality, which he later considers metaphysical and would prefer to see replaced by the concept of functional dependency.

Except for Mach, no one seems to have taken notice of Fechner's lecture. It is next mentioned in 1890 in a study on the history of the law of causality done by Edmund Koenig, a pupil of Wundt. Koenig interprets Fechner's "law of causality" as an attempt to avoid a complete determination of the mind by matter, without positing violation of the law of causality. Fechner represents the course of the world as a developmental process for which laws should not be taken as strict standards, coercing individual cases to comply with their necessity, but as descriptions of actual behavior, behavior which could be determined by entirely different motives. Although this view is logically feasible, one cannot support it, because there is no proof that the objects in the world themselves newly create the law governing their action.[137]

It is easy to explain why Fechner's ideas of 1849 gained little attention. They were either read superficially as a mere repetition of typical nineteenth-century determinism, which viewed natural laws as an expression of strict and necessary causal relations, or they were rejected, as Koenig rejected them, as a perhaps interesting, but otherwise empirically nonprovable theory.

But things were different for the doctrine of measuring collectives. Within a year after publication, the astronomer from Leipzig, Heinrich Bruns (1848–1919) published an article providing a general solution to Fechner's problem of characterizing the functions of distribution. He follows up one of Bessel's, the astronomer's, ideas from 1838 that had been left unelaborated and provides an analytic sketch of an arbitrary distribution function using Hermite's polynomials as an infinite series whose individual members represent successive derivations of the Gaussian formula. This ϕ-series (or Bruns's series) plays the same part in distribution functions that the Fourier-development does in treating periodical processes. Shortly thereafter, Bruns elaborated his new method in lectures given during the winter semester of 1898–1899, and published them in 1906.[138]

The editor of *Measuring Collectives*, Gottlob Friedrich Lipps (1865–1931) discussed Fechner's theory and Bruns's solution in 1898 in Wundt's *Philosophische Studien*. In later writings he tried to develop his own theory of collective objects, based on Fechner's and Bruns's theories.[139] Both Bruns and Lipps explicitly stated the opinion implicitly contained in every page of the *Theory of Measuring Collectives*, namely that the theory of collective concepts is not a science *alongside* the theory of probability, but is equal to it. Probability calculation, as a theory of frequencies, applies to the same objects as does the theory of collective objects, namely mass phenomena. The doctrine of measuring collectives is therefore nothing other than a probability theory and probability must be defined as relative frequency. For the first time this clearly marks Fechner's theory as an alternative to the Laplacian definition of probability.[140]

However, Bruns and Lipps simultaneously modified Fechner's definition of "collective object" in a way that entirely dilutes Fechner's achievement and cuts off its link to his notion of indeterminism. Namely, as Lipps puts it, they think that "it is absolutely unnecessary to consider chance when calculating probability."[141] Bruns, again, interprets chance entirely as meaning the subjective, epistemological standpoint advocated by astronomers from Bernoulli to Laplace: If processes seem to us to be random, it is an illusion. Randomness of an event merely reflects our "lack of knowledge."

True to this manner, Bruns distinguishes "chance due to complete ignorance" from "chance due to extreme complication." This is the case when we at least qualitatively know the effective conditions, but the complication of the circumstances does not allow our further pursuance of the individual process. For instance, we know the effective mechanistic causes involved with the lottery wheel, but we cannot pursue them into detail.[142]

But nonetheless, Bruns does see the possibility of an empirical theory of chance *alongside* science based on causal laws. If probability theory mentions chance, this is due to the abstraction that it must undertake, just like every other science that studies natural processes (in contrast to artificial processes created in the laboratory). Abstraction in probability theory means that one must "abstract the apparently random events of reality completely from the causal relations which actually prevail, even though we may not understand them."[143] These merely "imagined blind random events" become the object of mathematics. So whether the axioms of probability calculation are actually "valid for reality" cannot be decided a priori, but only empirically.

In addition, the axioms of probability calculation, being abstractions, are always only approximately valid for reality. Any "conflict between the principle of causality and the existence of a truly useful theory of chance is an illusion, because the blind random events of probability calculation are nothing other than a prerequisite set up for the purpose of investigation, or, in short, a working hypothesis."[144] Besides rejecting Fechner's concept of randomness in this way, Bruns particularly emphasizes the significance of two traits of Fechner's doctrine for measuring collectives: Fechner was the first to show the necessity of elaborating the subject of collective series "beginning with unified and comprehensive standpoints," and this aspect was no less important, although Fechner's mathematical apparatus was "quite primitive." And, Fechner got abolished the notion of symmetry in distribution curves: "Even a sharp and unprejudiced thinker like Fechner needed a certain moral jolt before he could admit that we actually have no valid reason for thinking, for instance, that the curve of a table of recruits must be symmetrical, and even that such symmetry, if it did actually prevail, would be a very strange phenomenon."[145] Bruns and Lipps also took notice of Karl Pearson's biometrical methods from the early 1890s, which mark the commencement of our present-day statistics. They compared Pearson's para-metrical method, based on five types of curves, with Fechner's method and de-cided against Pearson: While Pearson's approach did exhibit "greater flexibility" than that of Bruns, Bruns's series allows a more general treatment of the prob-lem. In 1905 Peason replied harshly regarding Fechner's method.[146] It is likely that the history of statistics would have been different, had Fechner's book ap-peared prior to Pearson's work on the topic.

In France, too, scholars took notice of Fechner's work. In a review of *Theory of Measuring Collectives* the mathematician and author of a widespread intro-duction to classical theory of probability, Joseph Bertrand (1822–1900) expressed profound doubt concerning Fechner's program. On the one hand, he did not believe that the existence of pure random variation is demonstrable (just as it is impossible to prove whether or not the series of the seventh digits in logarithms is random), on the other hand he believed that it is entirely improbable that there is a common law governing all kinds of variations.[147]

Around the turn of the century it seemed to be common among German-speaking scholars to use the Fechner-Bruns methods. The founder of energeti-cism, the physicist Georg Helm, advocated views similar to those of Lipps and Bruns and wrote of "collective forms of energy." In psychophysics, Lipps and his colleague in Leipzig, Wilhelm Wirth, made widespread use of the new methods, and Wundt followed suit in a more moderate manner. Fechner's approach was also tried out in anthropology. And the American philosopher Josiah Royce, pointed out—as some anthropologists had done—that Fechner's interpretation of collective objects contains an anti-essentialist definition of the concept of species.[148] After World War I this application-oriented tradition practically em-ployed only works coming from G. F. Lipps's school, which in the meantime had moved to Zurich.[149] After World War II this tradition vanished entirely. Ulti-mately, Pearson and Fisher's statistics were victorious.

It seems as if around the year 1908 Fechner's theory of collectives was standard knowledge for all German-speaking scholars working in probability theory and statistics. Even the traditionalist Emanuel Czuber, whose textbooks at that time dominated the field, adopted the term "doctrine of measuring collectives" for the title of his book. He refers explicitly to the element of chance involved in Fechner's definition of collective objects, but entirely neglects Bruns's and Lipps's interpretation of probability as frequency.[150]

In the year 1919 the mathematical quality of Fechner's approach was improved by Richard Von Mises.[151] A mathematician and philosopher, Richard Von Mises (1883–1953) had been a student under Czuber in Vienna from 1901 to 1906. This new phase of recognition for Fechner's work prompted by Von Mises had three main traits: It was the height of Fechner reception, the consolidation and breakthrough of frequency interpretation and the end of the Laplacian age in probability theory.[152]

As early as 1912, Von Mises needed only twelve printed pages to simply and elegantly describe the theory of measuring collectives, which in his opinion "differ from Bruns's elaborations not in the result, but merely in the way of thought." On the first page of this treatise he notes that he has borrowed "the concept and the designation 'theory of measuring collectives'" from Fechner "for the entirety of all the investigations included here."[153]

While this work actually disciplines Bruns's approach, the famous article of 1919 is foundational. Here Von Mises defines the term "collective" by returning to Fechner's original intuition. He begins by re-introducing Fechner's concept of chance. As we have seen, Fechner thought of the collective object as "an object varying randomly in terms of its quantitative definitions," whereby this variation was defined independently of the sequence exhibited by the specimens.[154] Von Mises also distinctly points out one of Fechner's thoughts that Bruns and Lipps had effaced, namely that "the ideal law of chance"—as Fechner put it— "is actually only [valid] for an infinite number," meaning that it is actually only applicable to infinitely large collectives. Talking about collective objects can therefore only mean deriving an estimate from a finite amount of observations and the laws of chance can therefore always only "be confirmed with sufficient approximation for the desired degree of precision."[155]

Von Mises formulates these thoughts in both of his renowned postulates demanding the existence of threshold values for relative frequencies and "irregularity" in assigning attributes to the components of a collective. Irregularity is defined as the threshold value of frequency in selections taken from the collective. (He obviously made use of Fechner's tests for randomness!) An infinite series of components that vary in one trait is a "collective" (according to Von Mises), if: 1) the relative frequencies of the occurrences of the traits in a countable infinite series tend toward a certain threshold value and 2) if, for the sequence of traits in ever infinite sequence, selected without using the trait difference of the components, the threshold value of relative frequencies are still the same as those for the entire collective.[156] Von Mises also calls this last demand "the principle of the impossibility of a gambling system": If we think of an infinite sequence of throws with a fair dice, even the most clever selection of games on which to place a bet cannot improve one's chances of winning in comparison to betting on all the tosses.

Von Mises calls any correlation between components of an infinite sequence and given traits that fulfills both of the aforementioned demands "random-like." Thus a collective can also be defined as an infinite sequence of components, whose traits are arranged in a random-like manner. 'Probability' then means only the 'probability that a trait will occur within a collective' and is defined as the threshold value of the relative frequency with which the trait occurs in the collective.

In 1928 Von Mises presented his ideas in detail in a book called *Probability, Statistics, and Truth* [Wahrscheinlichkeit, Statistik und Wahrheit], that is very readable, even for non-mathematicians. He investigates many areas of application for his empirical theory of probability and derives some philosophical conclusions from them. The book was reprinted many times and translated into many languages. But it also met with some critical opinions. His former teacher Czuber rejected explicitly the interpretation of probability as frequency: "The final line which V. Mises believes to have drawn under the calculation of probability as it has been developed over more than two centuries will probably not endure."[157] For a long time Von Mises's theory remained overshadowed by Kolmogorov's purely formal and, in terms of application, entirely neutral approach. In terms of the empirical application of probability and twentieth-century philosophy of science his theory was and remains epoch-making.[158] In the stormy course of the development of algorithmic cybernetics, however, Von Mises's approach did gain new actuality in formal mathematics.[159]

Why, in spite of Czuber's criticism, was Von Mises's notion of irregularity so successful, while Fechner's notions of chance and indetermination were so neglected and rejected that they were eventually entirely forgotten? Besides the fact that the times were just not "ripe" for Fechner, as we say, one reason for his lack of success may be that he expressed his ideas in such an unconventional way— in his lectures of 1849, in his publication on organic development in 1873, and in the *Theory of Measuring Collectives*. Von Mises's success is easier to explain. Two factors influenced how his work was received: The crisis of mechanical physics and Ernst Mach's philosophy.

Reacting to the downfall of the mechanistic worldview, Von Mises, in a lecture given in 1921 "On the Present Crisis in Mechanics," set down a list of problems that he held to be unsolvable by classical physics. These include the movement of liquids, elasticity of solid bodies, and of particular importance, Brownian motion. He makes reference to his own theory of Brownian motion, which, in contrast to Smoluchowski's and Einstein's approaches assumes initial probabilities and does without the infamous ergodic hypothesis. Von Mises sees the general structure of the problem exemplified in Galton's board. The distribution achieved with such an apparatus could not have been deducted by classical mechanics; in fact, we would not have a clue as to what such a deduction would look like.[160] Boltzmann's treatment of the problem could offer no alternative. It is simply contradictory to first give a classical description of the impact of gas molecules and then to "cross out" these calculations with purely statistical studies. If kinetic theory of gas has already familiarized us with statistical methods in physics, why should we refuse a purely statistical approach? We have all the more justification for such an approach, as new statistics in physics is concerned not only with the behavior of individual entities in individual cases, but as collectives. This way there is no conflict with deterministically formulated physics making statements about individual cases. So Von Mises advocates a purely probabilistic approach for solving the pressing problems of physics, and this at a time prior to quantum mechanics, before physicists had become accustomed to vagueness and indeterminacy.

All of his life, Von Mises was a loyal follower of Ernst Mach—perhaps more so than any of his other friends from the Vienna Circle. All of his writings not dealing with mathematics in the narrower sense were heavily influenced by Mach's philosophy. In the aforementioned article on the crisis in mechanics Von Mises says that it is unscientific to say that the path of a ball rolling across Galton's board is predetermined. Since we can never know all of the influences affecting the ball, we cannot decide whether or not such knowledge would put us in a position to predict the ball's exact path. Since assumptions that cannot be decided through experience are unscientific, it is also unscientific to assume that the ball's behavior is predetermined. In this Von Mises is entirely in accord with Ernst Mach's methodology (and therefore also with Fechner's).

Von Mises's critique of the a priori concept of causality is also, in a way, Machian (and therefore also Fechnerian). Causality is not the prerequisite for experience; it is a very general result of experience itself. "The principle of causality is variable and subject to the demands of physics." By borrowing this from

Mach's philosophy for his own theory of probability, Von Mises was sure to gain the approval of all those colleagues seeking out Mach's thought as a serious and perhaps even promising solution for the crisis in mechanics.

One such colleague might have been, for instance, the Viennese physicist and successor to Boltzmann's chair, Franz S. Exner. Around 1919 Exner wanted to combine Mach's and Boltzmann's conceptions of physics by assuming indeterministic behavior at the micro level that balanced out to an average state at the macro level. He thought that the causes of physical mass phenomena, like those that the determinist assumes, behave in a way as if there were no causes and pure chance ruled. Notice how high the pressure was on the concept of cause in physics at that time and how much impact Ernst Mach's position had.[161]

The indeterministic view developed later in quantum mechanics was naturally very agreeable to Von Mises. For centuries chance had been considered an illusion and determinism considered reality. Supported by the success of statistical approaches in physics, in 1930 Von Mises—in a Machian manner—turns the tables: "Strict determinism, normally attributed to the physics of differential equation, is *an illusion*; it does not hold up, if we say that basically a theory is valid only within the context of the experiments required to test it, meaning that its validity is restricted to *what can be perceived by the senses* or to what is 'in principle' observable. In the realm of microscopic things what is indeterministic is in part contained in the *objects* of observation, in part it is projected into them by *acts of measuring*; every microphysics, however, contains a statistical element, since it alone mediates the *transition to becoming a mass phenomenon* and *every measurement* is already just that."[162]

And finally, we must also consider how Von Mises saw himself in relation to Fechner. While he does admit in one or two inconspicuous passages that his own work was stimulated by Fechner's treatment of the concept of collectives, he strangely believes to have surpassed Fechner by demanding that collectives have an infinite number of components and by introducing irregularity as a property of collectives.[163] To this very day it is still claimed that the latter is what makes Von Mises's theory original and unique.[164]

Both of these claims, however, are simply wrong. The very first sentence of Fechner's work defines the collective object by using variation by chance. And in the same definition he says that collective objects consist of "an unknown quantity" of specimens, for which he later clarifies that this means: 'infinitely many.' Fechner goes to great lengths to develop methods for estimating the consistency of infinitely large collective objects by examining a finite number of specimens. "Ideal laws of chance" can "taken strictly, only apply to an infinite quantity."[165]

Perhaps it would have been somewhat embarrassing for a logical positivist like Von Mises to concede that he learned empiricism from someone whose philosophy had been called "Poetry in Concepts." (Wundt had used that expression, when seeking a comprehensive valuation of Fechner's work.[166]) This would mean that 'meaningless metaphysics'—the anathema for the Vienna Circle—led to one of the greatest triumphs of empiricism and scientific philosophy in the twentieth century.[167]

When we look back today over the line of indeterministic thought described here, following it from its 'half-Hegelian,' late-idealistic and epigenetic roots and the early debates on freedom in a world determined by natural laws, right down to Von Mises, we can join Stephen G. Brush (who used these words to refer to another neglected tradition of indeterministic thought) in saying: "The revolution was complete . . . but its history almost obliterated."[168] The history I have portrayed shows that twentieth-century probability theory is more closely connected to questions of the meaning of human existence and freedom than we might ever have imagined.

CONCLUSION

LOOKING BACK WITH a desire to comprehensively evaluate Fechner's work, we are first confronted with the irksome verdict that he allegedly escaped scientific and technological rationality and the modernity of the age by reverie in a literary counter-world. Max Horkheimer degraded Fechner's philosophy to "bourgeois fantasies" where Fechner contrasted his day view with the sober night view of natural science for reasons of personal "philosophical solace" and "edification."[1] This sort of opinion is obviously shortsighted. It fails to explain why Fechner pushed "rationality" to the limits in other domains and why his work gave such a prolific enduring impulse to the rise of "modern natural science" (understood in a parallel to "modern literature"). Revealing "inner conflicts" provides no explanation, but evades it. It is more appropriate to see in both sides of Fechner's work a struggle for emancipation, often expressed using mythical images: the reunion of natural science and direct human experience and emotion, making that union the foundation of the systematic human search for orientation in our world. While we cannot deny that Fechner's thought, particularly in his later years, was slightly tainted by piousness, we should remember that for all his religiousness, Fechner respected neither a power nor dogma, but always and only his own view of things. The abstruse elements of Fechner's thought are not bad metaphysics, but simply a result of entirely overestimating the quality of empirical evidence supporting his notions of panpsychism. Yet, if we accept functionalism as a solution to the mind-body problem, we must also concede that panpsychism is *logically tenable.*

Fechner's unwillingness to compromise and his need for independence considerably inhibited the dissemination of his works. When it did finally meet with broader interest (particularly around the turn of the century and up until the First World War), it was dominated by curiosity regarding the speculative and fantastical side of his efforts and neglected his scientific accomplishments. But this sort of interest petered quickly, since the public demand for worldviews was more satisfied by Eduard von Hartmann and Schopenhauer, David Friedrich Strauss, and Haeckel. Real interest in both sides of Fechner's work, and how they are related, was sustained by a much smaller circle of readers. Fechner's ideas were appreciated more by "revolutionary" scientists (using Thomas Kuhn's interpretation of the term) than by followers of "normal science."

In this context we must consider Fechner's scholarly life as an outsider and the conflicted role he endured. Though he did give lectures at the University of Leipzig, after 1843 he no longer—as we mentioned before—officially "belonged" there. He kept his distance from university life and was probably not authorized to certify examinations. This outsider role had two effects: On the one hand it led, via "hybridization" (Ben-David), to far-reaching scientific innovations. As Ben-David has shown, innovations often develop when a researcher, finding himself in a role conflict, uses the means and methods with which he would normally fulfill role A, to achieve the goals prescribed by role B.[2] This helps to describe how Fechner established psychophysics: He employed the means normally used by a physicist in order to render man's psychical side (and the "day view" in general) scientific.

But Fechner's outsider position simultaneously harbored the disadvantage that it had no really new professional *social* function and was doomed to remain a mere "hybridization of ideas." In order to achieve any noteworthy dissemination or continuance of his innovative ideas, he would have needed a well-defined institutional framework and a communication network among followers. Thus Fechner lacked anchorage for the novel social role (distinct from physics and traditional psychology) he intended for psychophysics, aesthetics, and the day view in general. But as Ben-David and Randall Collins have proven, such a role is an unconditionally necessary prerequisite for the social acceptance of scientific innovations originating in hybrid ideas and for their capacity to develop into autonomous traditions.[3]

It was Wilhelm Wundt, who, by founding a laboratory for psychology in Leipzig in 1879, and graced with the new role of the experimental psychologist, created the social framework for certain parts (of his own choosing) of Fechner's accomplishments. This is how Fechner's methods of measurement in psychophysics have survived to this day in textbooks for experimental psychology. It attests to the inner strength of Fechner's thought that it could, in many respects and in spite of lacking an appropriate social role, still remain vivacious and fascinating in several other domains. And it is certainly not wrong to interpret Fechner's doctrine of the world soul as meaning that the psychological element of mankind depends, come what may, on the earth's fate and the destiny of the entire universe.

APPENDIX

First Document: Fechner's handwritten biography from the Darmstadt Collection in the manuscript department of the National Library of Berlin—Foundation of Prussian Cultural Property.

I was born as the second of five siblings on April 19, 1801 in the little village of Gross-Sährchen in Lower Lusatia, situated between Muskau and Treibel. My father, Samuel Traugott Fechner and his father had been pastors in Gross-Sährchen. My mother, Johanna Dorothea F(echner), nee Fischer, came from Goltzen in Lower Lusatia, where her father had also been a pastor. Altogether, many in our family had taken holy orders and I myself was predestined to do so, too; but things went differently. Since my father died when I was age five, my maternal uncle, Gottlob Eusebius Fischer, at that time the arch deacon in Wurzen, later the pastor in Ranis, and finally superintendent in Sangershausen, took me and my brother Eduard, who was a year and a half older than I (and later died as a painter in Paris) to live with him. He raised and educated us until 1814, when he returned us to our mother; after that our ways diverged for further education appropriate to our dissimilar talents and inclinations. I was enrolled in a secondary school in Sorau, six hours away from Treibel; my brother was sent to the art academy in Dresden. A year later my mother took me and my three sisters to live there. After Easter of 1815 I went to the School of the Cross for a year and a half, then the academy for medicine and surgery for half a year. At Easter of 1817, at the age of sixteen, I left to study medicine in Leipzig. This was possible because the careful instruction in languages that I had enjoyed while living with my uncle, paired with my own diligence, enabled me to finish school faster than usual. I set about studying the theoretical and practical sides of medicine, inasmuch as this was possible considering the very imperfect institutional conditions in Leipzig, passing the doctoral examination, but never performing the disputation needed for a real promotion, only later being bestowed the title of honorary doctorate by the faculties of medicine at the universities of Leipzig and Breslau. In part, I had lost faith in the achievements of medicine, in part I felt that I had absolutely no practical talent for that profession, and I ultimately found myself gradually entering literary activity that led me away from medicine. During my studies I supported myself, besides getting a minimal and soon exhausted allowance from my mother at the onset, by receiving some grants and by giving private lessons. I finally established relations with the book businesses Baumgärtner and Voss and my literary jobs provided me not only with a means of sustenance, but also with enough leisure and finances to do experiments in physics, to which I had been inspired and—as it were—prepared by translating and editing Biot's textbook on physics. The scientific activity I did in this manner switched several times as external and internal circumstances dictated. And this is reflected in my writing. In 1823 I acquired a doctoral degree in philosophy (called a master's degree), and went through the process of habilitation. In 1831 and 1832 I was a university lecturer, and after Prof. Brandes died I was given a professorship in physics in 1834. But then I ruined my eyesight by doing experiments in subjective color perception, looking often at the sun through colored glass, and by frequently observing minute divisions well into the evenings, so that by Christmas 1839 I could no longer use my eyes and had to interrupt my lectures. When I finally could no longer bear daylight at all, I gave up my position. I also had headaches from previous intellectual strain. Then, in the autumn of 1843, peculiar circumstances led to a fairly quick, almost sudden considerable improvement, if not the complete convalescence of my eyesight and my mind and enabled me to return to my scientific activities. But I did not bother to regain my previous position; instead I gave a two-hour lecture every week (as of 1846), but not on physics. I had turned to other topics, such as the greatest good, the final things, anthropology, the mind-body relation, psychophysics, aesthetics and the philosophy of nature—issues with which the majority of my writings since then are concerned.

But then my eyesight dwindled anew and both eyes were affected by cataracts, requiring that I undergo surgery performed by Prof. Gräfe in Halle several times throughout the years from 1873 to 1877, with satisfactory results. Meanwhile I have requested to be relieved of lecture duties since 1873 and am now merely a supernumerary at the university, while I do continue to contribute to studies at the Royal Saxon Society. Next to memberships in a number of medical and scientific associations, I also became a correspondent member of the Berlin Academy in 1841, The Turin Academy in 1841, the Imperial Leopoldina Academy in 1859, the Vienna Academy in 1878, and an honorary citizen of the City of Leipzig in 1884.

My life is without any remarkable events; it took place mainly at my desk. I married Clara Maria, nee Volkmann, author of the popular storybook "The Black

Aunt," and the sister of the physiologist Alfred Wilhelm Volkmann in Halle, on April 18, 1833. Our otherwise very happy marriage remained childless.

Second document: Letter from Ernst Mach to William James, dated May 6, 1909, originally written in German. The English translation was published in Perry 1935, II, 591f. and Thiele 1978, 175f. The original is kept at the Houghton Library, Harvard University, Cambridge, Mass., shelf mark bMS AM 1092 (542).

Vienna, May 6, 1909

Most honored Professor,

Some days ago I received "A Pluralistic Universe" from you. Being quite ill in bed I could read only a little of the book. So far I have read the fourth lecture, on Fechner, which has interested me greatly, inasmuch as I knew Fechner, who was a fatherly friend to me personally as well as in his work. Your setting forth of his standpoint and way of thinking seems to me to be excellent; there is no German, so far as I know, who has achieved it. I came into closer contact with Fechner after the appearance of his "Psychophysik" in the year 1860, and was associated with him at intervals until near the time of his death. His ideas gain ground steadily in Germany, even his ideas concerning the soul-life of plants, with which the scientific botanists would have nothing to do in the years 1880–1885. People called Fechner a fool and a fanatic. He was himself in part responsible for this on account of his leaning towards spiritism, manifested in his association with Zöllner and Slade. And one must acknowledge that a little sobriety would have greatly furthered his always scientific thinking. The "Earth-Protecting Angel" makes a peculiar impression when one thinks, for example, of San Francisco or Messina.

I have also read the third lecture, on Hegel. I have constantly tried to read Hegel, supposing that I would find profound ideas in him, but I have never succeeded in arriving at a good understanding of him, perhaps because I approached him from the scientific side. Through your third lecture a first understanding of Hegel seems to dawn upon me. For this illumination I am very grateful to you.

To your books, of which I now possess a considerable series, I owe many points of view, and I hope and most eagerly desire that these writings may have their good effects in Germany.

With renewed and hearty thanks, and with best wishes,

Most respectfully yours,

Dr. Ernst Mach

NOTES

Archival materials frequently cited have been identified by the following abbreviations.

FN: Manuscript remnants by Gustav Theodor Fechner, kept at the University Library in Leipzig: *Nachlass* 36–41 (diaries), *Nachlass* 42 (letters to and from Fechner, diverse material).

UAL, PA 451: University Archive in Leipzig, *Personalakte* 451 (personal file on Fechner).

Notes to Introduction

1. *Naturphilosophie* denotes a kind of philosophy of nature peculiar to German philosophy, science, and literature from around 1790–1830, proposing an organic-dynamic worldview as opposed to the atomist-mechanistic outlook of modern science. *Naturphilosophie* generally favored Spinoza's dual aspect theory of mind and matter over French encyclopedist, mechanist materialism. A heterogeneous movement, *Naturphilosophie* had roots in German idealism, classicism, and romanticism. See the entry under *Naturphilosophie* in *Routledge Encyclopedia of Philosophy,* ed. Edward Craig, London 1998. For readability its rendition as philosophy of nature or nature philosophy is preferable in most cases. On *Naturphilosophie* as the alleged "plague of the century" see Liebig 1840, 24. Depictions of this confrontation can be found in Schnädelbach 1983, 25ff., 88ff., Schnabel 1934, Lange 1875, Vol. 2., Engelhardt 1981a. See also Richardson 1997.

2. Compare how the Vienna Circle narrated its own history: Neurath 1929, 82–84.

3. Schlick 1925, 368.

4. Schnädelbach 1983, 13.

5. Compare Kanitscheider's concept of a synthetic theory of science in Kanitscheider 1983.

6. A survey of variations in nonreductive materialism can be found in Bieri 1981, 51f. and Hastedt 1988, 257, 264–66. Kim offers a critical view in Kim 1989.

7. For example, in Feyerabend 1964.

8. Perry 1935, II, 416.

9. A scientific realist holds the reasons supporting an empirical theory also as reasons for believing that theoretical objects (such as atoms), events, and processes included in the theory factually exist.

10. For an explanation of the term "probabilistic revolution" see Krüger 1987, Heidelberger 1982, 1983a, Hacking 1983, 1990, Gigerenzer 1989.

11. See also Gundlach 1988, 17–30 for a good description of Fechner literature and definition of the epoch.

12. Kuntze 1892. See also Fechner's curriculum vitae in the appendix to this book.

13. Kuntze 1892, 82f.

14. Windelband 1910, 762f.

15. Wundt 1901.

16. Fechner 1890. Vogel 1988 describes how Vierordt and the Tübinger institute assimilated Fechner's psychophysics.

17. Fechner 1905. See also Wirth 1938.

18. Döring 1987.

19. Marshall 1982.

20. Marshall 1974, 1974a.

21. Adler 1977; Bringmann 1976; Ellenberger 1956; Scheerer 1987, 1987a, 1989, 1992—cf. also Murray 1990 on this point; Mattenklott 1986; Schreier 1979; Sprung 1987; Oelze 1989.

22. Lennig 1990.

23. Heidelberger 1993a.

24. Warren 1981, Krueger 1989, Murray 1993.

25. Boring 1921, 1942, 1950, 1961, 1961a.

26. For example, Stevens 1961.

27. James 1901, 1909a, chapter 4.

28. James 1890, I, 533–49; cf. Marshall 1974.

29. Foucault 1901.

30. Seydel 1880; Bölsche 1904; Hartung 1913; Siegel 1913; Wundt 1901.

31. Wundt 1908, chapter 9, and earlier editions; Titchener 1905; Foucault 1901.

32. Lasswitz 1896; Bölsche 1913, 109.

33. Lasswitz 1904.

34. Adolph 1923

35. For example, Friedrich Paulsen 1901, 192: "Move On to Fechner!" as an alternative to Ernst Haeckel and in opposition to the neo-Kantian slogan "Return to Kant!"

Notes to Chapter One

1. Today Gross-Särchen is called Zarki Wielkie and is in Poland, between Trzebiel (Triebel) and Leknica (Muskau). When Fechner was born it belonged to the Electorate of Saxony; after the Congress of Vienna in 1815 it was allotted to Prussia. For more details on Fechner's family background see Gundlach 1993.

2. The decision to study medicine may have been guided by the example of his paternal grandfather. Besides his incumbency duties, this pastor also took medical care of the members of his parish. For Fechner's origin and family see especially Gundlach 1993.

3. Kuntze 1892, 37. On the subject of Fechner's early studies and esp. on Mollweide see Brauns 1998.

4. The other two are the physicist Wilhelm Weber (1804–1891; 1831–1837 in Göttingen, 1843–1849 in Leipzig, then once more in Göttingen) and the anatomist Eduard Weber (1806–1871) in Leipzig. Kuntze 1892, 243f. vividly describes the three Weber brothers. See also Wiederkehr 1960, chapter I; Wiederkehr 1967, with reports particularly on Wilhelm Weber; and Schreier 1993.

5. Wilhelm Weber 1846. The dictionary entry originated in Weber 1834. See Murray 1983a, 149–152, and Murray and Ross 1988 on Weber's psychophysical approach. Recent work on E. H. Weber includes Brauns 1993, 1996, Hoffmann, Christoph 2001. See the translation of key texts of Weber bei Ross and Murray: Weber, Ernst Heinrich 1996.

6. Weber 1846, 115–118, in the original version 559ff.

7. The first products of this interest were two chapters inserted into Fechner 1828, IV, 471–88: "Subjective Optical Phenomena" and "A Brief Depiction of Goethe's Theory of Color," as well as an addendum in V, 465. Fechner seems to have been interested in color contrast very early—as far back as when he lived in Gross-Särchen, as he mentioned in a letter that is now lost; see Wirth 1938, 101.

8. Whistling 1887, Kuntze 1892, 38.

9. Cf. Fechner 1821 and 1822. For a historical literary interpretation of Fechner's satirical writings see Sengle 1972, 174f., Marshall 1969, 1988, 35–37, Rosencrantz 1933 and Gebhard 1984, 166–174. However, Gebhard's interpretation is dubitable.

10. Kuntze 1892, 39. Hartung 1913 discusses Fechner's relationship to Schelling and Oken. See also Heidelberger 1994, Heidelberger 1994a, and Lennig 1994a.

11. Kuntze 1892, 39f. Cf. also Fechner 1851, II, 351. Fechner notes there that the "titan boldness" of "some writing based on Schelling's views (Oken's philosophy of nature) " prodded him to go beyond the "common view of nature," but he could not find "*clear* convergencies" with Schelling's identity theory because "the entire view seemed obscure" to him. Cf. also Fechner 1855, p. XIV.

12. See the note 1 to the introduction. Philosophy of nature in general and the concept itself are covered in Gower 1973, Pearce Williams 1973, Snelders 1973, Knight 1975, Engelhardt 1981a, 1985, 1986, Heuser-Kessler 1986, Sandkühler 1984, Schmied-Kowarzik 1989 and Cunningham 1990, Mutschler 1990, Gloy 1993, Heuser-Kessler 1994, Poggi 1994, Snelders 1994, Poggi 1996, Bonsiepen 1997, Caneva 1997. For more recent literature see the overview in Heidelberger 1998b and in the introduction above.

13. Schelling 1797, 55f. and 39.

14. Schelling 1799, 33.

15. Engelhardt 1979, 79, Heuser-Kessler 1986, 28, Schnädelbach 1983, 105, Wundt 1901, 318. On Oken see J. von P. 1833, Guettler 1884, Hartung 1913, Schuster 1922, Strohl 1936, Richards 1990. Also Stallo 1848, Pross 1991, Mischer 1997 and Breidbach et al. 2001.

16. Oken 1809, I. p. VII.

17. Oken 1809, I, 6 (§7).

18. Oken 1809, I, 14f. (§34).

19. Oken 1809, I, 23 (§63f.). See ibid. I, 16f. (§39), I, 18f. (§47) and I, 22 (§59).

20. See Gould 1977 on the history of the idea that ontogenetic development follows the same pattern as that of phylogenetic development.

21. Oken 1809, I, 44.

22. Erdmann 1853, 285. [Translator's Note: Erdmann plays on the German word "Geist" (spirit) and "Melissengeist," mint liqueur (thus also "spirits").] In criticizing the second edition of Oken's *Lehrbuch* (textbook) someone writing under the abbreviation J. v. P. found that Oken remained stuck in the philosophy of 1809, unaware of events that "have outdated Schelling's system" (J. v . P. 1833, 161).

23. Schubert 1856, 659, quote (in German) taken from Engelhardt 1980, 29.

24. Lütgert 1925, 251 and Erdmann 1896, 753. F. A. Lange also views the materialists as progeny of philosophy of nature: Lange 1875, II, 544ff., 519, 536ff., although he also concedes that English materialism was equally important; Lange 1875, I, 335ff., II, 513.

25. Guettler 1884, 133f., 129. Haeckel claimed that Spinoza had been the forerunner of his own monistic "religion," namely "naturalistic monism," or "hylozoism"; Haeckel 1914, 35.

26. Kuntze 1892, 72.

27. Fechner 1823. See also Marshall 1974a and Marshall 1982, 66f.

28. Kuntze 1892, 106.

29. FN, Papers 41, 1878, Memorial Address for E. H. Weber, dated January 30, 1878.

30. On this matter see Kuntze 1892, 38f., 72, 106, Schreier 1987 and Jungnickel 1986, 58–62.

31. Fechner 1823a, 1823b.

32. Biot 1824, Fechner 1824 and 1828. The first edition (published in 1817) by Biot 1824 had been previously translated in 1818 by Friedrich Wolff (Wolff 1818). See also Schreier 1987.

33. Thénard 1813, Fechner 1825.

34. Whistling 1887.

35. For a general history of physics in Leipzig see Schreier 1985, Wislicenus 1897 and Jungnickel 1979, especially pp. 20–24 and 33–38.

36. UAL, PA 451, p. 5v, Report from the Faculty for Philosophy to the Minister (of Education), dated June 25, 1834.

37. Kuntze 1892, 39.

38. Kuntze 1892, 72 and Schreier 1985a, 60.

39. These are listed among the courses for the summer term 1833 offered by "Fechner Mg., Med. Bacc.": *Leipziger Literatur-Zeitung*, Intelligenzblatt, No. 14, April 1833, Column 116; see also Schreier 1985.

40. Kuntze 1892, 106. Wilhelm Wirth had seen a notebook folder of Fechner's covering the years 1823–1840 in which the first fourteen pages were full of extracts from Cauchy's writings (Wirth 1938, 106).

41. Cf. Heidelberger 1983 and 1979, Marshall 1982, 67f. and Hacking 1987. On Biot and his circle cf. Crosland 1967 and Frankel 1977.

42. Kuhn 1961, 288–98; Kuhn 1976.

43. Cf. Jungnickel 1986, 31, 61.

44. Kuntze 1892, 67 and 71f. Unfortunately Fechner's report on this trip has been lost.

45. Kuntze 1892, 72 and 82; UAL, PA 451, p. 7.

46. Whistling 1887.

47. Schreier 1985, 10, Schreier 1985a, 61.

48. Fechner 1831. (Wundt 1887, 353 praises Fechner's exemplary careful and precise work.) On Fechner's use of Ohm cf. Heidelberger 1979, especially 147ff., Winter 1948, Marshall 1982, 67f., Caneva 1978, Schreier 1985a. Jungnickel 1986, 124, 51–60, 75f. convincingly discusses the particular significance that Ohm's law had for theoretical physics in Germany.

49. Heidelberger 1980, 1979.

50. Fechner 1828, III, p. X; cf. also 11, 178ff., 190–223 and Fechner 1832, I, 354ff. ; II, 1ff.

51. Fechner 1830, cf. Bonitz 1977, 1983.

52. Fechner 1832; cf. also Fechner 1831a, where he deals in particular with Poisson's newest studies.

53. Jungnickel 1986, 110f.

54. Fechner 1828a, 257, 1826, 257; cf. also 1828, I, 396–411.

55. Helmholtz 1847, 3.

56. Fechner 1838, 1840; cf. also Fechner 1838a.

57. On Fechner's work concerning contrasting colors and how Helmholtz made use of it, cf. Turner 1988, Scheerer 1984. Kries 1905a, 206 spoke of the "Fechner-Helmholtz theory of . . . after images." See also Kremer 1992 and Turner 1993 and 1994.

58. Fechner 1879, 161.

59. Kügelgen 1925, 64. Kügelgen wrote in an entry dated July 5, 1860 about when he met

Fechner with friends in Leipzig's Rosental Park: "Our conversation was lively: Zezschwitz is a devout Lutheran theologian and Fechner is a pious disbeliever and witty beyond comparison." (Ibid. 293). Of course, Fechner may have advocated a materialist standpoint opposing Kügelgen for the mere pleasure of debate. Such esprit is frequently emphasized in secondary literature.

60. Kügelgen 1901, 637.

61. Kuntze 1892, 39.

62. Rosencrantz 1933, 30.

63. Fechner 1856, 13f. On the *Anatomy* see Marshall 1969, Gebhard 1984, 177–180 and Rosencrantz 1933, 28–30. In Fechner 1846a, 298ff. and 311ff. one finds a satire on Hegel and the dialectic method. (See also section 1.9 below.)

64. Drobisch 1956, 55.

65. On Herbart see Dorer 1932, 71–103, 160, Murray 1983a, 136–138, Lange 1865, Weiss 1928, Hatfield 1990, 117–128. Sachs-Hombach 1993, 1993a, Briese 1998, Boudewijnse et al. 1999, 2001.

66. Herbart 1824 [Psychologie als Wissenschaft], 185.

67. Herbart 1824, 201.

68. Herbart 1824, 285.

69. Herbart 1824, 250.

70. Herbart 1824, 292.

71. Herbart 1825, 173.

72. Herbart 1825, 174.

73. Fechner 1853a, 91.

74. Fechner 1853a, 76.

75. Fechner 1853a, 84.

76. Fechner 1851, II, 373.

77. Koschnitzke 1988, 88–91.

78. Erdmann 1879, 380, 379.

79. Koschnitzke 1988, 14–22, 93–98.

80. Kuntze 1892, 44. Pp. 435–50 contain Fechner's personal , detailed, and colorful acknowledgment of Schulze.

81. Kuntze 1892, 56.

82. Wetzels 1973, 8.

83. Goethe 1811, part 3, book 11, 490f.

84. Humboldt 1845, II, 45 and 400.

85. Kuntze 1892, 66. Lennig 1990, 33. Lennig discovered that this letter dated Oct. 6, 1825 was printed in Berlin's newspaper *National-Zeitung* (morning edition) No. 40, on Oct. 26, 1887, p. 622. Fechner's family was artistically talented: His brother became a painter in Paris, his parents and grandparents wrote personal literature (Kuntze 1892, 347), his nephew Theodor, the son of his sister Mathilde Kietz became a sculptor, his niece Marie, the daughter of his sister Clementine, was a virtuoso at the piano. Clementine's stepdaughter was Clara Schumann. Before she moved to Dresden in 1840 she was often seen among Fechner's acquaintances. Bettina von Arnim, née Brentano, was a frequent guest at Fechner's house after 1836. Kuntze 1892, 95ff.

86. Kuntze 1892, 143.

87. Kuntze 1892, 50.

88. [Translator's Note: A *Gewandhaus* (in the Middle Ages) was originally a drapery or

clothier's trade market building, reflecting the status of that guild. The Gewandhaus of classicist architecture in Leipzig was refurbished as a concert hall in 1781.]

On Fechner's humorous and belletristic compositions see Lasswitz 1896, 26–40. Freud read the *Riddle Booklet* (Buggle 1969, 177) and Fechner's poems impressed Hofmannsthal (Dorer 1986).

89. If we were to glance further back in Fechner's biography, we might possibly also want to include Fechner's childhood religion among the elements of conflict.

90. I have appropriated the concept of late idealism from Leese 1929, Horstmeier 1930, and Harmann 1937. See also J. Erdmann 1896, Schnädelbach 1983, 237, and Köhnke 1986, 88ff. on late idealism.

91. Kuntze 1892, 78 calls him Fechner's most intimate friend. On page 57 he writes that Weisse and the mathematician Ernst (?) Müller, a middle school director in Wiesbaden (whom Fechner knew from school and university days; Fechner in Fechner 1879, 101, entered with the initials I. H. T.), were the two friends "dearest to Fechner's heart." On Müller see also Döring 1987, 298, and Kuntze 1892, 198 and 35.

92. On the friendship between Fechner and Weisse see: Kuntze 1892, 169ff., and 10, 46, 141, 197, 249f. On pp. 253–55 Kuntze briefly characterizes Weisse. The latter was a grandson of the famous poet Christian Felix Weisse (1726–1804). Informative sources for Weisse's philosophy include: Seydel 1867, and J. E. Erdmann 1896, 643–48, 688 ff., 696, 799–803. Rosenkranz 1840 provided a cutting critique of Weisse's philosophy, spiced with irony reminiscent of Heinrich Heine, and a great pleasure to read. See also Fechner's newly available obituary on Weisse (Fechner 1866) and from recent literature Kruck 1994, Briese 1998, and Schneider 2001.

93. Kuntze 1892, 57.

94. Bähr 1894, 4 characterizes Weisse as a "metaphysician of the old school," "attracted to abstract speculation," and subscribing to the "aesthetic-theosophical bent of Schelling's last period."

95. Noack 1879, 921.

96. This point is made by Weisse's follower Seydel 1876; see Köhnke 1986, 61, 450.

97. Schnädelbach 1983, 237; Löwith 1978, 125ff. and 164ff.

98. Fichte 1855, 62; quote taken from Horstmeier 1930, 15.

99. For a detailed treatment of Weisse's concept of freedom see 8.2.

100. Weisse 1833, 6.

101. *System der Ethik*. On Weisse's aesthetics see Müller 1977.

102. Köhnke 1986, 61. On Kant's role in speculative theism cf. Köhnke 1986, 88–105, Horstmeier 1930, 11f., Lehmann 1963.

103. Erdmann 1896, 653f., Köhnke 1986, 94ff.

104. See Schnädelbach 1983, 110. On empiricism and Christian-theistic scientism in late idealism cf. Köhnke 1986, 97, 102f., 459.

105. Mach 1865, 1866a; Schlick in vol. 159, 1915; Reichenbach in vols. 161, 1916 and 162, 1917.

106. See Lotze 1857, 6, Wentscher 1913, 26ff., Leese 1929, 11, Heinze 1896, 593f. Lotze was an important teacher and forerunner of Gottlob Frege; cf. Sluga 1980, 5 and 52ff. Another of Frege's teachers, the mathematician and philosopher of nature in Jena Karl Snell was also one of Weisse's pupils. It is possible that the concept of "truth value" can be traced from Weisse via Lotze to Frege. On Lotze in general see Woodward 1975, Hall 1912, Schnädelbach 1983, Wentscher 1903, 1913, 1921, Lenoir 1982, 168–72, Orth 1986. Cp. the thorough and far-reaching study by Pester 1997; also Borgard 1999 and Gabriel 2000.

107. Kuntze 1892, 141, 242 and Wentscher 1913. Döring 1987, 291f. uses one of Fechner's diary entries to describe a typical meeting of such a circle.

108. See Lotze's reviews of Fechner's publications; Lotze 1847, 1850, 1855, 1879. On the relation between Lotze's and Fechner's philosophy see Siegel 1913, 331, Wentscher 1924, Simon 1894.

109. Kuntze 1892, 170.

110. Döring 1987, 291.

111. Kuntze 1892, 169.

112. Fechner 1861, 45.

113. Wundt 1901, 269.

114. For more about the Volkmann family consult Kuntze 1892, 222–25.

115. Fechner 1834.

116. UAL, PA 451, 15, 18.

117. Wirth 1938, 103.

118. FN Papers 41, 1875, 45–47. Breitkopf and Härtel still exists today, publishing sheet music.

119. Heidelberger 1979, 44–49. Cf. also Jungnickel 1986, 121.

120. Jungnickel 1986, 115.

121. *Büchlein vom Leben nach dem Tode.* On this booklet see Meyer 1937, 30–32, Gebhard 1984, 181–188, Lütgert 1925, 229f., J. E. Erdmann 1896, 690, Kuntze 1892, 145–148.

122. Sengle 1971, 74ff., Erdmann 1896, 688–91, Lütgert 1925, 228–49.

123. Fechner 1836, 131.

124. Fechner 1836, 130f. and 97.

125. Fechner 1836, 101f. and 103.

126. Fechner 1836, 97f.

127. According to Kuntze, Weisse is supposed to have used these words in one of the debates with Fechner. (Kuntze 1892, 254).

128. Gebhard 1984, 182f. In contrast see Mattenklott 1986, 155, who considers Fechner's "critique of reason" "not to be anti-enlightening" and not to be irrational.

129. Moravec 1988, 111.

130. Kuntze 1892, 139; cf. 107. On this section see Kuntze 1892, chapter V: "The Crisis. 1840–1843"; Lennig 1990, 38–49; Schröder 1991.

131. A detailed and vivid autobiographical report on this illness is made by Fechner, found in Kuntze 1982, 105–126, written shortly after his recovery in the sober style of a natural scientist.

132. Fechner 1840, 194. Fechner described his symptoms in detail in a letter to J. E. Purkynje on July 12th, 1840; cf. Hoskovec 1988, 196, 198–200.

133. Kuntze 1892, 111 and 128f.

134. Kuntze 1892, 114f.

135. Report from the dean of the philosophical faculty, A. Westermann, to the rector of the university, dated July 29, 1841. UAL, PA 451, p. 25–27.

136. See Jungnickel 1986, 136f. for a description of the circumstances surrounding Weber's appointment. [Translator's note: The Göttinger Seven were a group of university teachers: W. Albrecht, F. C. Dahlmann, H. Ewald, G. G. Gervinus, J. and W. Grimm and W. Weber in Göttingen, who were dismissed from their positions by King Ernst August on Dec. 12, 1837 because they accused him of impinging on the constitution of 1833 by repealing it. Their act was respected throughout Germany and encouraged a revival of liberalism

in many parts of the country. Almost all of the Göttinger Seven later became members of the Frankfurter National Parliament.]

137. UAL, PA 451, p. 38f., Report from the faculty to the government.

138. Kuntze 1892, 124f.

139. Kuntze, 1892, 140.

140. Kuntze 1892, 105–26.

141. The title of Bringmann's work 1976.

142. In a letter to Poggendorff in 1845 (or 1844) Fechner writes that his eyesight improved "quickly, as if miraculously" so that he "hoped it would be restored entirely." But then it "degenerated once more" to the extent that he had to dictate the letter and "once again feels condemned to inactivity." Testimony of his respect for Fechner is given by the fact that Poggendorff, as the editor of the *Annalen der Physik,* cited a longer passage taken from one of Fechner's letters in a footnote to Fechner's article in 1845 and also wished him the best of health. (Fechner 1845, 337.) Even then, personal remarks such as these were not common practice in scientific journals.

143. UAL, PA 451, p. 58 r.

144. Möbius 1894; Hermann 1925; Ellenberger 1970, 306; Ellenberger 1956, 203; Bringmann 1976, 46; Schröder 1991, 18. See also Lennig 1990, 41–46. On Hermann cf. Marshall 1980, 201–4.

145. Release from his obligation to teach physics seems to have been very important to Fechner. In reviewing his illness he writes that after recuperation he congratulated himself on being suspended from his position, "because giving many lectures always over-exerted me." (Kuntze 1892, 139). And while reviewing his life in later years he wrote that he accepted his eye illness as "an acceptable price" for gaining inner maturity and for losing his position, which he had been "increasingly unable to fulfill." (Kuntze 1892, 332.)

146. Kuntze 1892, 139.

147. Hartung 1913, 279. Döring 1987, 292 reports from Fechner's diary that Fechner was surprised when from one semester to the next the number of students in his course jumped from two to nine.

148. Fechner 1845.

149. Fechner 1846a, 270.

150. Fechner 1846, 298 and 311ff.

151. In 1879, 74f. Fechner lists all his writings as part of the day view in the narrowest sense.

152. Fechner 1846, 3.

153. Fechner 1846, 10, 13, 21–23.

154. Fechner 1846, 13.

155. Fechner 1846, 41f. Cf. also 17.

156. Oken 1809, 358f (§§3457 and 3459).

157. Fechner 1848a, 186. The reaction to Fechner 1846 is discussed in Lotze 1847. Traces of Fechner's influence are even to be found in the writings of the English economic and social scientist Francis Y. Edgeworth, who presented a mathematical ethics in 1881, entertaining the notion of the greatest happiness for the greatest number of people; see Wall 1978. Schönpflug 1998.

158. Fechner 1846, 66f. Fechner repeated this last phrase in 1848a, 193.

159. Fechner 1848a, 5. On Fechner 1848 cf. Nitzschke 1989.

160. Fechner 1848a, 22, 24.

161. Kuntze 1892, 131; Fechner 1848, 65 (89f.) and 294f. (391f.).

162. Putnam 1967, Fodor 1968.

163. Fechner 1848, chapter III, Fechner 1856, 27ff., 35f.

164. A history of the society is given in Wislicenus 1897, Jungnickel 1979, Lea 1984, Gundlach 1988a. Fechner held office from Dec. 2, 1848 until Dec. 12, 1850. He succeeded Moritz Wilhelm Drobisch as vice secretary and was followed by Wilhelm Gottlieb Hankel. Wilhelm Weber was secretary from 1846–48 and Ernst Heinrich Weber from 1848–74; see Wislicenus 1897, pp. xliv and 14.

165. Fechner 1851, I, p. xvi f.

166. Marshall 1988, 39. See chapter 5 for more on the identity view.

167. Fechner 1851, II, 373–86. Cf. Scheerer's commentary 1987a, 198 and Marshall 1982, 71–73. It is interesting that Wilhelm Fridolin Volkmann, one of Herbart's followers, took notice of Fechner's approach as early as 1856; Volkmann 1856, 55. (W. F. Volkmann was not related to Fechner's wife.)

168. Fechner 1860, II, 554.

169. Rosenzweig 1987, Stevens 1975, 7, 10.

170. Schaller 1852, 1036 and 1050. An anonymous review scoffs at Fechner's "passion" for page-long comparisons. The reviewer finds that the book lacks application and is worried that less experienced thinkers will be confused by Fechner's use of terminology. Nevertheless, he continues, when compared to its faults, the merits of the *Zend-Avesta* are obvious. (Anonymous 1852, 85).

171. Köhnke 1986, 121ff., Schnädelbach 1983, 123, Sluga 1980, 15ff., Gregory 1977, Ziegler 1899, chapter 11.

172. See Lenoir 1982, chapter 5, Cranefield 1957, Heidelberger 1993b.

173. Fechner 1856, 25.

174. Quote taken from Fechner 1856, 47. See Schleiden 1855. On the relationship between Schleiden and Fechner see Woodward 1972.

175. Kuntze 1892, 231.

176. Fechner 1860; cf. Gundlach 1988b and Brauns 2000. For the situation in psychology in the 1850s see Waitz 1852, Brauns 1994 and Reed 1994.

177. Fechner 1860, I, p. 8 and VII.

178. Wundt 1901, 296.

179. Boring 1950, 281; cf. also M. Müller 1991 and Titchener 1905.

180. Pelman 1862. Fichte 1861 mentions only a few aspects of Fechner's psychophysics.

181. FN, Papers 41, 1877, p. 97.

182. FN, Papers 41, 1875, p. 59 dated Nov. 16th. In Fechner 1882, p. iv Fechner attributes his failure to present a new edition of the *Elements* to being too old.

183. Fechner 1860, II, p. ix.

184. Ludwig Noack in Noack 1861 relays a review of Fechner 1861 that is almost as long as the reviewed book itself. Noack wanted to prove that Fechner's methods in *On the Soul* are unscientific, compared to those in *Elements of Psychophysics*.

185. On Fechner's aesthetics see Sprung 1988, Ebrecht 1988, Allesch 1988, Arnheim 1985 and Lalo 1908, also Jaensch 1932, Brentano 1959, Berlyne 1974, Müller 1977. For more recent literature see the overview in the introduction above.

186. Fechner 1876, 54, 82, 84f.

187. Fechner 1873, p. iii.

188. Fechner 1905, 311, No. 117 in the Fechner archive.

189. FN, Papers 41, 1878, p. 19, dated Oct. 18, 1878; Papers 40, 1873, p. 28f.

190. Breuer 1903, 312; Letter from Josef Breuer to Franz Brentano, in March 1903. On Breuer see also 7.3 below.

191. See Pernerstorfer 1912, Hartungen 1932, and Hemecker 1991. Lengauer 1989 is very comprehensive and instructive. For the mediating role of Lipiner between Fechner and Gustav Mahler see Barham 1998. See Meinong 1965, 3, 5 for a report on Lipiner by Thomas Masaryk at the occasion of his visit in Leipzig.

192. Fügner 1881, 39. Fügner also reports that Lipiner's *Prometheus* is in many ways modeled on Fechner's *Zend-Avesta*; ibid. 50.

193. On Lipiner's role in the Vienna Reading Club and Sigmund Freud's contact with Lipiner and this club, see Hemecker 1991, 59, 61f., 70ff. and Gödde 1991.

194. Lipiner 1878, 14; see also Hemecker 1991, 62.

195. Freud 1989, 85, 100, 190.

196. Freud 1989, 78, 83, and in letters from 22–23 of October 1874 and from Nov. 8, 1874.

197. This information was contributed by the Brentano Archive in Würzburg and the Research Center for Austrian Philosophy in Graz. Evidence of the letters is found in Döring 1987, 294. On the relationship between Freud and Brentano see Freud 1989, esp. 116f., as well as Brauns 1989. On Freud and Fechner see 7.3.

198. Scharlau 1990, 202.

199. Hemecker 1991, 11–13, 108ff., Karlick 1982, 13–22, Koschnitzke 1988, 90, Dorer 1932, 114. Franz Exner's sons, the physicist Franz Serafin Exner (1849–1926), and the physiologist Sigmund Exner (1846–1926) also participated in the debate on Fechner's psychophysics; Exner 1879. Franz S. Exner had been a teacher for Erwin Schrödinger, Marian von Smoluchkowski, and Friedrich von Hasenöhrl. See Karlik 1982, 130–41, Stöltzner 1999, 2002, 2003. On high-school education in Austria in the second half of the nineteenth century and Exner's role in it cp. Leitner 1998 and Wozniak 1998. See also Stachel 1998.

200. Lindner 1868, p. iv, 6, 9, 26f., 40, 52, 61, 73, 75.

201. Using the edition from 1872; Lindner 1872; from Hemecker 1991, 12, 110. See Hemecker 1991, 108–27 for Herbart's influence on Freud. Hemecker's work corrects Dorer's depiction in this respect; Dorer 1932, 103–6, 170, 175–78.

202. Fechner 1879, 15, 71.

203. Fechner 1879, 74.

204. This scenario reads like a belated (albeit less poetic) revival of "Christ's Oration from Atop the Highest Structure of the World, Proclaiming that There Is No God" found in Jean Paul's farcical novel *The Life of Lawyer Siebenkäs*. Cf. Jean Paul 1796, 274–85. At the age of twenty-three, Fechner had already written an essay reflecting "On the Definition of Life," which anticipates the day view and exhibits similarity to Jean Paul's protagonist's speech; cf. Fechner 1824a, 65–73.

205. Nitzschke 1987, 63. Nitzschke 1989 on the other hand, provides an adequate portrait of Fechner. Molella 1973, 176ff. also misconstrues the real nature of the day view.

206. Lotze 1879, 397. In the course of his life, Lotze became more prone to conservatism than Fechner.

207. See Ziegler's correct remarks in Ziegler 1899, 325f.

208. Lotze 1879, 437.

209. Wundt 1887, 360.

210. Kuntze 1892, 139, 194, 113.

211. UAL, PA 451, p. 54, 55; FN, papers 40, 1873, p. 33.

212. For more from this passage cf. chapter 10 in Kuntze 1892, and Wundt 1901.

213. Wundt 1901, 340.

214. On Zöllner in general see the revealing description given by Meinel 1991, also Molella 1973, chapters 6–9, Hermann 1982, Hall 1912, 144f., Buek 1905, 156–160; on Zöllner's spiritualism see Meinel 1991, 34–42. For spiritualism also see Hacking 1988 and Thiel 1987. On the importance that Zöllner's spiritualism had for the development of mathematical knot theory cf. Epple 1999.

215. Zöllner 1872, 299ff. See also Molella 1973, chapters 6 and 7, and Rosenberger 1887, 582–85.

216. On Fechner's role in the Leipziger spiritualist movement and his opinion on parapsychology cf. Lasswitz 1896, 105–107, Wundt 1901, 338–43, Bringmann 1988, and Meinel 1991. For Wundt's and Lasswitz's reaction to Fechner's spiritualism see Marshall 1980, 206f.

217. Fechner 1879, 269.

218. Mach 1896, 372. Mach keeps the circumstances anonymous, but there is no doubt that he is writing on events in Leipzig.

219. Köhnke 1986, 588, 591–93; Meinel 1991, 17.

220. Fechner 1879, chapter 22. I here contradict Molella's interpretation, see Molella 1973, 176–84, esp. 183.

221. Fechner 1879, 263, 272.

222. See Wise 1990, 353 and chapter 4 below.

223. Helmholtz 1874, 415, and Helmholtz 1874a (directed against Zöllner). Zöllner had also attacked Thomson's theory that life might have been transferred to earth by organic germs carried by meteors coming from the universe.

224. Ostwald 1927, 96f. On pages 80ff. Ostwald nicely depicts society in Leipzig during the late 1880s, proffering portraits of Karl Ludwig, Wundt, H. Bruns, Ratzel, Lamprecht, and others.

225. Kuntze 1892, 305.

226. Kuntze 1892, 313.

227. Although Hall reports that Fechner was fond of letter-writing; Hall 1912, 115.

228. Published in Wirth 1938a. The report is based on FN Papers 41, 1877, p. 18–23, dated July 13, 1877 and refers to a visit made by Masaryk a year earlier.

229. Kraus 1929, 61.

230. Hall stayed a year in Leipzig, around 1879, cf. Hall 1912, 114.

231. Fechner 1884, 110.

232. Camerer 1907, 236f. Some of Camerer's papers are to be found at the Institute for History of Medicine, University of Tübingen. The correspondence between Fechner and Camerer, reported by Wirth 1938, 111, 113 and also by Camerer himself, along with twenty-three Fechner manuscripts in Camerer's possession, are apparently lost. On Vierordt cf. Vogel 1988.

Notes to Chapter Two

1. Another possible designation would be: *non-eliminative naturalism*.

2. Fechner 1851, I, XXII.

3. Probably Külpe first applied this term to Fechner's work; Külpe 1901, 1908.

4. This is most obvious in Fechner's *Theory of Atoms*. As the full title of the book itself demonstrates, Fechner distinguishes a physical from a philosophical doctrine of atoms;

whereby the latter "is supported entirely by the first, and without the proofs of the former would have no foundation, while the reverse is not the case." (Fechner 1864, VII) The physical theory of atoms, however, is not actually to be understood as physics, but instead, as philosophy of science for atomic physics. It is meant to demonstrate that atomic theory is a view "representing a nexus of facts" (Ibid. VIII).

5. John Locke 1706, II, 8;8: "Whatsoever the mind perceives in itself, or is the immediate object of perception, thought, or understanding, that I call *idea*." (Cf. also Fechner 1851, I, 411 and Fechner's controversy with Herbart's devotee Drobisch concerning the concept of phenomenon: Drobisch 1856, 79f. and Fechner 1857, 182.)

6. Fechner 1861, 199. Cf. Fechner 1905, 257.

7. Fechner 1863, 21 and 22f.

8. Fechner 1851, I, 411.

9. Fechner 1863, 22.

10. Fechner 1863, 21. As early as 1823a, 20f. Fechner advocated a similar position.

11. Fechner 1851, II, 321.

12. Fechner discusses the perception of one's own body in Fechner 1851, II, 325 and 323.

13. Locke 1706, II, 1; 4.

14. Hume 1748, 14.

15. Berkeley 1710, §§28–30, 34.

16. Fechner 1851, II, 362.

17. Putnam 1962.

18. Putnam 1962, 374.

19. Fechner 1851, II, 321f. A similar passage exists in Fechner 1860, I, 4.

20. Stubenberg 1986, 201. Stubenberg and—by way of suggestion—Chisholm 1978 (whose comments Stubenberg employs) are the only authors who, to my knowledge, have dealt with Fechner's mind-body theory recently.

21. It is also very possible, that in other places Fechner sometimes uses "inner" and "outer" as synonymous with "self-phenomenon" and "foreign phenomenon." These are questions of style which change nothing for my solution to the difficulty addressed by Stubenberg.

22. This would dispense of one of "the two huge problems" that Stubenberg 1986, 198 sees attached to Fechner's mind-body theory. (The second problem concerns the question of what it means for something to appear in two manners. We shall return to this point later.)

23. For Fechner "belief" seldom carries a religious meaning. There are many parallels to David Hume's term "belief." As it is for Hume, so it is also for Fechner: belief is a prerequisite for normal human action and life. For Hume the content of belief is the connection of things by laws of nature. For Fechner belief consists in assuming the minds of others. This assumption, as we shall see, also rests on a connection of laws of nature.

24. Fechner 1863, 17.

25. Fechner 1863, 40.

26. Fechner 1863, 25.

27. Fechner 1863, 40.

28. Fechner 1863, 29.

29. Fechner 1863, 32.

30. Fechner 1879, 229.

31. Fechner 1861, 19.

32. Fechner 1863, 23.

33. Fechner 1863, 74.

34. Fechner 1863, 49

35. Fechner 1863, 137.

36. Fechner 1863, 45.

37. Fechner 1851, I, XXIV.

38. Fechner 1851a, 198. Cf. also Fechner 1863, 125. Similar to David Hume 1748, sect. V, part II, §44f. Here we have what we might call an "evolutionary theory of knowledge."

39. Fechner 1863, 71.

40. Fechner 1855, 119. Cf. also Fechner 1857, 73ff.

41. Fechner 1863, 122. This passage is also in Fechner 1861, II, 251f. and Fechner 1879, 78.

42. Fechner 1851, II, 251. This is an argument regarding the concerted action of the good and the true.

43. Fechner 1851a, 197.

44. Fechner 1861, 200.

45. Fechner 1851, I, p. XXI.

46. Fechner 1823a, 178f. Cf. Fechner 1879, 194–96.

47. Fechner 1879, 227.

48. Fechner 1879, 227 (Footnote).

49. Fechner 1851, II, 361.

50. Fechner 1879, 226f.

51. Fechner 1855, 95.

52. Carnap 1928.

53. Unless we count David Hume's critique of the concept of substance as such (Hume 1739). Classical examples of neutral monism are Bertrand Russell's and particularly William James's approaches (James 1904); the latter knew Fechner well and admired him (see Perry 1935, Marshall 1988, Scheerer 1988). James also was well acquainted with Fechner's *Zend-Avesta*; James 1909a.

54. Frank 1986, 59.

55. Cf., however, Carnap 1928, 226 (§163) and 88 (§65).

56. As, for example, Descartes does in his second meditation.

57. Fechner 1855, 96. Cf. also Fechner 1879, 244 and Fechner 1861, 208, 213.

58. Fechner 1855, 103.

59. Fechner 1861, 209.

60. Fechner 1855, 98.

61. Fechner 1855, 98. In the second edition of the *Theory of Atoms* Fechner omitted this passage in order to "snip" a "verbal dispute" he had with Herbart's devotee Drobisch on this matter (Fechner 1864, 114).

62. Fechner 1861, 202; cf. Fechner 1855, v f. and 98ff.

63. Fechner 1863, 205.

64. Fechner 1879, 244. Earlier, in the *Zend-Avesta*, Fechner had still allowed for the possibility of a substance existing "behind" the phenomena of body and soul, a substance that is unknown, although at the time he already believed he could do without this assumption. (Fechner 1851, I, 412f. and II, 367f.) Later he found the supposition of substance meaningless. This is one of the few cases where Fechner changed his opinion in the course of time. Cf. Fechner 1882, 256, for his comment on this change of mind.

65. Fechner 1863, 207; cf. also Fechner 1861, 9.

66. Fechner 1905, 282.

67. Fechner 1861, 213.

68. For instance, in Fechner 1861, 214.

69. Fechner 1879, 244. Cf. also Fechner 1861, 207.

70. Fechner 1861, 208.

71. Fechner 1855, 96. Cf. Fechner 1861, 201.

72. Fechner 1861, 16.

73. Fechner 1861, 221. Cf. also Fechner 1851, II, 350f. and 366. Fechner also calls his theory an "inter-relating (nature-philosophical) interpretation" (ibid., S. 361; cf. also Fechner 1860, I, 6). In the following, "identity view" as the term for Fechner's philosophy is strictly distinguished from the identity theory meant in the current mind-body debate. (Bieri 1981, 36–43) The difference between the two positions will be examined more closely below.

74. Fechner 1860, I, 3f.

75. The example using a coin does not apply for our tactile sense, since we do feel both sides of the coin at once.

76. Fechner 1879, 244.

77. Fechner 1851, 321f.

78. Fechner 1861, 210f.

79. Fechner 1851, II, 320f.

80. "Basic psychophysical law": in Fechner 1879, 203: "Most general law of psychophysics": in Fechner 1861, 211; "Functional principle" in Fechner 1860, II, 380; Fechner 1879, 233, Fechner 1882, 4. In his psychophysics Fechner later strives to find a quantitative expression for the relationship defined in the basic law.

81. Fechner 1861, 211.

82. Fechner 1879, 203. Cf. also Fechner 1882a, 221.

83. Davidson 1970, 214.

84. In contrast to Fechner, however, Davidson denies the possibility of psychophysical laws. A good survey of the debate on supervenience is given in Teller 1984.

85. Fechner 1851, II 330. Cf. also 347 and 365 and Fechner 1905, 255. The term "parallelism" can be used in both a monistic and a dualistic sense. Fechner's parallelism is monistic as opposed to the occasionalists' and Leibniz' dualistic parallelism. (See chapter 5.) Spinoza represents a special case, since his parallelism is dualistic for humans and monistic for God. I use "parallelism" here in Fechner's monistic meaning. It was used this way throughout the nineteenth century among German scholars.

86. Fechner 1905, 308.

87. For this distinction see Carnap 1928, §165. Functional relationships have nothing to do with modern mind-body "functionalism," with which we shall deal later on.

88. Fechner 1851, II, 347.

89. Fechner 1860, I, 8.

90. Or such a change can be presumed inside the body, without contradicting experience.

91. Fechner 1851, II, 347. Also Fechner 1860, I, 5. On the clock allegory see Du Bois-Reymond 1872, 472f., who notes that Leibniz did not overlook the possibility that the clocks might be identical—as Fechner had assumed—but decidedly denied it.

92. Fechner 1860, II, 415. Cf. Fechner 1905, 307.

93. Fechner 1857, 89.

94. Fechner 1851, II, 333. In Fechner 1882, 10 he sketches a physiological model for a psychophysical representation of self-reflection!

95. Fechner 1861, 212; see also Fechner 1851, II, 330ff., Fechner 1860, II, 388 and 526; Fechner 1879, 247; Fechner 1882, 15 and 259f. The existence of the synechiological principle excludes interpreting the parallelism of the mental and the physical as isomorphism.

96. Fechner 1882, 259 and Fechner 1860, II, 526.

97. It is possible that the expression "resultant" was common in Fechner's time; we find it used with this meaning in Lotze 1843, 144, 165 and in Wundt 1911, 755.

98. Fechner 1860, II, 415. Fechner attacks monadical doctrine particularly in Fechner 1879, 246ff. and Fechner 1864, 244ff.

99. Fechner 1860, I, 8f. Cf. also Fechner 1860, II, 388.

100. Cf. e.g. Fechner 1855, 104f. Fechner writes of the difference between the "experiential relation" of body and soul and the "representation" of this relation in a metaphysical view. (Fechner 1860, I, 6.)

101. Fechner 1856, 34.

102. See particularly Fechner 1860, I, 6f.; cf. Fechner 1879, 240ff. Fechner makes a list of the "basic facts" in Fechner 1855, II, 363ff.

103. Fechner, however, is convinced that in a later developmental phase of the science of psychophysics empirical aspects will be found to definitively support the identity view. (See chapter 6.)

104. Cf. Bieri 1981, 39ff. on this point, Hastedt 1988, 106ff., Metzinger 1985, 16ff.

105. Prior to World War II the expression *Zweiseitentheorie* was common in Germany: Carnap used it, e.g., Carnap 1928, 28 (§22). This expression was translated into English as "double (dual) aspect theory." Translated back into German, it became *Doppelaspekttheorie*, and more recently, *Aspektdualismus* (Hastedt 1988, 11). Prior to World War I the term *psychophysischer Parallelismus* was common (see chapter 5) (Hering 1878, 76; Wundt 1911, 745ff.) Klimke 1911, 311 lists *Identitätshypothese, Zweiseitentheorie, Parallelismustheorie,* and *psychophysischer Monismus* as names for Fechner's theory. Marshall 1988, 39 suggests the designation "dual perspective theory" to characterize Fechner's theory, since he seldom speaks of two sides, and usually speaks of "the difference of perspective" (standpoint). [Translator's note: Double aspect theory seems most appropriate, because the object is single, while options of perspective are double. It is not the aspect that is dual, but rather the number of aspects (perspectives) available is two.]

106. Popper 1977, 114; Armstrong 1987, 491; Armstrong 1968, 34.

107. For example, in Lennon 1984, 364.

108. Other expressions he uses are: The body as "support," "seat," "organ," "basis," "outer sheath," "tool," and "expression" of the mind (Fechner 1851, II, 332 and 345f). See also Fechner 1860, I, 10, 37; II, 380; Fechner 1879, 230, 243; Fechner 1905, 254, 287 and many other passages.

109. Fechner 1861, 221.

110. Fechner 1851, II, 348.

111. Fechner 1861, 221.

112. As we have seen, Fechner calls the supervening of the psychical over the physical the "functional principle." In Fechner 1851 II, 332 Fechner sketches the possibility that "higher mental functions" themselves exist in virtue of lower functions, such that there is an entire hierarchy of supervenience.

113. Fechner 1851, II, 357. Cf. also Fechner 1860, I, chapter V, dealing with the validity of the "principle of conservation of energy" for psychophysical activity and using it implicitly as an argument for the causal inclusiveness of the psychophysical realm. Similar ideas are also found in Fechner 1905, 280.

114. Fechner 1851, II, 318.

115. Fechner 1851, II, 342.

116. For "knowledge" in Fechner's sense of the word see above, section 2.1.

117. [Translator's note: As an apple drops from a tree, remains in the vicinity, and its seed grows.] Fechner 1855, 1864, XIV. On the relation of Fechner to Schelling see Hartung 1913. This dissertation is better than most rather shallow dissertations of that time. See also Lennig 1993, Heidelberger 1994, 1994a.

118. Fechner 1851, II, 350f.

119. Fechner 1905, 309.

120. Fechner 1905, 309. Cf. Fechner 1851, II, 351–55 and Fechner 1879, 245.

121. Hartung 1913, 280. Hartung quotes from one of the manuscripts (partially also printed in Fechner 1905) that Fechner used for his lectures in 1864, 1866–1867 and 1869.

122. Fechner 1851, II, 352.

123. For the same reason Fechner thinks that teleology compatible with natural science is possible; Fechner 1851, II, 352. We shall return to this point in chapter 7.

124. Schelling 1804, 501. (Thanks to Manfred Frank for pointing out this passage to me.) The manuscript containing this passage was published posthumously, so Fechner could not have known it when writing the *Zend-Avesta*.

125. "Anyone talking of polarity [as a nature-philosophic concept] in contexts other than those of magnetism and galvanism, is lost." (Fechner 1855, 48.)

126. Seydel 1875, 67 and 76.

Notes to Chapter Three

1. Fechner formulates his program for "nature philosophy" best in the preface to Fechner 1851, I, IXIf. Cf. also Fechner 1905, 275ff. For the term "philosophy of nature" see note 1 to the introduction to this book.

2. Fechner 1851, II, 326f. Cf. above, chapter 2.

3. For example: Du Bois-Reymond 1872, 440: "Knowledge of nature . . . is tracing the changes in the physical world back to the movements of atoms . . . or resolving natural processes into the mechanics of atoms."

4. Fechner 1879, 16f.

5. Fechner 1861, 19.

6. Fechner 1848, 10. On functionalism in the contemporary mind-body debate see Putnam 1967, Fodor 1968 and the summary in Bieri 1981, 47ff., Hastedt 1988, 142ff. In this context the term "functional" has nothing to do with the mind being functionally dependent on the body, as discussed above in chapter 2.

7. Fechner 1851, I, 179f., Fechner 1861, 49f., Fechner 1848, passim. The last two quotes in the following passage are taken from Fechner 1851a, 200. Fechner 1851 I, 71–178 lists and discusses 48 different functional similarities between the planet earth and the human body.

8. Fechner calls this type of inference "arguments of supplement, arguments of gradation, of relationship, of causality, of teleology." Fechner 1861, 66–105. Cf. also Fechner 1855, 78ff.

9. Fechner 1851, I, XXII; cf. also 327ff.

10. Fechner 1879, 29.

11. Du Bois-Reymond 1872, 464. Du Bois-Reymond does not mention Fechner here. But

it is clear that he meant him, particularly since he mentions him elsewhere. Even Moritz Schlick 1925, 368 raises this objection.

12. Fechner 1848, 10, 26ff., 35ff., Fechner 1851, I, 211ff., Fechner 1856, 69f., 77f., Fechner 1861, 27ff., Fechner 1863, 212 ff., Fechner 1879, 32, 87ff., Fechner 1882, 14, Fechner 1905, 291.

13. Fechner 1882, 14.

14. Fechner 1861, 27f.

15. Fechner 1861, 37f. and 1848, 132. Consider the distinction biologists make between homologous and analogous organs.

16. Fechner 1860, II, 262ff., 282ff.; see also Hering 1878, 75.

17. Fechner 1882a, 221.

18. Fechner 1856, 75, 100.

19. Fechner 1851, I, XIXf.

20. Fechner 1861, 150.

21. Fechner professes his belief in pantheism explicitly in 1851, I, XIf. and 333. The same is true for 1879, 84, however, here he notes that in contrast to so many other varieties of pantheism, his notion embraces the concept of a personal God. Wundt 1901, 320–23 also emphasizes the difference between Fechner's position and other pantheistic doctrines.

22. Fechner 1856, 22.

23. Fechner 1879, 240, 245. Cf. also Fechner 1855, 90.

24. For instance, in Fechner 1861, 221.

25. Fechner 1879, 240, 245. Also cf. Fechner 1855, 90.

26. Fechner 1857, 183.

27. Fechner 1861, 207. Cf. also Fechner 1855, 97f. and 1905, 310f.

28. Fechner 1879, 21. See also 273, where Fechner, remarking on this passage, reemphasizes that it is very difficult, if not impossible, to imagine "an independently existing perceptual sensation."

29. Fechner 1861, 211. Cf. also Fechner 1861, 202. Here Fechner says that belief in God makes the assumption of noumena superfluous, "because belief sees—as an effect of God existent without us and within us—those very things that we would like to view as an effect of dark things [meaning noumena and substances] behind us and God, by making visible in God those things which would remain invisible in the dark things behind the phenomena."

30. [Meaning: Christmas Eve.] Oken 1808, 45.

31. Berkeley 1710. Fechner does not seem to have read Berkeley. He never quotes him and does not use Berkeley's terminology.

32. Fechner 1879, 228.

33. Fechner 1879, 28.

34. Fechner 1879, 4.

35. Fechner 1879, 5 and 13.

36. Du Bois-Reymond 1872, 445. On Du Bois-Reymond see Ehrenfels 1886.

37. Fechner 1879, 237 and 4.

38. Fechner 1879, 4 and 7f.; cf. also 238 and other passages. Fechner himself notes that his book *Zend-Avesta* was "written with the needs of common sense in mind and should be judged accordingly." Fechner 1851a, 93.

39. Weizsäcker 1922, 8.

40. Fechner 230.

41. Fechner expounds this doctrine mainly in Fechner 1836, Fechner 1851, part III, and Fechner 1879, 38–47, 90–105, Fechner 1882, 7f.

42. Fechner 1860, II, 540.

43. Fechner 1879, 92.

44. Fechner 1879, 98.

45. Fechner 1857, 180. Bertrand Russell advocated a similar notion. In Russell 1912, 201 he discusses the problem of whether mechanism excludes teleology and concludes that it does not, using the same arguments brought forth by Fechner.

46. Fechner 1848a, 1.

47. Fechner 1855, II, 294ff.; 1879, 170ff.

48. For instance Wentscher 1921, Kuntze 1892; also Marshall 1982, 75 and Woodward 1972, 374–76.

49. Fechner 1846, 14.

Notes to Part III

1. Fechner 1879, 239.

2. Fechner 1857, 73.

3. Fechner 1855, 90. This last definition can be found in Poisson's textbook for mechanics, which was widespread at the time. The priority Fechner grants to the sense of touch is probably due to his own education in sensory physiology under Ernst Heinrich Weber. The physiology of the time was strongly influenced by Berkeley's *Theory of Vision*, 1709 (although Berkeley's immaterialism had no consequences for physiology). According to this theory, our concept of space rests primarily on our experience of touch. The fact that we can also *see* distances is the result of a learning process in which experiences of touch and sight are correlated. When we see space, we draw conclusions based on our learnt correlation between the sense of touch and that of sight.

4. Fechner 1857, 87.

5. Fechner 1855, 92.

6. Fechner 1855, 101.

7. Fechner 1879, 238.

8. Fechner 1879, 235.

9. Fechner 1879, 273f.

Notes to Chapter Four

1. See Leplin's collection, 1984, containing a good survey of the current debate. Further literature is listed there.

2. Presently the most interesting anti-realistic position is advocated by Fraassen 1980.

3. Fechner 1857, 165.

4. See Kaiser 1984, chapter 2 and Carrier 1990.

5. Fechner 1826, 257.

6. Fechner 1826, 269.

7. See Lauterborn 1934.

8. Fechner 1826, 274.

9. Fechner 1828a, 275.

10. Fechner 1828a, 290.

11. Fechner 1845. Cf. Rosenberger 1887, 503, Wiederkehr 1960, 122f., Wiederkehr

1967, 94ff., Molella 1973, 49f., Wise 1981, 23, Wise 1990, 346, Jungnickel 1986, 137, Archibald 1989.

12. On the connection between Fechner's and Weber's atomism cf. Wiederkehr 1960, chapter IV, 148ff., Wise 1981, 279, note 27, Wise 1982. Cf. also Molella 1973, 121f.

13. On Weber's electrodynamics see Wiederkehr 1960 and 1967; 90ff., Molella 1973, chapter V; Kaiser 1977 and 1981 and 1982a; Wise 1982, 1990; Archibald 1989.

14. For the history of atomism in the nineteenth century see Buchdahl 1959, Gardner 1979, Meyenn 1982, Priesner 1982, Kaiser 1984, Carrier 1990. Lange 1875 (sec. 2, II) is also readable and informative; in part also Buek 1905.

15. Cf. Köhnke 1986, 121–67, Schnädelbach 1983, chapter 3.

16. Immanuel Hermann Fichte played a major role: Fichte 1854.

17. On *The Theory of Atoms* see Molella 1973, chapters 3 and 4. Molella views Fechner's work as a continuation of the metaphysical atomism tradition; I do not share this assessment. Cf. also Kaiser 1984, 224–27, Wise 1981, 1982, Wiederkehr 1967, 148ff., Siegel 1913, Buek 1905, 92–94, 151–54, Lange 1875, II, 639–44.

18. Fechner 1855, III. Mach's conversion to anti-atomism occurred in the period between the first and second printing of Fechner's *Theory of Atoms*. For this reason I almost always refer to the first edition, although the second edition is more comprehensive.

19. Fechner 1855, IX.

20. Ibid (there are some differences in wording in the separate editions); Wiederkehr 1994.

21. Fechner 1855, chapters II and III. It is typical for Fechner to refer to Wilhelm Weber here.

22. Fechner 1855, 18ff.

23. Fechner 1855, 19.

24. Fechner 1855, 32.

25. Fechner 1854, 35.

26. Fechner 1855, 51f. Cf. also Fechner 1854, 41 and 43.

27. Fechner 1855, 28.

28. Fechner 1855, 29.

29. Fechner 1855, 28f. Cf. also 30, 88 and 92.

30. Fechner 1861, 216.

31. Fechner 1855, 52.

32. Fechner 1855, 105.

33. Fechner 1855, 119. Cf. also Fechner 1857, 73ff.

34. Fechner 1885, 33f.

35. Fechner 1855, 130, 161ff. The physicist Wilhelm Weber also advocated the view, as Fechner notes, that the concept of atomic mass must not necessarily imply spatial extension (ibid., 73; see also Lange 1875, 644). On the topic of point atomism in Germany in the nineteenth century see Snelders 1971, Rocke 1984, 128, 134, 144. As early as 1828 Fechner calls atoms "material points" (Fechner 1828a, 258).

36. Fechner 1855, 133. Cf. also Fechner 1864, 153, 239–44.

37. Fechner 1855, 80 and Fechner 1864, 95f.

38. Lange 1875, 640, 644. By obliterating the difference to dynamism Fechner's theory positions itself close to Schelling's "dynamic atomism"; Schelling 1799, 22f. Cf. also Bonsiepen 1988.

39. Fechner 1855, 126f.

40. Fechner 1855, 104.

41. Fechner 1857, 185.

42. Fechner 1855, 181ff. (chapter VII).

43. Fechner 1855, 193f.

44. Fechner 1855, 185.

45. For a history of short-distance molecular forces see Brush 1970. The most famous is probably Maxwell's force of repulsion with r^5, conceived of in 1867.

46. Fechner also considers the possibility of using atomism to find absolute units of measurement for physics (Fechner 1855, 197f.). In 1870 Maxwell attempted to do the same.

47. Fechner 1855, 113.

48. Fechner 1855, 94.

49. Fechner 1861, 207.

50. Lipps 1905, 282.

51. Fechner 1861, 213.

52. Fechner 1855, 111.

53. Fechner 1855, 107 (see Lange 1875, 664). Cf. also Fechner 1855, 90, 113, 179–81 and Fechner 1857, 177ff.: Here Fechner guards himself against attacks launched by Drobisch (a devotee of Herbart).

54. Fechner 1864, 117f., 135f. It is remarkable that even the physiologist Ernst Heinrich Weber identified force with volition, in a manner reminiscent of Schopenhauer. (Fechner 1864, 132).

55. Fechner 1857, 177f.

56. Fechner 1855, 112. "All forces are . . . in the end reducible to auxiliary concepts for describing the laws of balance and motion, according to which new positions and motion result from previous ones, or the previous ones are sustained. They are only relational concepts, that by definition presuppose entities, between which a relation exists." (Ibid. 72)

57. In Lotze 1857, 35–39, 25–29.

58. Fechner 1855, 180. Twenty-one years later Gustav R. Kirchhoff attracted considerable attention by making a similar remark in his *Lectures on Mathematical Physics*.

59. Fechner 1857, 190f.

60. Fechner 1861, 214.

61. The notion that non-perceived appearances may exist is not as strange as it may at first seem. Bertrand Russell (in *Mysticism and Logic*) wrote that *sensibilia* can exist without being perceived.

62. Fechner 1861, 125. The resemblance to Charles Sanders Peirce's pragmatism is striking. For similar reasons, Peirce sees "nominalism" as the *bête noire* that must be replaced by realism. Cf., for instance, Peirce's comments on Fraser's Berkeley Edition in Peirce 1960, vol. 8, §7ff.

63. Lange 1875, II, 639.

64. Fechner replies to eight different critical reviews (Fechner 1857, 64 and 190). Fechner 1858a deals with Schaller's critical comments. See Schaller 1857, Weisse 1855, Lotze 1855, Drobisch 1856. However, Lotze 1857, 17–50 defends Fechner's conception. Later Eduard von Hartmann joined Friedrich Albert Lange in criticizing Fechner; cf. Hartmann 1870.

65. On Mach's anti-atomism see Herneck 1956 and 1966, Seaman 1968, Brush 1968, Hiebert 1970, Laudan 1976, Curd 1978, chapter V, Feyerabend 1984, Blackmore 1988, Wolters 1988a. To my knowledge the only recent work on Mach that does consider Fechner more than merely marginally are Swoboda 1974, chapter V, Swoboda 1982, Wolters 1988. Newer

books on Mach in general include Haller 1988 and 1991, Hoffmann 1991, Blackmore 1992, 2001.

66. Mach 1910, 12, Herneck 1956, 212, Mach 1922, 24.

67. Mach 1863, 3, 12–18.

68. Mach 1863, VI.

69. Mach 1864, 69. Apparently Mach was moved to announce a new metaphysical view of matter by the reproach that his remarks were too close to philosophies of nature. See Swoboda 1982, 267f.

70. In light of these quotations I cannot see how anyone can claim that Mach, at that time, did not believe in the reality of atoms. For instance Blackmore, 1972, 33: "It has sometimes been alleged that Mach accepted the atomic theory during the period of his early scientific experiments, that is, between 1860 and 1863, but while there was a sense in which this was true, nonetheless, the assertion has been quite misleading. Mach never believed in the reality of atoms or in the indispensable value of the atomic theory in any normal sense." Cf. also ibid., p. 57, where Blackmore speaks of "Mach's 1863 rejection [!] of the atomic theory." Laudan 1981, 205 calls Mach's *Compendium* "striking for the reservations it voices about atomic doctrines." The only "reservation" I can discover here is Mach's reflection that metaphysical opinions about matter are susceptible to change.

71. Mach 1872, 26.

72. Mach 1872, 33.

73. Mach 1863a, 364.

74. Mach 1863a, 364.

75. Mach 1863a, 203f., 337; Blackmore 1992, 2001. Fechner is also skeptical of Herbart's theory of spatiality, cf. Fechner 1857, 184f. After critically commenting on Herbart's psychology of space, Mach discusses, as improved alternatives, Lotze's theory of local signs and how that theory was further developed by Wundt.

76. Mach 1872, 115.

77. Mach 1863a, 364.

78. Mach 1863a, 364.

79. Mach 1922, 24.

80. Herbart 1825, 286, 153.

81. Fechner 1860. On Fechner's effect on Mach in general, and particularly the influence of *Psychophysics* see Wolters 1988 and Swoboda 1982.

82. This was to become Mach 1865a.

83. Mach 1863a, 260, 169.

84. Mach 1886, 301: "*Fechner's* psychophysics, which had such an important effect, greatly stimulated me too, back then. Excited by this book, I gave fairly poor lectures." See also Swoboda 1982, 270 and Mach 1872, 55.

85. Letter from Mach to Fechner dated January 14, 1861; printed in Wolters 1988, 103.

86. Thiele 1966, 223ff. For the following see also Wolters 1988.

87. Mach 1886, VII (Foreword to the first edition).

88. Mach 1923, 562f. See Wolters 1988, 105. See also the letter from Mach to James in the appendix below.

89. Fechner 1855, 164. The entire chapter 5 in Fechner 1855 deals with the relationship between Herbart's monadology and atomism. Cf. also Fechner 1857, 184f.

90. Fechner 1854, 47.

91. On Grassmann: Fechner 1864, 237; on Lotze: Fechner 1857, 165ff.; on Hartmann: Hartmann 1870.

92. Mach 1872a. Zöllner's remark is in Zöllner 1872, 320.

93. Mach 1863a, 169. By proving the measurability of inner states, psychophysics had refuted the main objection to Herbart's mathematical psychology, namely that the psychical by its very nature is immensurable.

94. Mach 1910, 12.

95. Mach 1886, 24. Cf. also Mach 1910, 12.

96. Mach 1863a, 365. Cf. also Mach 1866, 1: "Today epoch-making combinations of apparently dissimilar sciences are no longer rare. Remember how mathematical, scientific methods were introduced into psychology by Herbart, how Mill, Quetelet, and others gave politics, statistics, and national economics a scientific treatment, and how Fechner combined physics and psychology to become psychophysics."

97. Mach 1863a, 170 and 202.

98. Mach 1863a, 364f.

99. Fechner 1860, I, 8.

100. Fechner 1860, I, 3 and 5.

101. Mach 1863a, 363.

102. Mach 1872, 29.

103. Mach 1872, 30. Cf. also ibid. 55. Cf. Lange's commentary on this: Lange 1875, 658. Mach's second attempt deals with the theory of heat. Heat must not necessarily be thought of as motion or matter. "Matter is possible appearance; it is the fitting expression for a conceptual gap." (Mach 1872, 25) Beyond this it is Mach's goal (in Mach 1872) to prove that the principle of the excluded perpetuum mobile does not depend on a mechanical worldview, but was discovered independently and before any mechanics by empirical method and is compatible with every alternative to mechanism that acknowledges the causal principle.

104. Helmholtz 1871, 45. Mach reports how he originally shocked his contemporaries with his ideas. "But now that some honorable scientists have begun to venture into the same region, they allow me to contribute my two bits towards clarifying the relevant issues." (Mach 1872, 2) Those "honorable scientists" were probably Zöllner and Riemann. In a note attached to this passage Mach emphasizes that he developed his own views without having read Riemann (Ibid., 55).

105. Fechner 1860, I, 7.

106. In Mach 1919, 12 Mach calls his non-metaphysical standpoint a "product of general cultural development."

107. Kraft 1918, 1210.

Notes to Chapter Five

1. See Feigl 1958, Place 1956 and Smart 1959.

2. Kim 1998, 1

3. Kim 1998, 1, Cp. also Kim 1997.

4. Jackson 1998, 395; see also Bieri 1997, 5–11.

5. Kim 1998, 2.

6. To my knowledge, the only author writing in English aware of identity theory's long anti-Cartesian prehistory is Milič Čapek. Cf. Čapek 1969, which provides valuable information.

7. More recent portrayals of nineteenth-century materialism are given in Wittkau-Horgby 1998 and Heidelberger 1998. The classic study is Gregory 1977. See also Schnädelbach 1983 and Ziegler 1899, chap. 11.

8. Vogt 1847, 1:206.

9. Virchow, as quoted by Büchner 1855, 274.

10. Carnap 1963, 11. For the history of monism cp. Gabriel 2000.

11. Lange 1873. For Lange see Teo 2002, Köhnke 1986, Krösche 1910.

12. The most important source for Fechner's psychophysical parallelism is the foreword and introduction to *Elements of Psychophysics*. See Fechner 1860, 1:vii–xiii, 1–20. For other accounts of Fechner's mind-body theory cp. Lennig 1994a, Heidelberger 1988 and Woodward 1972.

13. Case 1911, 235, 234.

14. Bain 1874. See Mischel 1970, 10.

15. This is probably partly due to Bertrand Russell's incorrect portrayal of psychophysical parallelism in *Analysis of Mind* (Russell 1921, 35). Therein he claimed that modern psychophysical parallelism is hardly distinguishable from Cartesian theory. For later authors expressing similar views, see Armstrong 1993, 8 and loci mentioned in that book's index; Heil 1998, 27; and (lacking all knowledge of the German tradition) Bieri 1997, 7. See also note 54 below.

16. The situation is, in fact, much more complicated. Occasionalism is logically *compatible* with (but not identical to) the first version of psychophysical parallelism, but not with the second version, which is, to a greater degree, of philosophical interest. (See the following discussion.)

17. Leibniz 1714, §78.

18. See chap. 2 above.

19. See above, ch. 2.

20. James 1890, 1:182.

21. See above, ch. 4.

22. For a detailed comparison of Fechner's and Mach's mind-body theories, see Heidelberger 2000a; 2000b; and below chap. 4, §4.4.

23. See above chap. 3, §3.2.

24. See, for example, Erdmann 1907.

25. See above chap. 7; Leibniz 1714, §79.

26. Leibniz 1714, §81. See Külpe 1898 for a sophisticated discussion of causality's role in the mind-body problem.

27. James 1904; Mach 1886, 14, 35, 50; Russell 1921, chap. 1. See also Hatfield 2003.

28. A contemporary's survey of the issues is given in Busse 1913, with an excellent appendix.

29. Riehl 1872, 128; 1887, 176–216; 1894, 167–205. Cf. also Riehl 1921, 112–46.

30. See Place 1990, 22.

31. Wundt 1894, 42.

32. Wundt 1863, 1:509, 513; see also 487.

33. Wundt 1894, 42.

34. Wundt 1880, 67.

35. Müller, Georg Elias 1896, §§1, 3, 5. Cp. also Müller, Georg Elias 1904 and Külpe, 1898, 114–120.

36. Peirce 1903, §15–126.

37. Bovet 1922, 902. Thomas Ryckman kindly drew my attention to this source.

38. See Favrholdt's introduction to Bohr 1999, xliii, 7. Favrholdt is wrong in claiming that Höffding's psychophysical parallelism is indebted to Leibniz (xliv).

39. Höffding 1893, 92. See also Höffding 1891; 1903, 26–30. On page 30 of this latter work, Höffding dismisses the term "parallelism" as actually being inappropriate and ambiguous and prefers "identity theory" instead.

40. Neumann 1932, 223–24, 262 n. 207. Neumann also feels obligated to discuss this topic with Leo Szilárd (262 n. 208).

41. See Heidelberger 1994a.

42. See Heidelberger 1993b, 493; 1997, 43–47.

43. The most notorious criticism in this respect might have come from Du Bois-Reymond 1872, 464.

44. H. A. Lorentz to theologian H. Y. Groenewegen, April 5, 1915. Inv.-No. 27, Archive H. A. Lorentz, Rijksarchief Noord-Holland, Haarlem, The Netherlands. Private information of Dr. L. T. G. Theunissen, Institute for the History of Science, Utrecht University. I would like to thank Bert Theunissen for permission to quote from Lorentz's letter, which he discovered while working on a project with his colleague Henk Klomp.

45. Vaihinger emphasized using Lotze's interpretation of "moderate occasionalism."

46. Lange 1968, 358.

47. See below chap. 6.

48. Sigwart 1911, 2:542–600.

49. Kronstorfer 1928, 173; see also 95.

50. Dilthey 1982, 281; see also 279.

51. Dilthey 1894.

52. Dilthey 1974, 142.

53. Ebbinghaus 1896. For recent treatments of the Dilthey-Ebbinghaus controversy see Harrington, Austin 2000, Gerhardt et al. 1999, 162–68, Schmidt, Nicole D. 1995, 37–60, Kusch 1995, 162–69 and Rodi 1987. See also Lessing 1985.

54. James 1890, 1:129, 138. James attributes the "conscious automaton theory" to Shadworth Holloway Hodgson, D. A. Spalding, Thomas Henry Huxley and William Kingdon Clifford. Ibid. 130. Chapter 5 in James's *Principles*, where these quotations are to be found, is titled "The Automaton-Theory" and appeared in almost the same wording in 1879 in *Mind*. A discussion of James's arguments is given in Čapek 1954. It is a serious mistake and highly misleading to characterize James's concept of automaton-theory as "logically identical to the sort of parallelism familiar from the writings of Leibniz and Malebranche," as Owen Flanagan puts it in his article (1997, 32; cf. also note 15 above).

55. Husserl 1911, 9.

56. Stumpf 1896.

57. See Pester 1997, chaps. 3.3 and 3.5 on Hermann Lotze's sophisticated methodical occasionalism. The Cambridge philosopher James Ward 1902, 66–69 and 1911, 600–603, who had studied with Lotze for some time, gives a remarkable defense of Lotze's view, containing one of the very rare presentations of psychophysical parallelism written in English.

58. Rickert 1900.

59. See Höfler 1897, 57–63. Höfler mentions a discussion he had with Boltzmann on this topic (58 n).

60. In dealing with the mind-body problem, Jodl appears to have been influenced by Alois Riehl. On the relationships of Breuer and Freud to Fechner, see above, chap. 1, sect.

1.11 and below, chap. 7, sect. 7.3. For further information on the situation in Austria, see below. chap. 6. On Mach's relationship to Fechner, see above, chap. 4.

61. Cp. Natsoulas 1984.

62. Jodl 1896, 74.

63. Taine 1870, pt. 1, bk. 4, chap. 2, §§4, 5.

64. Höffding 1893, 85.

65. Reininger 1916.

66. See Kronstorfer 1928, p. iv of the bibliography.

67. Schlick 1925, 336.

68. Schlick 1925, 351, 335, 336.

69. Schlick to Ernst Cassirer, Vienna, March 30, 1927, Moritz Schlick, Papers, Inv. No. 94. Used with kind permission from the Vienna Circle Stichting, Amsterdam, and the Philosophical Archive of the University at Constance, which let me see copies of the Schlick literary collection.

70. Schlick 1925, 324, 329.

71. Schlick 1925, 348–50.

72. Schlick 1925, 350.

73. Riehl 1887, 38; 1894, 40.

74. Riehl 1887, 196; 1894, 185.

75. Riehl 1887, 196; 1894, 185.

76. Feigl 1937–38, 413. Feigl also wrote that Schlick, "perhaps with greater clarity than all other monistic critical realists of the times[,] elaborated a physicalistic identity theory" worth being rediscovered "in modern semantic terms" (Feigl 1963, 261, 254). Once again, Riehl is explicitly mentioned as the scholar to whose views in these matters Schlick comes the closest. See also Feigl 1950, 614, where Schlick's view is called "double-language theory." In this article Feigl repeatedly characterizes his own theory as "identity or double-language view of mind and body" (617, 624, 626).

77. Carnap 1928, §22.

78. Carnap 1928, §166.

79. Carnap 1928, §22.

80. Carnap 1928, §169.

81. See Schlick 1925, 11.

82. Feigl 1964, 231.

83. Feigl 1964, 243.

84. Feigl 1964, 242.

85. Feigl 1934, 436. Cf. Feigl's own elucidation of his standpoint in 1934 in Feigl 1958, 23. In Feigl 1950 Feigl reports Felix Kaufmann's insistence that strict identity must be understood logically. Cf. also Sturma 1998.

86. Cf., for instance, Schlick 1925, 347.

87. Feigl 1958, v; see also 79 n.

88. Feigl 1958, 62, 109.

89. Feigl 1958, 94. Cf. also 95–97, 104. See also Feigl 1950, 616–17, where he refers to the "principle of parsimony."

90. Feigl 1958, 106, 109.

91. Feigl 1958, 138.

92. For the latter, see Place 1988, 1989, 1990, 1994.

93. Stubenberg 1997; see also Sturma 1998.

Notes to Chapter Six

1. Peirce 1900, 7.258.

2. The most comprehensive portrayals of Fechner's *Psychophysics* to date are: Foucault 1901, Gutberlet 1905, Titchener 1905, Wundt 1908, chapter 9. Titchener relies heavily on Foucault's exposition. A brief survey can be found in Boring 1942, 34–45, 50–52 and Boring 1950, chapter 14. Cp. Link 1992, 1994, 2001, 2002, Brauns 1996, 2000, and Nicolas 2002.

3. Fechner 1860, I, 8. In Fechner 1877, 2, psychophysics is additionally defined as the science that "preferably" deals with the "relations of measurement" holding between the psychical and the physical world.

4. Fechner 1860, I, 8.

5. Fechner 1851, 317; see also Fechner 1882, 257 and 1882a, 217.

6. Fechner 1860, I, 1f. Cf. also 1882, 257.

7. Marshall 1988, 40; cf. also Gregory 1981, 505f. Höffding 1891 and 1893, 65–93, especially 85. See also Bohr 1999.

8. Fechner 1858, 6; 1860, I, 55; 1887, 179; 1882, VII. The basic idea of the measurement principle and of measuring the mental was presented as early as 1851 in the second appendix to vol. 2 of *Zend-Avesta* (Fechner 1851, II, 373–86).

9. Fechner 1858, 2; 1887, 179.

10. Fechner 1887, 181.

11. Fechner 1858, 2; 1860, I, 56.

12. See Falmagne 1985.

13. According to Fechner 1887, 179.

14. Fechner 1858, 8 and 1860, I, 1.

15. Fechner 1858, 2.

16. Fechner 1858, 4.

17. Fechner 1858, 2.

18. Fechner 1855, 170–72.

19. Fechner 1887, 217, cf. also 215.

20. Fechner 1860, I, 45f. Cf. Fechner 1855, 180 and Fechner 1877, 142.

21. Fechner 1858, 4.

22. Fechner 1858, 4; cf. also Fechner 1887, 214f.

23. Fechner 1858, 4.

24. Fechner 1879, 200.

25. Fechner 1879, 201.

26. Fechner 1879, 201.

27. Fechner 1858, 4.

28. Fechner 1858, 7.

29. Fechner 1860, I, 58.

30. For more recent descriptions of Fechner's idea of measuring sensations see: Heidelberger 1993, Murray 1993, Gescheider 1985, Tiberghien 1984, 78–89, Murray 1983a, 153–58, Gregory 1981, 500–6; Falmagne 1974, Lowry 1971, 96–100. On Fechner's doctrine of method in psychophysics see Brauns 1990. Interesting descriptions are also in Mausfeld 1985, chapter 1 and Binswanger 1922, 76–90.

31. Fechner 1887, 180, 185, 199; Fechner 1877, 42f.

32. Fechner 1860, II, 10.

33. See Plateau 1872 and Laming/Laming 1996.

34. Fechner 1860, II, 191f.

35. Fechner 1860, II, 33ff.

36. Fechner 1860, I, 238.

37. Fechner 1860, II, 17.

38. Elsas made this claim in Elsas 1888. For general cases it was renewed in Luce 1958. In making this transition Fechner had made use of an auxiliary principle found in a mathematics textbook by Cournot. For Weber's Law it was admissible as an approximation, but it is not valid for general cases, where Weber's Law does not apply. Cf. Luce 1958 and a discussion in Mausfeld 1985, 63.

39. Read Fechner's reactions in Fechner 1877, 1882, 1887. Our contemporary controversies in psychophysics are stated clearly in Warren 1981, Krueger 1989, and Murray 1993. See also Laming 1994 and Norwich 1997.

40. Fechner 1851, II, 373; 1882a, 213f.

41. The interpretation addressed here is found, for instance, in Scheerer 1988, 274, who writes that for Fechner inner perception is at best semi-quantitative, and full quantification is reserved for external appearances. For the following paragraphs cf. Mausfeld 1985, chapter 1, and Mausfeld 1988, and also Murray 1983, 1983a, 1987, 1990.

42. "The Psychophysical Function: Harmonizing Fechner and Stevens." Wasserman 1979. Cf. also Warren 1981, Krueger 1989.

43. Newman 1974, 137. See Stevens 1946 and 1951.

44. Luce 1958, 225; Mausfeld 1985, 58ff.

45. Mausfeld 1985, 3. That renaissance is probably due to work done by Luce and Falmagne: See, for instance, Gescheider 1985 and the bibliography in Mausfeld 1985. For the development of psychophysics after Fechner cp. esp. Laming 1994 and Link 1992. See also the recent literature mentioned in the introduction above.

46. See, for instance, Gescheider 1985, 144ff.

47. Fechner 1882, 262; 1882a.

48. Fechner 1860, I, 67f. Observe that Fechner often calls his measurement formula, i.e. Fechner's Law, "Weber's Law" because he considers his own principle a simplified and tautological transformation of Weber's Law.

49. Fechner 1877, 14.

50. Fechner 1882, 262. See also 1882a, 218, 226. Cf. Scheerer 1987, 1987a, 1988, 1989, 1992 for informative discussions on Fechner's inner psychophysics.

51. Hering 1909 provides a comprehensive discussion of these three types of interpretation.

52. Ribot was also editor of the influential journal *Revue philosophique*, which for many years was the major medium of the French reception of Fechner's work. Ribot's series of articles was published later as a book that was widely read and also translated into German and English; Ribot 1879. Cf. Benno Erdmann's informative review; Erdmann 1879.

53. The articles and letters in their correct sequence are: Ribot 1874, 1875, Tannery 1875, Ribot 1875a, Wundt 1875, Tannery 1875a, Delboeuf 1875a. Ribot then continued his article on Wundt with 1875b and 1875c.

54. On Jules Tannery cf. Picard 1925. The *Revue scientifique*, 4 année, No. 29, 16 janvier 1875, 676f. reports on Tannery's dissertation. Jules Tannery is a brother of Paul Tannery (1843–1904), a scholar famous for his investigations into ancient history of science and also

for editing Descartes's works. Paul Tannery was also the one to reveal that it was his brother writing anonymously to the *Revue scientifique*; Tannery 1883. In Tannery 1888 he also reviewed Elsas's 1886 and Köhler's 1886 critiques of Fechner.

55. Tannery 1875, 876, cf. also 1875a, 1019 (Translation of the French text by MH.)

56. Tannery 1875a, 1019.

57. "I can assure you, that I have this logarithm on my mind." Tannery 1875, 877.

58. Boring 1921.

59. Tannery 1875, 876.

60. Bergson 1889, 61. His critique on Fechner is in ibid., 55–66; cf. the entire first chapter for Bergson's critique of contemporary psychology concerned with measurement. On Fechner and Bergson cf. Meissner 1962, Penna 1988, Romanos 1959.

61. Cf. Köhnke 1986, part 3, particularly 404ff., 319ff.

62. Titchener 1905, XLIX, Boring 1921. Neither Titchener nor Boring gave due credit to Tannery's key role in these events.

63. Boring 1921, 453.

64. Hornstein 1988, 6: "To take the quantity objection seriously implied retaining introspective data and ties with metaphysical philosophy."

65. Bernstein 1868.

66. Müller 1878, 835.

67. Fechner 1882, 256. On G. E. Müller cf. Blumenthal 1985, 53–61, Sprung 1997, 2000; Haupt 1998.

68. Liebmann, 1877, 515 and 517. Cf. also Liebmann's very positive portrayal of Fechner in Liebmann 1870. Liebmann had also heard lectures given by Fechner while he was a student in Leipzig.

69. Delboeuf 1878, 98; cf. 117. Joseph Rémy Léopold Delboeuf was a pupil of Plateau and a good friend of William James. Delboeuf had studied in Bonn from 1858–59 (see the *Zeitschrift für Philosophie und philosophische Kritik* 37, 1860, 153). See Fechner 1874b for a review of Delboeuf 1873. On Delboeuf see also Nicolas et. al. 1997.

70. Stadler 1878, 216. Cf. Holzhey, I, 25 and other places; Schulthess 1984, 10.

71. Stadler 1878, 222f. I have adapted Stadler's notation to match my own.

72. Stadler 1878, 219.

73. Stadler 1878, 223.

74. Stadler 1878, 223.

75. Stadler 1880, 583f.

76. This term was used by the professor for mathematics and physics in Brunswick, A. Wernicke, in a review written on Elsas 1886; Wernicke 1887, cf. Holzhey 1986, I, 25f. Cohen was professor in Marburg as of 1876. On Cohen cf. Edel 1988. On the whole Marburg movement of neo-Kantianism see Sieg 1994.

77. Holzhey 1986, I, 22.

78. Holzhey 1986, 381.

79. Müller 1882, p. Vf. On Müller see Holzhey 1986, I, 22; II, 171.

80. Müller 1882, 12.

81. Fechner 1860, II, chapter 31.

82. Müller 1882, 19.

83. Hering 1875.

84. Müller 1882, 33.

85. Müller 1882, 56.

86. Müller 1882, 117.

87. Cf. the declaration of 106 university teachers of philosophy dated February 12, 1912, protesting against the practice of filling philosophical chairs with representatives of experimental psychology (reprinted in Holzhey 1986, II, 519–21).

88. Cohen 1910, 411: "I am aware that I was encouraged to more fully and fundamentally dispute this problem by this [Stadler's] review." On Cohen's critique of psychology in general cf. Winrich de Schmidt 1976.

89. Ollig 1979, 1.

90. Cohen 1883, 129, §89.

91. Cohen 1883, 20, §25.

92. Cohen 1883, 107, §78.

93. Fechner 1860, I, 15.

94. Cohen 1883, 28, 26, 23, 107 (§§33, 31, 28, 78).

95. Cohen 1883, 17, §22.

96. Cohen 1883, 28, §33. Cf. also 26, §31.

97. Cohen 1883, 133, §92.

98. Cohen 1883, 144, §99.

99. Cohen 1883, 142, §97.

100. Funkenstein 1986 thinks he has identified Cohen as the first to have argued that observation is loaded with theory. "Science, as viewed by the School of Marburg, does not derive its validity . . . from principles outside itself. It creates of itself its methods and objects, facts, theories, and the criteria of discrimination between theories and facts." (Funkenstein 1986, 40) This may be the case, but for Cohen it was unthinkable that whatever science has brought forth can be *changed*, and this renders any real *history* of theories impossible. Thus it is absurd to want to make a methodologist for the history of science out of him.

101. Cohen 1883, 77, §60; 5, §7; 108, §78. Cf. also 28, §33; 153, §106.

102. Cohen 1883, 27, §32. Cf. also 151, §105.

103. Cohen 1883, 153, §106.

104. Cohen 1883, 40, §44.

105. Cohen 1883, 153, §106.

106. Cohen 1883, 109, §78.

107. Descartes, Third Meditation, 11.

108. Cohen 1883, 127, §88.

109. Cohen 1883, 155, §107.

110. Cohen 1883, 155, §108.

111. Cohen 1883, 157, §109.

112. Cohen 1883, 158, §110.

113. Cohen 1883, 161, §112. Cohen refers to Stadler's essay of 1878.

114. Cohen 1883, 152, §105.

115. Cohen 1883, 160, §111.

117. Cohen 1883, 162, §112.

118. It is interesting that three years after Cohen's piece was published F. A. Müller, the same who had been one of the first to criticize Fechner in a neo-Kantian manner, lost the favor of the Marburg School. He reproached Cohen for having started with the right idea, but then committing the same error of which he had accused Fechner, by attributing intensive value to the differential, when in reality it can only be characteristic of a real object. (Müller 1886, 96. See also Schulthess 1984, 41.)

119. Natorp 1888, 88 and 91. See also Natorp 1893, 611. More on Natorp's psychology can be found in Cramer 1974, Winrich de Schmidt 1976, and Binswanger 1922, 91–97.

120. Cf. Flach 1968, 34; Schulthess 1984, 37ff.

121. Kries 1925, 147.

122. Kries 1925, 166.

123. Kries 1925, 166.

124. Kries 1925, 155.

125. Kries 1882. See also the revealing discussion of this article in Kries 1925, 155–56, 168, 178f. In Kries 1905, 24 he sustains his criticism, making reference to the article dated 1882.

126. Kries 1882, 258.

127. Kries 1882, 260.

128. Kries 1882, 267f.

129. Kries 1882, 273.

130. Kries 1882, 275.

131. Kries 1882, 276.

132. Kries 1882, 274f.

133. Elsas was one of Helmholtz's students and a friend of Heinrich Hertz. Cf. *In memoriam* in Cohen 1895.

134. Holzhey 1986, I, 381.

135. Elsas 1886, p. VI; see Höfler 415.

136. Elsas 1886, 58.

137. Elsas, 1886, 61 and 62. Elsas's view is partially already apparent above in the last quotation of Tannery.

138. Elsas 1886, 65.

139. Elsas 1886, 67.

140. Elsas 1886, 70.

141. Lasswitz 1887, 3.

141. Elsas 1886, 70.

142. Natorp 1890, 152.

143. Carnap 1928, 91, §66. Cf. also 3 (§2), 139 (§100), and 148 (§49).

144. On the problem of accessibility to the Given as it was discussed in the Vienna Circle see Heidelberger 1985, particularly 151–57. On Carnap's neo-Kantian inheritance see Sauer 1985, Friedman 1987, 529f.

145. Zeller 1881, 1882. Paul Du Bois-Reymond (1882, 29–36), a brother of the more famous Emil, engaged in the debate at the same time.

146. Cf. Helmholtz 1887. Modern expressions of the axiom for extensive magnitudes can be found in Suppes 1980, 417 or Carnap 1966, part II. Cohen, incidentally, reacted to Helmholtz's essay (Cohen 1888), complaining that Helmholtz had not understood the importance of infinitesimal magnitude as the "generative magnitude and unit of reality."

147. Helmholtz 1887, 93.

148. Helmholtz 1903, 11.

149. Wundt 1887, 358; Wundt 1901, 308f.

150. Wundt 1904, 44f.

151. Wundt 1904, 42. In Wundt 1887, 358, the funeral oration for Fechner, Wundt had still claimed that Fechner had "become the founder of experimental psychology" by showing "that exact methods can be applied . . . not only to outside events in nature itself, but also to the facts of consciousness."

152. Wundt 1908.

153. Mach 1886, 1, 34.

154. See Mach 1896, 39–57. Cf. Ellis 1966, chapter VI on this topic.

155. Mach 1896, 39.

156. For example: We need not open the window in order to discover how cold it is outdoors. Instead, we peer through the glass, read the thermometer, and behave appropriately.

157. Mach 1896, 40.

158. Mach 1886, 157.

159. Mach 1896, 44.

160. Since for our purposes the main emphasis is on an exposition of the general idea of Mach's theory of measurement, I have not exhaustively described the criteria which intensive magnitudes must fulfill. An interested reader will find these in Carnap 1966, part II.

161. Mach 1896, 48 and 51.

162. Mach 1896, 51.

163. Mach 1896, 52. This is the only time that Mach mentions space and time in this chapter. For Einstein's reading of Mach's Science of Mechanics and Principles of the Theory of Heat cp. Einstein 1987, 229–31. (Document 54, letter to Mileva Marić of September 10, 1899, note 8).

164. Herneck 1966, 153f., Holton 1973, 232; 1988, 249f. On Einstein's appreciation of Mach cf. Einstein 1916.

165. Planck 1887, 171f.

166. Planck 1909, 47–51.

167. Fechner 1882, 325.

168. Mach 1896, 51.

169. Mausfeld 1985, Foreword.

170. Mach 1886, 282 and 67, Footnote 3. S. S. Stevens drew attention to this passage and suggested that Mach rejects measuring sensations in the way that our *present-day* meaning of 'measuring' is understood; Stevens 1975, 59; also Adler 1977.

171. Mach 1896, 281, almost literally repeated in Mach 1886, 67.

172. Weber 1851.

173. Ebbinghaus 1890.

174. Schlick 1910, 135f., 137. Cf. also Schlick 1925, 323, 315f.

175. Mach 1868, 12.

176. Hering 1875, 330f.; cf. also Hering 1878, 76f., a text that had been previously published in 1874. For Hering in general see Turner 1993, 1994, Busse/Bäumer 1996, Baumann 2002. For psychology, physiology and physics at Prague university of the time see Ühlein et al. 1994.

177. Hering 1875, 310.

178. Hering 1875, 310; cf. Mach 1922, 50f., 321.

179. Mach 1865b, 320.

180. Fechner 1877, chapter 8; cf. Scheerer 1988, 275f. and Scheerer 1992. It is interesting that later Sigmund Exner defended Fechner's view against Hering; Exner 1879, 240–44.

181. Brentano 1874, 101, cf. also 11f.

182. Fechner 1877, 24f. On the hypothesis of relation and the exponential law ensuing from it see above, section 6.3.

183. On this issue see Fechner 1882, 322f. and 1887, 184. In this latter passage Fechner once again stresses that measuring sensation his way is even still possible, if the empirical

hypothesis that people evaluate just noticeable differences of sensation as equal, is wrong. As proof he names photometry, in which claims about equal astral luminous intensities are used to infer physical regularities.

184. Mach 1886, 280f.

185. The one scholar who struggled most to escape this scheme was Cassirer, in his book *Substance and Function* [Substanzbegriff und Funktionsbegriff] (1910). He tried to show that in the history of the sciences the concept of substance was gradually replaced by the concept of relation. Indeed, this did not mean that for him the concept of substance had been abrogated, but that substance became something like a Platonic idea, the "shadowy core" of which Ernst Mach spoke. In general the Marburg School did have an inclination for Platonism. See Scheler 1922, 284f.

186. Scheler 1922, 284f. In Scheler's opinion the Marburg School's historical interpretation of the history of philosophy was "plainly ruinous."

187. Ellis 1967, 242.

Notes to Chapter Seven

1. A brief discussion of Fechner 1873 is given in Kamminga 1980, 132–36 and Kamminga 1986. A recent discussion is Köchy 1998.

2. Fechner 1873, chapters I and II.

3. Fechner 1873, 12.

4. Fechner 1873, p. IV. On the second idea see basically Fechner 1873, chapters III and XI.

5. Fechner 1873, 27f.

6. Fechner stated the principle variously. This description is probably the clearest. It is found in Fechner 1879, 209. Cf. a more technical formulation in Fechner 1873, 30: "For every system left to itself or any system of material parts which is subject to constant external conditions, and therefore also part of the material system of the world, inasmuch as we consider it a closed system, we find, if we exclude infinite movements, continuous progress from unstable towards stable states, to a full or approximately stable terminal state."

7. Cf. Fechner 1873, 37 and Fechner 1879, 207.

8. Fechner 1873, 31.

9. Fechner 1873, 34f. Complete inertia of all particles in the universe would mean that the "vis viva" transformed into heat would be lost. But since the principle of the conservation of energy holds, what remains at the end of the universe's development are the thermic oscillations of inorganic molecules. Cf. Fechner 1860, I, 35; Fechner 1879, 205 and 209.

10. Fechner 1873, 38 and 32.

11. Fechner 1873, 39.

12. Fechner 1873, chapter VI.

13. Fechner 1873, 56.

14. Fechner 1873, 64.

15. Fechner 1873, 89 and 64.

16. Fechner 1873, chapters V and VIII.

17. Fechner 1879, 221 and Fechner 1873, 55. Cf. also Fechner 1873, 68 and 75f.

18. Fechner 1873, 99.

19. Fechner 1873, 93.

20. As his biographer Kuntze 1892, 205f. reports.

21. Fechner 1848a, 1. On Fechner's hedonistic ethics see chapter 1.

22. Fechner 1873, 102.

23. Mayr 1974.

24. Fechner 1851, I, 468.

25. See above, chapter 1, section 1.2; and literature mentioned there. A comparison of Schelling, Oken and Fechner is given in Hartung 1913.

26. Oken 1809, vol. I, p. VIIf. and §34. (p. 14f.).

27. Oken 1809, vol. I, §63 (p. 23).

28. Oken 1809, vol. I, §70. (p. 24). Cf. also §71.

29. Oken 1809, vol. II, §817f. (p. 10), §906f. and 909f. (p. 15). Scheerer 1984a, 1336–38 deals with the organism-universe analogy in Schelling's philosophy and how it was received by, among others, Oken and Fechner.

30. Using this notion, Oken (and Fechner too) escapes Kant's verdict on hylozoism. Kant said that the "possibility of living matter" contains an inconceivable contradiction, "because lifelessness, inertia, constitutes the essential trait of the same [i.e. of matter]." (Kant 1790, A 323/B 327.)

31. Oken 1809, vol. I, §72 (p. 25). Cf. also §26f. (p. 12).

32. Irregularity and spontaneity of (cosmic)organic movement can be a part of Laplace determinism. No claim is made about violating the cause-effect relation, it is just a statement about the arbitrariness of the universe's initial conditions. If Fechner had not made the assumption of "deterministic chaos," he could not have assumed the reinstatement of previous conditions.

33. Fechner 1873, 25. On Zöllner, see above, chapter 1.

34. Zöllner 1870, 77.

35. Fechner 1873, 33f. Leibniz used this example in his clock allegory. For this and the history of this resonance phenomenon see Du Bois-Reymond 1872, 470f.

36. Zöllner 1872, p. 326; cf. 321.

37. Köhnke 1986, 404–33.

38. Du Bois-Reymond 1872, 451.

39. Smaasen 1846, 161.

40. Du Bois-Reymond 1872, 451.

41. Zöllner 1872, 320ff.

42. See Möbius 1901, Paulsen 1895, passim (on Paulsen's relation to Fechner and Schopenhauer see Fritsch 1910), on Bölsche and his dependence on Fechner and Darwin see Kelly 1981, 36–56. Bölsche was, throughout his life as an author, one of Fechner's most eager propagandists. He was a friend of Bruno Wille's, who was also a Fechner follower and who defended Fechner's work against the claim of being vain speculation; Wille 1905.

43. Schmidt 1874, 4. The journal *Das Ausland*, which carried the review in question, was a main organ of the Darwinians.

44. N-e [anonymous] 1874, 597.

45. Riehl 1875, Perty 1874, Snell 1874, Ulrici 1874.

46. Preyer 1890, 11. On Preyer's understanding of Fechner see Kamminga 1980, 136–46.

47. Printed in Preyer 1880, 52; cf. also 319.

48. Lange 1875, II, 744, 746f.

49. Hering mentions Fechner in Hering 1878, 73–77 and in Hering 1906, 135. On Hering see Hirschmüller 1978, 57f., Hillebrand 1918, and Schwartze 1984.

50. Hering 1884, 113–17, 119.

51. Hering 1888, 36.

52. Hering 1888, 37.

53. Hering 1888, 44. Hering and Breuer discovered the self-regulation of aspiration (1868), also called the Hering-Breuer reflex. See Rothschuh 1972, 103. On Breuer, see below.

54. See Dubos 1973 for a survey on Bernard and his tradition; cf. also Rothschuh 1972 and Holmes 1986.

55. Mach 1886, 40. The principle of economy states that the laws of empirical science should ideally be the most frugal, simple, convenient expression of the facts and not, as the mechanistic worldview demands, reduce phenomena to movements of atoms.

56. Mach 1896, 381. In 1886, 41, Mach notes the similarity between Hering's thoughts on the behavior of living substance and Avenarius's doctrine that the central nervous system "has, not only as a whole, but also in its parts, a tendency to maintain itself, a tendency to uphold its state of balance."

57. Mach 1896, 382.

58. Mach 1922, 305.

59. See Mach 1863, 234; Mach 1922, 41, 268.

60. This dynamics of theory can be found first in Mach 1871, 46 (cf. also 31, 55, 60) and resurfaces in almost all later writing. "Continuity, economy and perseverance are mutually conditional; they are actually only different sides of one and the same property of healthy thought." (Mach 1922, 268; cf. also 273). "The objective of scientific economy is a most complete, coherent, unified, serene, *worldview*, hardly disturbed by new events, a world-view of the greatest *stability*." (Mach 1896, 366.) Cf. Kaiser 1982 in general on the problem of mechanism in Mach's biology.

61. In 1912 Petzoldt founded the "*Gesellschaft für positivistische Philosophie*" [Society for Positivistic Philosophy] in Berlin, which existed until 1922. In 1927 it was revived as the Berlin Group of the "*Internationale Gesellschaft für empirische Philosophie*" [International Society for Empirical Philosophy], and in 1931 renamed the "*Gesellschaft für wissenschaftliche Philosophie*" [Society for Scientific Philosophy]. Petzoldt was elected chairman a few weeks before his death. Members of this society included such brilliant philosophers of science as C. G. Hempel, H. Reichenbach and R. von Mises. See Hentschel 1990, 15–36, particularly on Petzoldt 9–10.

62. See Petzoldt 1890, 1894, 1923.

63. Lampa 1915. Petzoldt 1923, 215f. Lampa also discusses a series of physical examples for which we could say that they exhibit a tendency toward stability, and among them are some that satisfy Le Chatelier's principle.

64. In his "Self-Portrait" Freud writes: "I was always susceptible to G. Th. Fechner's ideas and have followed his example on important points." (Freud 1925, 86.) Binswanger 1956, 89f. quotes further statements by Freud concerning how he was influenced by Fechner. See Nitzschke 1989, Scheerer 1988, 278ff., Buggle 1969, Ellenberger 1956, Dorer 1932, 106–12, 159, 176f. on the huge influence that Fechner had on Freud. Link 2002 is also very instructive. In a dubious and inadequate article by Riepe 2002 Fechner's influence on Freud is played down or altogether denied.

65. Freud 1920, 38.

66. Freud 1920, 67ff.

67. Freud 1920, 4f. Pertaining to this, Freud quotes Fechner's explications of his (as I call it) fifth nature-philosophical idea; Fechner 1873, 94. Cf. also Freud 1924, 371f., where he once again mentions that the principle of pleasure is a special case of Fechner's tendency

toward stability. Cf. also Bernfeld 1930 for an interesting early discussion of the Freud/ Fechner principle of stability and a comparison with the law of entropy.

68. Freud 1920, 45 and 40.

69. Freud 1920, 53. Freud also finds his instinct doctrine analogous to August Weismann's (1834–1914) theory, according to which organic matter consists of potentially immortal cell plasma, serving the purpose of reproduction, and mortal soma.

70. Freud 1920. 63.

71. On Breuer see Hirschmüller 1978, Dorer 1932, 71–103, on Freud and Breuer's cooperation see Hirschmüller 1978, 178–256.

72. Kann 1969, 41, letter dated October 2, 1901.

73. Breuer 1902, 34. This lecture is discussed in Hirschmüller 1978, 61–63.

74. Breuer 1902, 34.

75. Breuer 1903, 313.

76. Kann 1969, 42.

77. Gerland 1875, Ratzel 1882, quoted from Mühlmann 1967, 124. On Ratzel and how Fechner influenced him see Ratzel 1901, Steinmetzler 1956, 137–42.

78. Köhler 1920, 250. On Gestalt theory cf. Schurig 1985, who suggests that interest in the system property of "wholeness" can be traced back to W. Roux and the vitalist H. Driesch. But the roots go deeper, namely to Fechner.

79. Köhler 1925, 193, 201, 198f.

80. Metzger 1975, 209f.

81. Fechner 1873, 32.

82. Herzberg 1929, 257.

83. Peirce 1878; 1884; 1900; Behrens 1987, Hacking 1988; 1990, 205, Adler 1977, 23. However, Peirce does quote from Fechner's *Theory of Atoms*, i.e., Fechner 1855, or 1864 in his evolutionary metaphysics; Peirce 1892c, 6.240. On Peirce in general see Hacking 1990, chapter 23, Hacking 1990a, 699f., Hacking 1983, Apel 1967, Holmes 1964, Schulz 1988. On Peirce's evolutionary metaphysics see Short 1983, Schulz 1988, 224–40, Anderson 1987, chapter 4. See also Heidelberger 1994b, Wohlgemuth 1993, Hookway 1997, Reynolds 1997, Heidelberger 2003a.

84. Peirce 1891, 6.12.

85. Peirce 1893, 6.301.

86. Peirce 1892c, 6.246, Peirce 1893, 6.301 and Peirce 1892a, 6.59.

87. Peirce 1892c, 6.265.

88. Peirce 1898, 7.471.

89. Peirce 1891, 6.33.

90. Fechner 1851, I, 459f.

91. Peirce 1893, 6.295.

92. Peirce 1893, 6.299, 6.302.

93. Peirce 1892c, 6.268.

94. Peirce 1891, 6.24.

95. Peirce 1891, 6.25. Elsewhere Peirce speaks of "matter to be specialized and partially deadened mind." (Peirce 1892b, 6.102).

96. Peirce 1891, 6.24.

97. Fechner 1873, 92.

98. Simon 1962, 200–205.

99. Krohn 1987, 446.

100. Prigogine 1973, 675, reference to Eigen 1971.

Notes to Chapter Eight

1. Fechner 1849 and 1849a. In later writings Fechner varies the topics of these lectures several times. The *Zend-Avesta* contains entire passages taken from the lecture on causality. As much as 30 years later he repeats this theme in the *Day View*. See Fechner 1851, I, 207–22, II, 258–312, Fechner 1864, 125 and 170ff. and Fechner 1879, chapter XVII. Scharf 1996 has recently given a perceptive analysis of Fechner 1849.

2. Fechner 1849, 53.

3. Fechner 1849, 50. Fechner does not mention the origins of the claims and arguments he discusses. One possible source is the French physiologist Xavier Bichat (1771–1802), who distinguished invariable mathematical laws in physics from variable, incalculable laws in physiology; Bichat 1801, p. XXXV and LIII; quoted from Hacking 1990, 14. For a history of the concept of emergence see Stephan 1998, 1999. See also Heidelberger 1994b for a comparison between Fechner's and Peirce's concepts of emergence.

4. Fechner 1849, 52.

5. Fechner 1849, 52.

6. Fechner 1849, 53.

7. Fechner 1849, 56. Instead of "incommensurability" (one of Fechner's and Ch. H. Weisse's favorite terms), it would have been better to use the term "immensurability."

8. Fechner 1849, 61.

9. Fechner 1849, 61.

10. Fechner 1849, 63f.

11. Helmholtz 1847, 4 and 3. In *Prolegomena* Kant speaks of "empirical rules" (§29, Academy Edition VI, 312).

12. Fechner 1849, 56.

13. Fechner 1849a, 98.

14. Fechner 1849a, 100.

15. These examples are mine. Fechner mentions similar examples in Fechner 1823. See Marshall's (partial) translation, Marshall 1974a.

16. Fechner 1849a, 101.

17. Fechner 1849a, 102.

18. Fechner 1849a, 106.

19. Fechner 1849a, 109 and 108.

20. Woodward 1972, 374ff. and Marshall 1982, 75 do so.

21. Fechner 1849a, 110.

22. Fechner 1849a, 110f.

23. Fechner 1849a, 111.

24. Fechner 1849a, 111.

25. Fechner 1849a, 112.

26. Fechner 1851, II, 277.

27. Fechner 1849a, 113.

28. Fechner 1849a, 114.

29. Fechner 1849a, 114.

30. Weber 1846, 212.

31. Fechner 1849a, 114. Here Fechner refers to the Weber passage quoted. He also cites this passage in Fechner 1851, II, 287f.

32. Fechner 1849a, 115.

33. Here we find surprising similarity to theories of complexity from computer science. Howard 1983, 237, for example: "Our results show that the ability to calculate what happened in the past is not restricted to deterministic systems. A system can evolve in a very random way but still preserve complete information about its past. It is tempting to wonder whether our own universe is like this." Fechner would certainly have affirmed this last question.

34. Fechner 1849a, 118.

35. Fechner 1849a, 120.

36. Fechner 1849a, 116.

37. Weisse 1835, IV.

38. Weisse 1829, 139.

39. Weisse 1829, 120.

40. Weisse 1833, 290.

41. Weisse 1829, 149.

42. Weisse 1830, 1830, 92f.

43. Weisse 1841, 38.

44. Weisse 1833, 290.

45. Weisse 1833, 322.

46. Weisse 1829, 185f.

47. Wentscher 1913, 29 and 54. Cf. also Wentscher 1903. On Lotze see chapter 1, section 1.5.

48. Lotze 1842, 60.

49. Lotze 1848, 61.

50. Lotze 1843, 148.

51. Lotze 1848, 61.

52. Lotze 1838, 13.

53. Lotze 1848, 204.

54. Lotze 1843, 204. It comes as a great surprise to learn that Henderson was familiar with Fechner 1873. See Henderson 1917 where, in the second appendix, Henderson reprinted, without any further comment, excerpts from chapter 3 of Fechner 1873.

55. Lotze 1843, 204.

56. However, in Herbart's philosophy the monad gets disturbed, but only from the outside. On Lotze's interpretation the soul disturbs the body; it implies an inner disturbance within the organism.

57. Lotze 1843, 204.

58. Lotze 1843, 206.

59. Lotze 1843, 204.

60. Lotze 1843, 206.

61. Lotze 1846, 4 and 16.

62. Lotze 1846, 165.

63. Lotze 1846, 17. Lotze's emphasis on freedom in an indeterministic sense became weaker rather than stronger over the course of his life. See Wentscher 1903. Waitz 1852, 1021f. writes that "Lotze's views in psychology are essentially shaped by his interest in the freedom of will, which he takes for an unprovable moral postulate." Waitz goes on to reproach Lotze for insufficiently distinguishing his psychology from ethics.

64. See Cranefield 1957, Lenoir 1982, chapter 5.

65. Helmholtz 1847, 3f. On Helmholtz's philosophy of science see Heidelberger 1993b and 1997. For Helmholtz in general see Cahan 1993.

66. Helmholtz 1847, 4.

67. Helmholtz 1855, 116. Helmholtz 1867, 454. See also Köhnke 1986, 151ff.

68. In the bibliography to §26 in *Physiological Optics* Helmholtz also cites Lotze's *Medical Psychology* of 1852 and *Microcosmos* of 1856 (Helmholtz 1867, 456). In 1881 Helmholtz supported Lotze's call to Berlin for the university chair that formerly Trendelenburg and Harms had occupied (Wentscher 1913, 348; Pester 1987).

69. Du Bois-Reymond 1848, 9. *Investigations on Animal Electricity* were published over a period of thirty-six years! Vol. 1 in 1848, vol. 2, part 1, in 1849, part 2, pp. 1–384 in 1860 and finally, pp. 385–579 in 1884. On this opus and its failure cf. Cranefield 1957.

70. Du Bois-Reymond 1848, 23.

71. Temkin 1950.

72. Cf. Coleman 1971, chapter III, particularly 41ff., Gould 1977.

73. Temkin 1950, 387.

74. On Burdach cf. Kay 1970, Temkin 1950; on Von Baer cf. Lenoir 1982.

75. In 1851, III 180 and 240 Fechner cites Burdach's writings in physiology, but neither in Fechner 1849 nor 1848a.

76. Burdach 1842, 232f.

77. Fechner 1851, I, 350.

78. Burdach 1842, 235.

79. Burdach 1842, 237.

80. Burdach 1842, 245.

81. Burdach 1842, 248.

82. Burdach 1842, 257. Cf. also 238, 252, 258. On page 249 he quotes Quetelet's famous sentence on the budget of crime, which "is paid with frightful regularity." Quetelet was the first to reflect on the regularity of the statistical distribution of social facts. On Quetelet cf. below.

83. Burdach 1842, 259.

84. Burdach 1842, 255.

85. Cf. Mayr 1959, Sober 1980. An essentialist interpretation of Quetelet, which Burdach opposes, can be found in Carl Gustav Carus; Carus 1858, 71. Carus, like Fechner a member of the Royal Saxon Society of Science in Leipzig since 1846, seems to have sought proof for Goethe's notion of the primordial plant by using Quetelet's statistical averages. Fechner criticized this approach in *Centralblatt für Naturwissenschaft und Anthropologie*, which he edited (Fechner 1853, 505–19).

86. Burdach 1842, 248.

87. Lotze 1841, 92.

88. On this transition in general see Krüger 1987, Schneider 1987, Kamlah 1987, Krüger 1987a, Hacking 1983, 1987.

89. The only other names which Fechner mentions in his book are Poisson (once), Scheibner (a mathematician from Leipzig), the statisticians E. B. Elliott, Bodio, and Gould, the anthropologist Welcker and the meteorologists Dove and Schmid. We search in vain for names like Lexis, Venn, Galton, or Bienaymé.

90. On Quetelet cf. Porter 1981, 1983, 1985, 1986, 1987; Hacking 1990, chapter 13, Krüger 1987a; Lécuyer 1987; Stigler 1986; Buck 1981; Schramm 1982, Porter 1994, Toyoda 1997.

91. In the sequence of mention above: Drobisch 1834, 64; Quetelet 1848; Drobisch 1849, Drobisch 1867. In Lange 1873, II, 842–49 Albert Lange provides an excellent account of the debate on the consequences of Quetelet's theory for the freedom of will, which also includes Drobisch. The issue of combining statistics and freedom remained virulent into the

twentieth century, cf. Kaufmann 1913, chapter 6 (156–88), dedicated entirely to this topic. On the whole issue see Porter 1986.

92. Fechner 1897, 229.

93. Fechner 1897, 228

94. Fechner 1856, V.

95. See Singer 1979, 7. Singer's reference to Pearson must be corrected as follows: Pearson 1914–30, III, 464, 468.

96. Fechner 1860, I, chapter VIII. See also Stigler 1986, 242–54, Walker 1929, 24ff., 49ff.

97. Fechner 1860, II, 543.

98. Fechner 1861a, 1874, 1874a, 1876. Walker 1929, 57, 84ff. discusses Fechner's "central value" and Singer 1979, 6 deals with Fechner's method of "extreme ranks for subjective judgments."

99. Wundt 1901, 315.

100. Lipps 1898 provides a quick and adequate access to Fechner 1897, just as Czuber 1906 does also. See also Witting 1990, 782, 786–88 and Heidelberger 2001.

101. Fechner 1897, VI.

102. Fechner 1897, 3. This definition appears in similar wording as early as Fechner 1874a and Fechner 1871, 617, 619f. Ludwig 1988 provides a survey of various attempts during Fechner's time, particularly in biology, to use Gauss's law as a law of distribution. De Vries 1898 is also interesting on this issue.

103. Rümelin 1863, 213–15, 241 and 1874, 269–74. On Rümelin see also Porter 1987 and Kaufmann 1913, 12f.

104. Fechner 1897, chapter XXII.

105. Fechner 1897, 3.

106. Fechner 1897, 6.

107. Fechner 1897, 6.

108. Wundt 1901, 296, 255, 315.

109. Fechner 1897, VI.

110. Fechner 1897, 4.

111. Fechner 1897, 6.

112. Fechner 1987, 3.

113. Fechner 1897, 5.

114. Fechner 1897, 16.

115. Fechner 1897, 5.

116. Fechner 1897, 31.

117. Fechner 1897, 39.

118. Fechner 1897, 42.

119. Fechner 1897, 45. Cf. 229.

120. Fechner 1897, 46. The test reappears on page 365.

121. For instance, births, deaths, suicides during various seasons, male and female newborns, number of storms at different places, and so on.

122. Fechner 1897, 234ff.

123. Fechner 1897, 382–85. In chapters XXIII and XIV Fechner develops ranking methods similar to Kendall's 'tau.' See Fechner 1897, 372–75 and 386–98 and Singer 1979, 6.

124. Fechner 1897, 9, 20, 95f.

125. Fechner 1897, 85.

126. Walker 1929, 52.

127. Fechner 1897, chapter X, but also 12–26 and 84–98.

128. However, as early as 1838 the astronomer Friedrich Wilhelm Bessel had already re-marked that asymmetrical error distributions also exist and therefore normal distribution cannot cover all cases: "In both examples ... the law of the probability of error is very differ-ent from the frequently mentioned [Gauss] law ... [I have only to remark] that every at-tempt to see the law underlying the method of the smallest square as being *in general* the really prevailing law, will necessarily be in vain, because examples show that conditions, which are not merely mathematically possible, but can also be fulfilled empirically, lead to laws entirely different from it." (Bessel 1838, 377) Bessel's remark, however, was not fol-lowed up.

129. Quetelet 1846, 79 and 181. In discussing statistics on temperature medians Quetelet remarks: "... quand la moyenne n'est pas un résultat numerique purement abstrait et qu'il existe une température que des circonstances fortuites peuvent masquer plus ou moins, la courbe l'indique par sa régularité, pourvu que le nombre des observations soit suffisament grand." (79).

130. Fechner 1897, 64. Cf. also chapter XII.

131. There is another theoretical reason why Gauss's law cannot hold universally for dis-tributions: The probability of negative deviations from the mean is never zero according to Gauss's law. But in reality it cannot be larger than the mean. For this reason Fechner re-lates the law of distribution to ratio deviations or their logarithms, instead of to arithmeti-cal deviations. (Fechner 1897, 75).

132. Fechner 1897, 70ff.

133. De Vries 1894. See also De Vries 1898.

134. Wundt 1901, 330; Külpe 1901, 201; Bruns 1906, 309.

135. Mach 1872, 34f. I have been unable to find the "other place" where Mach claims to have shown that it is superfluous to include space and time in the law of causality.

136. Mach 1872, 36.

137. Koenig 1890, 478ff. Oettingen 1906 mentions Fechner's lecture on the law of causal-ity in the course of an essay on the energy interpretation of the law of causality. Wentscher 1921 then examines this in detail.

138. Bruns 1897, Bruns 1906. The latter was corrected twice: Bruns 1906a and 1906b. See also Bruns 1898. Bruns's series is also called "Charlier series" or "Charlier's A-series," and sometimes "Gram-Charlier series" or "Edgeworth series," in reference to work by Edge-worth 1905 and Charlier 1906. See Elderton 1906 for a review of Charlier and Edgeworth's methods. Edgeworth seems to have been the only Anglo-Saxon scholar to take note of Bruns's work: Edgeworth 1907. Cf. also Särndal 1971 and Cramér 1972 on the history of these distribution functions. On Bruns see Bauschinger 1921, which includes a bibliography of Bruns's writings and a list of eight dissertations which attempted to apply the theory of measuring collectives in various fields. Bauschinger also reports that Bruns left a finished manuscript behind for a second edition of *Calculating Probability* (Bruns 1906), as well as the manuscript for a book on error theory.

139. Lipps 1898, 1901, 1905a. Czuber 1906, 522ff., 525ff. comments on Lipps 1901 and Bruns 1906. G. F. Lipps was the younger brother of the renowned Theodor Lipps, founder of the Institute for Psychology at the University of Munich. On Bruns, Lipps, and Czuber see Witting 1990, 787–89.

140. Lipps 1901, 107; Bruns 1898, 341–43; Bruns 1906, 95. Felix Hausdorff was also engaged in this discussion (Hausdorff 1901). For Hausdorff's work on probablility see Girlich 1996.

141. Lipps 1901, 124; Bruns 1989, 346. Bruns was also aware of the fact that Fechner's theory of measuring collectives was based on his work of 1849. The first ideas for a general theory of the forms of collective objects "can be traced back very far in Fechner's work: we find them already stated in a lecture on 'mathematical treatment of organic designs and processes'... and they appear clearly in several later writings." (Bruns 1906, 309).

142. Bruns 1906, 6f.

143. Bruns 1906, 7.

144. Bruns 1906, 5.

145. Bruns 1906, 95. Cf. also 5.

146. Bruns 1906, 111; Lipps 1901, 155–62, Pearson 1905. Duncker 1899 contributed to the dissemination of Pearson's methods in Germany.

147. Bertrand 1899, particularly 11–17. In this review Bertrand provides a general survey of Fechner's publications, as well as a brief survey of Emile Dormoy's work on divergence coefficients. Dormoy's approach is very similar to that of Wilhelm Lexis; cf. Stigler 1986, 1987.

148. Helm 1902, 1907; Lipps 1906; Wundt 1908, 567–74; Wirth 1912. On Wirth's role in Leipzig see Schröder 1993. Regarding anthropology cf. Ranke 1904. In an essay, Ranke and Greiner arrived at the conclusion that Pearson's and Fechner's distributions are purely empirical descriptions and therefore of limited theoretical value. Pearson's reply (Pearson 1905) contains the aforementioned critique on Fechner. On Royce: Royce 1912, 72ff.

149. See Lipps 1933 for the names of some publications.

150. Czuber 1906, 1908. Regarding chance in Fechner's work: Czuber 1908, 350. On Czuber see Schneider 1987, 194; Kamlah 1987, 111.

151. Mises 1919. On Von Mises cf. Bernhardt 1984, 1990, Stadler 1990, Plato 1987, 395f., Witting 1990, 793f. For the history of twentieth-century probability, including von Mises, see Plato 1994, Stigler 1999, Hochkirchen 1999, Stöltzner 1999, 2002, 2003.

152. Von Mises himself understands his theory as marking the end of the rule of Laplace's definition of probability (Mises 1919, 66f.).

153. Mises 1912, 3.

154. Fechner 1897, 31 and 466.

155. Fechner 1897, 31.

156. Mises 1919, 60ff.

157. Czuber 1908 (fifth edition), 441.

158. Fine 1973, 83 views the contrast between Von Mises and Kolmogorov as follows: "Axiomatic probability hardly enables us to understand what probability is about or what its subject matter is. The Kolmogorov axioms, for example, provide neither a guide to the domain of applicability of probability nor a procedure for estimating probabilities nor appreciable insight into the nature of random phenomena." (See also Weatherford 1982, 144ff. for a similar judgement; also Krengel 1990, 461–66.) Fechner's (and thus also Von Mises's) approach was a first step toward fixing this.

159. See Bernhardt 1984, 213ff. and Fine 1973, chapter V for a pertinent discussion of work done by Kolmogorov, Loveland, Schnorr, and Martin-Löf. Work done by Chaitin 1974, 1975, 1982 is particularly promising. Schnorr 1971, 192 even sees a way to found the Kolmogorov approach on a Von Mises-like theory: "that the foundation of the theory of probability does not, as it was formerly assumed, lead to a conflict with Kolmogorov's axioms. Instead, the concept of random sequence must lead to Kolmogorov's axioms in a constructive way. Kolmogorov's axioms can be established using the model of frequency."

160. Mises 1921, 430. Mises sketched his theory of Brownian motion in Mises 1920.

161. See Exner 1922, particularly 680. On page 330 Exner quotes Fechner's definition of a collective object. On Exner see Hanle 1979 and Karlik 1982.

162. Mises 1930, 153. Although I here emphasize the proximity of Von Mises's and Mach's opinions, I am not saying that Mach advocated indeterminism. That cannot be easily claimed, although some of Mach's remarks about the uniqueness of events in the world come quite close to it. The proximity to Mach is one of verificationist-empiristic method.

163. Mises 1928 (First Edition) 26, 99; Mises 1912.

164. As, for example, in Geiringer 1973.

165. Fechner 1897, 3 and 31, also 94f., 60, 39, 234.

166. Wundt 1901, 310f. It is very likely that Carnap modeled the notorious designation for metaphysics as "poetry of concepts" on this.

167. By the way, Von Mises was familiar with Fechner's psychophysics, although he rejected Fechner's Law; Mises 1939, 337–40.

168. Brush 1976, 630.

Notes to Conclusion

1. Horkheimer 1926, 201f.

2. Ben-David 1960, 566; 1966, 459–61. Ben-David also uses the notion of role hybridization to explain the origins of Freudian psychoanalysis.

3. Ben-David 1960, 1966, 455, 459ff., also in Ben-David 1991.

BIBLIOGRAPHY

Achelis, Thomas. 1889. "Zur Würdigung G. Th. Fechners." *Zeitschrift für Völkerpsychologie und Sprachwissenschaft* 19:164–92.
———. 1891. "Gustav Theodor Fechner." *Nord und Süd* 56:272–96.
Adler, Helmut E. 1977. "The Vicissitudes of Fechnerian Psychophysics in America." *Annals of the New York Academy of Sciences* 291:21–32.
———. 1980. "Vicissitudes of Fechnerian Psychophysics in America." Pp. 11–23 in *Psychology: Theoretical-Historical Perspectives*, edited by Robert W. Rieber and Kurt D. Salzinger. New York: Academic Press. [2nd ed. 1998, 3–14. Washington, D.C.: American Psychological Association.]
———. 1992. "William James and Gustav Fechner: From Rejection to Elective Affinity." Pp. 253–61 in *Reinterpreting the legacy of William James*, edited by Margaret Donnely. Washington, D.C.: American Psychological Association.
———. 1992a. "Fechner's influence in America: Little-known work outside psychophysics." *Revista de Historia de la Psicologia* 13:85–91.
———. 1996. "Gustav Theodor Fechner: A German *Gelehrter*." In *Portraits of Pioneers in Psychology*, vol. 2, edited by Gregory A. Kimble, C. Alan Boneau, and Michael Wertheimer. Washington, D.C.: American Psychological Association. [Reprint, 1998, 1–13.]
Adolph, Heinrich (August Karl). 1923. *Die Weltanschauung Gustav Theodor Fechners*. Stuttgart: Strecker and Schröder.
Albert, Dietrich, and Horst Gundlach, eds. 1997. *Apparative Psychologie: Geschichtliche Entwicklung und gegenwärtige Bedeutung*. Lengerich: Pabst.
Aldrich, V. C., and Herbert Feigl. 1935. "Spatial Location and the Psychophysical Law." *Philosophy of Science* 2:256–61. [A letter to the editor by Aldrich and Feigl's response.]
Allesch, Christian G. 1987. *Geschichte der psychologischen Ästhetik. Untersuchungen zur historischen Entwicklung eines psychologischen Verständnisses ästhetischer Phänomene*. Göttingen: Hogrefe. [303–16, 352, 374, 397 on Fechner's aesthetics.]
———. 1988. "Gustav Theodor Fechner als Wegbereiter der psychologischen Ästhetik." Pp. 207–15 in *G. T. Fechner and Psychology*, edited by Josef Brožek and Horst Gundlach. Passau: Passavia.
———. 1988a. "100 Jahre Ästhetik von unten." *Musikpsychologie* 5:11–31.
Altmann, Irene. 1995. *Bibliographie Gustav Theodor Fechner*. Edited by Gustav-Theodor-Fechner-Gesellschaft e. V. Leipzig. Leipzig: Verlag im Wissenschaftszentrum Leipzig. [Includes Wolfram Meischner's plenary address to the *Leipziger Fechner Symposion*, July 6–10, 1987, 10–38.]
Anderson, Douglas R. 1987. *Creativity and the Philosophy of C. S. Peirce*. Dordrecht: Nijhoff.
[Anonymous.] 1849. [Review of Fechner 1848.] *Allgemeine Literatur-Zeitung* (Halle) (April): 617–40
———. 1852. [Review of Fechner 1851.] *Leipziger Repertorium der deutschen und ausländischen Literatur*. Edited by E. G. Gersdorf, 37 (1st vol. of 10th annual set):82–85.

Apel, Karl-Otto. 1967. *Der Denkweg von Charles S. Peirce. Eine Einführung in den amerikanischen Pragmatismus.* Frankfurt am Main: Suhrkamp.

Archibald, Thomas. 1989. "Energy and the Mathematization of Electrodynamics in Germany, 1845–1875." *Archives Internationales d'histoire des sciences* 39:276–308.

Arendt, Hans-Jürgen. 1994. *Gustav Theodor Fechner und sein Hauslexikon (1834–1838). Ein Beitrag zur Fechner-Biographie und zur Geschichte der enzyklopädischen Literatur in Deutschland.* Leipzig: Verlag im Wissenschaftszentrum Leipzig.

———. 1999. *Gustav Theodor Fechner. Ein deutscher Naturwissenschaftler und Philosoph im 19. Jahrhundert.* (Daedalus, vol. 12.) Frankfurt am Main: Lang.

———. 2001. "Gustav Theodor Fechner (1801–1887) und die Leipziger bürgerliche Gesellschaft im 19. Jahrhundert." *NTM—Schriftenreihe für Geschichte der Naturwissenschaften, Technik und Medizin* 9(1):2–14.

Arens, Katherine. 1989. *Structures of Knowing: Psychologies of the Nineteenth Century.* Dordrecht: Kluwer.

Argüelles, José A. 1972. *Charles Henry and the formation of a psychophysical aesthetic.* Chicago, University of Chicago Press.

Armstrong, David M. 1968. *A Materialist Theory of Mind.* London: Routledge. Revised ed., London: Routledge, 1993.

———. 1987. "Mind-Body Problem: Philosophical Theories." Pp. 490–91 in *The Oxford Companion to the Mind,* edited by Richard L. Gregory and O. L. Zangwill. Oxford: Oxford University Press.

Arnheim, Rudolf. 1985. "The Other Gustav Theodor Fechner." Pp. 856–65 in *A Century of Psychology as a Science,* edited by Sigmund Koch and David E. Leary. New York: McGraw-Hill. [First appeared 1980 in *Gestalt Theory* 2: 133–40. Also 1986, pp. 39–49 in *New Essays on the Psychology of Art,* edited by Rudolf Arnheim. Berkeley: University of California Press.]

Ash, Mitchell G. 1980. "Experimental Psychology in Germany before 1914: Aspects of an Academic Identity Problem." *Psychological Research* 42:75–86.

———. 1980a. "Academic Politics in the History of Science: Experimental Psychology in Germany, 1879–1941." *Central European History* 13:255–86.

———. 1995. *Gestalt Psychology in German Culture: 1890–1967: Holism and the Quest for Objectivity.* Cambridge: Cambridge University Press.

———. 2003. "Psychology." In *Cambridge History of Science,* edited by Dorothy Ross and Theodore Porter. Vol. 7: *The Modern Social and Behavioral Sciences.* Cambridge: Cambridge University Press.

Ash, Mitchell G., and William R. Woodward, eds. 1989. *Psychology in Twentieth-Century Thought and Society.* Cambridge: Cambridge University Press.

Aubert, Hermann. 1865. *Physiologie der Netzhaut.* Breslau: Morgenstern.

Baer, Karl Ernst von. 1864. *Reden gehalten in wissenschaftlichen Versammlungen und kleinere Aufsätze vermischten Inhalts,* Erster Theil: *Reden.* St. Petersburg: H. Schmitzdorff.

Bähr, Karl. 1894. *Gespräche und Briefwechsel mit Arthur Schopenhauer.* Aus dem Nachlasse von Karl Bähr. Edited by Ludwig Schemann. Leipzig: Brockhaus.

Bain, Alexander. 1874. *Geist und Körper: Die Theorien über ihre gegenseitigen Beziehungen.* Authorized transl. Leipzig: Brockhaus. [English original 1873. *Mind and Body: The Theories of Their Relation.* New York: Appleton.]

Baird, John C. 1997. *Sensation and Judgment: Complementarity Theory of Psychophysics.* Mahwah, N.J.: Erlbaum.

Baird, John C., and Elliot Noma. 1978. *Fundamentals of Scaling and Psychophysics*. New York: Wiley.

Barham, Jeremy. 1998. "Mahler's Third Symphony and the Philosophy of Gustav Fechner: Interdisciplinary Approaches to Criticism, Analysis and Interpretation." Ph.D. diss. in Music, University of Surrey.

Batschmann, Oskar. 1996. "Der Holbein-Streit. Eine Krise der Kunstgeschichte." *Jahrbuch der Berliner Museen* 38 (Suppl.):87–100.

Baumann, Christian. 2002. *Der Physiologe Ewald Hering (1834–1918)*. (Deutsche Hochschulschriften, vol. 1216) Frankfurt am Main: Hänsel-Hohenhausen.

Baumgartner, Hans Michael. 1994. *Wissenschaftshistorischer Bericht zu Schellings Naturphilosophischen Schriften 1797–1800. Ergänzungsband zu Werke Band 5 bis 9*. Supplementary volume to vols. 5–9 of series I: *Werke* of the *Historisch-kritische Ausgabe* of Schelling's writings. Edited on behalf of Bayerische Akademie der Wissenschaften by Hans Michael Baumgartner, Wilhelm G. Jacobs, and Hermann Krings. Stuttgart: Frommann-Holzboog. [Collection of articles on the state of chemistry, physics and physiology around 1800, as related to Schelling's *Naturphilosophie*.]

Bauschinger, Julius. 1921. "Nekrolog. Heinrich Bruns." *Vierteljahrsschrift der Astronomischen Gesellschaft* 56:59–69.

Becher, Erich. 1911. *Gehirn und Seele*. (Die Psychologie in Einzeldarstellungen, vol. 5) Heidelberg: Winter.

———. 1929. "Gustav Theodor Fechner." Pp. 31–45 in *Deutsche Philosophen*. Munich/ Leipzig: Duncker and Humblot.

Beckermann, Ansgar, Hans Flohr and Jaegwon Kim, eds. 1992. *Emergence or Reduction? Essays on the Prospects of Non-reductive Physicalism*. Berlin: De Gruyter.

Behrens, Peter J. 1987. "Fechnerian Psychophysics in America: The Contribution of Charles Peirce and Joseph Jastrow, 1884." Paper given at the Leipziger Fechner Symposion, Karl-Marx-Universität Leipzig, July 6–10.

Ben-David, Joseph. 1960. "Roles and Innovations in Medicine." *American Journal of Sociology* 65:557–68.

———. 1991. *Scientific growth: Essays on the Social Organization and Ethos of Science*. Edited by Gad Freudenthal. Berkeley: University of California Press.

Ben-David, Joseph, and Randall Collins. 1966. "Social Factors in the Origins of a New Science: The Case of Psychology." *American Sociological Review* 31:451–65.

Benjafield, John G. 2001. "The Psychology of Mathematical Beauty in the 19th Century: The Golden Section." Pp. 87–105 in *The Transformation of Psychology: Influences of 19th-century Philosophy, Technology, and Natural Science*, edited by Christopher D. Green, Marlene Shore, and Thomas Teo. Washington, D.C.: American Psychological Association.

Bergmann, Gustav, and Kenneth W. Spence. 1944. "The Logic of Psychophysical Measurement." *Psychological Review* 51:1–24.

Bergson, Henri. 1889. *Essai sur les données immédiates de la conscience*. (Bibliothèque de philosophie contemporaine) Paris: Félix Alcan. [Cited according to reprint, 1945. Genève: Albert Skira. Authorized transl. 1913 by Frank L. Pogson as *Time and Free Will: An Essay on the Immediate Data of Consciousness*. New York: George Allen.]

———. 1896. *Matière et Mémoire. Essai sur la relation du corps à l'esprit*. (Bibliothèque de philosophie contemporaine) Paris: Félix Alcan. [Authorized transl. 1980 by Nancy Margaret Paul and W. Scott Palmer as *Matter and Memory*. New York: Zone Books.]

———. 1901. "Le parallélisme psycho-physique et la métaphysique positive." *Bulletin de la*

Societé française de philosophie. 33–34, 43–57. [Cited according to reprint, 1957, pp. 139–67 in Henri Bergson, *Écrits et paroles.* Edited by R.-M. Mossé-Bastide. Vol. 1. Paris: Presses Universitaires de France.]

Berkeley, George. 1710. *A Treatise Concerning the Principles of Human Knowledge.* Dublin.

Berlyne, D. E., ed. 1974. *Studies in the New Experimental Aesthetics: Steps Toward an Objective Psychology of Aesthetic Appreciation.* Washington, D.C.: Hemisphere Pub. Corp.

Bernfeld, Siegfried. 1944. "Freud's Earliest Theories and the School of Helmholtz." *Psychoanalytic Quarterly* 13:341–62.

Bernfeld, Siegfried, and Sergei Feitelberg. 1930. "Der Energiesatz und der Todestrieb." *Imago* 16:187–206.

Bernhardt, Hannelore. 1984. *Richard von Mises und sein Beitrag zur Grundlegung der Wahrscheinlichkeitsrechnung im 20. Jahrhundert.* Habilitation thesis (Dissertation B), Humboldt-Universität zu Berlin.

———. 1990. "Der Berliner Mathematiker R. v. Mises und sein wahrscheinlichkeitstheoretisches Konzept." *Wissenschaftliche Zeitschrift der Humboldt-Universität zu Berlin,* Reihe Mathematik/Naturwissenschaften, 39:205–9.

Bernstein, Julius. 1868. "Zur Theorie des Fechner'schen Gesetzes der Empfindung." [Reichert's und du Bois-Reymond's] *Archiv für Anatomie, Physiologie und wissenschaftliche Medizin:* 388–93.

Bertalanffy, Ludwig von. 1926. *Fechner und das Problem der Integrationen höherer Ordnung. Ein Versuch zur induktiven Metaphysik.* Ph.D. diss., phil. Fak., Universität Wien.

Bertrand, Joseph. 1899. [Review of] "Kollektiv-Masslehre von Gustav Theodor Fechner." *Journal des Savants* January: 5–17.

Berzelius, Jacob. 1836. "Einige Ideen über eine bei der Bildung organischer Verbindungen in der lebenden Natur wirksame, aber bisher nicht bemerkte Kraft." *Jahres-Bericht über die Fortschritte der physischen Wissenschaften von Jacob Berzelius* 15: 237–45.

Bessel, Friedrich Wilhelm. 1838. "Untersuchungen über die Wahrscheinlichkeit der Beobachtungsfehler." *Astronomische Nachrichten* 15: 369ff. [Cited according to reprint, 1876, pp. 372–91 in *Abhandlungen von Friedrich Bessel,* edited by Rudolf Engelmann. Vol. 2. Leipzig: Q. Engelmann.]

Bichat, Xavier. 1801. *Anatomie générale. Appliquée à la physiologie et à la médecine.* 4 vols. Paris: Brosson. [Nouvelle édition, 1812, 4 vols. Translation 1822 by George Hayward as *General Anatomy, Applied to Physiology and Medicine.* 3 vols. Boston: Richardson and Lord.]

Bieri, Peter, ed. 1997. *Analytische Philosophie des Geistes.* Königstein: Hain, 1997. [1st ed. 1981.]

———. 1982. "Nominalismus und innere Erfahrung." *Zeitschrift für philosophische Forschung* 36 (1982): 3–24.

Binswanger, Ludwig. 1922. *Einführung in die Probleme der allgemeinen Psychologie.* Heidelberg: Springer, 1922. Reprint Amsterdam: Bonset, 1965.

———. 1956. *Erinnerungen an Sigmund Freud.* Bern: Francke, 1956. [Transl. by Norbert Guterman as *Reminiscenses of a friendship.* New York: Grune and Stratton, 1957.]

Biot, Jean-Baptiste. 1818. *Anfangsgründe der Erfahrungs-Naturlehre.* Transl. by Friedrich Wolff. 2 parts. Berlin: Voß, 1818–19. [German transl. of the 1st French ed. 1817 of Biot 1824.]

———. 1824. *Précis élémentaire de physique expérimentale.* 2 vols. 3rd ed. Paris: Deterville, 1824. [For German transl. of this ed. see Fechner 1824 and 1828. 1st French ed. 1817, 2nd ed. 1821.]

Blackmore, John T. 1972. *Ernst Mach. His Work, Life, and Influence.* Berkeley: University of California Press, 1972.

———. 1988. "Mach über Atome und Relativität – neueste Forschungsergebnisse." *Ernst Mach—Werk und Wirkung.* Eds. Rudolf Haller and Friedrich Stadler. Vienna: Hölder, 1988, 463–83.

Blackmore, John T. ed. 1992. *Ernst Mach—A deeper look: Documents and new perspectives.* (Boston Studies in the Philosophy of Science, vol. 143) Dordrecht: Kluwer, 1992.

Blackmore, John T. et al. eds. 2001. *Ernst Mach's Vienna, 1895–1930, or, phenomenalism as philosophy of science.* (Boston Studies in the Philosophy of Science, vol. 218) Dordrecht: Kluwer, 2001.

Blumenthal, Arthur L. 1985. "Shaping a Tradition: Experimentalism Begins." *Points of View in the Modern History of Psychology.* Edited by Claude E. Buxton. Orlando, Florida: Academic Press, 1985, 51–83.

Boas, Franz. 1881. "Ueber eine neue Form des Gesetzes der Unterschiedsschwelle." [Pflüger's] *Archiv für die gesammte Physiologie* 26 (1881): 493–500.

———. 1882. "Ueber die verschiedenen Formen des Unterschiedsschwellenwerthes." [Pflüger's] *Archiv für die gesammte Physiologie* 27 (1882): 214–22.

———. 1882a. "Die Bestimmung der Unterschiedsempfindlichkeit nach der Methode der übermerklichen Unterschiede." [Pflüger's] *Archiv für die gesammte Physiologie* 28 (1882): 562–66.

———. 1882b. "Ueber die Grundaufgabe der Psychophysik." [Pflüger's] *Archiv für die gesammte Physiologie* 28 (1882): 566–76.

Bohr, Niels. 1999. *Complementarity Beyond Physics (1928–1962).* Edited by David Favrholdt. (Vol. 10 of *Niels Bohr Collected Works.* General editor Finn Aaserud) Amsterdam: Elsevier, 1999.

Bolland, Gerardus Johannes Petrus Josephus. 1910. *Schelling, Hegel, Fechner en de nieuwere theosophie. Eene geschiedkundige voor-en inlichting.* Leiden: Adriani, 1910.

Bölsche, Wilhelm. 1892. "Neues zur Erinnerung an Gustav Theodor Fechner." [Review of Kuntze 1892.] *Freie Bühne* 3 (1892): 358–66.

———. 1897. "Fechner. Ein Charakterbild." *Deutsche Rundschau* 92 (1897): 344–69.

———. 1901. "Herolde der 'Tagesansicht'. Ein Wort zu Fechner." *Ethische Kultur* 9(32) (1901): 249–52.

———. 1904. "Fechner." In Wilhelm Bölsche, *Hinter der Weltstadt. Friedrichshagener Gedanken zur ästhetischen Kultur.* Jena and Leipzig: Eugen Diederichs, 1904, 259–347.

———. 1913. *Stirb und werde. Naturwissenschaftliche und kulturelle Plaudereien.* Jena: Diederichs, 1913.

Bonitz, Manfred. 1977. "Notes on the Development of Secondary Periodicals from the 'Journal des Sçavans' to the 'Pharmaceutisches Central-Blatt'." *International Forum on Information and Documentation* 2:26–31.

———. 1983. "Gustav Theodor Fechner—Der anonyme erste Redakteur des 'pharmaceutischen Central-Blatts'." *Zentralblatt für Bibliothekswesen* 97:70–72.

Bonsiepen, Wolfgang. 1988. "Die Ausbildung einer dynamischen Atomistik bei Leibniz, Kant und Schelling und ihre aktuelle Bedeutung." *Allgemeine Zeitschrift für Philosophie* 13(1):1–20.

———. 1997. *Die Begründung einer Naturphilosophie bei Kant, Schelling, Fries und Hegel. Mathematische versus spekulative Naturphilosophie.* Frankfurt am Main: Klostermann.

Borgard, Thomas. 1999. *Immanentismus und konjunktives Denken. Die Entstehung eines modernen Weltverständnisses aus dem strategischen Einsatz einer "psychologia prima"*

(1830–1880). (Studien und Texte zur Sozialgeschichte der Literatur, vol. 63) Tübingen: Niemeyer.

Boring, Edwin G. 1920. "The Logic of the Normal Law of Error in Mental Measurement." *American Journal of Psychology* 31:1–33.

———. 1921. "The Stimulus-Error." *American Journal of Psychology* 32(4):449–71. [Also in Boring 1963, 255–73.]

———. 1928. "Did Fechner Measure Sensation?" *Psychological Review* 35(443):445.

———. 1933. *The Physical Dimensions of Consciousness.* New York: Dover. [Reprint 1963.]

———. 1942. *Sensation and Perception in the History of Experimental Psychology.* New York: Appleton.

———. 1950. *A History of Experimental Psychology.* 2nd ed. New York: Appleton. [Chapter 14: Gustav Theodor Fechner.]

———. 1961. "Fechner: Inadvertent Founder of Psychophysics." *Psychometrika* 26:3–8. [Cited according to Boring 1963, 126–31.]

———. 1961a. "The Beginning and Growth of Measurement in Psychology." *Isis* 52:238–57. [Cited according to Boring 1963, 140–58.]

———. 1963. *History, Psychology, and Science: Selected Papers.* Eds. Robert I. Watson and Donald T. Campbell. New York: Wiley.

Boudewijnse, Geert-Jan A., David J. Murray and Christina A. Bandomir. 1999. "Herbart's Mathematical Psychology." *History of Psychology* 2:163–93.

Boudewijnse, Geert-Jan A., David J. Murray, and Christina A. Bandomir. 2001. "The fate of Herbart's Mathematical Psychology." *History of Psychology* 4(2):107–32.

Bovet, E. 1922. "Die Physiker Einstein und Weyl. Antworten auf eine metaphysische Frage." *Wissen und Leben* 15(19):901–6.

Brandt, Richard and Jaegwon Kim. 1967. "The Logic of the Identity Theory." *The Journal of Philosophy* 64:515–37.

Brasch, Moritz. 1888. "Fechner und die Psychophysik." Pp. 613–23 in *Die Philosophie der Gegenwart. Ihre Richtungen und Hauptvertreter.* Leipzig: Greßner und Schramm.

———. 1888a. "Gustav Theodor Fechner, der Begründer der Psychophysik." *Westermanns illustrirte deutsche Monats-Hefte* 32(64):80–85.

———. 1894. "Gustav Theodor Fechner, der Begründer der Psychophysik. Eine philosophische Charakteristik." Pp. 1–13 in *Leipziger Philosophen. Portraits und Studien aus dem wissenschaftlichen Leben der Gegenwart.* Leipzig: Adolf Weigel.

Brauns, Horst-Peter. 1990. "Fechners experimentelle Versuchsplanung in 'Elemente der Psychophysik' im Lichte heutiger Methodenlehre des psychologischen Experiments." *Psychologie und Geschichte* 1(4):10–23.

———. 1993. "Ernst Heinrich Weber." Pp. 27–28 in *Illustrierte Geschichte der Psychologie,* edited by Helmut E. Lück and Rudolf Miller. München: Quintessenz.

———. 1994. "Zur Lage der Psychologie um das Jahr 1850." Pp. 207–218 in *Arbeiten zur Psychologiegeschichte* edited by Horst Gundlach. Göttingen: Hogrefe.

———. 1996. "Ernst Heinrich Weber und das Paradigma seiner sensumetrischen Experimente." Pp. 25–44 in *Untersuchungen zur Geschichte der Psychologie und der Psychotechnik,* edited by Horst Gundlach. (Passauer Schriften zur Psychologiegeschichte, vol. 11) Munich: Profil.

———. 1998. "Über einige Beziehungen des jungen Fechner zu K. B. Mollweide und W. T. Krug sowie ihre individualhistorische Relevanz." Pp. 121–49 in *Psychologiegeschichte— Beziehungen zu Philosophie und Grenzgebieten,* edited by Jürgen Jahnke, Jochen Fahrenberg, et al. (Passauer Schriften zur Psychologiegeschichte, Bd. 12.) Munich: Profil.

———. 2000. "Gustav Theodor Fechner: Elemente der Psychophysik I/II (1860)." Pp. 36–45 in *Klassiker der Psychologie*, edited by Helmut E. Lueck, Rudolf Miller, and Gabi Sewz-Vosshenrich. Stuttgart: Kohlhammer.

Brauns, Horst-Peter and Alfred Schöpf. 1989. "Freud und Brentano. Der Medizinstudent und der Philosoph." Pp. 40–79 in *Freud und die akademische Philosophie*, edited by Bernd Nitzschke. Munich: Psychologie-Verlag.

Breidbach, Olaf, ed. 1997. *Natur der Ästhetik—Ästhetik der Natur*. Vienna: Springer.

Breidbach, Olaf and Paul Ziche, ed. 2001. *Naturwissenschaften um 1800: Wissenschaftskultur in Jena-Weimar*. Weimar: Böhlau.

Breidbach, Olaf, Hans-Joachim Fliedner and Klaus Ries, eds. 2001. *Lorenz Oken (1779–1851): Ein politischer Naturphilosoph*. Weimar: Verlag Hermann Böhlaus Nachfolger.

Brentano, Franz. 1874. *Psychologie vom empirischen Standpunkt*. 2 vols. in one. Leipzig: Duncker und Humblot, 1874. [Cited according to reprint, 1955, 2 vols. Hamburg: Meiner, 1955. Translation 1973 by A. C. Rancurello, D. B. Terrell, and L. L. McAlister as *Psychology from an empirical standpoint*. London: Routledge.]

———. 1959. *Grundzüge der Ästhetik*. Edited by Franziska Mayer-Hillebrand from the Brentano estate. Bern: Francke. [On Fechner 19–29 and 62–65.]

Breuer, Josef. 1902. *Die Krisis des Darwinismus und die Teleologie*. Speech held on May 2, 1902. [Reprint ed. Gerd Kimmerle. Tübingen: ed. diskord, 1986.]

———. 1903. [Letter by Josef Breuer to Franz Brentano, March] Document 11, in Albrecht Hirschmüller, *Physiologie und Psychoanalyse in Leben und Werk Josef Breuers*. Bern: Huber, 1978, 303–14.

Briese, Olaf. 1998. *Konkurrenzen. Philosophische Kultur in Deutschland 1830–1850. Porträts und Profile*. Würzburg: Königshausen and Neumann, 1998. [Essays on Ch. H. Weisse, Herbart, Schelling, Fechner, and others.]

Bringmann, Wolfgang G. and William D. G. Balance. 1976. "Der Psychologe, der sich selbst geheilt hat. Das Leben und Werk von Gustav Theodor Fechner." *Psychologie heute* 3(9) (1976): 43–48.

Bringmann, Wolfgang G., Norma J. Bringmann and Eberhard Bauer. 1990. "Fechner und die Parapsychologie." *Zeitschrift für Parapsychologie und Grenzgebiete der Psychologie* 32(1–2) (1990): 19–43.

Bringmann, Wolfgang G., Norma J. Bringmann and Norma L. Medway. 1988. "Fechner and Psychical Research." Pp. 243–55 in *G. T. Fechner and Psychology*, edited by Josef Brožek and Horst Gundlach. Passau: Passavia.

Brown, William. 1913. "Are the Intensity Differences of Sensation Quantitative?" *The British Journal of Psychology* 6:184–89.

Brožek, Josef and Horst Gundlach, eds. 1988. *G. T. Fechner and Psychology*. (Passauer Schriften zur Psychologiegeschichte, vol. 6) Passau: Passavia.

Brožek, Josef and Jiří Hoskovec. 1987. *J. E. Purkyně and Psychology, with a Focus on Unpublished Manuscripts*. Prague: Academia.

Bruchmann, Kurt. 1887. "Gustav Theodor Fechner, ein deutscher Metaphysiker." *Preußische Jahrbücher* 59: 293–309.

Bruns, Heinrich. 1897. "Über die Darstellung von Fehlergesetzen." *Astronomische Nachrichten* 143:329–49.

———. 1898. "Zur Collectiv-Masslehre." *Philosophische Studien* 14:339–75.

———. 1906. *Wahrscheinlichkeitsrechnung und Kollektivmasslehre*. (Teubners Sammlung von Lehrbüchern auf dem Gebiete der mathematischen Wissenschaften mit Einschluß ihrer Anwendungen, vol. 17) Leipzig: B. G. Teubner.

———. 1906a. "Beiträge zur Quotenrechnung." *Berichte der Königlich Sächsischen Gesellschaft der Wissenschaften* 58:571–613.

———. 1906b. "Das Gruppenschema für zufällige Ereignisse." *Abhandlungen der Königlich Sächsischen Gesellschaft der Wissenschaften* 29:577–628.

Brush, Stephen G. 1968. "Mach and Atomism." *Synthese* 18:192–215.

———. 1970. "Interatomic Forces and Gas Theory from Newton to Lennard-Jones." *Archive for Rational Mechanics and Analysis* 39:1–29.

———. 1976. "Irreversibility and Indeterminism: Fourier to Heisenberg." *Journal of the History of Ideas* 37:603–30.

———. 1978. *The Temperature of History: Phases of Science and Culture in the Nineteenth Century.* New York: Bart Franklin.

Buchdahl, Gerd. 1959. "Sources of Scepticism in Atomic Theory." *British Journal for the Philosophy of Science* 10:120–34.

———. 1987. "Stadien der begrifflichen Entwicklung von Atomtheorien." Pp. 101–29 in *Begriffswandel und Erkenntnisfortschritt in den Erfahrungswissenschaften,* edited by Friedrich Rapp and Hans-Werner Schütt. (TUB-Dokumentation Heft 32) Berlin: Technische Universität Berlin.

Büchner, Ludwig (Louis). 1855. *Kraft und Stoff. Empirisch-naturphilosophische Studien. In allgemein-verständlicher Darstellung.* Frankfurt am Main: Meidinger. [5th ed. 1858. Frankfurt am Main: Meidinger. 21st ed. 1904. Leipzig: Thomas.]

Buck, Peter. 1981. "From Celestial Mechanics to Social Physics: Discontinuity in the Development of the Sciences in the Early Nineteenth Century." Pp. 19–33 in *Epistemological and Social Problems of the Sciences in the Early Nineteenth Century,* edited by Hans Nils Jahnke and Michael Otte. Dordrecht: Reidel.

Buek, Otto. 1905. "Die Atomistik und Faradays Begriff der Materie. Eine logische Untersuchung." *Archiv für Geschichte der Philosophie* 18:65–110, 139–65.

Buggle, Franz and Paul Wirtgen. 1969. "Gustav Theodor Fechner und die psychoanalytischen Modellvorstellungen Sigmund Freuds. Einflüsse und Parallelen." *Archiv für die gesamte Psychologie* 121:148–201.

Bunge, Mario. 1977. "Emergence and the Mind." *Neuroscience* 2:501–9.

———. 1980. *The Mind-Body Problem. A Psychobiological Approach.* Oxford: Pergamon.

Burdach, Karl Friedrich. 1842. "Die persönliche Besonderheit." *Blicke ins Leben.* Vol. 2: *Comparative Psychologie, zweiter Theil.* Leipzig: Leopold Voß.

Busse, Ludwig. 1900. "Die Wechselwirkung zwischen Leib und Seele und das Gesetz der Erhaltung der Energie." Pp. 89–126 in *Philosophische Abhandlungen. Christoph Sigwart zu seinem siebzigsten Geburtstage am 28. März 1900 gewidmet.* Tübingen: Mohr.

———. 1913. *Geist und Körper, Seele und Leib.* 2nd ed. (with supplementary appendix summarizing the newest literature by Ernst Dürr). Leipzig: Meiner, 1913. [1st ed. 1903. Leipzig: Dürr.]

Busse, Michael, and Änne Bäumer-Schleinkofer. 1996. "Ewald Hering und die Gegenfarbentheorie." *NTM—Schriftenreihe für Geschichte der Naturwissenschaften, Technik und Medizin* 4:159–72.

Cadwallader, T. C. 1974. "Charles S. Peirce (1839–1914): The First American Experimental Psychologist." *Journal of the History of the Behavioral Sciences* 10:291–98.

Cahan, David, ed. 1993. *Hermann von Helmholtz and the Foundations of Nineteenth-Century Science.* Edited by David Cahan. Berkeley: University of California Press.

Camerer, [Johann] Wilhelm. 1881. "Versuche über den Raumsinn der Haut bei Kindern, angestellt an der obern Extremität nach der Methode der richtigen und falschen Fälle." *Zeitschrift für Biologie* 17:1–22.

———. 1883. "Versuche über den Raumsinn der Haut nach der Methode der richtigen und falschen Fälle." *Zeitschrift für Biologie* 19:280–300.

———. 1885. "Die Methode der richtigen und falschen Fälle angewendet auf den Geschmackssinn." *Zeitschrift für Biologie* 21:570–602.

———. 1887. "Die Methode der Äquivalente, angewandt zur Maassbestimmung der Feinheit des Raumsinnes." *Zeitschrift für Biologie* 23:509–59.

———. 1907. "Das Verhältnis von Seele und Leib nach dem heutigen Stand der Wissenschaft." *Besondere Beilage des Staats-Anzeigers für Württemberg* 13–14:193–215; 15–16: 225–40.

———. 1908. *Philosophie und Naturwissenschaft.* 4th ed. Stuttgart: Kosmos-Franckh, no date. [ca. 1908]

Caneva, Kenneth L. 1974. *Conceptual and Generational Change in German Physics: The Case of Electricity, 1800–1846.* Ph.D. diss., Princeton University, 1974.

———. 1978. "From Galvanism to Electro-Dynamics: the Transformation of German Physics and its Social Context." *Historical Studies in the Physical Sciences* 9:63–159.

———. 1993. *Robert Mayer and the Conservation of Energy.* Princeton: Princeton University Press.

———. 1997. "Physics and *Naturphilosophie*: A Reconnaissance." *History of Science* 35:35–107.

Čapek, Milič. 1954. "James's Early Criticism of the Automaton Theory." *Journal of the History of Ideas* 15:260–79.

———. 1961. "*The Philosophical Impact of Contemporary Physics.* Princeton: Van Nostrand.

———. 1969. "The Main Difficulties of the Identity Theory." *Scientia* 104:388–404.

Carnap, Rudolf. 1928. *Der logische Aufbau der Welt.* Berlin-Schlachtensee: Weltkreis-Verlag. [Reprint 1967. Frankfurt am Main: Ullstein, 1979. Translation 1967 by R. George as *The Logical Structure of the World.* Berkeley: University of California Press.]

———. 1963. "Intellectual Autobiography." Pp. 1–84 in *The Philosophy of Rudolf Carnap,* edited by Paul Arthur Schilpp. La Salle, Ill: Open Court.

———. 1966. *Philosophical Foundations of Physics.* Edited by Martin Gardner. New York: Basic Books. [Corrected reprint, with new foreword, of 1974 ed., 1995, retitled *An Introduction to the Philosophy of Science.* New York: Basic Books.]

Carrier, Martin. 1990. "Kants Theorie der Materie und ihre Wirkung auf die zeitgenössische Chemie." *Kant-Studien* 81: 170–210.

Carriere, Moriz. 1852. "Zend-Avesta von G. Th. Fechner." *Blätter für literarische Unterhaltung.* 2 (40):937–41.

———. 1882. "Zur Psychophysik." *Deutsche Revue* 7(2):397–98.

Carus, Carl Gustav. 1858. *Symbolik der menschlichen Gestalt.* 2nd expanded ed. Leipzig: Brockhaus.

Case, Thomas. 1911. "Metaphysics." Pp. 224–53 of vol. 18 in *Encyclopedia Britannica.* 11th ed. Cambridge: Cambridge University Press.

Caspari, Otto. 1869. *Die psychophysische Bewegung in Rücksicht der Natur ihres Substrats. Eine kritische Untersuchung als Beitrag zur empirischen Psychologie.* Leipzig: Leopold Voß.

Chaitin, Gregory J. 1974. "Information-Theoretic Computational Complexity." *IEE Transactions on Information Theory* IT-20(1):10–15. [Cited according to reprint, 1985, pp.

285–99 in *New Directions in the Philosophy of Mathematics*, edited by Thomas Tymoczko. Boston: Birkhäuser.]

———. 1975. "Randomness and Mathematical Proof." *Scientific American* 232:47–52.

———. 1982. "Gödel's Theorem and Information." Pp. 300–11 in *New Directions in the Philosophy of Mathematics*, edited by Thomas Tymoczko. Boston: Birkhäuser.

Cheng, Chung-ying, ed. 1975. *Philosophical Aspects of the Mind-Body Problem*. Honolulu: University Press of Hawaii.

Chisholm, Roderick M. 1978. "Is There a Mind-Body Problem?" *Philosophic Exchange* 2:25–34.

Chuseau, Hans. 1905. *Eduard von Hartmanns Stellung zum psychophysischen Parallelismus*. Ph.D. diss., phil. Fak., Universität Königsberg.

Clifford, William Kingdon. 1878. "On the Nature of Things-in-themselves." *Mind*. [Also 1879, pp. 71–88 in vol. 2 of *Lectures and Essays*, edited by Leslie Stephen and Frederick Pollock. London: Macmillan.]

Cohen, Hermann. 1883. *Das Princip der Infinitesimal-Methode und seine Geschichte. Ein Kapitel zur Grundlegung der Erkenntnisskritik*. Berlin: Dümmler. [Reprint 1968 Frankfurt am Main: Suhrkamp, 1968. Cited according to 4th ed., 1984 (reprint of the 1st ed.) *Werke von Hermann Cohen*. Vol. 5. Hildesheim: Olms.]

———. 1884. [Review of Fechner 1882.] Pp. 476–77 in Hermann Cohen, 1928. *Schriften zur Philosophie und Zeitgeschichte*, edited by. Albert Görland and Ernst Cassirer. Vol 2. Berlin: Akademieverlag.

———. 1885. *Kants Theorie der Erfahrung*. 2nd rev. ed. Berlin: Dümmler, 1885. [1st ed. 1871; 3rd ed. 1918. Berlin: Bruno Cassirer.]

———. 1888. "Jubiläums-Betrachtungen." [Review of *Philosophische Aufsätze, Eduard Zeller zu seinem fünfzigjährigen Doctor-Jubiläum gewidmet*. Leipzig: Fues, 1887.] *Philosophische Monatshefte* 24:257–91.

———. 1895. "Worte an der Bahre von Adolf Elsas." [Obituary of 1895.] Pp. 396–97 in Hermann Cohen, 1928. *Schriften zur Philosophie und Zeitgeschichte*, edited by Albert Görland and Ernst Cassirer. Vol 2. Berlin: Akademieverlag.

Coleman, William. 1971. *Biology in the Nineteenth Century*. Cambridge: Cambridge University Press.

Collins, James. 1947. "Religious Thoughts of a Scientist." *Modern Schoolman* 24:235–38. [On Fechner's philosophy of religion.]

Corso, John F. 1963. "A Theoretico-historical Review of the Threshold Concept." *Psychological Bulletin* 53:371–93.

Cramer, Konrad. 1974. "Erlebnis." Pp. 537–603 in *Stuttgarter Hegel-Tage 1970*, edited by Hans-Georg Gadamer. (Hegel-Studien, Beiheft 11) Bonn: Bouvier.

Cramér, Harald. 1972. "On the History of Certain Expansions Used in Mathematical Statistics." *Biometrika* 59:205–7.

Cranefield, Paul F. 1957. "The Organic Physics of 1847 and the Biophysics of Today." *Journal of the History of Medicine and Allied Sciences* 12:407–23.

———. 1966. "Freud and the School of Helmholtz." *Gesnerus* 23:35–39.

Crosland, Maurice. 1967. *The Society of Arcueil: A View of French Science at the Time of Napoleon I*. London: Heinemann.

Cummins, Robert. 1983. *The Nature of Psychological Explanation*. Cambridge, Mass.: Bradford Books.

Cunningham, Andrew and Nicholas Jardine, eds. 1990. *Romanticism and the Sciences*. Cambridge: Cambridge University Press.

Curd, Martin V. 1978. *Ludwig Boltzmann's Philosophy of Science: Theories, Pictures and Analogies.* Ph.D. diss., University of Pittsburgh.

Czaja, Johannes. 1993. *Psychophysische Grundperspektive und Essayismus. Untersuchungen zu Robert Musils Werk mit besonderem Blick auf Gustav Theodor Fechner und Ernst Mach.* Ph.D. diss., phil. Fak., Universität Tübingen.

Czuber, Emanuel. 1906. "Die Kollektivmaßlehre." *Zeitschrift für das Realschulwesen* (Vienna) 31(9):513–32.

———. 1908. *Wahrscheinlichkeitsrechnung und ihre Anwendung auf Fehlerausgleichung, Statistik und Lebensversicherung.* Vol. 1: *Wahrscheinlichkeitstheorie, Fehlerausgleichung, Kollektivmasslehre.* 2nd ed. Leipzig: Teubner. [Reprint 1968. New York: Johnson. 4th ed. 1924. 5th ed. 1938.]

Danziger, Kurt. 1979. "The Positivist Repudiation of Wundt." *Journal of the History of the Behavioural Sciences* 15:205–30.

———. 1980. "Wundt and the two traditions of Psychology." Pp. 73–88 in *Wilhelm Wundt and the Making of a Scientific Psychology,* edited by Robert W. Rieber. New York: Plenum.

———. 1990. *Constructing the Subject: Historical Origins of Psychological Research.* Cambridge: Cambridge University Press.

———. 1997. *Naming the Mind: How Psychology Found its Language.* London: Sage Publications.

Davidson, Donald. 1970. "Mental Events." Pp. 79–101 in *Experience and Theory,* edited by L. Foster and J. W. Swanson. Amherst: University of Massachussetts Press. [Cited according to reprint 1987. Pp. 207–55 in Donald Davidson, *Action and Events.* Oxford: Clarendon.]

Delbœuf, Joseph (Rémy Léopold). 1873. *Étude psychophysique. Recherches théoriques et expérimentales sur la mesure des sensations et spécialement des sensations de lumière et de fatigue.* (Mémoires couronnés et autres mémoires publiés par l'Académie royale des sciences, des lettres et des beaux-arts de Belgique, tome XXIII) Brussels: Hayez.

———. 1873a. "La mesure des sensations." *La Revue Scientifique,* 2e série, 3e année, No. 3:66–68.

———. 1875. "La mesure des sensations. Réponses à propos du logarithme des sensations." *La Revue Scientifique,* 2e série, 4e année, No. 43:1014–17.

———. 1875a. "La mesure des sensations (1)." *La Revue Scientifique,* 2e série, 4e année, No. 46:1089–90.

———. 1876. *Théorie générale de la sensibilité.* (Mémoires couronnés et autres mémoires publiés par l'Académie royale des sciences, des lettres et des beaux-arts de Belgique, tome XXVI) Brussels: Hayez.

———. 1877. "La loi psychophysique. Hering contre Fechner." *Revue philosophique* 3: 225–63.

———. 1878. "La loi psychophysique. Fechner contre ses adversaires." *Revue philosophique* 5:34–63, 127–57.

———. 1883. *Examen critique de la loi psychophysique, sa base et sa signification.* Paris: Germer Baillière. [Consists of Delboeuf 1877 and 1878.]

———. 1883a. *Eleménts de psychophysique générale et spéciale.* Paris: Germer Baillière.

Delezenne, Charles (Edouard Joseph) 1826. "Mémoire sur les valeurs numériques des notes de la gamme." *Recueil des travaux de la société des sciences, de l'agriculture et des arts de Lille* 4:1–56.

Dennert, Eberhard. 1902. *Fechner als Naturphilosoph und Christ. Ein Beitrag zur Kritik des Pantheismus.* Gütersloh: C. Bertelsmann.

————. 1903. "Gustav Theodor Fechner und das Christentum." *Reformation* 2(7):98–117.

Dilthey, Wilhelm. 1894. "Ideen über eine beschreibende und zergliedernde Psychologie." *Sitzungsberichte der Königlich Preußischen Akademie der Wissenschaft*, philosoph-histor. Klasse, 54. [Cited according to reprint 1974. Pp. 136–240 in Wilhelm Dilthey, *Gesammelte Schriften*. Vol. 5. Edited by Georg Misch. Stuttgart: Teubner.]

————. 1982. "Ausarbeitungen und Entwürfe zum zweiten Band der Einleitung in die Geisteswissenschaften. Viertes bis Sechstes Buch (ca. 1880–1890)." Pp. 78–295 in Wilhelm Dilthey, 1982. *Gesammelte Schriften*. Vol. 19. Eds. Helmut Johach and Frithjof Rodi. Göttingen: Vandenhoeck.

Dittenberger, Wilhelm. 1896. "Über das psychophysische Gesetz." *Archiv für systematische Philosophie* 2:71–102.

Dittmar, Carl. 1884. [Review of Fechner 1882.] *Allgemeine Zeitschrift für Psychiatrie* 40:19-23.

Dorer, Lore Muerdel. 1986. "Hofmannsthals Terzinen und G. Th. Fechner. Hinweise auf bisher unbeachtete Bezüge in Leben und Werk." *Modern Austrian Literature* 19(2):33–46.

Dorer, Maria. 1932. *Historische Grundlagen der Psychoanalyse*. Leipzig: Meiner.

Döring, Detlef and Annelies Plätzsch. 1987. "Kommentiertes Verzeichnis des handschriftlichen Nachlasses Gustav Theodor Fechners in der Universitätsbibliothek Leipzig." Pp. 286–305 in *Psychophysische Grundlagen mentaler Prozesse. In memoriam G. Th. Fechner (1801–1887)*, edited by Hans-Georg Geissler and Konrad Reschke. Leipzig: Karl-Marx-Universität Leipzig.

Driesch, Hans. 1916. *Leib und Seele. Eine Prüfung des psycho-physischen Grundproblems*. Leipzig: E. Reinicke.

Drobisch, Moritz Wilhelm. 1834. *Beiträge zur Orientierung über Herbart's System der Philosophie*. Leipzig: Leopold Voß.

————. 1849. [Review of Quetelet 1848.] *Leipziger Repertorium der deutschen und ausländischen Literatur* 25:28–39.

————. 1856. "Synechologische Untersuchungen. (Dritter Artikel.)." *Zeitschrift für Philosophie und philosophische Kritik*, 28:52–91.

————. 1864. "Ueber den neuesten Versuch die Psychologie naturwissenschaftlich zu begründen." [Review of Wundt 1862 and 1863.] *Zeitschrift für exacte Philosophie* 4:313–48.

————. 1867. *Die moralische Statistik und die Willensfreiheit*. Leipzig: Leopold Voß.

Du Bois-Reymond, Emil. 1848. "Vorrede." In Emil Du Bois-Reymond, *Untersuchungen über thierische Elektricität*. 2 vols. Berlin: Reimer, 1848–49, vol. 1. [Cited according to reprint 1912. "Über die Lebenskraft. Aus der Vorrede zu den 'Untersuchungen über thierische Electrizität', vom März 1848." Pp. 1–26 in *Reden von Emil Du Bois-Reymond*, edited by Estelle Du Bois-Reymond. 4th ed., vol. 1. Leipzig: Veit.]

————. 1872. *Über die Grenzen des Naturerkennens*. (Ein Vortrag in der zweiten öffentlichen Sitzung der 45. Versammlung Deutscher Naturforscher und Ärzte zu Leipzig am 14. August 1872) Leipzig: Veit, 1872. [Cited according to reprint 1912. Pp. 441–73 in *Reden von Emil Du Bois-Reymond*, edited by Estelle Du Bois-Reymond. 4th ed., vol. 1. Leipzig: Veit.]

Du Bois-Reymond, Paul. 1882. *Die allgemeine Functionentheorie*. Part 1: *Metaphysik und Theorie der mathematischen Grundbegriffe: Größe, Grenze, Argument und Function*. Tübingen: Laupp 1882. Reprint 1968. Detlef Laugwitz, ed. Darmstadt: Wissenschaftliche Buchgesellschaft.

Dubos, René. 1973. "Environment." Pp. 120–27 in *Dictionary of the History of Ideas*, edited by Philip P. Wiener. Vol. 1. New York: Scribner's.

Duncker, Georg. 1899. "Die Methode der Variationsstatistik." *Archiv für Entwicklungsmechanik der Organismen* 8:112–83.

Dupéron, Isabelle. 2000. *G. T. Fechner: Le Parallélisme psychophysiologique*. Paris: Presses Universitaires de France.

Durner, Manfred. 1994. "Theorien der Chemie." Pp. 1–161 in Baumgartner 1994. [Chemistry at the time of Schelling's philosophy of nature, ca. 1797–1800.]

Ebbinghaus, Hermann. 1890. "Über negative Empfindungswerte." *Zeitschrift für Psychologie* 1:320–34, 463–85.

———. 1896. "Über erklärende und beschreibende Psychologie." *Zeitschrift für Psychologie* 6:161–205.

Ebrecht, Angelika. 1988. "'Verkehrte Welt'—Strukturmomente einer psychologischen Ästhetik im Werk Fechners." Pp. 229–41 in *G. T. Fechner and Psychology*, edited by Josef Brožek and Horst Gundlach. Passau: Passavia.

———. 1992. *Das individuelle Ganze. Zum Psychologismus der Lebensphilosophie*. Stuttgart: Metzler.

Edel, Geert. 1988. *Von der Vernunftkritik zur Erkenntnislogik. Die Entwicklung der theoretischen Philosophie Hermann Cohens*. Freiburg i. Br.: Alber.

Edgeworth, Francis Ysidro. 1877. *New and Old Methods of Ethics*. Oxford: Parker.

———. 1907. "On the Representation of Statistical Frequency by a Series." *Journal of the Royal Statistical Society* 70:102–6.

Ehrenfels, Christian von. 1886. "Metaphysische Ausführungen im Anschlusse an Emil du Bois-Reymond." *Sitzungsberichte der Kaiserlichen Akademie der Wissenschaften (Wien)*, phil.-hist. Klasse, 112:429–503. [Cited according to reprint 1990. Pp. 11–67 in Christian Ehrenfels, *Metaphysik. Philosophische Schriften*. 4th ed., edited by Reinhard Fabian. Munich: Philosophia.]

Eigen, Manfred. 1971. "Self Organization of Matter and the Evolution of Biological Macromolecules." *Naturwissenschaften* 58:465–523.

Einstein, Albert. 1916. "Ernst Mach." *Physikalische Zeitschrift* 17(7):101–4. [Facsimile reprint 1996. Pp. 278–81 in *The Collected Papers of Albert Einstein*. Vol. 6: *The Berlin Years: Writings, 1914-1917*, edited by A. J. Kox, Martin J. Klein, and Robert Schulman. Princeton: Princeton University Press. Also in Ernst Mach 1987. Pp. 641–46 in *Populär-wissenschaftliche Vorlesungen*. Vienna: Böhlau.]

———. 1987. *The Collected Papers of Albert Einstein*. Vol. 1: *The Early Years, 1879–1902*, edited by John Stachel. Princeton: Princeton University Press.

Eisler, Rudolf. 1893. *Der psychophysische Parallelismus. Eine philosophische Skizze*. Leipzig: Wilhelm Friedrich.

Elderton, W. Palin. 1906. "Notices of Memoires and Books." *Biometrika* 5:206–10.

Ellenberger, Henri F. 1956. "Fechner and Freud." *Bulletin of the Menninger Clinic* 20:201–14.

———. 1970. *The Discovery of the Unconscious: The History and Evolution of Dynamic Psychiatry*. London: Fontana. Reprint 1994. [Cited according to the German translation 1985. *Die Entdeckung des Unbewußten. Geschichte und Entwicklung der dynamischen Psychiatrie von den Anfängen bis zu Janet, Freud, Adler und Jung*. Bern: Diogenes.]

Ellis, Brian D. 1966. *Basic Concepts of Measurement*. Cambridge: Cambridge University Press.

———. 1967. "Measurement." Pp. 241–50 in *Encyclopedia of Philosophy*, edited by Paul Edwards. Vol. 5. New York: Macmillan.

Elsas, Adolf. 1886. *Ueber die Psychophysik. Physikalische und erkenntnisstheoretische Betrachtungen.* Marburg: Elwert.

———. 1888. "Die Deutung der psychophysischen Gesetze." *Philosophische Monatshefte* 24:129–55.

———. 1888a. "Zum Andenken Gustav Theodor Fechners." *Die Grenzboten* 47:73–80, 113–24.

Engelhardt, Dietrich von. 1980. "Schuberts Stellung in der romantischen Naturforschung." Pp. 11–36 in *Gotthilf Heinrich Schubert. Gedenkschrift zum 200. Geburtstag des romantischen Naturforschers.* (Erlanger Forschungen, Reihe A, vol. 25) Erlangen: Verlag Universitätsbund Erlangen-Nürnberg.

———. 1981. "Du Bois-Reymond im Urteil der zeitgenössischen Philosophie." Pp. 187–205 in *Naturwissen und Erkenntnis im 19. Jahrhundert: Emil Du Bois-Reymond,* edited by Gunter Mann. Mainz: Gerstenberg.

———. 1981a. "Prinzipien und Ziele der Naturphilosophie Schellings—Situation um 1800 und spätere Wirkungsgeschichte." Pp. 77–98 in *Schelling. Seine Bedeutung für eine Philosophie der Natur und der Geschichte.* (1st Internat. Schelling Conference, Zürich 1979), edited by Ludwig Hasler. Stuttgart: Frommann-Holzboog.

———. 1985. "Die organische Natur und die Lebenswissenschaften in Schellings Naturphilosophie." Pp. 39–57 in *Natur und Subjektivität. Zur Auseinandersetzung mit der Naturphilosophie des jungen Schelling.* (2nd Internat. Schelling Conference, Zürich 1983), edited by Reinhard Heckmann, Hermann Krings, and Rudolf W. Meyer. Stuttgart: Frommann-Holzboog.

———. 1986. "Der Entwicklungsbegriff zwischen Naturwissenschaft und Naturphilosophie um 1800." *Annalen der internationalen Gesellschaft für dialektische Philosophie. Societas Hegeliana* (Cologne) III:309–16.

Erdmann, Benno. 1879. "Zur zeitgenössischen Psychologie in Deutschland." [Review of Ribot 1879.] *Vierteljahrsschrift für wissenschaftliche Philosophie* 3:377–407.

———. 1907. *Wissenschaftliche Hypothesen über Seele und Leib. Vorträge gehalten an der Handelshochschule zu Köln.* Cologne: Dumont-Schauberg.

Erdmann, Johann Eduard. 1853. *Die Entwicklung der deutschen Spekulation seit Kant III.* Riga. (Vol. 7 of *Versuch einer wissenschaftlichen Darstellung der Geschichte der neuern Philosophie.* 7 vols. Riga 1848–53) [Cited according to facsimile reprint 1982. Stuttgart: Frommann.]

———. 1896. *Grundriß der Geschichte der Philosophie.* 4th ed., rev. by Benno Erdmann. Vol. 2. Berlin. [The appendix: *Die deutsche Philosophie seit Hegels Tode,* 639–916, is cited according to the separate facsimile reprint, introduced by Hermann Lübbe, 1964. Stuttgart: Frommann.]

Erhardt, Franz. 1897. *Die Wechselwirkung zwischen Leib und Seele. Eine Kritik der Theorie des psychophysischen Parallelismus.* Leipzig: Reisland.

Eshleman, Cyrus H. 1909. "Professor James on Fechner's Philosophy." *The Hibbert Journal* 7(3):671–73.

Exner, Franz. 1919. *Vorlesungen über die physikalischen Grundlagen der Naturwissenschaften.* 2nd expanded ed. Leipzig/Vienna: Deuticke. [Cited according to the 2nd expanded ed. 1922.]

Exner, Sigmund. 1879. "Physiologie der Großhirnrinde." Pp. 189–350 in *Handbuch der Physiologie,* edited by Ludimar Hermann. Vol. 2. Leipzig: Vogel. [On Fechner's Psychophysics see chapter 2, esp. 215–46.]

Faggi, Adolfo. 1895. "Fechner e la sua costruzione psicofisica." *Rivista Italiana* 10.

Falckenberg, Richard. 1897. "Aus Hermann Lotzes Briefen an Theodor und Clara Fechner." *Zeitschrift für Philosophie und philosophische Kritik* 111:177–90.

Falmagne, Jean-Claude. 1974. "Foundations of Fechnerian Psychophysics." Pp. 127–159 in *Contemporary Developments in Mathematical Psychology*, edited by David H. Krantz, R. C. Atkinson, and R. Duncan Luce. Vol. 2: *Measurement, Psychophysics, and Neural Information Processing*. San Francisco: Freeman.

———. 1977. "Weber's Inequality and Fechner's Problem." *Journal of Mathematical Psychology* 16:267–71.

———. 1985. *Elements of Psychophysical Theory*. Oxford Psychology Series No. 6. Oxford: Clarendon. Reprint 2002. Oxford: Oxford University Press.

Fechner, Gustav Theodor. 1821. *Beweis, daß der Mond aus Jodine bestehe*. Germanien [*sic*]: Penig. 2nd ed. Leipzig: Leopold Voß. 1832. [Cited according to Fechner 1875, *Kleine Schriften*, 1-20. This article appeared under the pseudonym of "Dr. Mises."]

———. 1822. *Panegyrikus der jezigen Medicin und Naturgeschichte*. Leipzig: C. H. F. Hartmann. [Cited according to Fechner 1875, *Kleine Schriften*, 21–68. This article appeared under the pseudonym of "Dr. Mises."]

———. 1823. *Praemissae ad theoriam organismi generalem*. Leipzig: Staritz. [Habilitation thesis, Universität Leipzig. Partial translation in Marilyn E. Marshall 1974. "G. T. Fechner: Premises Toward a General Theory of Organisms (1823)." *Journal of the History of the Behavioral Sciences* 10:438–47.]

———. 1823a. *Katechismus der Logik oder Denklehre bestimmt zum Selbst- und Schulunterricht*. Leipzig: Baumgärtner.

———. 1823b. *Katechismus oder Examinatorium über die Physiologie des Menschen*. Leipzig: Baumgärtner.

———. 1824. *Lehrbuch der Experimental-Physik oder Erfahrungs-Naturlehre*. By Jean-Baptiste Biot. 4 vols. Leipzig: Leopold Voß. [Translation of Biot 1824 by Fechner with many additions.]

———. 1824a. *Stapelia mixta*. Leipzig: Leopold Voß, 1824. [Partial reprint in Fechner 1875, *Kleine Schriften*. Reprint Eschborn: Klotz, 1994. This collection of articles appeared under the pseudonym of "Dr. Mises."]

———. 1825. *Lehrbuch der theoretischen und praktischen Chemie*. By Louis-Jacques Thénard. Transl. and supplemented by Gustav Theodor Fechner according to the 5th ed., 7 vols. in 9 sections. Leipzig: Leopold Voß. [Translation of Thénard 1813 and 1827. With many additions by Fechner. Vols. 4 and 5 are written almost completely by Fechner himself.]

———. 1825a. "Carl Brandan Mollweide." [Obituary] *Zeitung für die elegante Welt*. Leipzig 25:545–47. [Also at http://www.uni-leipzig.de/~fechner/mollweid.htm. Not in the bibliography of Fechner's works by Müller, Rudolph 1892.]

———. 1825b. *Vergleichende Anatomie der Engel. Eine Skizze*. Leipzig: Baumgärtner, 1825. [Also in Fechner 1875, *Kleine Schriften*,. 195–240. Reprinted in Breidbach 1997. Translation Marilyn E. Marshall, "Gustav Fechner, Dr. Mises, and the Comparative Anatomy of Angels." *Journal of the History of the Behavioral Sciences* 5(1):39–58.]

———. 1825c. [Letter, Fechner to Jean Paul, October 10.] See P. 1887; Kuntze 1892, 66; or Lennig 1994, 47.

———. 1826. "Ueber die Möglichkeit, scheinbare Abstossungen auf Anziehungskräfte zurückzuführen." [Kastner's] *Archiv für die gesammte Naturlehre* 9:257–83. [Not in the bibliography of Fechner's works by Müller, Rudolph 1892.]

————. 1828. *Lehrbuch der Experimental-Physik oder Erfahrungs-Naturlehre.* By Jean-Baptiste Biot. 2nd German ed. with the addition of new discoveries by Gustav Theodor Fechner. 5 vols. Leipzig: Leopold Voß, 1st vol. 1828, the other vols. 1829. [Translation of Biot 1824. Vol. 3 also titled: *Lehrbuch des Galvanismus und der Elektrochemie.* Revised according to new original sources by Gustav Theodor Fechner. Although it appeared under Biot's name, this vol. of 564 pp. was completely written by Fechner himself.]

————. 1828a. "Ueber die Anwendung des Gravitationsgesetzes auf die Atomenlehre." [Kastner's] *Archiv für die gesammte Naturlehre* 15: 257–90. [With diagram of a model of a planetary atom. Not in the bibliography of Fechner's works by Müller, Rudolph 1892.]

————. 1829. See Ohm 1938.

————. 1830a. *Elementbuch des Elektromagnetismus nebst Beschreibung der hauptsächlichen elektromagnetischen Apparate.* Leipzig: Leopold Voß.

————. 1831. *Maaßbestimmungen über die galvanische Kette.* Leipzig: Brockhaus.

————. 1831a. "Resultate verschiedener physikalisch-mathematischer Abhandlungen." [Schweigger's] *Journal für Chemie und Physik* 63:352–62, 437–44; and 64:38–49. [Short presentation of several works by S.-D. Poisson and A.-L. Cauchy. Not in the bibliography of Fechner's works by Müller, Rudolph 1892.]

————. 1832. *Repertorium der Experimental-Physik, enthaltend eine vollständige Zusammen-stellung der neueren Fortschritte dieser Wissenschaft. Als Supplement zu neuern Lehr- und Wörterbüchern der Physik.* 3 vols. Leipzig: Leopold Voß.

————. 1834. *Das Hauslexicon. Vollständiges Handbuch praktischer Lebenskenntnisse für alle Stände.* Redacteur G. Th. Fechner. 8 vols. Leipzig: Breitkopf und Härtel. [2nd ed. 1841.]

————. 1836. *Das Büchlein vom Leben nach dem Tode.* Dresden: Grimmer. [This 1st ed. appeared under the name of "Dr. Mises." Subsequent eds. appeared in 1866, 1887, 1900, 1903, 1906, 1911, 1919. Reprint 2001. Schutterwald/Baden: Wiss. Verl. Cited according to Fechner 1984. *Das unendliche Leben.* Munich: Matthes and Seitz. This ed. includes passages from Fechner 1848 and 1851 and an afterword by Gert Mattenklott, which also appears abbreviated in Mattenklott 1986. 1st ed. translated 1977 by Mary C. Wadsworth as *The Little Book of Life after Death.* Reprint 1977 of 1904 ed. New York: Arno Press.]

————. 1838. "Ueber die subjectiven Complementarfarben." *Annalen der Physik und Chemie* 44:221–45, 513–35.

————. 1838a. "Ueber eine Scheibe zur Erzeugung subjectiver Farben." *Annalen der Physik und Chemie* 45:227–32.

————. 1840. "Ueber die subjectiven Nachbilder und Nebenbilder." *Annalen der Physik und Chemie* 50:193–221, 427–70.

————. 1841. *Gedichte.* Leipzig: Breitkopf und Härtel, 1841. [Appeared under the name of "Dr. Mises."]

————. 1845. "Ueber die Verknüpfung der Faraday'schen Inductions-Erscheinungen mit den Ampère'schen elektro-dynamischen Erscheinungen." *Annalen der Physik und Chemie* 64:337–45.

————. 1846. *Ueber das höchste Gut.* Leipzig: Breitkopf und Härtel. [Reprint 1925. Frankfurt am Main: Diesterweg.]

————. 1846a. *Vier Paradoxa.* Leipzig: Leopold Voß, 1846. [Cited according to Fechner 1875, *Kleine Schriften*, 241–321.]

————. 1848. *Nanna oder über das Seelenleben der Pflanzen.* Leipzig: Leopold Voß. [Reprint

1992. Eschborn: Klotz. Reprint 1997. Karben: Wald. Cited according to the 4th ed. 1908. Hamburg: Leopold Voß.]

―――. 1848a. "Ueber das Lustprincip des Handeln." *Zeitschrift für Philosophie und philosophische Kritik* 19:1–30, 163–94.

―――. 1849. "Die mathematische Behandlung organischer Gestalten und Processe." *Berichte über die Verhandlungen der Königlich Sächsischen Gesellschaft der Wissenschaften zu Leipzig*, Mathematisch-physische Classe (Jahrgang 1849):50–64. [Reprint 1996. Pp. 15–24 in *Abstand und Nähe. Vorträge im Rückblick. Sächsische Akademie der Wissenschaften zu Leipzig*, edited by Helga Bergmann im Auftrag des Präsidiums. Berlin: Akademie. See Scharf 1996 for a commentary on this speech.]

―――. 1849a. "Ueber das Causalgesetz." *Berichte über die Verhandlungen der Königlich Sächsischen Gesellschaft der Wissenschaften zu Leipzig*, Mathematisch-physische Classe (Jahrgang 1849): 98–120.

―――. 1850. *Räthselbüchlein*. Leipzig: Wigand. [Appeared under the name of "Dr. Mises."]

―――. 1851. *Zend-Avesta oder über die Dinge des Himmels und des Jenseits. Vom Standpunkt der Naturbetrachtung*. 3 Theile. Leipzig: Leopold Voß. [Reprint 1998. Eschborn: Klotz. The 3 parts are cited with I, II, III. "Kurze Darlegung eines Princips mathematischer Psychologie." II, 373–86, translated by Eckart Scheerer 1987 as "Outline of a New Principle of Mathematical Psychology." *Psychological Research* 49:203–7.]

―――. 1851a. "Ueber die Erkenntniß Gottes in der Natur aus der Natur." *Zeitschrift für Philosophie und philosophische Kritik* 21:193–209.

―――. 1853a. "Zur Kritik der Grundlagen von Herbart's Metaphysik." *Zeitschrift für Philosophie und philosophische Kritik* 25:70–102.

―――. 1854. "Ueber die Atomistik." *Zeitschrift für Philosophie und philosophische Kritik* 25:25–57.

―――. 1855. *Ueber die physikalische und philosophische Atomenlehre*. Leipzig: Hermann Mendelssohn, 1855. [For the 2nd ed. see Fechner 1864.]

―――. 1856. *Professor Schleiden und der Mond*. Leipzig: Adolf Gumprecht.

―――. 1857. "In Sachen der Atomistik." *Zeitschrift für Philosophie und philosophische Kritik* 30: 61–89, 165–91.

―――. 1858. Das psychische Maß." *Zeitschrift für Philosophie und philosophische Kritik* 32: 1–24.

―――. 1858a. "Ueber den Punct." *Zeitschrift für Philosophie und philosophische Kritik* 33:161–83.

―――. 1860. *Elemente der Psychophysik*. Erster und zweiter Theil [Part 1 and 2]. Leipzig: Breitkopf und Härtel, 1860. [Reprint 1964. Amsterdam: Bonset. Reprint 1998. Bristol: Thoemmes. 2nd unchanged ed., 1889, edited by Wilhelm Wundt. Leipzig: Breitkopf und Härtel. 3rd ed. 1903. Leipzig: Breitkopf und Härtel. Translation of vol. 1 1966 by Helmut E. Adler as *Elements of Psychophysics*, edited by Davis H. Howes and Edwin G. Boring. New York: Holt, Rinehart and Winston. Translation of vol. 2 forthcoming.]

―――. 1861. *Ueber die Seelenfrage. Ein Gang durch die sichtbare Welt, um die unsichtbare zu finden*. Leipzig: C. F. Amelang. [3rd ed. 1928, edited by Eduard Spranger. Leipzig: Voß. Reprint 1992. Eschborn: Klotz.]

―――. 1861a. "Ueber die Correctionen bezüglich der Genauigkeitsbestimmung der Beobachtungen." *Berichte über die Verhandlungen der Königlich Sächsischen Gesellschaft der Wissenschaften*, math.-phys. Klasse.

———. 1863. *Die drei Motive und Gründe des Glaubens.* Leipzig: Breitkopf und Härtel, 1863. [Cited according to reprint 1923, edited by Wilhelm Platz. Stuttgart: Strecker and Schröder.]

———. 1864. *Ueber die physikalische und philosophische Atomenlehre.* 2nd expanded ed. Leipzig: Hermann Mendelssohn. [Reprint 1982. Frankfurt am Main: Minerva. Reprint 1995, edited by Ecke Bonk. Vienna: Springer. Compare Fechner 1855.]

———. 1866. "Erinnerung an Christian Hermann Weiße (geb. d. 10 August 1801, gest. d. 19 Sept. 1866)." *Leipziger Tageblatt* (October 7). [Available at http://www.uni-leipzig.de/~fechner/weisse.htm]

———. 1871. "Zur experimentalen Ästhetik." *Abhandlungen der Königlich Sächsischen Gesellschaft der Wissenschaften* 9:555–635.

———. 1873. *Einige Ideen zur Schöpfungs- und Entwickelungsgeschichte der Organismen.* Leipzig: Breitkopf und Härtel. [Cited according to reprint 1985. Tübingen: ed. diskord.]

———. 1874. "Ueber die Bestimmung des wahrscheinlichen Fehlers eines Beobachtungsmittels durch die Summe der einfachen Abweichungen." *Annalen der Physik und Chemie* (Jubelband):66–81.

———. 1874a. "Ueber den Ausgangswerth der kleinsten Abweichungssumme." *Abhandlungen der Königlich Sächsischen Gesellschaft der Wissenschaften,* math.-phys. Klasse, 11:1–76. [Usually cited with the year 1874. The volume I inspected bears 1878 as date.]

———. 1874b. [Review of Delboeuf 1873.] *Jenaer Literaturzeitung,* 1 (28):421–23.

———. 1875. *Kleine Schriften.* Leipzig: Breitkopf und Härtel, 1875. [Appeared under the name of "Dr. Mises."]

———. 1875. "Nachruf für Hermann Härtel." *Leipziger Tageblatt* (August 10). [Available at http://www.uni-leipzig.de/~fechner/haertel.htm]

———. 1876. *Vorschule der Aesthetik.* Zwei Theile. Leipzig: Breitkopf und Härtel. [Reprint 1978. Hildesheim: Olms. Chapter XIV translated by Monika Niemann, Julia Quehl, and Holger Höge as "Various Attempts to Establish a Basic Form of Beauty: Experimental Aesthetics, Golden Section, and Square." *Empirical Studies of the Arts* 15(2):115–30.]

———. 1877. *In Sachen der Psychophysik.* Leipzig, Breitkopf und Härtel, 1877. [Reprint Amsterdam: Bonset, 1968.]

———. 1879. *Die Tagesansicht gegenüber der Nachtansicht.* Leipzig: Breitkopf und Härtel. [Reprint 1994. Eschborn: Klotz.]

———. 1882. *Revision der Hauptpuncte der Psychophysik.* Leipzig: Breitkopf und Härtel, 1882. [Reprint 1965. Amsterdam: Bonset. Chapter XXI translated by Eckart Scheerer as "Some Thoughts on the Psychophysical Representation of Memories." *Psychological Research* 49:209–12.]

———. 1882a. "Über die Aufgabe der Psychophysik." *Allgemeine Zeitung,* München, Beilage Nr. 339:4993-95 and 5010. [Cited according to reprint 1890. Pp. 204–26 in *Wissenschaftliche Briefe von Gustav Theodor Fechner und W. Preyer,* edited by William Thierry Preyer. Hamburg: Leopold Voß, 1890.]

———. 1884. "Ueber die Methode der richtigen und falschen Fälle in Anwendung auf die Maaßbestimmungen der Feinheit oder extensiven Empfindlichkeit des Raumsinns." *Abhandlungen der Königlich Sächsischen Gesellschaft der Wissenschaften,* mathematisch-physische Classe, 13:109–312.

———. 1887. "Ueber die psychischen Massprincipien und das Weber'sche Gesetz. Discussion mit Elsas und Köhler." [Wundt's] *Philosophische Studien* 4: 161–230. [Translation of pp. 178–98 by Eckart Scheerer as "My Own Viewpoint on Mental Measurement." *Psychological Research* 49:213–19.]

————. 1888. "Zöllners mediumistische Experimente. Aufzeichnungen aus dem Tagebuche von Gustav Theodor Fechner." *Sphinx* 3(5):217–21.

————. 1890. *Wissenschaftliche Briefe von Gustav Theodor Fechner und W. Preyer. Nebst einem Briefwechsel zwischen K. von Vierordt und Fechner und 9 Beilagen.* Edited by William Thierry Preyer. Hamburg: Leopold Voß.

————. 1890a. "Über negative Empfindungswerte." [Excerpts from Fechner's letters to William Preyer] *Zeitschrift für Psychologie und Physiologie der Sinnesorgane* 1:29–46, 108–20.

————. 1897. *Kollektivmasslehre.* Posthumously published on behalf of the Royal Saxon Society of the Sciences. Edited by Gottlob Friedrich Lipps. Leipzig: W. Engelmann.

————. 1905. Gottlob Friedrich Lipps, "Bericht über das Fechner-Archiv." *Berichte über die Verhandlungen der Königlich Sächsischen Gesellschaft der Wissenschaften zu Leipzig,* mathematisch-physische Klasse, 57:247–312. [With many excerpts from the destroyed Fechner estate.]

————. 1946. *Religion of a Scientist: Selections from Gustav Th. Fechner.* Translated and edited by Walter Lowrie. New York: Pantheon.

————, ed. 1830. *Pharmaceutisches Centralblatt* 1–5.

————, ed. 1853. *Centralblatt für Naturwissenschaften und Anthropologie.* 1. Jahrgang 1853; 2. Jahrgang 1854. Leipzig: Avenarius und Mendelssohn. [No further vols.]

Feigl, Herbert. 1934. "Logical Analysis of the Psychophysical Problem. A Contribution of the New Positivism." *Philosophy of Science* 1:420–45.

————. 1935. See Aldrich 1935.

————. 1937. "Moritz Schlick." *Erkenntnis* 7:393–419. [Reprint 1979. Pp. xv–xvii in Moritz Schlick, *Philosophical Papers.* Vol. 1. Edited by Henk L. Mulder and Barbara F. B. van de Velde-Schlick. Dordrecht: Reidel. Reprint 1982. Pp. 55–82 in *Rationality and Science: A Memorial Volume for Moritz Schlick in Celebration of the Centennial of His Birth,* edited by Eugene T. Gadol. Vienna: Springer.]

————. 1950. "The Mind-Body Problem in the Development of Logical Empiricism." *Revue Internationale de Philosophie* 4:64–83. [Cited according to reprint 1953. Pp. 612–26 in *Readings in the Philosophy of Science,* edited by Herbert Feigl and May Brodbeck. New York: Appleton.]

————. 1958. "The 'Mental' and the 'Physical.'" Pp. 370–497 in *Minnesota Studies in the Philosophy of Science.* Vol. 2. Edited by Herbert Feigl, Michael Scriven and Grover Maxwell. Minneapolis: University of Minnesota Press. [Cited according to separate ed. 1967, *The "Mental" and the "Physical." The Essay and a Postscript.* Minneapolis: University of Minnesota Press.]

————. 1960. "Mind-Body, Not a Pseudoproblem." Pp. 33–44 in *Dimensions of Mind,* edited by Sidney Hook. New York: Collier. [Cited according to the 5th ed. 1973.]

————. 1963. "Physicalism, Unity of Science and the Foundations of Psychology." Pp. 227–67 in *The Philosophy of Rudolf Carnap,* edited by Paul Arthur Schilpp. La Salle, Ill.: Open Court.

————. 1964. "Logical Positivism after Thirty-Five Years." *Philosophy Today* 8(4):228–45.

————. 1971. "Some Crucial Issues of Mind-Body Monism." *Synthese* 22:295–312.

————. 1975. "Russell and Schlick. A Remarkable Agreement on a Monistic Solution of the Mind-Body Problem." *Erkenntnis* 9:11–34.

Feuchtmüller, Rupert. 1946. *Dichtung und Naturwissenschaft von Gustav Theodor Fechner bis Friedrich Schnack.* Ph.D. diss., phil. Fak., Universität Wien.

Feyerabend, Paul K. 1964. "Realismus und Instrumentalismus: Bemerkungen zur Logik der Unterstützung durch Tatsachen." Pp. 79–112 in *Ausgewählte Schriften*. Vol. 1. Braunschweig: Vieweg.

———. 1984. "Mach's Theory of Research and its Relation to Einstein." *Studies in History and Philosophy of Science* 15:1–22.

Feyerabend, Paul K., and Grover Maxwell, eds. 1966. *Mind, Matter, and Method: Essays in Philosophy and Science in Honor of Herbert Feigl*. Minneapolis, University of Minnesota Press.

Fichte, Immanuel Hermann. 1854. "Ueber die neuere Atomenlehre und ihr Verhältniß zur Philosophie und Naturwissenschaft." *Zeitschrift für Philosophie und philosophische Kritik* 24:24–46.

———. 1854a. "Die Seelenlehre des Materialismus." *Zeitschrift für Philosophie und philosophische Kritik* 25:58–77; 169–179.

———. 1855 [1834]. *Die Idee der Persönlichkeit und der individuellen Fortdauer*. Elberfeld: Büschler. [Cited according to the reprint 1855. Leipzig: Dyk'sche.]

———. 1861. "Beiträge zur Lehre vom 'Seelenorgan'. Mit Bezug auf G. Th. Fechner's 'Elemente der Psychophysik' [...], so wie auf R. Wagner's und J. M. Schiff's neueste Untersuchungen über Nervenphysiologie." *Zeitschrift für Philosophie und philosophische Kritik* 38:157–82; 39:67–101.

———. 1869. "Seele, Geist, Bewußtseyn vom Standpunkte der Psychophysik." *Zeitschrift für Philosophie und philosophische Kritik* 55:237–59; 56:47–86.

———. 1878. *Der neuere Spiritualismus, sein Werth und seine Täuschungen. Eine anthropologische Studie*. Leipzig: Brockhaus.

Fick, Monika. 1993. *Sinnenwelt und Weltseele. Der psychophysische Monismus in der Literatur der Jahrhundertwende*. (Studien zur deutschen Literatur, vol. 125) Tübingen: Niemeyer.

Fiedler, Johann Kuno. 1918. *Die Motive der Fechner'schen Weltanschauung*. Ph.D. diss., phil. Fak., Universität Leipzig. [Doctoral advisors: Johannes Volkelt, Eduard Spranger.]

Fine, Terence L. 1973. *Theories of Probability. An Examination of Foundations*. New York: Academic Press.

Finkelstein, L. 1982. "Theory and Philosophy of Measurement." Pp. 1–30 in *Handbook of Measurement Science*. Vol. 1. Edited by P. H. Sydenham. New York: Wiley.

Fitch, Florence Mary. 1903. *Der Hedonismus bei Lotze und Fechner*. Ph.D. diss., phil. Fak., Friedrich-Wilhelms-Universität Berlin.

Fix, Ulla, ed. 2003. *Fechner und die Folgen außerhalb der Naturwissenschaften*. Tübingen: Niemeyer.

Flach, Werner. 1968. "Einleitung." In *Das Princip der Infinitesimal-Methode und seine Geschichte. Ein Kapitel zur Grundlegung der Erkenntnisskritik*, edited by Hermann Cohen, Frankfurt am Main: Suhrkamp. [See Cohen 1883.]

Flanagan, Owen. 1997. "Consciousness as a Pragmatist Views It." Pp. 25–48 in *The Cambridge Companion to William James*, edited by Ruth Anna Putnam. Cambridge: Cambridge University Press.

Florey, Ernst and Olaf Breidbach, eds. 1993. *Das Gehirn—Organ der Seele? Zur Ideengeschichte der Neurobiologie*. Berlin: Akademie Verlag.

Fodor, Jerry A. 1968. *Psychological Explanation*. New York: Random House.

Foerster, Heinz von. 1984. "Erkenntnistheorien und Selbstorganisation." *Delfin* 4:6–19. [Cited according to reprint 1987. Pp. 133–158 in *Der Diskurs des Radikalen Konstruktivismus*, edited by Siegfried J. Schmidt. Frankfurt am Main: Suhrkamp.]

Foucault, Marcel. 1901. *La Psychophysique*. Ph.D. diss. (Thèse pour le doctorat ès lettres), Université de Paris. Paris: Félix Alcan.

Fraassen, Bas C. van. 1980. *The Scientific Image.* Oxford: Clarendon.

Frank, Manfred. 1985. *Eine Einführung in Schellings Philosophie.* Frankfurt am Main: Suhrkamp.

———. 1986. *Die Unhintergehbarkeit von Individualität.* Frankfurt am Main: Suhrkamp.

Frankel, Eugene. 1977. "J. B. Biot and the Mathematization of Experimental Physics in Napoleonic France." *Historical Studies in the Physical Sciences* 8:33–72.

Freud, Sigmund. 1920. "Jenseits des Lustprinzips." *Internationale Zeitschrift für Psychoanalyse.* Beiheft 2. Leipzig: Internationaler Psychoanalytischer Verlag. [Cited according to reprint 1972. Pp. 1–69 in Sigmund Freud, *Gesammelte Werke.* 7th ed. Edited by Anna Freud et al. Vol. 13. Frankfurt am Main: S. Fischer. Translated as "Beyond the Pleasure Principle." *Standard Edition* 18:1–64.]

———. 1924. "Das ökonomische Problem des Masochismus." *Internationale Zeitschrift für Psychoanalyse* 10:121–31. [Cited according to reprint 1972. Pp. 369–83 in Sigmund Freud, *Gesammelte Werke.* 7th ed. Edited by Anna Freud et al. Vol. 13. Frankfurt am Main: S. Fischer. Translated as "The Economic Problem of Masochism." *Standard Edition* 19:155–70]

———. 1925. "Selbstdarstellung." Pp. 1–58 in *Die Medizin der Gegenwart in Selbstdarstellungen.* Vol. 4. Edited by Louis Ruyter Grote. Leipzig: Meiner. [Cited according to reprint 1972. Pp. 33–96 in Sigmund Freud, *Gesammelte Werke.* 7th ed. Edited by Anna Freud et al. Frankfurt am Main: S. Fischer. Translated as "An Autobiographical Study." *Standard Edition* 20:7–70.]

———. 1989. *Jugendbriefe an Eduard Silberstein 1871-1881.* Edited by Walther Boehlich. Frankfurt am Main: S. Fischer.

Freudenberg, Gotthold. 1883. "Über das psychophysische Grundgesetz." Pp. 3–8 in *Programm der Lehr- und Erziehungs-Anstalt für Knaben von Dr. Ernst Zeidler, früher Albani, Dresden.* Dresden: Lehmann.

Freudenreich, Hans. 1904. *Fechners psychologische Anschauungen.* Ph.D. diss., phil. Fak., Universität Leipzig. [Doctoral advisors: Wilhelm Wundt, Max Heinze.]

Friedlein, Curt. 1912. *Das Verhältnis der Naturanschauung Fechners zu derjenigen Örsteds.* Ph.D. diss., phil. Fak., Universität Leipzig. [Doctoral advisors: Paul Barth, Johannes Volkelt.]

Friedman, Michael. 1987. "Carnap's *Aufbau* Reconsidered." *Nous* 21:521–45.

Fritsch, Paul. 1910. *Friedrich Paulsens philosophischer Standpunkt. Insbesondere sein Verhältnis zu Fechner und Schopenhauer.* (Abhandlungen zur Philosophie und ihrer Geschichte, Heft 17) Leipzig: Quelle and Meyer. [Also Ph.D. diss., phil. Fak., Universität Erlangen, 1910. Doctoral advisor: Richard Falckenberg.]

Fritzsch, T. 1924. "Von Herbart bis Fechner." *Literarische Berichte aus dem Gebiete der Philosophie* 4:33–35.

Fügner, F. 1881. "Siegfried Lipiner, ein Dichter unserer Zeit." *Blätter für Handel, Gewerbe und sociales Leben, (Beiblatt zur Magdeburgischen Zeitung).* Montagsblätter 5 (January 31):38–39; 6 (February 7):42–45; 7 (February 14):49–51.

Funkenstein, Amos. 1986. "The Persecution of Absolutes: On the Kantian and Neo-Kantian Theories of Science." Pp. 39–63 in *The Kaleidoscope of Science*, edited by Edna Ullmann-Margalit. Dordrecht: Reidel.

Fürnrohr, A. E. 1849. [Review of Fechner 1848.] *Flora oder allgemeine botanische Zeitung* 32(5) (N. R. 7):88–91.

Füßl, Wilhelm and Margrit Prussat. 2001. *Der wissenschaftliche Nachlaß von Ernst Mach (1838–1916).* (Veröffentlichungen aus dem Archiv des Deutschen Museums, vol. 4) Munich: Deutsches Museum.

Gabriel, Gottfried. 2000. "Einheit in der Vielheit. Der Monismus als philosophisches Programm." Pp. 23–39 in *Monismus um 1900: Wissenschaftskultur und Weltanschauung*, edited by Paul Ziche. Berlin: Verlag für Wissenschaft und Bildung.

Gardner, Michael R. 1979. "Realism and Instrumentalism in 19th-Century Atomism." *Philosophy of Science* 46:1–34.

Gebhard, Walter. 1984. *"Der Zusammenhang der Dinge." Weltgleichnis und Naturverklärung im Totalitätsbewußtsein des 19. Jahrhunderts.* (Hermaea, N. F., vol. 47) Tübingen: Niemeyer. [164–221; 349–56 on Fechner.]

Geiringer, Hilda. 1973. "Probability: Objective Theory." Pp. 605–23 in *Dictionary of the History of Ideas*, edited by Philip P. Wiener. Vol. 3. New York: Scribner's.

Geissler, Hans-Georg, ed. 1990. *Psychophysical explorations of mental structures*. In collaboration with Martin H. Müller and Wolfgang Prinz. Lewiston, N.Y.: Hogrefe and Huber, 1990.

Geissler, Hans-Georg, Stephen W. Link, and James T. Townsend, eds. 1992. *Cognition and Psychophysics: Basic Issues*. Hillsdale: Erlbaum.

Geissler, Hans Georg, and Konrad Reschke, eds. 1987. *Psychophysische Grundlagen mentaler Prozesse. In memoriam G. Th. Fechner (1801–1887)* (Wissenschaftliche Beiträge der Karl-Marx-Universität Leipzig) Leipzig: Karl-Marx-Universität.

Gerhardt, Volker, Reinhard Mehring, and Jana Rindert. 1999. *Berliner Geist. Eine Geschichte der Berliner Universitätsphilosophie bis 1946*. Berlin: Akademie Verlag.

Gerland, Georg. 1875. *Anthropologische Beiträge*. Halle: Max Niemeyer.

Gescheider, George A. 1985. *Psychophysics. Method, Theory, and Application*. 2nd ed. Hillsdale, N.J.: Erlbaum.

———. 1997. *Psychophysics: The Fundamentals*. Mahwah, N.J.: Erlbaum.

Gigerenzer, Gerd and David J. Murray. 1987. *Cognition as Intuitive Statistics*. Hillsdale, N.J.: Erlbaum.

Gigerenzer, Gerd, Zeno Swijtink, Theodore Porter, Lorraine Daston, John Beatty, and Lorenz Krüger. 1989. *The Empire of Chance: How probability changed science and everyday life*. Cambridge: Cambridge University Press.

Giovanni, George di. 1979. "Kant's Metaphysics of Nature and Schelling's *Ideas for a Philosophy of Nature*." *Journal of the History of Philosophy* 17:197–215.

Girlich, Hans-Joachim. 1981. "Felix Hausdorff und die angewandte Mathematik." Pp. 134–46 in *100 Jahre mathematisches Seminar der Karl-Marx-Universität Leipzig*, edited by Herbert Becker and Horst Schumann. Berlin: Deutscher Verlag der Wissenschaften.

———. 1996. "Hausdorffs Beiträge zur Wahrscheinlichkeitstheorie." Pp. 31–69 in *Felix Hausdorff zum Gedächtnis*, edited by E. Brieskorn. Vol 1. Braunschweig: Vieweg.

Globus, Gordon. 1989. "The 'Strict Identity' Theory of Schlick, Russell, Feigl, and Maxwell." Pp. 257–284 in *Science, Mind, and Psychology: Essays in Honor of Grover Maxwell*, edited by Mary Lou Maxwell and C. Wade Savage. Lanham, Mad.: University Press of America.

Gloy, Karen and Paul Burger, eds. 1993. *Die Naturphilosophie im Deutschen Idealismus*. Stuttgart: Frommann-Holzboog.

Gödde, Günter. 1991. "Freuds philosophische Diskussionskreise in der Studentenzeit." *Jahrbuch der Psychoanalyse*. Eds. Friedrich-Wilhelm Eickhoff and Wolfgang Loch. Stuttgart: Frommann, 27:73–113.

———. 1999. *Traditionslinien des "Unbewußten". Schopenhauer—Nietzsche—Freud*. Tübingen: ed. diskord, 1999.

Goethe, Johann Wolfgang von. 1811. *Aus meinem Leben. Dichtung und Wahrheit*. Reprint 1989. *Goethes Werke*. (Hamburger Ausgabe) *Autobiographische Schriften*. Vol. 9, 11th ed. Munich: Beck.

Goldschmidt, Alfred. 1902. *Fechners metaphysische Anschauungen*. Ph.D. diss., phil. Fak., Universität Würzburg.

Gould, Stephen Jay. 1977. *Ontogeny and Phylogeny*. Cambridge, Mass.: Harvard University Press.

Gower, Barry. 1973. "Speculation in Physics: The History and Practice of *Naturphilosophie*." *Studies in History and Philosophy of Science* 3:301–56.

Gregory, Frederick. 1977. *Scientific Materialism in Nineteenth Century Germany*. Dordrecht: Reidel.

Gregory, Richard L. 1981. *Mind in Science. A History of Explanations in Psychology and Physics*. London: Weidenfeld and Nicolson.

Grotenfelt, Arwid. 1888. *Das Webersche Gesetz und die psychische Relativität*. Helsingfors: Frenckell.

Grün, Klaus-Jürgen. 1993. *Das Erwachen der Materie. Studie über die spinozistischen Gehalte der Naturphilosophie Schellings*. Zürich: Olms.

Guettler, Carl. 1884. *Lorenz Oken und sein Verhältnis zur modernen Entwicklungslehre. Ein Beitrag zur Geschichte der Naturphilosophie*. Langensalza: Hermann Beyer.

Gundlach, Horst. 1986. "Dr. Mises and Mr. Fechner—How Early Was Their Identity Uncovered?" *American Psychologist* 41:582–83.

———. 1988. "Fechner scholarship—Aspects of its history." Pp. 17–30 in *G. T. Fechner and Psychology*, edited by Josef Brožek and Horst Gundlach. Passau: Passavia.

———. 1988a. "Verwandte, Freunde, Vertraute. Die mathematisch-physische Abteilung der Königlich Sächsischen Gesellschaft der Wissenschaften zu Leipzig als Treibhaus der Psychophysik." Pp. 87–102 in *G. T. Fechner and Psychology*, edited by Josef Brožek and Horst Gundlach. Passau: Passavia.

———. 1988b. *Index Psychophysicus. Bio- und bibliographischer Index zu Fechners Elementen der Psychophysik und den Parerga*. (Passauer Schriften zur Psychologiegeschichte, vol. 7) Passau: Passavia Universitätsverlag.

———. 1989. "Zur Verwendung physiologischer Analogien bei der Entstehung der experimentellen Psychologie." *Berichte zur Wissenschaftsgeschichte* 12:167–76.

———. 1991. "Über den Anfang der Zeitschrift für Psychologie und Physiologie der Sinnesorgane." *Zeitschrift für Psychologie. Supplement* 11:13–24.

———. 1992. "Psychologie und Experiment um die Mitte des 19. Jahrhunderts." Pp. 105–19 in *Psychologische Forschung und Methode: Das Versprechen des Experiments. Festschrift für Werner Traxel*, edited by Horst Gundlach. Passau: Passavia.

———. 1993. *Entstehung und Gegenstand der Psychophysik*. (Lehr-und Forschungstexte Psychologie, vol. 45) Berlin: Springer.

Gundlach, Horst, ed. 1994. *Arbeiten zur Psychologiegeschichte*. Göttingen: Hogrefe.

———. 1996. *Untersuchungen zur Geschichte der Psychologie und der Psychotechnik*. (Passauer Schriften zur Psychologiegeschichte, vol. 11) Munich: Profil.

Gutberlet, Constantin. 1879. "Die Psychophysik." *Natur und Offenbarung* 25:665–77, 729–41; 26:97–117, 166–77, 281–90, 603–12, 717–27.

———. 1880. [Review of Langer 1876.] *Natur und Offenbarung. Organ zur Vermittlung zwischen Naturforschung und Glauben für Gebildete aller Stände* 25:437–45.

————. 1880a. "In Sachen des Spiritismus." *Natur und Offenbarung. Organ zur Vermittlung zwischen Naturforschung und Glauben für Gebildete aller Stände* 26:505–9. [Review of Zöllner 1879.]

————. 1898. "Der psychophysische Parallelismus." *Philosophisches Jahrbuch der Görresgesellschaft* 11:369–96.

————. 1905. *Psychophysik. Historisch-kritische Studien über Experimentelle Psychologie.* Mainz: Kirchheim.

Haarhaus, Julius R. 1901. "Gustav Theodor Fechner." *Blätter für Bücherfreunde* 1:7–9.

Hacking, Ian. 1975. *The Emergence of Probability.* Cambridge: Cambridge University Press.

————. 1981. "How Should We Do the History of Statistics?" *I and C* 8:15–26.

————. 1983. "Nineteenth-Century Cracks in the Concept of Determinism." *Journal of the History of Ideas* 44:455–76.

————. 1987. "Was There a Probabilistic Revolution 1800–1930?" Pp. 45–55 in *The Probabilistic Revolution.* Vol. 1: *Ideas in History.* Edited by Lorenz Krüger, Lorraine J. Daston, and Michael Heidelberger. Cambridge, Mass.: MIT Press/Bradford Books.

————. 1987a. "Prussian Numbers 1860–1882." Pp. 377–394 in *The Probabilistic Revolution.* Vol. 1: *Ideas in History.* Edited by Lorenz Krüger, Lorraine J. Daston, and Michael Heidelberger. Cambridge, Mass.: MIT Press/Bradford Books.

————. 1988. "Telepathy: Origins of Randomization in the Design of Experiments." *Isis* 79(298):427–51.

————. 1990. *The Taming of Chance.* Cambridge: Cambridge University Press.

————. 1990a. "Probability and Determinism, 1650–1900." Pp. 690–701 in *Companion to the History of Modern Science,* edited by Robert C. Olby et al. London: Routledge.

Haeckel, Ernst. 1914. *Gott-Natur (Theophysis). Studien über monistische Religion.* 2nd ed. Leipzig: Kröner.

Hagner, Michael und Bettina Wahrig-Schmidt, eds. 1992. *Johannes Müller und die Philosophie.* Berlin: Akademie Verlag.

Halbfass, Wilhelm. 1887. "Gustav Theodor Fechner als Naturphilosoph I. Ein Beitrag zur Geschichte des Positivismus." Pp. 12–46 in *Programm des Gymnasiums zu Neuhaldensleben.* Neuhaldensleben: Besser.

Hall, (Granville) Stanley. 1912. *Founders of Modern Psychology (Hartmann, Zeller, Lotze, Gustav Theodore Fechner, Helmholtz, Wundt).* New York: Appleton [Reprint 1924. Cited according to the abbreviated German translation 1914 by Raymund Schmidt *Die Begründer der Modernen Psychologie (Lotze, Fechner, Helmholtz, Wundt).* (Wissen und Forschen, vol. 7) Leipzig: Meiner.]

Haller, Rudolf. 1991. "Ernst Mach: Das unrettbare Ich." Pp. 210–44 in *Grundprobleme der großen Philosophen.* Edited by Josef Speck, *Philosophie der Neuzeit V.* Göttingen: Vandenhoeck.

Haller, Rudolf and Friedrich Stadler, eds. 1988. *Ernst Mach—Werk und Wirkung.* Vienna: Hölder.

Hanle, Paul A. 1979. "Indeterminacy before Heisenberg: The Case of Franz Exner and Erwin Schrödinger." *Historical Studies in the Physical Sciences* 10:225–69.

Harrington, Anne. 1996. *Reenchanted Science: Holism in German Culture from Wilhelm II to Hitler.* Princeton: Princeton University Press.

Harrington, Austin. 2000. "In Defence of Verstehen and Erklaeren: Wilhelm Dilthey's Ideas Concerning a Descriptive and Analytical Psychology." *Theory and Psychology* 10(4):435–51.

Hartmann, Albert. 1937. *Der Spätidealismus und die Hegelsche Dialektik.* (Neue deutsche Forschungen, Abt. Philosophie, vol. 26) Berlin: Juncker and Dünnhaupt.

Hartmann, Eduard von. 1869. *Philosophie des Unbewussten. Versuch einer Weltanschauung.* 1st ed. 2 vols. Berlin: Duncker. [Cited according to the 12th ed., 1923. 3 vols. Leipzig: Kröner. Vol. 3 containing writings from 1872–77. Translation 1931 by William Chatterton Coupland as *Philosophy of the Unconscious: Speculative Results According to the Inductive Method of Physical Science.* Preface by C. K. Ogden. 3 vols. New ed. London: Paul, Trench, Trübner, and New York: Harcourt, Brace and Company.]

———. 1870. "Dynamismus und Atomismus (Kant, Ulrici, Fechner)." *Philosophische Monatshefte* 6:187–205.

———. 1891. "Fechners Universalbewusstsein." *Sphinx* 11:321–30.

———. 1901. *Die moderne Psychologie. Eine kritische Geschichte der deutschen Psychologie in der 2. Hälfte des 19. Jahrhunderts.* (Ausgewählte Werke, vol. 13) Leipzig: Haacke.

Hartung, Walter. 1913. "Die Bedeutung der Schelling-Okenschen Lehre für die Entwicklung der Fechnerschen Metaphysik." *Vierteljahrsschrift für wissenschaftliche Philosophie und Soziologie* 37 (N. F. 12): 253–89 and 371–422. [Also separately as 1912 Ph.D. diss., phil. Fak., Universität Bonn.]

Hartungen, Hartmut von. 1932. *Der Dichter Siegfried Lipiner (1856–1911).* Ph.D. diss., typescript, phil. Fak., Universität München.

Hastedt, Heiner. 1988. *Das Leib-Seele-Problem. Zwischen Naturwissenschaft des Geistes und kultureller Eindimensionalität.* Frankfurt am Main: Suhrkamp.

Hatfield, Gary. 1990. *The Natural and the Normative. Theories of Spatial Perception from Kant to Helmholtz.* Cambridge, Mass.: Bradford.

———. 1995. "Remaking the Science of Mind: Psychology as Natural Science." Pp. 184–231 in *Inventing Human Science: Eighteenth-Century Domains,* edited by Christopher Fox, Roy Porter, and Robert Wokler. Berkeley: University of California Press.

———. 1997. "Wundt and Psychology as Science: Disciplinary Transformations." *Perspectives on Science* 5(3):349–82.

———. 2003. "Sense Data and the Mind-Body Problem: Mach, James, and Russell." *Principia* 6.

Haupt, Edward J. 1998. "Origins of American Psychology in the Work of G. E. Müller: Classical Psychophysics and Serial Learning." Pp. 17–75 in *Psychology: Theoretical-Historical Perspectives,* edited by Robert W. Rieber and Kurt D. Salzinger. 2nd ed. Washington, D.C.: American Psychological Association.

Hausdorff, Felix. 1897. "Das Risiko bei Zufallsspielen." *Berichte über die Verhandlungen der Königlich Sächsischen Gesellschaft der Wissenschaften* 49:497–548.

———. 1901. "Beiträge zur Wahrscheinlichkeitsrechnung." *Berichte über die Verhandlungen der Königlich Sächsischen Gesellschaft der Wissenschaften* 53:152–78.

Hayungs, Heino. 1912. *Die Lehre von der Beseeltheit der Pflanzen von Fechner bis zur Gegenwart.* Ph.D. diss., phil. Fak., Universität Kiel. [Doctoral advisor: Martius.]

Heidelberger, Michael. 1979. *Der Wandel der Elektrizitätslehre zu Ohms Zeit. Eine methodengeschichtliche Untersuchung und logische Rekonstruktion.* Ph.D. diss., phil. Fak., Universität München.

———. 1980. "Towards a Logical Reconstruction of Revolutionary Change: The Case of Ohm as an Example." *Studies in History and Philosophy of Science* 11(2):103–21.

———. 1981. "Some Patterns of Change in the Baconian Sciences of the early 19th-Century Germany." Pp. 3–18 in *Epistemological and Social Problems of the Sciences in the*

Early Nineteenth Century, edited by Hans Niels Jahnke and Michael Otte. Dordrecht: Reidel.

———. 1983. "Wandlungstypen in den Baconischen Wissenschaften im Deutschland des frühen 19. Jahrhunderts." *Philosophia Naturalis* 20(1):112–26.

———. 1985. "Zerspaltung und Einheit: Vom logischen Aufbau der Welt zum Physikalismus." Pp. 144–89 in *Philosophie, Wissenschaft, Aufklärung. Beiträge zur Geschichte und Wirkung des Wiener Kreises,* edited by Hans-Joachim Dahms. Berlin: Walter de Gruyter.

———. 1986. "Zur Philosophie der Messung im 19. Jahrhundert." Pp. 159–68 in *Die historische Metrologie in den Wissenschaften,* edited by Harald Witthöft et al., St. Katharinen: Scripta Mercaturae Verlag.

———. 1987. "Fechner's Indeterminism: From Freedom to Laws of Chance." Pp. 117–56 in *The Probabilistic Revolution.* Vol. 1: *Ideas in History.* Edited by Lorenz Krüger, Lorraine J. Daston, and Michael Heidelberger. Cambridge, Mass.: MIT Press/Bradford Books.

———. 1988. "Fechners Leib-Seele-Theorie." Pp. 61–77 in *G. T. Fechner and Psychology,* edited by Josef Brožek and Horst Gundlach. Passau: Passavia.

———. 1988a. "Fechner und Mach zum Atomismus in der Physik." P. 75–112 in *Die geschichtliche Perspektive in den Disziplinen der Wissenschaftsforschung,* edited by Hans Poser and Clemens Burrichter. (TUB-Dokumentation, vol. 39) Berlin: Technische Universität Berlin.

———. 1989. "Probabilismus, probabilistisch." In *Historisches Wörterbuch der Philosophie.* Eds. Joachim Ritter and Karlfried Gründer. Vol. 7. Basel: Schwabe.

———. 1990. "Selbstorganisation im 19. Jahrhundert." Pp. 67–104 in *Selbstorganisation— Aspekte einer wissenschaftlichen Revolution,* edited by Wolfgang Krohn and Günter Küppers. Braunschweig: Vieweg.

———. 1990a. "Concepts of Self-Organization in the 19th Century." Pp. 170–80 in *Selforganization: Portrait of a Scientific Revolution.* (Sociology of the Sciences, Yearbook 1990, Vol. 14), edited by Wolfgang Krohn, Günter Küppers, and Helga Nowotny. Dordrecht: Kluwer.

———. 1993. *Die innere Seite der Natur. Gustav Theodor Fechners wissenschaftlich-philosophische Weltauffassung.* Frankfurt am Main: Klostermann. [Original German version of this book.]

———. 1993a. "Fechner's Impact for Measurement Theory." [Open peer-commentary to Murray 1993] *Behavioral and Brain Sciences* 16:146–48.

———. 1993b. "Force, Law, and Experiment: The Evolution of Helmholtz's Philosophy of Science." Pp. 461–97 in *Hermann von Helmholtz and the Foundations of Nineteenth-Century Science,* edited by David Cahan. Berkeley: University of California Press.

———. 1994. "The Unity of Nature and Mind: Gustav Theodor Fechner's Non-Reductive Materialism." Pp. 215–36 in *Romanticism in Science: Science in Europe, 1790–1840,* edited by Stefano Poggi and Maurizio Bossi. (Boston Studies in the Philosophy of Science, vol. 152) Dordrecht: Kluwer.

———. 1994a. "Fechners Verhältnis zur Naturphilosophie Schellings." Pp. 201–18 in *Schelling und die Selbstorganisation. Neue Forschungsperspektiven,* edited by Marie-Luise Heuser-Kessler and Wilhelm G. Jacobs (*Selbstorganisation. Jahrbuch für Komplexität in den Natur-, Sozial- und Geisteswissenschaften,* vol. 5) Berlin: Duncker and Humblot.

———. 1994b. "Die Wirklichkeit emergenter Eigenschaften." Pp. 340–58 in *Kreativität und Logik. Charles S. Peirce und das philosophische Problem des Neuen,* edited by Helmut Pape. Frankfurt am Main: Suhrkamp.

————. 1995. "Helmholtz als Philosoph." *Deutsche Zeitschrift für Philosophie* 43(5):835–44.

————. 1997. "Beziehungen zwischen Sinnesphysiologie und Philosophie im 19. Jahrhundert." Pp. 37–58 in *Philosophie und Wissenschaften. Formen und Prozesse ihrer Interaktion.* Edited by Hans Jörg Sandkühler. Frankfurt am Main: Peter Lang.

————. 1998. "Büchner, Friedrich Karl Christian Ludwig (Louis) (1824–1899)." Pp. 48–51 in *Routledge Encyclopedia of Philosophy*, edited by Edward Craig. Vol. 2. London: Routledge.

————. 1998a. "From Helmholtz's Philosophy of Science to Hertz's Picture-Theory." Pp. 9–24 in *Heinrich Hertz (1857–1894): Classical Physicist, Modern Philosopher*, edited by Davis Baird, R. I. G. Hughes, and Alfred Nordmann. (Boston Studies in the Philosophy of Science, vol. 198) Dordrecht: Kluwer.

————. 1998b. "Naturphilosophie." Pp. 737–43 in *Routledge Encyclopedia of Philosophy*, edited by Edward Craig. Vol. 6. London: Routledge.

————. 1999. See Schiemann, Gregor, and Michael Heidelberger. 1999.

————. 2000. "Psychophysics." Pp. 608–9 in *Reader's Guide to the History of Science*, edited by Arne Hessenbruch. Chicago: Fitzroy Dearborn Publishers.

————. 2000a. "Der Psychophysische Parallelismus: Von Fechner und Mach zu Davidson und wieder zurück." pp. 91–104 in *Elemente moderner Wissenschaftstheorie. Zur Interaktion von Philosophie, Geschichte und Theorie der Wissenschaften*, edited by Friedrich Stadler. Vienna: Springer.

————. 2000b. "Fechner und Mach zum Leib-Seele Problem." Pp. 53–67 in *Materialismus und Spiritualismus. Philosophie und Wissenschaften nach 1848*, edited by Andreas Arndt and Walter Jaeschke. Hamburg: Meiner.

————. 2001. "Gustav Theodor Fechner." Pp. 142–47 in *Statisticians of the Century*, edited by Christopher Charles Heyde and Eugene Seneta. New York: Springer.

————. 2002. "Wie das Leib-Seele Problem in den Logischen Empirismus kam." Pp. 40–72 in *Phänomenales Bewusstsein—Rückkehr der Identitätstheorie?* edited by Michael Pauen and Achim Stephan. Paderborn: Mentis. [This is the German original of ch. 5 of this book. Translated into French 2003 by Fabien Schang as "Les racines de la théorie de l'identité de Feigl dans la philosophie et dans la psychophysiologie du 19ème siècle." *Herbert Feigl, De la Physique au Mental*, edited by Bernard Andrieu. Paris.]

————. 2003. "The Mind-Body Problem in the Origin of Logical Empiricism: Herbert Feigl and Psychophysical Parallelism." Pp. 233–62 in *Logical Empiricism: Historical and Contemporary Perspectives*, edited by Paolo Parrini, Wesley C. Salmon, and Merrilee H. Salmon. Pittsburgh.: University of Pittsburgh Press. [Translation of Heidelberger 2002.]

————. 2003a. "Fechners wissenschaftlich-philosophische Weltauffassung." Pp. 25–42 in *Fechner und die Folgen außerhalb der der Naturwissenschaften*, edited by Ulla Fix. Tübingen: Niemeyer.

Heidelberger, Michael and Lorenz Krüger, eds. 1982. *Probability and Conceptual Change in Scientific Thought.* (Report Wissenschaftsforschung, vol. 22) Bielefeld: B. Kleine.

Heidelberger, Michael, Lorenz Krüger and Rosemarie Rheinwald, eds. 1983a. *Probability Since 1800.* (Report Wissenschaftsforschung, vol. 25) Bielefeld: B. Kleine.

Heil, John. 1998. *Philosophy of Mind: A Contemporary Introduction.* London: Routledge.

Heinrich, W. 1899. *Die moderne Physiologische Psychologie in Deutschland. Eine historisch-kritische Untersuchung mit besonderer Berücksichtigung des Problems der Aufmerksamkeit.* 2nd ed. Zürich: E. Speidel.

Heinze, Max. 1896. "[Christian Hermann] Weiße." *Allgemeine Deutsche Biographie* (Munich) 41:590–94. Reprint 1971. Berlin: Duncker and Humblot.

Hellige, Reiner. 1982. *Einige philosophische Aspekte der Mathematisierung der Psychologie im 19. Jahrhundert.* Ph.D. diss., Humboldt-Universität zu Berlin.

Helm, Georg. 1902. "Die Wahrscheinlichkeitslehre als Theorie der Collectivbegriffe." *Annalen der Naturphilosophie* 1:364–81.

———. 1907. "Die kollektiven Formen der Energie." *Annalen der Naturphilosophie* 6:366–72.

Helmholtz, Hermann von. 1847. *Über die Erhaltung der Kraft. Eine physikalische Abhandlung.* Berlin: G. Reimer. [Cited according to reprint 1902. In *Klassiker der Naturwissenschaften*, edited by Ostwald. Vol. 1. Leipzig: Engelmann. Translation 1995 by David Cahan as "On the Conservation of Force." Pp. 96–126 in *Science and Culture. Popular and Philosophical Essays*, edited by David Cahan. Chicago: University of Chicago Press.]

———. 1855. *Ueber das Sehen des Menschen.* Leipzig: Leopold Voß. [Cited according to reprint 1903. Pp. 87–117 in *Vorträge und Reden von Hermann von Helmholtz.* 5th ed. Vol. 1. Braunschweig: Vieweg.]

———. 1867. *Handbuch der physiologischen Optik.* Leipzig: Leopold Voß. [Translation of 3rd ed. 1909–10. *Helmholtz's Treatise on Physiological Optics*, edited by James P. C. Southall. Reprint 2000 of 1924–25 ed. 3 vols. Bristol: Thoemmes.]

———. 1871. "Gedächtnisrede auf Gustav Magnus." *Abhandlungen der Königlichen Akademie der Wissenschaften zu Berlin* (Jahrg):1–17. [Cited according to retitled reprint 1903. "Zum Gedächtniss an Gustav Magnus." Pp. 33–51 in *Vorträge und Reden von Hermann von Helmholtz.* 5th ed. Vol. 2. Braunschweig: Vieweg. Translation 1881 by Edmund Atkinson as "Gustav Magnus: In Memoriam." Pp. 1–26 in Hermann Helmholtz, *Popular Lectures on Scientific Subjects.* Vol. 2. London: Longmans, Green and Co. Reprint 1996, edited by Andrew Pyle. Bristol: Thoemmes.]

———. 1874. "Vorrede." In William Thomson and Peter Guthrie Tait, *Handbuch der theoretischen Physik.* Braunschweig: Fr. Vieweg und Sohn, 1874, v–xiv. [Preface to the German transl. of *Treatise on Natural Philosophy.* 2nd part of 1st vol. Oxford: Clarendon, 1873. Cited according to reprint "Induction und Deduction. Vorrede zum zweiten Theile des ersten Bandes der Uebersetzung von William Thomson's und Tait's 'Treatise on Natural Philosophy.'" *Vorträge und Reden von Hermann von Helmholtz.* 2 vols. 5th ed. Braunschweig: Vieweg, 1903. Vol. 2, 411–21.]

———. 1874a. "Vorrede." Pp. v–xxv in John Tyndall, *Wissenschaftliche Fragmente.* Braunschweig: Fr. Vieweg und Sohn. [Preface to the German translation of John Tyndall 1871. *Fragments of Science for Unscientific People: A Series of Detached Essays, Lectures, and Reviews.* London: Longmans, Green and Co. Cited according to reprint 1903. "Ueber das Streben nach Popularisirung der Wissenschaft. Vorrede zu der Uebersetzung von Tyndall's 'Fragments of Science' 1874." Pp. 422–34 in *Vorträge und Reden von Hermann von Helmholtz.* 5th ed. Vol. 2. Braunschweig: Vieweg.]

———. 1878. *Die Thatsachen in der Wahrnehmung.* Berlin: Kgl. Akademie der Wissenschaften, 1878. [Cited according to 1903. Pp. 213–47 in *Vorträge und Reden von Hermann von Helmholtz.* 5th ed. Vol. 2 Braunschweig: Vieweg. Translation 1995 by David Cahan as "The Facts of Perception." Pp. 342–80 in *Science and Culture. Popular and Philosophical Essays*, edited by David Cahan. Chicago: University of Chicago Press.]

———. 1887. "Zählen und Messen, erkenntnistheoretisch betrachtet." Pp. 17–52 in *Philosophische Aufsätze, Eduard Zeller zu seinem fünfzigjährigen Doctor-Jubiläum gewidmet.* Leipzig: Fues. [Cited according to Hermann von Helmholtz 1921. Pp. 70–108 in *Schriften zur Erkenntnistheorie*, edited by Paul Hertz and Moritz Schlick. Berlin: J. Springer. Translated 1977 by Malcolm F. Lowe as "Numbering and Measuring from an

Epistemological Viewpoint." Pp. 72–103 in Hermann von Helmholtz, *Epistemological Writings. The Paul Hertz/Moritz Schlick Centenary Edition of 1921*, edited by Robert S. Cohen and Yehuda Elkana (*Boston Studies in the Philosophy of Science*, vol. 37) Dordrecht: Reidel.]

———. 1903. *Vorlesungen über theoretische Physik*. Edited by Arthur König et al. Vol. 6: *Theorie der Wärme*. Edited by Franz Richarz. Leipzig: Barth.

———. 1921. *Schriften zur Erkenntnistheorie*. Edited by Paul Hertz and Moritz Schlick. Berlin: J. Springer. [Reprint 1998. Ecke Bonk. New York: Springer. Translation 1977 by Malcolm F. Lowe as *Epistemological Writings: the Paul Hertz/ Moritz Schlick Centenary Edition of 1921*. Edited by Robert S. Cohen and Yehuda Elkana. (Boston Studies in the Philosophy of Science, vol. 37.) Dordrecht: Reidel.]

Hemecker, Wilhelm W. 1991. *Vor Freud. Philosophiegeschichtliche Voraussetzungen der Psychoanalyse*. Munich: Philosophia, 1991.

Henderson, Lawrence J. 1917. *The Order of Nature: An Essay*. Cambridge, Mass.: Harvard University Press. Reprint 1925. [The appendix is titled "Fechner and the Tendency to Stability" and reprints parts of the 3rd chap. of Fechner 1873.]

Hensel, Paul. 1897. [Review of Lasswitz 1896.] *Historische Zeitschrift* 79 (N. F. 43):300–3.

Hentschel, Klaus. 1990. *Die Korrespondenz Petzoldt-Reichenbach: Zur Entwicklung der "wissenschaftlichen Philosophie" in Berlin*. (Berliner Beiträge zur Geschichte der Naturwissenschaften und der Technik, vol. 12.) Berlin: Sigma.

Herbart, Johann Friedrich. 1813. *Lehrbuch zur Einleitung in die Philosophie*. Königsberg: Unzer. [Reprinted 1989. *Joh. Fr. Herbart's Sämtliche Werke*, edited by Karl Kehrbach and Otto Flügel. Vol. 4 (2nd reprint of the 1891 ed. Langensalza: H. Beyer.) Aalen: Scientia. Translation 1891 by M. K. Smith as *A Text-Book in Psychology: An Attempt to Found the Science of Psychology in Experience, Metaphysics, and Mathematics*. New York: Appleton.]

———. 1816. *Lehrbuch zur Psychologie*. Königsberg: Unzer. [Reprinted 1989. *Joh. Fr. Herbart's Sämtliche Werke*, edited by Karl Kehrbach and Otto Flügel. Vol. 4 (2nd reprint of the 1891 ed. Langensalza: H. Beyer.) Aalen: Scientia.]

———. 1824/25. *Psychologie als Wissenschaft. Neu gegründet auf Erfahrung, Metaphysik und Mathematik*. Königsberg: Unzer. [Reprinted 1989. *Joh. Fr. Herbart's Sämtliche Werke*, edited by Karl Kehrbach and Otto Flügel. Vol. 6 (2nd reprint of the 1891 ed. Langensalza: H. Beyer.) Aalen: Scientia.]

———. 1828/29. *Allgemeine Metyphysik, nebst den Anfängen der philosophischen Naturlehre*. Erster, historisch-kritischer Theil, Königsberg: Unzer, 1828; zweyter, systematischer Theil, Königsberg: 1829. [Reprinted 1989. *Joh. Fr. Herbart's Sämtliche Werke*, edited by Karl Kehrbach and Otto Flügel. Vols. 7 (part 1) and 8 (part 2) (2nd reprint of the 1891 ed. Langensalza: H. Beyer.) Aalen: Scientia.]

Hering, Ewald. 1861. *Beiträge zur Physiologie* 1–5. Heft. Leipzig: W. Engelmann.

———. 1868. *Die Lehre vom binocularen Sehen*. Leipzig: W. Engelmann. [Translated and edited 1977 by Bruce Bridgman and Lawrence Stark as *The Theory of Binocular Vision*. New York: Plenum.]

———. 1870. "Über das Gedächtnis als eine allgemeine Funktion der organisierten Materie." (Vortrag gehalten in der feierlichen Sitzung der Kaiserlichen Akademie der Wissenschaften in Wien am 30. Mai). Reprint 1921. Pp. 5–31 in *Fünf Reden von Ewald Hering*, edited by H. E. Hering. Leipzig: Engelmann. [Reprint 1912. In *Klassiker der Naturwissenschaften*, edited by Ostwald. Vol. 1. Nr. 148. 2nd ed. Leipzig: W. Engelmann. Translation 1880 by Samuel Butler as "Professor Ewald Hering 'On Memory'." In

Samuel Butler, *Unconscious Memory*. London: A. C. Fifield. New ed. 1910 by Marcus Hartog. London: A. C. Fifield. Also 1897 by Paul Carus as *On Memory and the Specific Energies of the Nervous System*. Chicago: Open Court.]

———. 1875. "Zur Lehre von der Beziehung zwischen Leib und Seele. I. Mittheilung. Über Fechner's psychophysisches Gesetz." *Sitzungsberichte der Kaiserlichen Akademie der Wissenschaften (Wien)*, math.-natw. Klasse, 72(III. Abt.):310–48. [Only this part appeared.]

———. 1878. *Zur Lehre vom Lichtsinne. Sechs Mittheilungen an die Kaiserl. Akademie der Wissenschaften in Wien*. 2nd unchanged ed. Vienna: Carl Gerold. [Originally in *Sitzungsberichte der Kaiserlichen Akademie der Wissenschaften (Wien)*, math.-natw. Klasse, III. Abt. (1872–74). Translation 1964 by Leo M. Hurvich and Dorothea Jameson as *Outlines of a Theory of the Light Sense*. Cambridge, Mass.: Harvard University Press.]

———. 1884. "Ueber die specifischen Energieen des Nervensystems." [Inaugurational Speech as Rector in 1882] *Lotos* (Prague) 33 (N. F. 5):113–26. [Also pp. 33–51 in Hering 1921.]

———. 1888. "Zur Theorie der Vorgänge in der lebendigen Substanz." *Lotos* (Prague) 37 (N. F. 9):35–70. [Speech from 1888; also pp. 53–103 in Hering 1921.]

———. 1906. "Antwortrede [bei Verleihung der goldenen Graefemedaille, 1906]." Pp. 133–40 in *Fünf Reden von Ewald Hering*, edited by Heinrich Ewald Hering. Leipzig: Engelmann, 1921.

———. 1909. *Die Deutungen des Psychophysischen Gesetzes*. Tübingen: G. Schnürlen.

———. 1921. *Fünf Reden von Ewald Hering*. Edited by Heinrich Ewald Hering. Leipzig: Engelmann. [Includes Hering 1870 and 1888.]

Hermann, Dieter B. 1982. *Karl Friedrich Zöllner*. (Biographien hervorragender Naturwissenschaftler, Techniker und Mediziner, vol. 57) Leipzig: Teubner.

Hermann, Imre. 1925. "Gustav Theodor Fechner." *Imago. Zeitschrift für Anwendung der Psychoanalyse auf die Geisteswissenschaften* 11(4):371–419.

———. 1980. *Parallélismes: Bolyai, Fechner, Darwin, Brouwer, Hilbert, Cantor, Russell*. Preface Jean Petitot-Cocorda and Claude Lorin. Text assembled and revised by Eva Fuzessery and Nicole Sels. Translated from the German and Hungarian into French by Marie-José Amrhein-Locquet, Eva Fuzessery, Jean-Jacques Gorog, and Suzanne Hommel. Paris: Denoël.

Herneck, Friedrich. 1956. "Über eine unveröffentlichte Selbstbiographie Ernst Machs." *Wissenschaftliche Zeitschrift der Humboldt-Universität zu Berlin*, Math.-Nat. Reihe, 6:209–19.

———. 1966. "Die Beziehungen zwischen Einstein und Mach, dokumentarisch dargestellt." *Wissenschaftliche Zeitschrift der Friedrich-Schiller-Universität Jena*, math.-naturw. Reihe, 15:1–14. [Cited according to Friedrich Herneck 1976. *Einstein und sein Weltbild*. Berlin: Der Morgen.]

Herzberg, Alexander. 1913. *Über die Unterscheidung zwischen Physischem und Psychischem und über den Sinn der Wechselwirkungslehre und des Parallelismus*. Ph.D. diss., phil. Fak., Friedrich-Wilhelms-Universität Berlin. [Doctoral advisors: Alois Riehl, Benno Erdmann.]

———. 1929. "Das Stabilitätsprinzip in der modernen Psychologie." *Annalen der Philosophie* 8:238–58.

Heuser-Kessler, Marie-Luise. 1986. *Die Produktivität der Natur. Schellings Naturphilosophie und das neue Paradigma der Selbstorganisation in den Naturwissenschaften*. (Erfahrung und Denken, vol. 69) Berlin: Duncker and Humblot.

Heuser-Kessler, Marie-Luise and Wilhelm G. Jacobs, eds. 1994. *Schelling und die Selbstorganisation. Neue Forschungsperspektiven.* (*Selbstorganisation. Jahrbuch für Komplexität in den Natur-, Sozial- und Geisteswissenschaften,* vol. 5) Berlin: Duncker and Humblot.

Hicks, Dawes. 1913. "Are the intensity differences of sensation quantitative?" *The British Journal of Psychology* 6:155–74.

Hiebert, Erwin N. 1970. "The Genesis of Mach's Early Views on Atomism." Pp. 79–106 in *Ernst Mach. Physicist and Philosopher.* (Boston Studies in the Philosophy of Science, vol. 6.) Edited by Robert S. Cohen and R. J. Seeger. Dordrecht: Reidel.

Hiebsch, Hans. 1977. *Wilhelm Wundt und die Anfänge der experimentellen Psychologie. Sitzungsberichte der sächsischen Akademie der Wissenschaften zu Leipzig,* philolog.-histor. Klasse, 118(4).

Hill, J. Arthur. 1915. "Fechner's Theory of Life After Death." *The Hibbert Journal* 14(1):156–66.

Hillebrand, Franz. 1918. *Ewald Hering. Ein Gedenkwort der Psychophysik.* Berlin: Springer.

Hirschberger, Egon Maximilian. 1951. *Die Sinnesqualitäten und die Fechnersche Tagesansicht.* Ph.D. diss., phil. Fak., Universität München.

Hirschmüller, Albrecht. 1978. *Physiologie und Psychoanalyse im Leben und Werk Josef Breuers.* (Jahrbuch der Psychoanalyse, Beiheft 4) Huber: Bern.

Hochfeld, Sophus. 1908. *Fechner als Religionsphilosoph.* Ph.D. diss., phil. Fak., Universität Erlangen.

Hochkirchen, Thomas. 1999. *Die Axiomatisierung der Wahrscheinlichkeitsrechnung und ihre Kontexte. Von Hilberts sechstem Problem zu Kolmogoroffs Grundbegriffen.* (Studien zur Wissenschafts-, Sozial- und Bildungsgeschichte der Mathematik, vol. 13) Göttingen: Vandenhoeck and Ruprecht. [Also 1998 Ph.D. diss., Fachbereich Mathematik, Universität Wuppertal. Doctoral advisor: Erhard Scholz.]

Hofer, Veronika. 1996. *Organismus und Ordnung. Zu Genesis und Kritik der Systemtheorie Ludwig von Bertalanffys.* Ph.D. diss., phil. Fak., Universität Wien.

———. 2002. "Philosophy of Biology around the Vienna Circle: Ludwig von Bertalanffy, Joseph Henry Woodger and Philipp Frank." Pp. 325–33 in *History of Philosophy of Science: New Trends and Perspectives.* Edited by Michael Heidelberger and Friedrich Stadler. (Vienna Circle Institute Yearbook, vol. 9) Dordrecht: Kluwer.

Höffding, Harald. 1891. "Psychische und physische Aktivität." *Vierteljahrsschrift für wissenschaftliche Philosophie* 15:233–50.

———. 1893. *Psychologie in Umrissen auf Grundlage der Erfahrung.* Transl. from Danish by F. Bendixen. 2nd German ed. according to the 3rd Danish ed. Leipzig: Reisland. [Translation 1893 by Mary E. Lowndes as *Outlines of Psychology.* London: Macmillan.]

———. 1896. *Geschichte der neueren Philosophie.* 2 vols. Transl. from Danish by F. Bendixen. Leipzig: Reisland. [Translation 1924 by B. E. Meyer as *A History of Modern Philosophy: A Sketch of the History of Philosophy from the Close of the Renaissance to Our Own Day.* London: Macmillan. Reprint 1955. New York: Dover.]

———. 1903. *Philosophische Probleme.* Leipzig: Reisland. [Translation 1906 by Galen M. Fisher as *Problems of Philosophy.* London: Macmillan.]

———. 1905. *Moderne Philosophen. Vorlesungen, gehalten an der Universität in Kopenhagen im Herbst 1902.* Transl. from Danish by F. Bendixen, assisted by the author. Leipzig: Reisland. [Translation 1915 by Alfred C. Mason as *Modern Philosophers: Lectures Delivered at the University of Copenhagen during the Autumn of 1902, and Lectures on Bergson Delivered in 1913.* London: Macmillan.]

Hoffmann, Christoph. 2001. "Haut und Zirkel. Ein Entstehungsherd: Ernst Heinrich Webers Untersuchungen 'Ueber den Tastsinn.'" Pp. 191–223 in *Ansichten der Wissenschaftsgeschichte*. Frankfurt am Main: S. Fischer.

Hoffmann, Dieter and Hubert Laitko, eds. 1991. *Ernst Mach. Studien und Dokumente zu Leben und Werk*. Berlin: Deutscher Verlag der Wissenschaften.

Höfler, Alois. 1887. [Review of Elsas 1886.] *Vierteljahrsschrift für wissenschaftliche Philosophie* 11:351–71.

———. 1897. *Psychologie*. Vienna/ Prague: Tempsky.

Höge, Holger. 1995. "Fechner's Experimental Aesthetics and the Golden Section Hypothesis Today." *Empirical Studies of the Arts* 13:131–48.

———. 1996. "The Golden Section Hypothesis—A Funeral, but Not the Last One." *Visual Arts Research* 22:79–89.

———. 1997. "Fechner in Context: Aesthetics from Below, Inner and Outer Psychophysics: A Reply to Pavel Machotka." *Empirical Studies of the Arts* 15(1):91–97.

———. 1997a. "The Golden Section Hypothesis—Its Last Funeral." *Empirical Studies of the Arts* 15(2):233–55.

Hölder, Otto. 1901. "Die Axiome der Quantität und die Lehre vom Mass." *Berichte über die Verhandlungen der Königlich Sächsischen Gesellschaft der Wissenschaften zu Leipzig*, Mathem.-phys. Klasse, 53:1–46. [Translation 1996 by Joel Michell and C. Ernst as "The Axioms of Quantity and the Theory of Measurement." *Journal of Mathematical Psychology* 40:235–52; and 41:345–56.]

Hollander, Karl von. 1908. *Über die Bedeutung von Fechners "Nanna" für die Gegenwart*. Ph.D. diss., phil. Fak., Universität Göttingen, 1908. [Doctoral advisor: Baumann.]

Holmes, Larry. 1964. "Prolegomena to Peirce's Philosophy of Mind." Pp. 359–81 in *Studies in the Philosophy of Charles Sanders Peirce*. 2nd series. Edited by Edward C. Moore and Richard S. Robin. Amherst, Mass.: The University of Massachusetts Press.

Holmes, Frederic L. 1986. "Claude Bernard, The Milieu Intérieur, and Regulatory Physiology." *History and Philosophy of the Life Sciences* 8:3–25.

Holton, Gerald. 1973. *Thematic Origins of Scientific Thought*. Cambridge, Mass.: Harvard University Press. Revised ed. 1988.

Holzhey, Helmut. 1986. *Cohen und Natorp*. 2 vols. Vol. 1: *Ursprung und Einheit. Die Geschichte der 'Marburger Schule' als Auseinandersetzung um die Logik des Denkens*. Vol. 2: *Der Marburger Neukantianismus in Quellen*. Basel/Stuttgart: Schwabe.

Hookway, Christopher. 1997. "Design and Chance: The Evolution of Peirce's Evolutionary Cosmology." *Transactions of the Charles Sanders Peirce Society* 33(1):1–34.

Horkheimer, Max. 1926. *Gesammelte Schriften*. Vol. 10: *Nachgelassene Schriften 1914-1931*. Edited by Alfred Schmidt. Frankfurt am Main: S. Fischer, 1990. [On pp. 200–12: "Fechner, Lotze und die Psychophysik." Part of the lecture-series of 1926: "Einführung in die Philosophie der Gegenwart."]

Hornstein, Gail A. 1988. "Quantifying Psychological Phenomena: Debates, Dilemmas, and Implications." Pp. 1–34 in *The Rise of Experimentation in American Psychology*, edited by Jill G. Morawski. New Haven: Yale University Press.

Horstmeier, Marie. 1930. *Die Idee der Persönlichkeit bei Immanuel Hermann Fichte und Christian Hermann Weiße*. Göttingen: Vandenhoeck.

Hoskovec, Jiří. 1988. "Das Echo des Werkes Fechners in den west- und ostslawischen Ländern." Pp. 193–200 in *G. T. Fechner and Psychology*, edited by Josef Brožek and Horst Gundlach. Passau: Passavia.

Howard, J. V. 1983. "Random Sequences of Binary Digits in Which Missing Values Can Almost Certainly be Restored." *Statistics and Probability Letters* 1:233–38.

Hubbeling, H. G. 1980. "De Fundamenten Van Heymans' Esthetica." *Algemeen Nederlands Tijdschrift voor Wijsbegeerte* 72:174–92.

Humboldt, Alexander von. 1845. *Kosmos. Entwurf einer physischen Weltbeschreibung.* 4 vols. Stuttgart: Cotta. Vol. 1: 1845; vol. 2: 1847; vol. 3: 1851; vol. 4: 1858.

Hume, David. 1739. *A Treatise of Human Nature.* In 1964. *David Hume. The Philosophical Works,* edited by T. H. Green and T. H. Grose. Vol. 4. Aalen: Scientia.

―――. 1748. *An Enquiry Concerning Human Understanding.* In 1964. *David Hume. The Philosophical Works,* edited by T. H. Green and T. H. Grose. Vol. 1. Aalen: Scientia.

Husserl, Edmund. 1911. "Philosophie als strenge Wissenschaft." *Logos* (Tübingen) 1:289–341. [Cited according to reprint 1987. Pp. 3–62 in Edmund Husserl, *Aufsätze und Vorträge (1911–1921).* (Husserliana 25) Dordrecht: Nijhoff.]

Itelson, Gregorius. 1890. "Zur Geschichte des psychophysischen Problems." *Archiv für Geschichte der Philosophie* 3:282–90.

Jackson, Frank. 1998. "Mind, Identity Theory of." Pp. 395–99 in *Routledge Encyclopedia of Philosophy,* edited by Edward Craig. Vol. 6. London: Routledge.

Jaeger, Siegfried. 1988. "Fechners Psychophysik im Kontext seiner Weltanschauung." Pp. 49–59 in *G. T. Fechner and Psychology,* edited by Josef Brožek and Horst Gundlach. Passau: Passavia.

Jaensch, Erich R. 1932. "Fechners Grundlegung der Aesthetik und die Wendung von Psychophysik zu organischer Psychologie." Pp. 98–109 in *Wissenschaft am Scheidewege von Leben und Geist, Festschrift für Ludwig Klages zum 60. Geburtstag,* edited by Hans Prinzhorn. Leipzig: Barth.

James, William. 1890. *The Principles of Psychology.* 2 vols. New York: Holt. Reprint 1981, Cambridge, Mass.: Harvard University Press. Reprint 1998. Bristol: Thoemmes. [On Fechner see esp. vol. 1, 533–49 and more generally chs. 6, 9, 13, and 20.]

―――. 1904. "Does 'Consciousness' Exist?" *Journal of Philosophy, Psychology, and Scientific Methods,* 1:477–91. [Cited according to William James 1976. Pp. 3–19 in *Essays in Radical Empiricism.* (Vol. 3 of *The Works of William James*) Cambridge, Mass.: Harvard University Press.]

―――. 1909. "The Doctrine of the Earth-Soul and of Beings Intermediate between Man and God. An Account of the Philosophy of G. T. Fechner." *The Hibbert Journal* 7(2):278–94. [Cited according to reprint with alterations in William James, *A Pluralistic Universe.* See James 1909a, 63–82.]

―――. 1909a. *A Pluralistic Universe: Hibbert Lectures at Manchester College on the Present Situation in Philosophy.* London: Longmans, Green and Co. [Cited according to Willam James 1977. Vol. 4 of *The Works of William James.* Cambridge, Mass.: Harvard University Press. Lecture IV: "Concerning Fechner."]

Jammer, Max. 1973. "Indeterminacy in Physics." Pp. 586–94 in *Dictionary of the History of Ideas,* edited by Philip P. Wiener. Vol. 2. New York: Scribner's.

Jantzen, Jörg. 1994. "Physiologische Theorien." Pp. 373–668 in Baumgartner 1994. [Physiology at the time of Schelling's philosophy of nature, ca. 1797–1800.]

Jastrow, Joseph. 1884. See Peirce 1884.

―――. 1888. "Critique of Psychophysic Methods." *American Journal of Psychology* 1:271–309.

Jean Paul [pseudonym of Johann Paul Friedrich Richter]. 1987 [1796]. *Siebenkäs.* Frankfurt am Main: Insel.

Jodl, Friedrich. 1896. *Lehrbuch der Psychologie.* Stuttgart: Cotta.

Jonas, Hans. 1986. "Parallelism and complementarity: The Psycho-Physical Problem in Spinoza and in the Succession of Niels Bohr." Pp. 237–47 in *Spinoza and the Sciences.* Dordrecht: Reidel.

Jungnickel, Christa. 1979. "Teaching and Research in the Physical Sciences and Mathematics in Saxony, 1820–1850." *Historical Studies in the Physical Sciences* 10:3–47.

Jungnickel, Christa and Russell McCormmach. 1986. *Intellectual Mastery of Nature. Theoretical Physics from Ohm to Einstein.* Vol. 1: *The Torch of Mathematics 1800–1870.* Chicago: University of Chicago Press.

Kaiser, Walter. 1977. "Operative Gesichtspunkte bei der Diskussion des Weberschen Gesetzes." *Zeitschrift für allgemeine Wissenschaftstheorie* 8:39–47.

———. 1981. *Theorien der Elektrodynamik.* Hildesheim: Gerstenberg.

———. 1982. "Ernst Mach und das Mechanismusproblem in der Biologie." *Medizinhistorisches Journal* 17:179–94.

———. 1982a. "Einleitung." Pp. 5–32 in Ludwig Boltzmann, *Vorlesungen über Maxwells Theorie der Elektrizität und des Lichtes.* Graz: Akademische Druck- und Verlagsanstalt: Vieweg.

———. 1984. *Zur Struktur wissenschaftlicher Kontroversen.* Unpublished habilitation thesis, mathematics faculty. Universität Mainz.

Kamlah, Andreas. 1987. "The Decline of the Laplacian Theory of Probability: A Study of Stumpf, von Kries, and Meinong." Pp. 91–116 in *The Probabilistic Revolution.* Vol. 1: *Ideas in History.* Edited by Lorenz Krüger, Lorraine J. Daston, and Michael Heidelberger. Cambridge, Mass.: MIT Press/Bradford Books.

Kamminga, Harmke. 1980. *Studies in the History of Ideas on the Origin of Life from 1860.* Ph.D. diss., dept. of History and Philosophy of Science, Chelsea College, University of London. [Chapter IV, pp. 132–47: "Spontaneous Degeneration: The Living Origins of Inorganic Matter,"references to Fechner.]

———. 1982. "Life from Space—A History of Panspermia." *Vistas in Astronomy* 26:67–86.

———. 1988. "Historical Perspective: The Problem of the Origin of Life in the Context of Developments in Biology." *Origins of Life and Evolution of the Biosphere* 18:1–11.

Kanitscheider, Bernulf. 1983. "Naturphilosophie und analytische Tradition." Pp. 63–79 in *Naturphilosophie,* edited by Bernulf Kanitscheider. Würzburg: Königshausen and Neumann.

Kann, Robert A. 1969. *Marie von Ebner-Eschenbach—Dr. Josef Breuer. Ein Briefwechsel, 1889–1916.* Vienna: Bergland.

Kant, Immanuel. 1790. *Kritik der Urteilskraft.* 2nd ed. 1793.

Kanz, Kai Torsten, ed. 1994. *Philosophie des Organischen in der Goethezeit. Studien zu Werk und Wirkung des Naturforschers Carl Friedrich Kielmeyer (1765–1844).* Stuttgart: Steiner.

Karlik, Berta and Erich Schmid. 1982. *Franz Serafin Exner und sein Kreis. Ein Beitrag zur Geschichte der Physik in Österreich.* Vienna: Verlag der Österreichischen Akademie der Wissenschaften.

Kaufmann, Alexander. 1913. *Theorie und Methoden der Statistik. Ein Lehr- und Lesebuch für Studierende und Praktiker.* Tübingen: Mohr.

Kay, Alan S. 1970. "Burdach, Karl Friedrich." Pp. 594–97 in *The Dictionary of Scientific Biography.* Vol. 2. New York: Scribner.

Kelly, Alfred. 1981. *The Descent of Darwin: The Popularization of Darwinism in Germany, 1860–1914.* Chapel Hill: University of North Carolina Press.

Kiesewetter, Carl. 1891. *Geschichte des Neueren Occultismus.* Leipzig: Friedrich. [See pp. 666–702 on Zöllner's spiritistic experiments, including two letters by Fechner referring to them.]

Kim, Jaegwon. 1989. "The Myth of Nonreductive Materialism." *Proceedings and Addresses of the American Philosophical Association* 63:31–47.

———. 1997. "The Mind-Body Problem: Taking Stock After Forty Years." Pp. 185–207 in *Philosophical Perspectives.* Vol. 11: *Mind, Causation and World.*

———. 1998. *Mind in a Physical World: An Essay in the Mind-Body Problem and Mental Causation.* Cambridge, Mass.: MIT Press/Bradford Books.

Klimke, Friedrich. 1911. *Der Monismus.* Freiburg i. Br.: Herder.

Knight, David M. 1975. "German Science in the Romantic Period." Pp. 161–78 in *The Emergence of Science in Western Europe,* edited by Maurice Crosland. London: Macmillan.

Köchy, Kristian. 1995. "Organische Ganzheit. Die maßgeblichen Prinzipien des romantischen Organismus-Konzeptes." *Biologisches Zentralblatt* 114(2).

———. 1996. "Perspektiven der Welt. Vielfalt und Einheit im Weltbild der deutschen Romantik." *Philosophia Naturalis* 33:317–42.

———. 1997. *Ganzheit und Wissenschaft. Das historische Fallbeispiel der romantischen Naturforschung.* (Epistemata, Reihe Philosophie, vol. 180) Würzburg: Königshausen and Neumann.

———. 1998. "Der 'Grundwiderspruch der Naturwissenschaften' mit umgekehrten Vorzeichen. Fechners Kritik an Darwin als Fallbeispiel für den verschlungenen Entwicklungsgang der biologischen Theorien." *Jahrbuch für Geschichte und Theorie der Biologie* 5:55–70.

Koenig, Edmund. 1890. *Die Entwicklung des Causalproblems in der Philosophie seit Kant.* Leipzig: Wigand. Reprint Leipzig: Zentralantiquariat, 1972.

Koenigsberger, Leo. 1903. *Hermann von Helmholtz.* Vol. 2. Braunschweig: Vieweg. [Pp. 61–64: Two of Fechner's letters to Helmholtz from 1869.]

Köhler, Alfred. 1886. "Ueber die hauptsächlichen Versuche einer mathematischen Formulirung des psychophysischen Gesetzes von Weber." *Philosophische Studien* 3:572–642.

Köhler, Wolfgang. 1920. *Die physischen Gestalten in Ruhe und im stationären Zustand.* Braunschweig: Vieweg.

———. 1925. "Gestaltprobleme und Anfänge einer Gestalttheorie." *Jahresberichte für die gesamte Physiologie und experimentelle Pharmakologie für das Jahr 1922* 3:512–39. [Cited according to reprint 1983 in *Gestalt Theory* 5:178–205.]

———. 1938. *The Place of Value in a World of Facts.* (The William James Lectures 1934–35) New York: Liveright. [Cited according to the German translation 1968. *Werte und Tatsachen.* Edited by Mira Koffka. Berlin: Springer.]

Köhnke, Klaus Christian. 1986. *Entstehung und Aufstieg des Neukantianismus. Die deutsche Universitätsphilosophie zwischen Idealismus und Positivismus.* Frankfurt am Main: Suhrkamp. [Translation 1991 (without the copious footnotes) by R. J. Hollingdale as *The Rise of Neo-Kantianism: German Academic Philosophy between Idealism and Positivism.* Cambridge: Cambridge University Press.]

Koschnitzke, Rudolf. 1988. *Herbart und Herbartschule.* Aalen: Scientia.

Kössler, Karl. 1901. *Gustav Theodor Fechner. Gedächtnisrede zur Säcularfeier seines Geburtstages gehalten im "Naturwissenschaftlichen Verein" an der k.k. Universität in Wien.* Leipzig: Franz Deuticke.

Kraft, Viktor. 1918. "Ein österreichischer Denker: Ernst Mach." *Donauland* 2(11):1209–13.

Krantz, David H. 1972. "Measurement Structures and Psychological Laws." *Science* 175:1427–35.

Kraus, Oskar. 1929. *Franz Brentano, Über die Zukunft der Philosophie*. Leipzig: Meiner. [On Fechner p. 61.]

Kremer, Richard L. 1992. "From Psychophysics to Phenomenalism: Mach and Hering on Color Vision." Pp. 147–73 in *The Invention of Physical Science*, edited by Mary Jo Nye. Dordrecht: Kluwer.

Krengel, Ulrich. 1990. "Wahrscheinlichkeitstheorie." Pp. 457–89 in *Ein Jahrhundert Mathematik 1890–1990. Festschrift zum Jubiläum der DMV*. (Dokumente zur Geschichte der Mathematik, vol. 6) Edited by Gerd Fischer et al. Braunschweig: Vieweg.

Kries, Johannes von. 1882. "Ueber die Messung intensiver Grössen und über das sogenannte psychophysische Gesetz." *Vierteljahrsschrift für wissenschaftliche Philosophie* 6:257–94. [Translation 1995 by K. K. Niall as "Conventions of Measurement in Psychophysics. Von Kries on the So-called Psychophysical Law." *Spatial Vision* 9:275–305.]

———. 1886. *Prinzipien der Wahrscheinlichkeitsrechnung*. Freiburg i. Br.: Mohr. [Reprint 1927, with a new introduction. Tübingen: Mohr.]

———. 1905. "Zur Psychologie der Sinne." Pp. 16–29 in *Handbuch der Physiologie des Menschen*, edited by Willibald Nagel. Vol. 3: *Physiologie der Sinne*, edited by Johannes von Kries et al. Braunschweig: Vieweg.

———. 1905a. "Die Gesichtsempfindungen." Pp. 109–282 in *Handbuch der Physiologie des Menschen*, edited by Willibald Nagel. Vol. 3: *Physiologie der Sinne*, edited by Johannes von Kries et al. Braunschweig: Vieweg.

———. 1925. "Johannes von Kries." Pp. 125–87 in *Die Medizin der Gegenwart in Selbstdarstellungen*. Vol. 4, edited by Louis Ruyter Grote. Leipzig: Meiner.

Krohn, Wolfgang, Günther Küppers and Rainer Paslack. 1987. "Selbstorganisation—Zur Genese und Entwicklung einer wissenschaftlichen Revolution." Pp. 441–65 in *Der Diskurs des Radikalen Konstruktivismus*, edited by Siegfried J. Schmidt. Frankfurt am Main: Suhrkamp.

Kronenberg, Moritz. 1925. "Fechner und Lotze." *Die Naturwissenschaften* 13(48):957–64.

Kronstorfer, Richard. 1928. *Drei typische Formen des Psychophysischen Parallelismus (Spinoza, Fechner, Mach)*. Ph.D. diss., phil. Fak., Universität Wien.

Krösche, Kurt. 1910. *Wie weit stimmt die Lehre Spinozas vom Parallelismus der göttlichen Attribute überein mit der Theorie vom psychisch-physischen Parallelismus bei Fechner und Fr. Alb. Lange?* Ph.D. diss., phil. Fak., Universität Erlangen.

Kruck, Günter. 1994. *Hegels Religionsphilosophie der absoluten Subjektivität und die Grundzüge des spekulativen Theismus Christian Hermann Weißes*. (Philosophische Theologie, vol. 4) Wien: Passagen.

Krueger, Lester E. 1989. "Reconciling Fechner and Stevens: Toward a Unified Psychophysical Law." *Behavioral and Brain Sciences* 12:251–67.

Krüger, Lorenz. 1987a. "The Slow Rise of Probabilism: Philosophical Arguments in the Nineteenth Century." Pp. 59–89 in *The Probabilistic Revolution*. Vol. 1: *Ideas in History*. Edited by Lorenz Krüger, Lorraine J. Daston, and Michael Heidelberger. Cambridge, Mass.: MIT Press/Bradford Books.

Krüger, Lorenz et al., eds. 1987. *The Probabilistic Revolution*. Vol. 1: *Ideas in History*. Eds. Lorenz Krüger, Lorraine J. Daston, and Michael Heidelberger. Vol. 2: *Ideas in the Sciences*. Eds. Lorenz Krüger, Gerd Gigerenzer, and Mary S. Morgan. Cambridge, Mass.: MIT Press/Bradford Books.

Kügelgen, Wilhelm von. 1909. *Jugenderinnerungen eines alten Mannes.* Leipzig: Max Hesse, no date. [ca. 1909]

―――. 1925. *Erinnerungen 1802–1867.* Vol. 3: *Lebenserinnerungen des Alten Mannes in Briefen an seinen Bruder Gerhard 1840–1867.* Eds. Paul Siegwart von Kügelgen and Johannes Werner. Leipzig: Koehler.

Kuhn, Thomas S. 1959. "Energy Conservation as an Example of Simultaneous Discovery." Pp. 321–56 in *Critical Problems in the History of Science,* edited by Marshall Clagett. Madison, Wisc.: The University of Wisconsin Press. [Cited according to *The Essential Tension: Selected Studies in Scientific Tradition and Change.* Chicago: University of Chicago Press, 1977, 66–104.]

―――. 1961. "The Function of Measurement in Modern Physical Science." *Isis* 52:161–90. [Cited according to *The Essential Tension: Selected Studies in Scientific Thought and Change.* Chicago: University of Chicago Press, 1977, 178–224.]

―――. 1962. *The Structure of Scientific Revolutions.* Chicago: University of Chicago Press. 2nd ed., enlarged, 1970.

―――. 1976. "Mathematical vs. Experimental Traditions in the Development of Physical Science." *Journal of Interdisciplinary History* 7:1–31. [Cited according to *The Essential Tension: Selected Studies in Scientific Thought and Change.* Chicago: University of Chicago Press, 1977, 31–65.]

Külpe, Oswald. 1893. "Anfänge und Aussichten der experimentellen Psychologie." *Archiv für Geschichte der Psychologie* 6:170–89.

―――. 1896. [Review of Lasswitz 1895.] *Zeitschrift für Psychologie und Physiologie der Sinnesorgane* 12:279–80.

―――. 1898. "Ueber die Beziehungen zwischen körperlichen und seelischen Vorgängen." *Zeitschrift für Hypnotismus, Psychotherapie sowie andere psychophysiologische und psychopathologische Forschungen* 7:97–120.

―――. 1901. "Zu Gustav Theodor Fechners Gedächtnis." *Vierteljahrsschrift für wissenschaftliche Philosophie und Soziologie* 25:191–217.

―――. 1908. *Die Philosophie der Gegenwart in Deutschland.* 4th ed. Leipzig: B. G. Teubner. [1st ed. 1902. On Fechner: 74–80.]

Kuntze, Johannes Emil. 1892. *Gustav Theodor Fechner (Dr. Mises). Ein deutsches Gelehrtenleben.* Leipzig: Breitkopf und Härtel. [Müller, Rudolph 1892 and Wundt 1887 as appendices.]

Kusch, Martin. 1995. *Psychologism: A Case Study in the Sociology of Knowledge.* London: Routledge.

Lalo, Charles. 1908. *L'esthétique expérimentale de Fechner.* Ph.D. diss. (Thèse pour le doctorat ès lettres) Université de Paris. Paris: Félix Alcan.

Laming, Donald R. J. 1994. "Psychophysics." Pp. 251–77 in *Companion Encyclopedia of Psychology,* edited by Andrew M. Colman. London: Routledge.

―――. 1997. *The Measurement of Sensation.* Oxford: Oxford University Press.

Laming, Janet and Donald R. J. Laming. 1996. "On the Measurement of Physical Sensations and on the Law which Links the Intensity of These Sensations to the Intensity of the Source." *Psychological Research/Psychologische Forschung* 59(2):134–44. [Discusses two papers published in 1872 by the Belgian physicist Joseph Plateau that are here translated into English. See Plateau 1872.]

Lampa, Anton. 1915. "Über die Tendenz zur Stabilität." Pp. 147–53 in *Festschrift für Wilhelm Jerusalem zu seinem 60. Geburtstag, von Freunden, Verehrern und Schülern.* With contributions by Max Adler et al. [no editor given] Vienna: Wilhelm Braumüller.

Lange, Friedrich Albert. 1865. *Die Grundlegung der mathematischen Psychologie. Ein Versuch zur Nachweisung des fundamentalen Fehlers bei Herbart und Drobisch.* Duisburg: Falk and Volmer.

———. 1873. *Geschichte des Materialismus und Kritik seiner Bedeutung in der Gegenwart.* 2nd revised and expanded ed. 2 vols. Iserlohn: Baedeker. Vol. 1: 1873; vol. 2: 1875. [Cited according to reprint 1974 in 2 vols, edited by Alfred Schmidt. Frankfurt am Main: Suhrkamp. Translation 1877–79 of the 2nd German ed. by Ernest Chester Thomas as *The History of Materialism and Criticism of its Present Importance.* London: The English and Foreign Philosophical Library. 3 vols. Reprint 1925 of the 3rd ed. of 1892 in one volume with introd. by Bertrand Russell. London: Kegan, Paul. Also 1950, New York: Humanities; and 2000, London: Routledge. The 1st German ed. appeared Iserlohn: Baedeker, 1866.]

———. 1887. "Seelenlehre (Psychologie)." *Encyklopädie des gesamten Erziehungs- und Unterrichtswesens.* Edited by Karl Adolf Schmid. 2nd corr. ed. 10 vols. Gotha: Besser. From vol. 5 Leipzig: Fues, 1876–87. Vol. 8, 1887, 521–613. [On Fechner 588–94.]

———. 1968. *Über Politik und Philosophie. Briefe und Leitartikel 1862 bis 1875.* (Duisburger Forschungen, 10. Beiheft) Edited by Georg Eckert. Duisburg: Walter Braun.

Langer, Paul. 1876. *Die Grundlagen der Psychophysik. Eine kritische Untersuchung.* Jena: Dufft.

———. 1893. *Psychophysische Streitfragen.* (Separatabdruck aus dem Programm des Gräflich Gleichenschen Gymnasiums zu Ohrdruf) Ohrdruf: Grapenthin.

Lasswitz, Kurd. See also Roob 1981.

———. 1885. [Review of Cohen 1883.] *Vierteljahrsschrift für wissenschaftliche Philosophie* 9(4):494–503.

———. 1887. [Review of F. A. Müller 1886.] *Deutsche Litteraturzeitung* Nr. 28 (July 9): 1004.

———. 1887a. [Review of Elsas 1886.] *Deutsche Litteraturzeitung* Nr. 1 (January 1): 3.

———. 1887b. "Die psychische Elle." *Die Nation* 4(19):286–88.

———. 1893. "Humor und Glauben bei Gustav Theodor Fechner (Dr. Mises)." *Vossische Zeitung.* Sonntagsbeilage (Sunday supplement) (February 5 and 12). [Cited according to 1920. Kurd Lasswitz. Pp. 197–32 in *Empfundenes und Erkanntes. Aus dem Nachlasse.* Leipzig: Elischer.]

———. 1894. "Fechner's 'Aesthetik von unten.'" *Die Nation* 11(36):539–42.

———. 1896. *Gustav Theodor Fechner.* (Frommanns Klassiker der Philosophie. Edited by Richard Falckenberg, vol. 1) Stuttgart: Frommann. [3rd ed. 1910. Reprint 1992. Eschborn: Klotz.]

———. 1904. "Verirrte Naturforschung." [Review of Pastor 1903.] *Die Zukunft* 47, 12. Jg.: 139–47.

Laudan, Larry. 1981. "The Methodological Foundations of Mach's Opposition to Atomism and Their Historical Roots." Pp. 202–25 in *Science and Hypothesis: Historical Essays on Scientific Methodology.* Dordrecht: Reidel. [Also 1976. Pp. 390–417 in *Motion and Time, Space and Matter: Interrelations in the History and Philosophy of Science,* edited by Peter K. Machamer and Robert G. Turnbull. Columbus: Ohio State University Press.]

Lauterborn, Robert. 1934. "Gustav Theodor Fechner und sein Atombild vom Jahre 1828." *Natur und Volk. Senckenbergische Naturforschende Gesellschaft* 64:439–42.

Lea, Elisabeth and Gerald Wiemers. 1984. *Planung und Entstehung der Sächsischen Akademie der Wissenschaften zu Leipzig.* Unpublished ms. Dresden.

Lécuyer, Bernard-Pierre. 1987. "Probability in Vital and Social Statistics." Pp. 317–35 in *The*

Probabilistic Revolution. Vol. 1: *Ideas in History.* Edited by Lorenz Krüger, Lorraine J. Daston, and Michael Heidelberger. Cambridge, Mass.: MIT Press/Bradford Books.

Leese, Kurt. 1929. *Philosophie und Theologie im Spätidealismus. Forschungen zur Auseinandersetzung von Christentum und idealistischer Philosophie im 19. Jahrhundert.* Berlin: Juncker and Dünnhaupt.

Lehmann, Gerhard. 1931. *Geschichte der nachkantischen Philosophie. Kritizismus und kritisches Motiv in den philosophischen Systemen des 19. und 20. Jahrhunderts.* Berlin: Junker und Dünnhaupt.

———. 1963. "Kant im Spätidealismus und die Anfänge der neukantischen Bewegung." *Zeitschrift für philosophische Forschung* 17:438–56.

Lehner, Kurt M. 2002. "Von der Anschauung der Welt als beseeltes Wesen. Mensch und Natur in der Sicht Gustav Theodor Fechners." *Grenzgebiete der Wissenschaft* 51(2):119–32.

Leibniz, Gottfried Wilhelm. 1714. *Monadologie.* [First German ed. Jena 1720. Latin 1737, French 1840.]

Leihkauf, H. 1983. "K. F. Zöllner und der physikalische Raum." *NTM—Schriftenreihe für Geschichte der Naturwissenschaften, Technik und Medizin* 20(1):29–33.

Leisering, Bruno. 1907. "Studien zu Fechners Metaphysik der Pflanzenseele." *Wissenschaftliche Beilage zum Jahresbericht der Elften Realschule zu Berlin.* Easter 1907. Programm Nr. 146. Berlin: Weidmann.

Leitner, Rainer. 1998. "Das Reformwerk von Exner, Bonitz und Thun: Das österreichische Gymnasium in der zweiten Hälfte des 19. Jahrhunderts." Pp. 17–70 in *Zwischen Orientierung und Krise. Zum Umgang mit Wissen in der Moderne.* (Studien zur Moderne, Bd. 2) Edited b. Sonja Rinofner-Kreidl. Vienna: Böhlau.

Lengauer, Hubert. 1989. "Siegfried Lipiner. Biographie im Zeichen des Prometheus." Pp. 1227–46 in *Die österreichische Literatur.* Edited by Herbert Zeman. Vol. 4, part 2: *Ihr Profil von der Jahrhundertwende bis zur Gegenwart (1880–1980).* Graz: Akademische Verlagsanstalt.

Lennig, Petra [see also Stach, Petra]. 1990. *Gustav Theodor Fechner. Biographisch-ideengeschichtliche Studie zur Rolle der Fechnerschen Philosophie bei der Entstehung der Psychophysik.* Ph.D. diss., Gesellschaftswissenschaftliche Fakultät, Humboldt-Universität zu Berlin.

———. 1993. "Gustav Theodor Fechner und die Naturphilosophie." Pp. 240–23 in *Die Naturphilosophie im Deutschen Idealismus,* edited by Karen Gloy and P. Burger. Stuttgart: Frommann-Holzboog.

———. 1994. *Von der Metaphysik zur Psychophysik: Gustav Theodor Fechner (1801–1887). Eine ergobiographische Studie.* (Beiträge zur Geschichte der Psychologie, vol. 8) Frankfurt am Main: Lang.

———. 1994a. "Die Entwicklung des Grundkonzeptes der Psychophysik durch Gustav Theodor Fechner. Eine spezielle Lösungsvariante des philosophisch tradierten Leib-Seele-Problems?" *NTM—Schriftenreihe für Geschichte der Naturwissenschaften, Technik und Medizin* 31:159–74.

Lennon, Kathleen. 1984. "Anti-Reductionist Materialism." *Inquiry* 27:363–80.

Lenoir, Timothy. 1982. *The Strategy of Life. Teleology and Mechanics in Nineteenth Century German Biology.* Dordrecht: Reidel.

Leplin, Jarrett, ed. 1984. *Scientific Realism.* Berkeley: University of California Press.

Lessing, Hans-Ulrich. 1985. "Briefe an Dilthey anläßlich der Veröffentlichung seiner 'Ideen über eine beschreibende und zergliedernde Psychologie.'" *Dilthey-Jahrbuch* 3:193–232.

Lexis, Wilhelm. 1877. *Zur Theorie der Massenerscheinungen in der menschlichen Gesellschaft.* Programm zur Uebernahme des Lehrstuhles der cameralistischen Fächer an der Grossherzoglich Badischen Universität Freiburg. Freiburg i. Br.: Wagner.

Ley, Michael. 1993. "Wahrnehmung in Bewegung—Ernst Machs 'manieristische' Psychophysik." Pp. 67–70 in *Illustrierte Geschichte der Psychologie,* edited by Helmut E. Lück and Rudolf Miller. München: Quintessenz.

———. 1994. "Ernst Mach und die Gestaltpsychologie." Pp. 123–31 in Arbeiten *zur Psychologiegeschichte,* edited by Horst Gundlach. Göttingen: Hogrefe.

———. 1995. "Eine sinnesphysiologische Untersuchung Ernst Machs als Beispiel für die Übergangssituation der Psychologie in der Mitte des 19. Jahrhunderts." Pp. 229–37 in *Psychologie im soziokulturellen Wandel—Kontinuitäten und Diskontinuitäten,* edited by Siegfried Jaeger, Irmingard Staeuble, Lothar Sprung and Horst-Peter Brauns. Frankfurt am Main: Peter Lang.

———. 1995a. "Der Stellenwert des Isomorphie-Gedankens im System der Gestalttheorie." *Psychologie und Geschichte* 7(3):200–209.

Liebe, Reinhard. 1903. *Fechners Metaphysik. Im Umriß dargestellt und beurteilt.* Ph.D. diss., phil. Fak., Universität Straßburg.

Liebmann, Otto. 1870. "Ueber eine moderne Anwendung der Mathematik auf die Psychologie." *Philosophische Monatshefte* 5:1–24.

———. 1877. [Review of Fechner 1877.] *Philosophische Monatshefte* 13:515–22.

Lienhard, Fritz. 1919. *Der Gottesbegriff bei Gustav Theodor Fechner.* Ph.D. diss., phil. Fak., Universität Bern.

Lindner, Gustav Adolph. 1868. *Lehrbuch der empirischen Psychologie als inductiver Wissenschaft. Für den Gebrauch an höheren Lehranstalten und zum Selbstunterricht.* 2nd, completely revised and expanded ed. Vienna: Gerold. [1st ed. 1858. *Lehrbuch der empirischen Psychologie nach genetischer Methode.* Vienna: Cilli. Translation 1889 by Charles De Garmo as *Manual of Empirical Psychology as an Inductive Science: A Textbook for Highschools and Colleges.* Boston: Heath.]

Link, Stephen W. 1992. *The Wave Theory of Difference and Similarity.* Hillsdale, N.J.: Erlbaum.

———. 1994. "Rediscovering the Past: Gustav Fechner and Signal Detection Theory." *Psychological Science* 5(6):335–40.

———. 2001. "Psychophysical Theory and Laws, History of." Pp. 12439–444 in *International Encyclopedia of the Social and Behavioral Sciences,* edited by Neil J. Smelser and Paul B. Baltes. Amsterdam: Elsevier.

———. 2002. "Fechner, Gustav Theodor." Pp. 126–29 in *Encyclopedia of Cognitive Science,* edited by Lynn Nadel. Vol. 2. New York: Nature Publishing Group/Macmillan.

———. 2002. "Fechner, Gustav Theodor." *The Freud Encyclopedia: Theory, Therapy and Culture,* edited by Edward Erwin. New York: Routledge.

Lipiner, Siegfried. 1878. *Über die Elemente einer Erneuerung religiöser Ideen in der Gegenwart. Vortrag gehalten im Lesevereine der deutschen Studenten Wiens am 19. Januar 1878.* Vienna: im Selbstverlage des Vorstandes des Lesevereins der deutschen Studenten Wiens.

Lipps, Gottlob Friedrich, ed. 1897. See Fechner 1897.

———. 1898. "Über Fechner's Collectivmasslehre und die Vertheilungsgesetze der Collectivgegenstände." *Philosophische Studien* 13:579–612.

———. 1901. "Die Theorie der Collectivgegenstände." *Philosophische Studien* 17:78–183; 467–575.

————. 1903. *Grundriß der Psychophysik.* Leipzig: Göschen.

————. 1905. See Fechner 1905.

————. 1905a. "Die Bestimmung der Abhängigkeit zwischen Merkmalen eines Gegenstandes." *Berichte über die Verhandlungen der Königlich Sächsischen Gesellschaft der Wissenschaften* 57:1–32.

————. 1905b. See Weber, Wilhelm 1850.

————. 1906. *Die psychischen Massmethoden.* Braunschweig: Vieweg.

————. 1912. "Psychophysik. Fundamentale Psychophysik." Pp. 1145–55 in *Handwörterbuch der Naturwissenschaften,* edited by E. Korschelt et al. Vol. 7. Jena: Gustav Fischer.

Lipps, Gottlob Friedrich and J. Witzig. 1933. "Psychophysik." Pp. 111–21 in *Handwörterbuch der Naturwissenschaften,* edited by R. Dittler, G. Joos et al. 2nd ed. Vol. 8. Jena: Gustav Fischer.

Locke, John. 1706. *An Essay Concerning Human Understanding.* 1st ed. 1690, 5th ed. 1706.

Lockwood, Michael. 1989. *Mind, Brain and the Quantum: The Compound 'I'.* Oxford: Blackwell. Reprint 1991.

Lotze, (Rudolph) Hermann. 1838. *De futurae biologiae principiis philosophicis.* Ph.D. diss., med. Fak., Universität Leipzig. Leipzig: Breitkopf und Härtel. [Cited according to 1885. Hermann Lotze, *Kleine Schriften.* Vol. 1. Edited by David Peipers. Leipzig: Hirzel, 1–25.]

————. 1841. *Metaphysik.* Leipzig: Weidmann. New ed. 1879. [Translation 1884 and 1887 by Bernard Bosanquet as *Metaphysic, in Three books, Ontology, Cosmology, and Psychology.* Oxford: Clarendon.]

————. 1842. *Allgemeine Pathologie und Therapie als mechanische Naturwissenschaften.* Leipzig: Weidmann. [2nd ed. see Lotze 1848.]

————. 1843. "Leben. Lebenskraft." Pp. ix–lxvii in *Handwörterbuch der Physiologie mit Rücksicht auf physiologische Pathologie,* edited by Rudolph Wagner. Vol. 1. Braunschweig: Vieweg. [Cited according to Hermann Lotze 1885. Pp. 139–220 in *Kleine Schriften.* Vol. 1. Edited by David Peipers. Leipzig: Hirzel.]

————. 1846. "Seele und Seelenleben." Pp. 142–264 in *Handwörterbuch der Physiologie mit Rücksicht auf physiologische Pathologie,* edited by Rudolph Wagner. Vol. 3. Braunschweig: Vieweg. [Cited according to Hermann Lotze 1886. Pp. 1–204 in *Kleine Schriften.* Vol. 2. Edited by David Peipers. Leipzig: Hirzel.]

————. 1847. "Recension von Gustav Theodor Fechner, *Über das höchste Gut.*" *Göttinger gelehrte Anzeigen* Stück 3–5:28–43. [Cited according to Hermann Lotze 1886. Pp. 272–84 in *Kleine Schriften.* Vol. 2. Edited by David Peipers. Leipzig: Hirzel.]

————. 1848. *Allgemeine Pathologie und Therapie als mechanische Naturwissenschaften.* 2nd improved ed., Leipzig: Weidmann.

————. 1850. "Recension von Gustav Theod. Fechner, Nanna, oder über das Seelenleben der Pflanzen." *Göttinger gelehrte Anzeigen* Stück 167:1661–70. [Cited according to Hermann Lotze 1886. Pp. 505–12 in *Kleine Schriften.* Vol. 2. Edited by David Peipers. Leipzig: Hirzel.]

————. 1851. *Allgemeine Physiologie des körperlichen Lebens.* Leipzig: Weidmann.

————. 1852. *Medicinische Psychologie oder Physiologie der Seele.* Leipzig: Weidmann. [Reprint 1966. Amsterdam: Bonset.]

————. 1852a. "Recension von Gustav Theodor Fechner, Zend-Avesta oder über die Dinge des Himmels und des Jenseits. Vom Standpunct der Naturbetrachtung (3 Theile, Leipzig 1851)." *Literarisches Centralblatt für Deutschland* (Jan. 17): cols. 39–40. [Also in Richard Falckenberg 1901. *Hermann Lotze.* Part 1: *Das Leben und die Entstehung der Schriften nach den Briefen.* Stuttgart: Frommann.]

————. 1855. "Recension von Gustav Theodor Fechner, Über die physikalische und philosophische Atomenlehre. (Leipzig 1855)." *Göttinger gelehrte Anzeigen* Stück 109–12:1081–12. [Cited according to Hermann Lotze 1891. Pp. 215–38 in *Kleine Schriften.* Vol. 3, 1. Abtheilung. Edited by David Peipers. Leipzig: Hirzel.]

————. 1857. *Streitschriften: Erstes Heft in Bezug auf Prof. I. H. Fichte's Anthropologie.* Leipzig: Hirzel.

————. 1879. "Alter und neuer Glaube, Tagesansicht und Nachtansicht." *Deutsche Revue über das gesammte nationale Leben der Gegenwart,* Jg. 3, 3(8) (May): 175–201. [Review of Fechner 1879 and Strauss 1872. Cited according to Hermann Lotze 1891. Pp. 396–497 in *Kleine Schriften.* Vol. 3, 1. Abtheilung. Edited by David Peipers. Leipzig: Hirzel.]

————. 1989. *Kleine Schriften zur Psychologie.* Introduction and comments by Reinhardt Pester. Berlin: Springer. [Reprints of works originally published 1840–52.]

Löwith, Karl. 1978. *Von Hegel zu Nietzsche. Der revolutionäre Bruch im Denken des bürgerlichen Geistes.* 7th ed. Hamburg: Meiner.

Lowry, Richard. 1971. *The Evolution of Psychological Theory. 1650 to the Present.* Chicago: Aldine.

Lübbe, Hermann. 1963. *Politische Philosophie in Deutschland. Studien zu ihrer Geschichte.* [Cited according to reprint 1974. Munich: dtv.]

————. 1972. *Bewußtsein in Geschichten. Studien zur Phänomenologie der Subjektivität. Mach, Husserl, Schapp, Wittgenstein.* Freiburg i. Br.: Rombach.

Luce, R. Duncan and Ward Edwards. 1958. "The Derivation of Subjective Scales from Just Noticeable Differences." *Psychological Review* 65 (1958): 222–37.

Luce, R. Duncan and E. Galanter. 1963. "Discrimination." Pp. 191–244 in *Handbook of Mathematical Psychology.* Vol. 1. Edited by R. Duncan Luce, Robert R. Bush, and Eugene Galanter. New York: Wiley.

Ludwig, Friedrich. 1898. "Die Variabilität der Lebewesen und das Gausssche Fehlergesetz." *Zeitschrift für Mathematik und Physik* 43:230–42.

Lülmann, Christian. 1917. *Monismus und Christentum bei G. Th. Fechner.* Berlin: Schwetschke.

Lütgert, Wilhelm. 1925. *Die Religion des deutschen Idealismus und ihr Ende.* 4 vols. Gütersloh: Bertelsmann, 1923–30. Vol. 3: *Höhe und Niedergang des Idealismus.* Reprint 1967. Hildesheim: Olms.

————. 1930. *Die Religion des deutschen Idealismus und ihr Ende.* 4 vols. Gütersloh: Bertelsmann, 1923–30. Vol. 4: *Das Ende des Idealismus im Zeitalter Bismarcks.* Reprint 1967. Hildesheim: Olms.

Luttenberger, F. 1977. "Friedrich Zöllner, der Spiritismus und der vierdimensionale Raum." *Zeitschrift für Parapsychologie und Grenzgebiete der Psychologie* 19:195–214.

Mach, Ernst. 1863. *Compendium der Physik für Mediciner.* Vienna: Wilhelm Braumüller.

————. 1863a. "Aus Dr. Mach's Vorträgen über Psychophysik." *Oesterreichische Zeitschrift für praktische Heilkunde* 9: columns 146–48, 167–70, 202–4, 225–28, 242–45, 260–61, 277–79, 294–98, 316–18, 335–38, 352–54, 362–66.

————. 1864. "Vorläufige Bemerkung über das Licht glühender Gase." *Zeitschrift für Mathematik und Physik* 9:69–70.

————. 1865. "Bemerkungen zur Lehre vom räumlichen Sehen." *Zeitschrift für Philosophie und philosophische Kritik* 46:1–5. [Also in Mach 1923, 117–23.]

————. 1865a. *Untersuchungen über den Zeitsinn des Ohres. Sitzungsberichte der Kaiserlichen Akademie der Wissenschaften (Wien),* math.-naturw. Klasse, II. Abth., 51:133–50.

———. 1865b. "Über die Wirkung der räumlichen Vertheilung des Lichtreizes auf die Netz-haut." *Sitzungsberichte der Kaiserlichen Akademie der Wissenschaften (Wien)*, math.-naturw. Klasse, II, Abth., 52:303–22.

———. 1866. *Einleitung in die Helmholtzsche Musiktheorie. Populär für Musiker dargestellt.* Graz: Leuschner and Lubensky.

———. 1866a. "Bemerkungen über die Entwicklung der Raumvorstellungen." *Zeitschrift für Philosophie und philosophische Kritik* 49:227–32.

———. 1868. "Über die physiologische Wirkung räumlich vertheilter Lichtreize, 4. Ab-handlung." *Sitzungsberichte der Kaiserlichen Akademie der Wissenschaften (Wien)*, math.-naturw. Klasse, II, Abth., 57:11–19.

———. 1872. *Die Geschichte und die Wurzel des Satzes von der Erhaltung der Arbeit. Vortrag gehalten in der k. böhm. Gesellschaft der Wissenschaften am 15. Nov. 1871.* Prague: Calve. [Cited according to reproduction of the second reprint of 1909, with a preface and a bibliography of Mach's writings. Edited by Joachim Thiele. Amsterdam: E. J. Bonset, 1969. Translation 1911 by P. E. B. Jourdain as *History and Root of the Principle of the Conservation of Energy.* Chicago: Open Court.]

———. 1872a. "Ueber einige Hauptfragen der Physik. Sommer 1872." In *Notizbuch Nr. 20* from the Mach-Archive in the Deutsches Museum, Munich. Sign.: NL 174/0449. 39 pp. Unpublished, handwritten, no page numbers.

———. 1883. *Die Mechanik in ihrer Entwicklung historisch-kritisch dargestellt.* Leipzig: F. A. Brockhaus. [Cited according to reprint 1991 of 9th ed. (Leipzig: Brockhaus, 1933) Darmstadt: Wissenschaftliche Buchgesellschaft. Translation 1960 by Thomas J. McCormack as *The Science of Mechanics. A Critical and Historical Account of its Development.* 6th ed. with revisions through the 9th German ed. La Salle, Ill.: Open Court; Paperback 1974.]

———. 1886. *Beiträge zur Analyse der Empfindungen.* Jena. [Later ed.s titled *Die Analyse der Empfindungen und das Verhältnis des Physischen zum Psychischen.* Cited according to the reprint 1991 of the 9th ed. (Jena: Gustav Fischer, 1922) Darmstadt: Wissenschaftliche Buchgesellschaft. Translation 1996 of the 5th ed. (1914) by C. M. Williams and Sydney Waterlow as *The Analysis of Sensations.* (The Origins of Modern Philosophy of Science, 1830–1914) London: Routledge/Thoemmes.]

———. 1896. *Die Principien der Wärmelehre. Historisch-kritisch entwickelt.* Leipzig: J. Ambrosius Barth. [Cited according to the almost identical 3rd ed. 1919. Translation 1986 as *Principles of the Theory of Heat: Historically and Critically Elucidated.* Edited by Brian McGuinness. (Vienna Circle Collection, vol. 17) Dordrecht: Reidel. Translation of chapter 3, "Critique of the Concept of Temperature," by M. J. Scott-Taggart and Brian D. Ellis appears as an appendix to Brian D. Ellis 1996. *Basic Concepts of Measurement.* Cambridge: Cambridge University Press.]

———. 1905. *Erkenntnis und Irrtum. Skizzen zur Psychologie der Forschung.* Leipzig: J. A. Barth. [Cited according to the reprint 1980 of the 5th ed. (ibid. 1926) Darmstadt: Wissenschaftliche Buchgesellschaft. Translation 1976 by Thomas J. McCormack and Paul Foulkes as *Knowledge and Error: Sketches on the Psychology of Enquiry.* (Vienna Circle Collection, vol. 3) Edited by Brian McGuinness. Dordrecht: Reidel.]

———. 1919. "Die Leitgedanken meiner naturwissenschaftlichen Erkenntnislehre und ihre Aufnahme durch die Zeitgenossen." Leipzig: J. A. Barth, 1919. [Also in *Scientia* 7 (1910): 225–40 and in *Physikalische Zeitschrift* 9 (1910): 599–606.]

———. 1923. *Populär-wissenschaftliche Vorlesungen.* 5th ed. Leipzig: J. A. Barth. [1st ed. 1894. Cited according to reprint 1987 with introduction by Adolf Hohenester and preface by Friedrich Herneck. Vienna: Böhlau. 1st ed. translated 1895 by Thomas J. McCormack as *Popular Scientific Lectures.* Chicago: Open Court. Reprint 1986. La Salle, Ill.: Open Court.]

———. 1988. "Auszüge aus den Notizbüchern 1871–1910." Pp. 167–211 in *Ernst Mach— Werk und Wirkung,* edited by Rudolf Haller and Friedrich Stadler. Vienna: Hölder.

Machotka, Pavel. 1995. "Aesthetics: If Not from Below, Whence?" *Empirical Studies of the Arts* 13:105–18. [A critique of Fechner's aesthetics. See response by Höge 1997.]

Marquard, Odo. 1987. *Transzendentaler Idealismus, Romantische Naturphilosophie, Psychoanalyse.* Cologne: Dinter.

Marks, Lawrence E. 1974. *Sensory Processes: the New Psychophysics.* New York: Academic Press.

Marshall, Marilyn E. 1969. See Fechner 1825b.

———. 1974. "William James, Gustav Fechner, and the Question of Dogs and Cats in the Library." *Journal of the History of the Behavioral Sciences* 10:304–12.

———. 1974a. "G. T. Fechner: Premises Toward a General Theory of Organisms (1823)." *Journal of the History of the Behavioral Sciences* 10:438–47.

———. 1980. "Biographical Genre and Biographical Archetype: Five Studies of Gustav Theodor Fechner." *Storia e Critica della Psicologia* 1(2):197–210. [Also in Ludy T. Benjamin, Jr. ed. 1997. Pp. 120–67 in *A History of Psychology: Original Sources and Contemporary Research.* 2nd ed. New York: McGraw-Hill.]

———. 1982. "Physics, Metaphysics, and Fechner's Psychophysics." Pp. 65–87 in *The Problematic Science: Psychology in Nineteenth-Century Thought,* edited by William R. Woodward and Mitchell G. Ash. New York: Praeger.

———. 1987. "G. T. Fechner: In Memoriam (1801–1887)." *History of Psychology Newsletter* 19:1–9.

———. 1988. "Gustav Theodor Fechner—Psychological Acrobat." Pp. 31–43 in *G. T. Fechner and Psychology,* edited by Josef Brožek and Horst Gundlach. Passau: Passavia.

———. 1990. "The Theme of Quantification and the Hidden Weber in the Early Work of Gustav Theodor Fechner." *Canadian Psychology/ Psychologie Canadienne* 31(1):45–53.

———. 2000. "Fechner, Gustav Theodor." Pp. 344–47 in *Encyclopedia of Psychology,* edited by Alan E. Kazdin. Vol. 3. Washington, D.C.: American Psychological Association, and Oxford: Oxford University Press.

Marshall, Marilyn E. and Barbara Rodway. 1987. "Hermann Ebbinghaus and the Legacy of Gustav Fechner." Pp. 304–12 in *Ebbinghaus-Studien 2. Beiträge zum Internationalen Hermann-Ebbinghaus-Symposion.* (Passauer Schriften zur Psychologiegeschichte, vol. 5), edited by Werner Traxel and Horst Gundlach. Passau: Passavia.

Martinelli, Riccardo. 1999. *Misurare l'anima: Filosofia e psicofisica da Kant a Carnap.* Macerata: Quodlibet.

Mattenklott, Gert. 1984. "Nachwort." Pp. 169–92 in Gustav Theodor Fechner, *Das unendliche Leben.* Munich: Matthes and Seitz.

———. 1986. *Blindgänger. Physiognomische Essays.* Frankfurt am Main: Suhrkamp. ["Exkurs 1: Gustav Theodor Fechner." 148–56.]

Mausfeld, Rainer. 1985. *Grundzüge der Fechner-Skalierung. Prinzipien der Diskrimination psychophysikalischer Diskriminationsskalen.* Frankfurt am Main: Peter Lang.

———. 1988. "Die Metamorphose der Fechnerschen Psychophysik um die Jahrhundertwende in Deutschland und in den USA." Pp. 45–48 in *G. T. Fechner and Psychology,* edited by Josef Brožek and Horst Gundlach. Passau: Passavia.

———. 1994. "Methodologische Grundlagen und Probleme der Psychophysik." Pp. 137–98 in *Methodologische Grundlagen der Psychologie*. (Encyklopädie der Psychologie, B, I, 1), edited by Theo Hermann and Werner H. Tack. Göttingen: Hogrefe.

Mayr, Ernst. 1974. "Teleological and Teleonomic, A New Analysis." Pp. 91–117 in *Methodological and Historical Essays in the Natural and Social Sciences*. (Boston Studies in the Philosophy of Science, vol. 14), edited by Robert Cohen and Marx Wartofsky. Dordrecht: Reidel.

———. 1976. "Typological versus Population Thinking." Pp. 26–29 in Ernst Mayr, *Evolution and the Diversity of Life*. Cambridge, Mass. Harvard University Press.

Maywald, Friedrich. 1907. *Wechselwirkung oder Parallelismus? Versuch einer Darstellung der Ansicht Eduard v. Hartmanns über diese Frage nebst einigen Bemerkungen über die Ansichten Paulsens, Busses und Wentschers über dieselbe*. Breslau: Selbstverlag des Verfassers.

McGregor, Douglas. 1935. "Scientific Measurement and Psychology." *Psychological Review* 42:246–66.

Meinel, Christoph. 1991. *Karl Friedrich Zöllner und die Wissenschaftskultur der Gründerzeit*. (Berliner Beiträge zur Geschichte der Naturwissenschaften und der Technik, vol. 13) Berlin: Sigma.

Meinong, Alexius. 1896. "Über die Bedeutung des Weberschen Gesetzes. Beiträge zur Psychologie des Vergleichens und Messens." *Zeitschrift für Psychologie und Physiologie der Sinnesorgane* 11:81–133.

———. 1965. *Philosophenbriefe. Aus der Wissenschaftlichen Korrespondenz von Alexius Meinong*. Edited by Rudolf Kindinger. Graz: Akademische Druck- und Verlagsanstalt.

Meischner, Wolfram. 1987. See Altmann 1995.

———. 1993. "Wilhelm Wundt." Pp. 35–40 in *Illustrierte Geschichte der Psychologie*, edited by Helmut E. Lück and Rudolf Miller. München: Quintessenz.

Meischner, Wolfram and Erhard Eschler. 1977. *Wilhelm Wundt*. Leipzig: Urania.

Meischner, Wolfram and Anneros Metge. 1985. *Zur Geschichte des Psychologischen Denkens an der Universität Leipzig*. (Psychologiehistorische Manuskripte 1) Leipzig: Karl-Marx-Universität Leipzig, Sektion Psychologie.

Meischner-Metge, Anneros. 1993. "Gustav Theodor Fechner." Pp. 32–34 in *Illustrierte Geschichte der Psychologie*, edited by Helmut E. Lück and Rudolf Miller. München: Quintessenz.

Meissner, W. W. 1962. "The Problem of Psychophysics: Bergson's Critique." *The Journal of General Psychology* 66:301–9.

Merkel, Julius. 1896. "Die Abhängigkeit zwischen Reiz und Empfindung." *Zeitschrift für Psychologie und Physiologie der Sinnesorgane* 12:226–42.

Merz, John Theodore. 1896. *A History of European Thought in the Nineteenth Century*. 4 vols. Edinburgh/ London: Blackwood, 1896–1914. [Reprint 2000 with new introduction by Giuseppe Micheli. Bristol: Thoemmes.]

Metge, Anneros. 1977. *Zur Herausbildung der Experimentalpsychologie unter besonderer Berücksichtigung des Beitrages von Wilhelm Wundt*. Ph.D. diss., Sektion Psychologie, Universität Leipzig.

Metzger, Wolfgang. 1961. "Ansprache von Professor Dr. Wolfgang Metzger." [Welcome address as president of the 16th International Congress for Psychology in Bonn, 1960; on Fechner] *Acta Psychologica* 19:12–22.

———. 1968. "Fechner, Gustav Theodor." Pp. 350–53 in *International Encyclopedia of the Social Sciences*, edited by David L. Sills. Vol. 5. New York: Macmillan.

———. 1975. *Psychologie. Die Entwicklung ihrer Grundannahmen seit der Einführung des Experiments*. 5th ed. Darmstadt: Steinkopff.

Metzinger, Thomas. 1985. *Neuere Beiträge zur Diskussion des Leib-Seele-Problems.* Frankfurt am Main: Lang.

Meyenn, Karl von. 1982. "Engpässe in der Atomtheorie des frühen 19. Jahrhunderts." Pp. 35–55 in *Atomvorstellungen im 19. Jahrhundert*, edited by Charlotte Schönbeck. Paderborn: Schöningh.

Meyer, Friedrich August Eberhard. 1937. *Philosophische Metaphysik und christlicher Glaube bei Gustav Theodor Fechner.* Ph.D. diss., theol. Fak., Universität Tübingen.

Michell, Joel. 1986. "Measurement Scales and Statistics: A Clash of Paradigms." *Psychological Bulletin* 100:398–407.

————. 1990. *An Introduction to the Logic of Psychological Measurement.* Hillsdale, N.J.: Erlbaum.

————. 1993. "The Origins of the Representational Theory of Measurement: Helmholtz, Hölder, and Russell." *Studies in History and Philosophy of Science* 24:183–206.

————. 1999. *Measurement in Psychology: Critical History of a Methodological Concept.* Cambridge: Cambridge University Press.

————. 2001. "Measurement Theory: History and Philosophy." Pp. 9451–54 in *International Encyclopedia of the Social and Behavioral Sciences*, edited by Neil J. Smelser and Paul B. Baltes. Amsterdam: Elsevier.

Mill, John Stuart. 1865. *An Examination of Sir William Hamilton's Philosophy and of the Principal Philosophical Questions Discussed in His Writings.* Boston: W. V. Spencer. [Cited according to John Stuart Mill 1979. *Collected Works.* Vol. 9. Edited by John M. Robson. Toronto: Toronto University Press.]

Mischel, Theodor. 1970. "Wundt and the Conceptual Foundations of Psychology." *Philosophy and Phenomenological Research* 31(1):1–26.

Mischer, Sibille. 1997. *Der verschlungene Zug der Seele. Natur, Organismus und Entwicklung bei Schelling, Steffens und Oken.* (Epistemata. Reihe Philosophie, vol. 222) Würzburg: Königshausen and Neumann.

Mises, Richard von. 1912. "Über die Grundbegriffe der Kollektivmasslehre." *Jahresberichte der Deutschen Mathematiker-Vereinigung* 21:9–20. [Reprint in Mises 1964, 3–14.]

————. 1919. "Grundlagen der Wahrscheinlichkeitsrechnung." *Mathematische Zeitschrift* 5:52–99. [Reprint in Mises 1964, 57–105.]

————. 1920. "Ausschaltung der Ergodenhypothese in der physikalischen Statistik." *Physikalische Zeitschrift* 21:225–32, 256–64.

————. 1920a. "Berichtigung zu meiner Arbeit 'Grundlagen der Wahrscheinlichkeitsrechnung'." *Mathematische Zeitschrift* 7:323.

————. 1921. "Über die gegenwärtige Krise der Mechanik." *Zeitschrift für angewandte Mathematik und Mechanik* 1:425–31. [Also in *Naturwissenschaften* 10:25–29.]

————. 1928. *Wahrscheinlichkeit, Statistik und Wahrheit. Einführung in die neue Wahrscheinlichkeitslehre und ihre Anwendung.* (Schriften zur wissenschaftlichen Weltauffassung. Eds. Philipp Frank and Moritz Schlick, vol. 3) Vienna: Springer. [2nd ed. ibid. 1936. 4th ed. ibid. 1972. Cited according to the 3rd revised ed. Vienna: Springer, 1951. Translated as *Probability, Statistics and Truth.* 2nd rev. English ed. 1957 prepared by Hilda Geiringer. London: Allen and Unwin. Reprint 1981. New York: Dover.]

————. 1930. "Über kausale und statistische Gesetzmässigkeit in der Physik." *Die Naturwissenschaften* 18:145–53.

————. 1939. *Kleines Lehrbuch des Positivismus. Einführung in die empiristische Wissenschaftsauffassung.* (Library of Unified Science, vol. 1, Book Series. Edited by Otto

Neurath) Den Haag: W. P. van Stockum and Zoon. [Cited according to reprint 1990 by Friedrich Stadler. Frankfurt am Main: Suhrkamp. Translation 1968 as *Positivism: A Study in Human Understanding*. New York: Dover.]

———. 1964. *Selected Papers of Richard von Mises*. Edited by Philipp Frank, S. Goldstein, M. Kac, et al. Vol. 2: *Probability and Statistics, General*. Providence: American Mathematical Society.

Möbius, Paul Julius. 1894. *Neurologische Beiträge*. 2. Heft. Leipzig: Abel.

———. 1901. *Stachyologie. Weitere vermischte Aufsätze*. Leipzig: J. A. Barth.

Moiso, Francesco. 1994. "Magnetismus, Elektrizität, Galvanismus." In Baumgartner 1994, 163–372. [Electricity and Magnetism at the Time of Schelling's Philosophy of Nature, ca. 1797-1800.]

Molella, Arthur Philip. 1973. *Philosophy and Nineteenth-Century German Electrodynamics: The Problem of Atomic Action at a Distance*. Ph.D. thesis, Cornell University.

Moore, Edward C. and Richard S. Robin. 1964. *Studies in the Philosophy of Charles Sanders Peirce*. 2nd series. Amherst, Mass.: University of Massachusetts Press.

Moravec, Hans. 1988. *Mind Children: The Future of Robot and Human Intelligence*. Cambridge: Harvard University Press.

Mühlmann, Wilhelm E. 1967. *Geschichte der Anthropologie*. 2nd revised and expanded ed. Frankfurt am Main: Athenäum. [Cited according to the 3rd ed. 1984. Wiesbaden: Aula.]

Müller, Ferdinand August. 1882. *Das Axiom der Psychophysik und die psychologische Bedeutung der Weber'schen Versuche. Eine Untersuchung auf Kantischer Grundlage*. Marburg: Elwert. [Pages 1–56 also as Ph.D. diss., phil. Fak., Universität Marburg.]

———. 1886. *Das Problem der Continuität in Mathematik und Mechanik*. Marburg: Elwert.

Müller, Georg Elias. 1878. *Zur Grundlegung der Psychophysik. Kritische Beiträge*. (Bibliothek für Wissenschaft und Literatur, vol. 23) Berlin: Grieben. [Reprint 1981. Ann Arbor, Mich.: University Microfilms.]

———. 1878a. [Review of Fechner 1877.] *Göttingische gelehrte Anzeigen* Stück 26 (June 26):801–32; Stück 27 (July 3):833–37.

———. 1878b. [Précis of Georg Elias Müller 1878.] *Göttingische gelehrte Anzeigen* Stück 6 (February 6):161–68.

———. 1891. [Review of Münsterberg 1889.] *Göttingische gelehrte Anzeigen* Nr. 11 (June 1):393–429.

———. 1896. "Zur Psychophysik der Gesichtsempfindungen." *Zeitschrift für Psychologie und Physiologie der Sinnesorgane* 10:1–82, 321–413; and 11:1–76, 161–93.

———. 1904. *Die Gesichtspunkte und Tatsachen der psychophysischen Methodik*. Wiesbaden: Bergmann.

Müller, Götz. 1977. "Zur Bedeutung Jean Pauls für die Ästhetik zwischen 1830 und 1848 (Weisse, Ruge, Vischer)." *Jahrbuch der Jean-Paul-Gesellschaft* 12:105–36.

Müller, Johann Jakob. 1870. "Über eine Ableitung des Hauptsatzes der Psychophysik." *Berichte über die Verhandlungen der Königlich Sächsischen Gesellschaft der Wissenschaften zu Leipzig*, Mathem.-physische Classe, 22:562–67.

Müller, Martin. 1991. "On the Interdisciplinary Genesis of Experimental Methods in Nineteenth-Century German Psychology." Pp. 129–40 in *World Views and Scientific Discipline Formation. Science Studies in the German Democratic Republic*. (Boston Studies in the Philosophy of Science, vol. 134) Edited by William R. Woodward and Robert S. Cohen. Dordrecht: Kluwer.

Müller, Rudolph. 1892. "Chronologisches Verzeichniß der Werke und Abhandlungen

G. Th. Fechner's." Pp. 363–72 in Johannes Emil Kuntze, *Gustav Theodor Fechner (Dr. Mises). Ein deutsches Gelehrtenleben.* Leipzig: Breitkopf und Härtel. [Also 1889 in vol. 1 of the 2nd ed. of Fechner 1860, 337–46.]

Münsterberg, Hugo. 1889. *Beiträge zur Experimentellen Psychologie.* Heft 1: *Einleitung: Bewusstsein und Gehirn. Willkürliche und unwillkürliche Vorstellungsverbindung.* Freiburg i. Br.: Mohr.

Murphy, Gardner. 1926. "A Brief Interpretation of Fechner." *Psyche* 7(1):75–80.

Murray, David J. 1983. "The Use of Probability Theory in Psychology prior to 1930." Pp. 141–63 in *Probability and Conceptual Change in Scientific Thought.* Edited by Michael Heidelberger and Lorenz Krüger. Bielefeld: B. Kleine.

———. 1983a. *A History of Western Psychology.* Englewood Cliffs, N.J.: Prentice Hall. 2nd ed. 1988. [149–58 on the beginnings of psychophysics.]

———. 1987. "A Perspective for Viewing the Integration of Probability Theory into Psychology." Pp. 73–100 in *The Probabilistic Revolution.* Vol. 2: *Ideas in the Sciences.* Edited by Lorenz Krüger, Gerd Gigerenzer, and Mary Morgan. Cambridge, Mass.: MIT Press/Bradford Books..

———. 1990. "Fechner's Later Psychophysics." *Canadian Psychology/Psychologie Canadienne* 31(1):54–60.

———. 1993. "A Perspective for Viewing the History of Psychophysics." [Target-article with open peer-commentaries] *Behavioral and Brain Sciences* 16 (March):115–37. [Commentaries: 137–86.]

———. 1995. *Gestalt Psychology and the Cognitive Revolution.* New York/London: Harvester Wheatsheaf.

———. 1996. See Weber, Ernst Heinrich 1996.

Murray, David J. and Helen E. Ross. 1988. "E. H. Weber and Fechner's Psychophysics." Pp. 79–86 in *G. T. Fechner and Psychology.* Eds. Josef Brožek and Horst Gundlach. Passau: Passavia.

N-e. 1874. [Reviews of works on Darwinism, among them Fechner 1873.] *Literarisches Centralblatt* (Leipzig) No. 18 (May 2):594–98.

Nagel, Thomas. 1974. "What is it Like to Be a Bat?" *Philosophical Review* 83:435–50.

———. 1979. "Panpsychism." In *Mortal Questions.* Cambridge: Cambridge University Press.

———. 1986. *The View From Nowhere.* New York: Oxford University Press.

Narens, Louis. 2002. *Theories of meaningfulness.* (Scientific psychology series). Mahwah, N.J.: Erlbaum.

Natorp, Paul. 1888. *Einleitung in die Psychologie nach kritischer Methode.* Freiburg i. Br.: Mohr.

———. 1891. "Quantität und Qualität in Begriff, Urtheil und gegenständlicher Erkenntnis." *Philosophische Monatshefte* 27:1–32, 129–60.

———. 1893. "Zu den Vorfragen der Philosophie." *Philosophische Monatshefte* 29:581–611.

Natsoulas, Thomas. 1984. "Gustav Bergmann's Psychophysical Parallelism." *Behaviorism* 12:41–69.

Neumann, Johann von. 1932. *Mathematische Grundlagen der Quantenmechanik.* (Die Grundlehren der mathematischen Wissenschaften in Einzeldarstellungen, vol. 38) Berlin: Springer. [Cited according to reprint 1996. Berlin: Springer. Translation by Robert T. Beyer as *Mathematical Foundations of Quantum Mechanics.* Princeton: Princeton University Press, 12th printing 1996.]

Neurath, Otto. 1979. "Wissenschaftliche Weltauffassung—der Wiener Kreis." Pp. 81–101 in

Wissenschaftliche Weltauffassung, Sozialismus und logischer Empirismus, edited by Rainer Hegselmann. Frankfurt am Main: Suhrkamp.

Newman, Edwin B. 1974. "On the Origin of 'Scales of Measurement.'" Pp. 137–45 in *Sensation and Measurement: Papers in Honor of S. S. Stevens* edited by Howard R. Moskowitz, Bertram Scharf and Joseph C. Stevens. Dordrecht: Reidel.

Nicolas, Serge. 2001. "Histore de la psychophysique en France." *Teorie et Modelli* 6:5–28.

———. 2001a. "G. T. Fechner et les précurseurs français de la psychophysique: Pierre Bouguer (1727, 1760) et Charles Delezenne (1827)." *Psychologie et Histoire* 2: 86–130.

———. 2002. "La Fondation de la psychophysique de Fechner: Des présupposés de métaphysique aux écrits scientifiques de Weber." *Année Psychologique* 102(2):255–98.

Nicolas, Serge, David J. Murray and B. Farahmand. 1997. "The Psychophysics of J.-R.-L. Delboeuf, 1831–1896." *Perception* 26(10):1297–315.

Nitzschke, Bernd. 1987. "Die helle und die dunkle Vernunft. Der Stoff, aus dem die Theorien sind—Zum 100. Todestag des Philosophen Gustav Theodor Fechner." *Die Zeit* Nr. 43 (October 16):63.

———. 1989. "Freud und Fechner. Einige Anmerkungen zu den psychoanalytischen Konzepten 'Lustprinzip' und 'Todestrieb.'" Pp. 80–96 in *Freud und die akademische Psychologie. Beiträge zu einer historischen Kontroverse,* edited by Bernd Nitzschke. Munich: Psychologie-Verlags-Union.

Noack, Ludwig. 1861. "Die Weltperspective des Seelenscheines." [Review of Fechner 1861.] *Psyche. Populär-wissenschaftliche Studien, Kritiken und Forschungen zur Erkenntnis des menschlichen Geisteslebens* (Leipzig: Wiegand) 4:6–97.

———. 1879. *Philosophie-geschichtliches Lexikon. Historisch-biographisches Handwörterbuch zur Geschichte der Philosophie.* Leipzig: Erich Koschny.

Norwich, K. H. and W. Wong. 1997. "Unification of psychophysical phenomena. The complete form of Fechner's law." *Perception and Psychophysics* 59:929–40.

Oberkofler, Gerhard, 1986. "Aus Briefen von Ewald Hering an Franz Hillebrand." Pp. 183–203 in *Aufsätze zur Geschichte der Naturwissenschaften und Geographie.* Edited by Günther Hamann. (Sitzungsberichte der Österreichische Akademie der Wissenschaften, Philosophisch-historische Klasse, vol. 475) Vienna: Verlag der Österreichischen Akademie der Wissenschaften.

Oelze, Berthold. 1989. *Gustav Theodor Fechner. Seele und Beseelung.* Münster: Waxmann.

Oettingen, Arthur von. 1906. "Das Kausalgesetz." *Berichte über die Verhandlungen der Königlich Sächsischen Gesellschaft der Wissenschaften* 58:454–70.

Ohm, Georg Simon. 1938. *Das Grundgesetz des elektrischen Stromes. Drei Abhandlungen von Georg Simon Ohm (1825 und 1826) und Gustav Theodor Fechner (1829).* Edited by Carl Piel. (Ostwald's Klassiker der exakten Wissenschaften, vol. 244) Leipzig: Akademische Verlagsgesellschaft. [Repr. Thun: Deutsch, 1996.]

Oken, Lorenz. 1808. *Über das Universum als Fortsetzung des Sinnensystems. Ein pythagoräisches Fragment.* Jena: Frommann.

———. 1809. *Lehrbuch der Naturphilosophie.* Jena: Friedrich Frommann, vol. 1 (Erster und zweiter Theil) 1809; vol. 2 (Dritter Theil. Erstes und zweites Stück) 1810; vol. 3 (Dritter Theil. Drittes und letztes Stück) 1811. [2nd reworked ed.. 1831. The 3rd ed. of 1843 is reprinted Hildesheim: Olms, 1991 and translated by Alfred Tulk 1847 as *Elements of Physiophilosophy.* London: Ray Society.]

Ollig, Hans Ludwig. 1979. *Der Neukantianismus.* Stuttgart: Metzler.

Orth, Ernst Wolfgang. 1986. "R. H. Lotze: Das Ganze unseres Welt- und Selbstverständnisses." Pp. 9–51 in *Grundprobleme der großen Philosophen, Philosophie der Neuzeit IV*, edited by Josef Speck. Göttingen: Vandenhoeck and Ruprecht.

Ostwald, Wilhelm. 1905. [Review of] *"Die Tagesansicht gegenüber der Nachtansicht* von G. Th. Fechner. 2nd ed. 1904." *Annalen der Naturphilosophie* 4:499–501.

———. 1926. *Lebenslinien. Eine Selbstbiographie.* Vol. 1: *Riga—Dorpat—Riga 1853–1887.* vol. 2: *Leipzig 1887–1905.* Vol. 3: *Gross-Bothen und die Welt 1905–1927.* Berlin: Klasing. [On Fechner esp. vol 2: 96–97 and vol 3: 363–65, 395, 409–10.]

P. 1887. [Obituary of Fechner.] *Nationalzeitung.* Morgen-Ausgabe. (Berlin) 622 (October 26): 3. [With reprod. of a letter by Fechner to Jean Paul from October 10, 1825, repr. in Kuntze 1892, 66 and Lennig 1994, 46.]

P., J. v. 1833. [Review of] *"Lehrbuch der Naturphilosophie* von Oken (jetzt Prof. in Zürich). Zweyte, umgearbeitete Auflage. Jena, Frommann, 1831." *Leipziger Literatur-Zeitung* Nr. 203 (August 24):1617–24. [See Oken 1809.]

Pap, Arthur. 1954. "Das Leib-Seele-Problem in der Analytischen Philosophie." *Archiv für Philosophie* 5(2):113–29.

Pastor, Willy. 1901. *Gustav Theodor Fechner und die durch ihn erschlossene Weltanschauung.* Speech at the occasion of the philosopher's 100th birthday. Leipzig: Georg Heinrich Meyer.

———. 1901a. *Im Geiste Fechners: Fünf naturwissenschaftliche Essays. Hrsg. zur Feier des 100. Geburtstags Gustav Theodor Fechners.* Leipzig: Meyer.

———. 1903. *Lebensgeschichte der Erde.* Leipzig: Diederichs.

———. 1904. "Im Geiste Fechners." *Die Zukunft* 47, 12. Jg.:191–93. [Response to Lasswitz 1904.]

———. 1905. "Gustav Theodor Fechner und die Weltanschauung der Alleinslehre." *Vorträge und Aufsätze aus der Comenius-Gesellschaft* 13(1):1–19.

Pauen, Michael. 1995. "Die Wissenschaft vom Schönen: Kunstpsychologie und die Ästhetik der Moderne." *Zeitschrift für philosophische Forschung* 49(1):54–75.

Paul, Jean. 1796. see: Jean Paul.

Paul, Julius. 1891. "Zur Lehre Fechners." *Sphinx* 12:146–48.

Pauli, Richard. 1920. *Über psychische Gesetzmäßigkeit.* Jena: Fischer.

Pauli, Richard and Aloys Wenzl. 1925. "Experimentelle und theoretische Untersuchungen zum Weber-Fechnerschen Gesetz." *Archiv für die gesamte Psychologie* 51.

Paulsen, Friedrich. 1895. *Einleitung in die Philosophie.* 3rd ed. Berlin: Hertz.

———. 1901. "Ernst Haeckel als Philosoph." Pp. 119–92 in *Philosophia militans. Gegen Klerikalismus und Naturalismus. Fünf Abhandlungen.* 2nd ed. Berlin: Reuther and Reichard.

———. 1907. "Gustav Theodor Fechner." *Internationale Wochenschrift für Wissenschaft, Kunst und Technik* 1:25–32, 55–64.

Paulsen, Johannes. 1907. *Der Begriff der Empfindung bei Fechner.* Phil. Diss, phil. Fak., Universität Marburg. [Doctoral advisors: Hermann Cohen, Paul Natorp. Also in *Philosophische Arbeiten* 1(4):241–99 as "Der Begriff der Empfindung in der Psychophysik (Fechner)."]

Pearce, Williams, L. 1973. *Kant, Naturphilosophie and Scientific Method. Foundations of Scientific Method: The Nineteenth Century.* Eds. Ronald N. Giere and Richard S. Westfall. Bloomington: Indiana University Press. 2nd printing 1974, 3–22.

Pearson, Karl. 1905. "'Das Fehlergesetz und seine Verallgemeinerungen durch Fechner und Pearson'. A Rejoinder." *Biometrika* 4:169–212.

Peirce, Charles Sanders. 1878. *Photometric Researches*. (Annals of the Astronomical Observatory of Harvard College, vol. 9) Leipzig: Engelmann.

———. 1884. "On small differences of sensation." [Together with Joseph Jastrow.] *Memoirs of the National Academy of Sciences* 3:73–83. [Cited according to 1958. *Collected Papers of Charles Sanders Peirce*. Vol. 7. Edited by Arthur W. Burks. Cambridge, Mass.: Harvard University Press. Reprint 1960, §§ 21–48.]

———. 1891. "The Architecture of Theories." *The Monist* 1:161–76. [Cited according to 1935. *Collected Papers of Charles Sanders Peirce*. Vol. 6. Edited by Charles Hartshorne and Paul Weiss. Cambridge, Mass.: Harvard University Press. Reprint 1960, §§ 7–34.]

———. 1892a. "The Doctrine of Necessity Examined." *The Monist* 2:321–37. [Cited according to 1935. *Collected Papers of Charles Sanders Peirce*. Vol. 6. Edited by Charles Hartshorne and Paul Weiss. Cambridge, Mass.: Harvard University Press. Reprint 1960, §§ 35–65.]

———. 1892b. "The Law of Mind." *The Monist* 2:533–59. [Cited according to 1935. *Collected Papers of Charles Sanders Peirce*. Vol. 6. Edited by Charles Hartshorne and Paul Weiss. Cambridge, Mass.: Harvard University Press, 1935. Reprint 1960, §§ 102–63.]

———. 1892c. "Man's Glassy Essence." *The Monist* 3:1–22. [Cited according to 1935. *Collected Papers of Charles Sanders Peirce*. Vol. 6. Edited by Charles Hartshorne and Paul Weiss. Cambridge, Mass.: Harvard University Press. Reprint 1960, §§ 238–71.]

———. 1893. "Evolutionary Love." *The Monist* 3:176–200. [Cited according to 1935. *Collected Papers of Charles Sanders Peirce*. Vol. 6. Edited by Charles Hartshorne and Paul Weiss. Cambridge, Mass.: Harvard University Press. Reprint 1960, §§ 287–317.]

———. 1898. "Habit." In 1958. *Collected Papers of Charles Sanders Peirce*. Vol. 7. Edited by Arthur W. Burks. Cambridge, Mass.: Harvard University Press. Reprint 1960, §§ 468–528.

———. 1900. "Notes on Science." In 1958. *Collected Papers of Charles Sanders Peirce*. Vol. 7. Edited by Arthur W. Burks. Cambridge, Mass.: Harvard University Press. Reprint 1960, §§ 256–312.

———. 1903. "Lowell Lectures of 1903." In 1935. *Collected Papers of Charles Sanders Peirce*. Vol. 1. Edited by Charles Hartshorne and Paul Weiss. Cambridge, Mass.: Harvard University Press. Reprint 1960.

———. 1960. *Collected Papers of Charles Sanders Peirce*. 8 vols. Edited by Charles Hartshorne and Paul Weiss (vols. 1–6) and Arthur W. Burks (vols. 7–8). 2nd ed. Cambridge, Mass.: Harvard University Press.

———. 1988. *Naturordnung und Zeichenprozeß. Schriften über Semiotik und Naturphilosophie*. Edited by Helmut Pape. Aachen: Alano. [Reprint 1991. Frankfurt am Main: Suhrkamp.]

Pelman, Carl (Georg) Wilhelm. 1862. [Review of Fechner 1860.] *Allgemeine Zeitschrift für Psychiatrie* 19:231–56.

———. 1878. "Gehirnphysiologie und Psychologie." *Archiv für Psychiatrie und Nervenkrankheiten* 8(7):713–21.

———. 1879. "Ideen zu einer allgemeinen Psychiatrie." *Allgemeine Zeitschrift für Psychiatrie und psychiatrisch-gerichtliche Medicin* 35(5):463–85.

Pendlebury, D. L. 1968. "The Day View of Gustave T. Fechner." *Systematics* 6:1–26.

Penna, Antonio Gomes. 1988. "Bergson's Critique of Fechner's Psychophysics." Pp. 151–55 in *G. T. Fechner and Psychology*, edited by Josef Brožek and Horst Gundlach. Passau: Passavia.

Pernerstorfer, Engelbert. 1912. "Nekrolog. Siegfried Lipiner." *Zeitschrift des österreichischen Vereines für Bibliothekswesen* 16 (Neue Folge 3):121–25.

Perry, Ralph Barton. 1935. *The Thought and Character of William James.* 2 vols. London: H. Milford.

Perty, Maximilian. 1874. "Fechner's neueste Schrift." *Blätter für literarische Unterhaltung*:88–90.

Pester, Reinhardt. 1987. "Lotzes Berufung an die Philosophische Fakultät." *Deutsche Zeitschrift für Philosophie* 35:806–14.

———. 1997. *Hermann Lotze. Wege seines Denkens und Forschens: Ein Kapitel deutscher Philosophie- und Wissenschaftsgeschichte im 19. Jahrhundert.* Würzburg: Königshausen and Neumann.

Petzoldt, Joseph. 1890. "Maxima, Minima und Oekonomie." *Vierteljahrsschrift für wissenschaftliche Philosophie und Soziologie* 14:206–39, 354–442.

———. 1894. "Ueber den Begriff der Entwicklung und einige Anwendungen desselben." *Naturwissenschaftliche Wochenschrift* 9 (7 and 8) (February 18 and 25): 77–81, 89–93.

———. 1923. *Das allgemeinste Entwicklungsgesetz.* Munich: Rösl, and Berlin: Paetel.

Philippi, E. 1884. [Review of Fechner 1882.] *Philosophische Monatshefte* 20:140–44.

Picard, Émile. 1925. *La vie et l'œuvre de Jules Tannery.* Institut de France, Académie des Sciences. Paris: Gauthier-Villars.

Piéron, Henri. 1960. "La psychophysique de Fechner, son rôle dans l'évolution de la psychologie et sa place dans la science actuelle." *Schweizerische Zeitschrift für Psychologie und ihre Anwendungen* 19:5–25.

Place, Ullin T. 1956. "Is Consciousness a Brain Process?" *British Journal of Psychology* 47:44–50.

———. 1988. "Thirty Years on—Is Consciousness still a Brain Process?" *Australasian Journal of Philosophy* 66:208–19.

———. 1989. "Thirty Five Years on—Is Consciousness Still a Brain Process?" *Grazer Philosophische Studien* 36:17–29.

———. 1990. "E. G. Boring and the Mind-Brain Identity Theory." *British Psychological Society, History and Philosophy of Psychology Newsletter* 11:20–31.

———. 1994. "'Is Consciousness Still a Brain Process?' Some Misconceptions about the Article." Pp. 9–15 in *Consciousness at the Crossroads of Philosophy and Cognitive Science.* (Selected proceedings of the final meeting of the Tempus Project 'Phenomenology and Cognitive Science', Maribor, Slovenia) Edited by Borstner and Shawe-Taylor. Thorverton: Imprint Academic.

Planck, Max. 1887. *Das Prinzip der Erhaltung der Energie.* (Wissenschaft und Hypothese, vol. 6) Leipzig: Teubner. [Cited according to the 3rd ed. 1913.]

———. 1909. "Die Einheit des physikalischen Weltbildes." *Physikalische Zeitschrift* 10:62–75. [Cited according to Max Planck 1983. Pp. 28–51 in *Vorträge und Erinnerungen.* Reprint of 5th ed. Darmstadt: Wissenschaftliche Buchgesellschaft.]

Plateau, Joseph (Antoine Ferdinand). 1872. "Sur la mesure des sensations physiques et sur la loi qui lie l'intensité de ces sensations à l'intensité de la cause excitante." *Bulletin de l'Académie royale des sciences, des lettres et des beaux-arts de Belgique,* 2e série, 33:376–88. [See Laming 1996]

Plato, Jan von. 1987. "Probabilistic Physics the Classical Way." Pp. 379–409 in *The Probabilistic Revolution.* Vol. 2: *Ideas in the Sciences.* Edited by Lorenz Krüger, Gerd Gigerenzer, and Mary Morgan. Cambridge, Mass.: MIT Press/Bradford Books.

———. 1994. *Creating Modern Probability: Its Mathematics, Physics and Philosophy in His-*

torical Perspective. (Cambridge Studies in Probability, Induction, and Decision Theory) Cambridge: Cambridge University Press. Reprint 1995.

Poggi, Stefano. 1996. "La 'Naturphilosophie' dell' idealismo e del romanticismo: lo stato della ricerca." *Rivista di Filosofia* 87(1):111–28.

Poggi, Stefano and Maurizio Bossi, eds. 1994. *Romanticism in Science: Science in Europe, 1790–1840.* (Boston Studies in the Philosophy of Science, vol. 152) Dordrecht: Kluwer.

Poincaré, Henri. 1902. *La Science et l'hypothèse.* Paris: Flammarion, 1902. Reprint ibid. 1968. [Cited according to the authorized German translation 1906 by F. and L. Lindemann as *Wissenschaft und Hypothese.* 2nd ed. Leipzig: Teubner.]

Popper, Karl R. and John C. Eccles. 1977. *The Self and its Brain.* New York: Springer. [Cited according to German translation 1982. *Das Ich und sein Gehirn.* Munich: Piper.]

Porter, Theodore M. 1981. "A Statistical Survey of Gases: Maxwell's Social Physics." *Historical Studies in the Physical Sciences* 12:77–116.

———. 1983. "Private Chaos, Public Order: The Nineteenth-Century Statistical Revolution." Pp. 27–40 in *Probability Since 1800.* Edited by Michael Heidelberger, Lorenz Krüger, and Rosemarie Rheinwald. Bielefeld: B. Kleine.

———. 1985. "The Mathematics of Society: Variation and Error in Quetelet's Statistics." *British Journal for the History of Science* 18:51–69.

———. 1986. *The Rise of Statistical Thinking, 1820–1900.* Princeton: Princeton University Press.

———. 1987. "Lawless Society: Social Science and the Reinterpretation of Statistics in Germany, 1850–1880." Pp. 351–75 in *The Probabilistic Revolution.* Vol. 1: *Ideas in History.* Eds. Lorenz Krüger, Lorraine J. Daston and Michael Heidelberger. Cambridge, Mass.: MIT Press/Bradford Books.

———. 1994. "From Quetelet to Maxwell: Social Statistics and the Origins of Statistical Physics." Pp. 345–62 in *The Natural Sciences and the Social Sciences. Some Critical and Historical Perspectives,* edited by I. Bernard Cohen. Dordrecht: Kluwer.

Preyer, William Thierry. 1877. *Elemente der reinen Empfindungslehre.* Jena: Dufft. [Also in *Sammlung physiologischer Arbeiten.* Vol. 1. Edited by William Preyer. Jena: Dufft, 1877, 537–638.]

———. 1880. *Naturwissenschaftliche Thatsachen und Probleme. Populäre Vorträge.* Berlin: Paetel.

———. 1890. See Fechner 1890.

Priesner, Claus. 1982. "Zur Entwicklung der Atom- und Moleküldefinition in der Chemie im 19. Jahrhundert." Pp. 7–33 in *Atomvorstellungen im 19. Jahrhundert,* edited by Charlotte Schönbeck. Paderborn: Schöningh.

Prigogine, Ilya and P. Glansdorff. 1973. "L'écart à l'équilibre interprété comme une source d'ordre des structures dissipatives." *Académie royale de Belgique, Bulletin de la Classe des Sciences* Sér. 5, vol. 59:672–702.

Pross, Wolfgang. 1991. "Lorenz Oken. Naturforschung zwischen Naturphilosophie und Naturwissenschaft." In *Die deutsche literarische Romantik und die Wissenschaften,* edited by Nicholas Saul. München: Iudicium.

Prytulak, Lubomir S. 1975. "Critique of S. S. Stevens' Theory of Measurement Classification." *Perceptual and Motor Skills* 41:3–28.

Putnam, Hilary. 1962. "The Analytic and the Synthetic." Pp. 358–97 in *Minnesota Studies in the Philosophy of Science.* Vol. 3: *Scientific Explanation, Space, and Time,* edited by Herbert Feigl and Grover Maxwell. Minneapolis: University of Minnesota Press.

———. 1967. "The Nature of Mental States." Pp. 37–48 in *Art, Mind and Religion,* edited by

W. H. Capitan and D. D. Merrill. Pittsburgh.: University of Pittsburgh Press. [Cited according to Hilary Putnam 1975. *Mind, Language and Reality*. Vol. 2 of *Philosophical Papers*. Cambridge: Cambridge University Press.]

Quetelet, Adolphe (Jacques Lambert). 1835. *Sur l'homme et le développement de ses facultés, ou Essai de physique sociale*. 2 vols. Paris: Bachelier. [2nd ed. 1869 in 2 vols. titled *Physique sociale, ou Essai sur le développement des facultés de l'homme*. Reprint 1997. (Mémoires de la classe des lettres, 3e sér., vol. 15.) Brussels: Académie royale de Belgique. Translation 1842 of 1st ed. as *A Treatise on Man and the Development of his Faculties*. Edinburgh: Chambers.]

———. 1846. *Lettres à S. A. R. le Duc Régnant de Saxe-Cobourg et Gotha, sur la théorie des probabilités, appliquée aux sciences morales et politiques*. Brussels: Hayez. [Translation 1849 as *Letters addressed to H. R. H. the Grand Duke of Saxe Coburg and Gotha, on the Theory of Probability*. London: Layton. Reprint 1981. New York: Arno.]

———. 1848. "Sur la statistique morale et les principes qui doivent en former la base." *Mémoires de l'Académie royale des sciences et belles-lettres de la Belgique* 21.

Radakovic, M. 1890. "Ueber Fechner's Ableitungen der psychophysischen Massformel." *Vierteljahrsschrift für wissenschaftliche Philosophie und Soziologie* 14:1–26.

Ranke, Karl Ernst and Richard Greiner. 1904. "Das Fehlergesetz und seine Verallgemeinerungen durch Fechner und Pearson in ihrer Tragweite für die Anthropologie." *Archiv für Anthropologie* 30 (N. F. 2):295–332.

Rath, Martin. 1994. *Der Psychologismusstreit in der deutschen Philosophie*. Freiburg i. Br.: Alber.

Ratzel, Friedrich. 1882. *Anthropo-Geographie, oder Grundzüge der Anwendung der Erdkunde auf die Geschichte*. Vol. 1. Stuttgart: Engelhorn.

———. 1901. "Die Tagesansicht Gustav Theodor Fechners." *Die Grenzboten* 60(2):169–78. [Also in Friedrich Ratzel 1905. Pp. 497–501 in *Glücksinseln und Träume. Gesammelte Aufsätze aus den Grenzboten*. Leipzig: Grunow.]

Reed, Edward S. 1994. "The Separation of Psychology from Philosophy: Studies in the Sciences of Mind 1815–1879." Pp. 297–356 in *Routledge History of Philosophy*. Vol. 7: *The Nineteenth Century*. Edited by C. L. Ten. London: Routledge.

Reichenbach, Carl Freiherr von. 1856. *Odische Erwiederungen an die Herren Professoren Fortlage, Schleiden, Fechner und Hofrath Carus*. Vienna: Braumüller.

Reichlin-Meldegg, Karl-Alexander von. 1866. [Review of Fechner 1863.] *Zeitschrift für Philosophie und philosophische Kritik* 48:101–21.

Reincke, Ernst. 1886. "Zweck und Methode der Psychophysik." Pp. 5–30 in *Programm des Realgymnasiums zu Malchin*. Nr. 601. Malchin: Heese.

Reininger, Robert. 1916. *Das psycho-physische Problem. Eine erkenntnistheoretische Untersuchung zur Unterscheidung des Physischen und Psychischen überhaupt*. Vienna: Braumüller. [2nd ed. 1930.]

Revers, Wilhelm Josef. 1960. "Gustav Theodor Fechners unbeabsichtigte Gründung." *Schweizerische Zeitschrift für Psychologie und ihre Anwendungen* 19:26–38.

Reynolds, Andrew. 1997. "Peirce's Cosmology and the Laws of Thermodynamics." *Transactions of the Charles Sanders Peirce Society* 33:403–23.

Rezazadeh-Schafagh, Sadeg. 1928. *Mystische Motive in Fechners Philosophie*. Ph.D. diss., phil. Fak., Universität Berlin. [Doctoral advisors: Eduard Spranger, Heinrich Maier.]

Ribot, Théodule. 1874. "La psychologie physiologique en Allemagne. La mesure des sensations." *La Revue Scientifique* 2e série, 4e année, tome VII, No 24 (12 decembre):553–63.

———. 1875. "La psychologie allemande contemporaine. M. Wilhelm Wundt." *La Revue Scientifique* 2e série, 4e année, tome VIII, No 31 (30 janvier): 723–31; 751–60.

———. 1875a. "Correspondance. A propos du logarithme des sensations." *La Revue Scientifique* 2e série, 4e année, tome VIII, No 37 (13 mars):877–78.

———. 1875b. "La psychologie physiologique en Allemagne. M. W. Wundt." *La Revue Scientifique* 2e série, 5e année, tome IX, No 22 (27 novembre):505–16.

———. 1875c. "La psychologie physiologique en Allemagne. M. W. Wundt (1)." *La Revue Scientifique* 2e série, 5e année, tome IX, No 23 (4 décembre):544–49.

———. 1879. *La Psychologie allemande contemporaine (École expérimentale)*. Paris: Germer Baillière. [Cited according to German translation 1881. *Die experimentelle Psychologie der Gegenwart in Deutschland.* Authorized German ed. Braunschweig: Vieweg. Translation 1885 (from the 2nd French ed. 1885, corrected and augmented, Paris: Félix Alcan) by James Mark Baldwin as *German Psychology of To-Day: The Empirical School.* New York: Macmillan. Reprint 1998. Bristol: Thoemmes.]

Richards, Evelleen. 1990. "'Metaphorical Mystifications': The Romantic Gestation of Nature in British Biology." Pp. 130–43 in *Romanticism and the Sciences*, edited be Andrew Cunningham and Nicholas Jardine. Cambridge: Cambridge University Press.

Rickert, Heinrich. 1900. "Psychophysische Causalität und psychophysischer Parallelismus." Pp. 59–87 in *Philosophische Abhandlungen. Christoph Sigwart zu seinem siebzigsten Geburtstage am 28. März 1900 gewidmet.* Tübingen: Mohr.

Riehl, Alois. 1872. *Über Begriff und Form der Philosophie.* Leipzig: Haacke. [Cited according to reprint 1925. Pp. 91–174, 332–39 in *Philosophische Studien aus vier Jahrzehnten.* Leipzig: Quelle and Meyer.]

———. 1875. [Review of Fechner 1873.] *Wissenschaftliche Monatsblätter* 3:7–9.

———. 1876. *Der philosophische Kriticismus und seine Bedeutung für die positive Wissenschaft.* Vol. 1: *Geschichte des philosophischen Kritizismus.* Leipzig: Engelmann.

———. 1879. *Der philosophische Kriticismus und seine Bedeutung für die positive Wissenschaft.* Vol. 2, Part 1: *Die sinnlichen und logischen Grundlagen der Erkenntnis.* Leipzig: Engelmann.

———. 1887. *Der philosophische Kriticismus und seine Bedeutung für die positive Wissenschaft.* Vol. 2, Part 2: *Wissenschaftstheorie und Metaphysik.* Leipzig: Engelmann.

———. 1894. *The Principles of the Critical Philosophy: Introduction to the Theory of Science and Metaphysics.* Translated by Arthur Fairbanks. London: Kegan, Paul, Trench, Trübner and Co. [Translation of Riehl 1887. Part II, chapter II of this volume bears the title: "On the Relation of Psychical Phenomena to Material Processes."]

———. 1921. *Zur Einführung in die Philosophie der Gegenwart.* 6th ed. Leipzig: Teubner. [1st ed. 1903. The 5th lecture on the mind-body problem is titled: "Der naturwissenschaftliche und der philosophische Monismus" (Scientific and Philosophical Monism) 112–46.]

Riepe, Manfred. 2002. "Das Ornament der Maße. Gustav Theodor Fechners Bedeutung für die Psychoanalyse Sigmund Freuds." *Psyche. Zeitschrift für Psychoanalyse und ihre Anwendungen* 56(8):756–89.

Rocke, Alan J. 1979. "The Reception of Chemical Atomism in Germany." *Isis* 70:519–36.

———. 1984. *Chemical Atomism in the Nineteenth Century.* Columbus, Ohio: Ohio State University Press.

Rodi, Frithjof. 1987. "Die Ebbinghaus-Dilthey-Kontroverse. Biographischer Hintergrund und sachlicher Ertrag." Pp. 146–54 in *Ebbinghaus-Studien 2. Beiträge zum Internationalen*

Hermann-Ebbinghaus-Symposion. (Passauer Schriften zur Psychologiegeschichte, vol. 5) Edited by Werner Traxel and Horst Gundlach. Passau: Passavia.

Rohracher, Hubert. 1926. *Die Erkenntnistheorie und Methodenlehre Gustav Theodor Fechners.* Ph.D. diss., phil. Fak., Universität München.

Romanos, Konstantin P. 1991. "Henri Bergsons Kritik der Quantität als allgemeine Entfremdungstheorie der Gegenwart." *Revue internationale de philosophie*:151–84.

Roob, Helmut. 1981. *Kurd Laßwitz. Handschriftlicher Nachlaß und Bibliographie seiner Werke.* (Veröffentlichungen der Forschungsbibliothek Gotha, Heft 19) Gotha.

Rorty, Richard. 1979. *Philosophy and the Mirror of Nature.* Princeton: Princeton University Press.

Rosenberger, Ferdinand. 1887. *Die Geschichte der Physik.* Vol. 3. Braunschweig: Vieweg.

Rosencrantz, Gerhard. 1933. *G. Th. Fechners Stellung in der Geschichte der gelehrten Satire.* Ph.D. diss., phil. Fak., Universität Königsberg. [Doctoral advisors: Nadler, Ziesemer.]

Rosenkranz, Karl. 1840. *Kritische Erläuterungen des Hegel'schen Systems.* Königsberg: Bornträger.

Rosenzweig, Saul. 1987. "The Final Tribute of E. G. Boring to G. T. Fechner: Concerning the Date of October 22, 1850." *American Psychologist* 42:787–90.

Ross, Helen E. 1987. "Die Arbeiten von Weber und Fechner über Händigkeit und Gewichtsunterscheidung: Ein Vergleich mit aktuellen Forschungen." Pp. 269–85 in *Psychophysische Grundlagen mentaler Prozesse. In Memoriam G. Th. Fechner (1801–1887)*, edited by Hans-Georg Geissler and Konrad Reschke. Leipzig: Karl-Marx-Universität Leipzig.

———. 1996. See Weber, Ernst Heinrich 1996.

Rothschuh, Karl E. 1972. "Historische Wurzeln der Vorstellung einer selbsttätigen informationsgesteuerten biologischen Regelung." *Nova Acta Leopoldina* (Edited by Joachim-Hermann Scharf) 37(1):91–106.

Royce, Josiah. 1912. "Prinzipien der Logik." Pp. 61–136 in *Encyclopädie der Philosophischen Wissenschaften*, edited by Arnold Ruge, in collaboration with Wilhelm Windelband. Vol. 1: *Logik.* Tübingen: Mohr. [Translation 1913 by E. Bethel Meyer as *Encyclopaedia of the Philosophical Sciences.* Vol. 1: *Logic.* London: Macmillan.]

Rümelin, Gustav. 1863. "Zur Theorie der Statistik I. 1863." Pp. 208–64 in *Reden und Aufsätze.* Freiburg i. Br.: Mohr.

———. 1874. "Zur Theorie der Statistik II. 1874." Pp. 265–84 in *Reden und Aufsätze.* Freiburg i. Br.: Mohr.

Rüscher, Alois. 1902. *Göttliche Notwendigkeits-Weltanschauung, Teleologie, mechanische Naturansicht und Gottesidee. Mit besonderer Berücksichtigung von Haeckel, Wundt, Lotze und Fechner.* Zürich: Müller. [Also 1902 Ph.D., phil. Fak., Universität Zürich.]

Russell, Bertrand. 1912. "On the Notion of Cause." *Proceedings of the Aristotelian Society* 13:1–26. [Cited according to Bertrand Russell 1951. Pp. 180–208 in *Mysticism and Logic.* 10th ed. London: Allen and Unwin.]

———. 1921. *The Analysis of Mind.* London: George Allen and Unwin. Reprint 1995. London: Routledge.

Sachs-Hombach, Klaus. 1993. *Philosophische Psychologie im 19. Jahrhundert. Ihre Entstehung und Problemgeschichte.* Freiburg i. Br.: Alber.

———. 1993a. "Der Geist als Maschine—Herbarts Grundlegung der naturwissenschaftlichen Psychologie." Pp. 91–111 in *Das sichtbare Denken. Modelle und Modellhaftigkeit in der Philosophie und den Wissenschaften*, edited by Jörg F. Mass. Amsterdam: Rodopi.

Salmon, Wesley C. 1984. *Scientific Explanation and the Causal Structure of the World.* Princeton: Princeton University Press.

Sandkühler, Hans Jörg, ed. 1984. *Natur und geschichtlicher Prozeß. Studien zur Naturphilosophie F. W. J. Schellings.* Frankfurt am Main: Suhrkamp.

Särndal, Carl-Erik. 1971. "The Hypothesis of Elementary Errors and the Scandinavian School in Statistical Theory." *Biometrika* 58:375–91.

Sauer, Werner. 1985. "Carnaps 'Aufbau' in Kantianischer Sicht." *Grazer philosophische Studien* 23:19–35.

Savage, C. Wade. 1970. *The Measurement of Sensation: A Critique of Perceptual Psychophysics.* Berkeley: University of California Press.

Savage, C. Wade and Philip Ehrlich, eds. 1991. *Philosophical and Foundational Issues in Measurement Theory.* Hillsdale, N.J.: Erlbaum.

Schaller, Julius. 1852. "Ueber die Beseeltheit der Gestirne." [Review of Fechner 1851.] *Allgemeine Monatsschrift für Wissenschaft und Literatur*:1035–58.

———. 1857. "Zur Kritik des Atomismus." *Zeitschrift für Philosophie und Philosophische Kritik* 31:1–48.

———. 1878. "Psychophysische Fragen und Bedenken." [Review of Fechner 1877 and G. E. Müller 1878.] *Zeitschrift für Philosophie und Philosophische Kritik* 72:281–310.

Scharf, Joachim-Hermann. 1996. "Die anonymen Privatheilande und die junge Biomathematik." Pp. 25–38 in *Abstand und Nähe. Vorträge im Rückblick. Sächsische Akademie der Wissenschaften zu Leipzig,* edited by Helga Bergmann im Auftrag des Präsidiums. Berlin: Akademie. [A commentary on Fechner 1849.]

Scharlau, Winfried. 1990. *Mathematische Institute in Deutschland 1800–1945.* (Dokumente zur Geschichte der Mathematik, vol. 5) Braunschweig: Vieweg.

Scheerer, Eckart. 1984. "Nachbild." Cols. 341–48 in *Historisches Wörterbuch der Philosophie,* edited by Joachim Ritter and Karlfried Gründer. Vol. 6. Basel: Schwabe.

———. 1984a. "Organismus. II. Kosmologie, Soziologie und Psychologie." Cols. 1336–48 in *Historisches Wörterbuch der Philosophie,* edited by Joachim Ritter and Karlfried Gründer. Vol. 6. Basel: Schwabe.

———. 1987. "Was wurde aus der inneren Psychophysik?" Pp. 8–26 in *Psychophysische Grundlagen mentaler Prozesse. In Memoriam G. Th. Fechner (1801–1887),* edited by Hans-Georg Geissler and Konrad Reschke. Leipzig: Karl-Marx-Universität.

———. 1987a. "The Unknown Fechner." *Psychological Research* 49:197–202.

———. 1989. "Conjuring Fechner's spirit." [Open peer-commentary to Krueger 1989] *Behavioral and Brain Sciences* 12:288–90.

———. 1989a. "Psychologie." Cols. 1599–1653 in *Historisches Wörterbuch der Philosophie,* edited by Joachim Ritter and Karlfried Gründer. Vol. 7. Basel: Schwabe.

———. 1992. "Fechner's Inner Psychophysics: Its Historical Fate and Present Status." Pp. 3–21 in *Cognition and Psychophysics: Basic Issues,* edited by Hans-Georg Geissler, Stephen W. Link, and James T. Townsend. Hillsdale, N.J.: Erlbaum.

———. 1993. "Gustav Theodor Fechner und die Neurobiologie: 'Innere Psychophysik' und 'tierische Elektrizität.'" Pp. 259–86 in *Das Gehirn—Organ der Seele?* edited by Ernst Florey and Olaf Breidbach. Berlin: Akademie.

———. 1994. "Psychoneural Isomorphism: Historical Background and Current Relevance." *Philosophical Psychology* 7(2):183–210.

Scheerer, Eckart and Helmut Hildebrandt. 1988. "Was Fechner an Eminent Psychologist?" Pp. 269–81 in *G. T. Fechner and Psychology,* edited by Josef Brožek and Horst Gundlach. Passau: Passavia.

Scheler, Max. 1922. *Die deutsche Philosophie der Gegenwart.* Pp. 261–330 in *Gesammelte Werke,* edited by Max Ferdinand Scheler. Vol. 7. Bern: Francke.

Schelling, Friedrich Wilhelm Joseph von. 1797. *Ideen zu einer Philosophie der Natur als Einleitung in das Studium dieser Wissenschaft.* Leipzig: Breitkopf und Härtel. 2nd ed. 1803. Landshut: Krüll. [Cited according to Schelling, ed. 1857. *F. W. J. Schelling's Sämmtliche Werke.* Vol. 2. Stuttgart: Cotta. Reprint 1985 as *Ausgewählte Schriften.* Vol. 1. Frankfurt am Main: Suhrkamp. 2nd ed. 1995, 245–94. Translation 1988 by Errol E. Harris and Peter Heath as *Ideas for a Philosophy of Nature as Introduction to the Study of this Science.* (Texts in German Philosophy) Introd. by Robert Stern. Cambridge: Cambridge University Press.]

———. 1799. *Erster Entwurf eines Systems der Naturphilosophie. Zum Behuf seiner Vorlesungen.* Jena/Leipzig: Gabler. [Cited according to Schelling, ed. 1857. *F. W. J. Schelling's Sämmtliche Werke.* Vol. 3. Stuttgart: Cotta. Reprint 1985 as *Ausgewählte Schriften.* Vol. 1. Frankfurt am Main: Suhrkamp, 2nd ed. 1995, 317–36.]

———. 1804. *System der gesammten Philosophie und der Naturphilosophie insbesondere. (Aus dem handschriftlichen Nachlaß.) 1804.* [Cited according to Schelling, ed. 1860. *F. W. J. Schelling's Sämmtliche Werke.* Vol 6. Stuttgart: Cotta. Reprint 1985 as *Ausgewählte Schriften.* Vol. 3. Frankfurt am Main: Suhrkamp. 2nd ed. 1995, 141–587.]

———. 1976. Friedrich W. J. Schelling, *Historisch-kritische Ausgabe.* Edited on behalf of the Schelling Commission of the Bavarian Academy of the Sciences by Hans Michael Baumgartner et al. Stuttgart: Frommann-Holzboog, 1976ff. (1st series: *Werke:* 7 vols. so far and a supplementary volume. 3rd series: *Briefe:* 1 vol. so far.)

———. 1994. See Baumgartner 1994.

Schiemann, Gregor and Michael Heidelberger, 1999. "Philosophie IX: Naturphilosophie." Pp. 1127–38 in *Enzyklopädie Philosophie,* edited by Hans Jörg Sandkühler. Vol. 2. Hamburg: Meiner.

Schleiden, Matthias Jakob. 1845. "Pflanzenchemie." [Review of books on agricultural chemistry.] *Jenaische allgemeine Literatur-Zeitung* 4(162):645–47.

———. 1855. "Vierte Vorlesung. Die Beseelung der Pflanzen. Gespräch und Rechtfertigung." Pp. 131–82 in Matthias Jakob Schleiden, *Studien. Populäre Vorträge.* Leipzig: W. Engelmann.

Schlick, Moritz. 1910. "Die Grenzen der naturwissenschaftlichen und philosophischen Begriffsbildung." *Vierteljahrsschrift für wissenschaftliche Philosophie* 34:121–42. [Translation by Peter Heath as "The Boundaries of Scientific and Philosophical Concept Formation." Pp. 25–40 in Moritz Schlick 1979. *Philosophical Papers,* edited by Henk L. Mulder and Barbara F. B. van de Velde-Schlick. Vol. 1: 1909–1922. Dordrecht: Reidel.]

———. 1916. "Idealität des Raumes, Introjektion und psychophysisches Problem." *Vierteljahrsschrift für wissenschaftliche Philosophie und Soziologie* 40:230–54. [Pp. 190–206 in Moritz Schlick 1979. *Philosophical Papers,* edited by Henk L. Mulder and Barbara F. B. van de Velde-Schlick. Vol. 1: 1909–1922. Dordrecht: Reidel.]

———. 1925. *Allgemeine Erkenntnislehre.* 2nd rev. ed. Berlin: Springer. Reprint 1979. Frankfurt am Main: Suhrkamp. [Translation 1974 by Albert E. Blumberg as *General Theory of Knowledge.* Wien: Springer-Verlag. Also 1985. La Salle, Ill.: Open Court. 1st German ed. 1918. Berlin: Springer.]

———. 1935. "De la relation entre les notions psychologiques et les notions physiques." *Revue de Synthèse* 10:5–26. [German original first in Moritz Schlick 1938. *Gesammelte Aufsätze 1926-1036.* Wien: Gerold. Reprint Hildesheim: Olms, 267–87. Translation by Wilfrid Sellars as "On the Relation between Psychological and Physical Concepts." Pp.

420–36 in Moritz Schlick 1979. *Philosophical Papers*, edited by Henk L. Mulder and Barbara F. B. van de Velde-Schlick. Vol. 2: 1925–1936. Dordrecht: Reidel.]

Schmidt, Nicole D. 1995. *Philosophie und Psychologie. Trennungsgeschichte, Dogmen und Perspektiven.* Reinbek: Rowohlt.

Schmidt, Oscar. 1874. "Fechners Ideen zur Schöpfungs- und Entwicklungsgeschichte der Organismen." *Das Ausland* 47(8) (February 23):141–44.

Schmidt, Siegfried J., ed. 1987. *Der Diskurs des Radikalen Konstruktivismus.* Frankfurt am Main: Suhrkamp.

Schmidt, Walther. 1921. *Zur Geschichte von Mass und Zahl in der Psychologie.* (Sammlung wissenschaftlicher Arbeiten, Heft 62) Langensalza: Wendt und Klauwell.

Schmidt, Winrich de. 1976. *Psychologie und Transzendentalphilosophie. Zur Psychologierezeption bei Hermann Cohen und Paul Natorp.* Bonn: Bouvier.

Schmied-Kowarzik, Wolfdietrich. 1989. "Friedrich Wilhelm Joseph Schelling (1775–1854)." Pp. 241–62 in *Klassiker der Naturphilosophie*, edited by Gernot Böhme. Munich: Beck.

———. 1996. *"Von der wirklichen, von der seyenden Natur." Schellings Ringen um eine Naturphilosophie in Auseinandersetzung mit Kant, Fichte und Hegel.* Stuttgart: Frommann-Holzboog.

Schnabel, Franz. 1934. *Deutsche Geschichte im 19. Jahrhundert.* Vol. 3: *Erfahrungswissenschaften und Technik.* Freiburg i. Br.: Herder.

Schnädelbach, Herbert. 1983. *Philosophie in Deutschland 1831–1933.* Frankfurt am Main: Suhrkamp. [Translation 1984 by E. Matthews as *Philosophy in Germany 1831–1933.* Cambridge: Cambridge University Press.]

Schneider, Anatol. 2001. *Person und Wirklichkeit. Nachidealistische Schellingrezeption bei Immanuel Hermann Fichte und Christian Hermann Weiße.* Würzburg: Königshausen and Neumann.

Schneider, Ivo. 1987. "Laplace and Thereafter: The Status of Probability Calculus in the Nineteenth Century." Pp. 191–214 in *The Probabilistic Revolution.* Vol. 1: *Ideas in History.* Edited by Lorenz Krüger, Lorraine J. Daston, and Michael Heidelberger. Cambridge, Mass.: MIT Press/Bradford Books.

Schnorr, C.-P. 1971. *Zufälligkeit und Wahrscheinlichkeit. Eine algorithmische Begründung der Wahrscheinlichkeitstheorie.* (Lecture Notes of Mathematics, 218) Berlin: Springer.

Schönpflug, Wolfgang. 1998. "Mathematical Psychics—das Glückskalkül des Francis Y. Edgeworth." Pp. 230–49 in *Psychologische Methoden und soziale Prozesse*, edited by Christoph Klauer and Hans Westmeyer. Lengerich: Pabst.

Schramm, Matthias. 1982. "Some Remarks on Quetelet." Pp. 115–127 in *Probability and Conceptual Change in Scientific Thought*, edited by Michael Heidelberger and Lorenz Krüger. Bielefeld: B. Kleine.

Schreier, Wolfgang. 1974. *Zu den Wechselbeziehungen von Physik, insbesondere Elektrophysik, und entstehender Elektrotechnik in der Zeit der Ausbreitung der Industriellen Revolution (1820–1870).* Habilitation thesis (Dissertation B), Universität Leipzig.

———. 1977. "Fechners experimentelle Methoden in der Psychophysik. Ein Beitrag über Wechselbeziehungne zwiscehn Physik und Psychologie." Pp. 22–29 in *Beiträge zur Wundt-Forschung.* Vol. 2. Leipzig: Karl-Marx-Universität, Sektion Psychologie.

———. 1979. "Über historische Wurzeln von Fechners Psychophysik." Pp. 61–71 in *Zur Geschichte der Psychologie*, edited by Georg Eckardt. Berlin: VEB Deutscher Verlag der Wissenschaften.

————. 1985. "Die Physik an der Leipziger Universität bis zum Ende des 19. Jahrhunderts." *Wissenschaftliche Zeitschrift der Karl-Marx-Universität Leipzig*, Mathematisch-naturwissenschaftliche Reihe, 34(1):5–11.

————. 1985a. "Gustav Theodor Fechner (1801-1887)." *Wissenschaftliche Zeitschrift der Karl-Marx-Universität Leipzig*, Mathematisch-naturwissenschaftliche Reihe, 34(1):60–62.

————. 1987. "Gustav Theodor Fechner als Physiker. Anläßlich des 100. Jahrestages seines Todes am 18. November 1987." *NTM—Schriftenreihe für Geschichte der Naturwissenschaften, Technik und Medizin* 24(2):81–85.

————. 1993. "Die drei Brüder Weber und Gustav Theodor Fechner—Untersuchungen zur medizinischen, Psycho- und technischen Physik." *NTM—Schriftenreihe für Geschichte der Naturwissenschaften, Technik und Medizin* 2 (N. S.):111–16.

Schröder, Christina. 1993. "Wilhelm Wirth und das psychophysische Seminar der Universität Leipzig." Pp. 41–46 in *Illustrierte Geschichte der Psychologie*, edited by Helmut E. Lück and Rudolf Miller. München: Quintessenz.

Schröder, Christina and Harry Schröder. 1991. "Gustav Theodor Fechner (1801–1887) in seiner Lebenskrise—Versuch der pathophysiologischen Rekonstruktion eines komplexen Krankheitsgeschehens." *Psychologie und Geschichte* 3:9–23.

Schubert, Gotthilf Heinrich. 1856. *Der Erwerb aus einem vergangenen und die Erwartungen aus einem zukünftigen Leben. Eine Selbstbiographie.* Vol. 3 in 2 parts. Erlangen: Paln und Enke. [Vol. 1: 1854; vol. 2: 1855.]

Schuekarew, A. 1907. "Über die energetischen Grundlagen des Gesetzes von Weber-Fechner und der Dynamik des Gedächtnisses." *Annalen der Naturphilosophie* 6:139–49.

Schulthess, Peter. 1984. "Einleitung." Pp. 7–46 in Hermann Cohen, *Das Princip der Infinitesimal-Methode und seine Geschichte. Ein Kapitel zur Grundlegung der Erkenntnisskritik. (Werke von Hermann Cohen*, vol. 5) 4th ed. as reprint of the 1st ed. (Berlin: Dümmler, 1883) Hildesheim: Olms.

Schulz, Lorenz. 1988. *Das rechtliche Moment der pragmatischen Philosophie von Charles Sanders Peirce.* Ebelsbach: Rolf Gremer.

Schurig, Volker. 1985. "Die Entdeckung der Systemeigenschaft 'Ganzheit'." *Gestalt Theory* 7: 208–27.

Schuster, Julius. 1922. *Oken, der Mann und sein Werk. Vortrag auf der Jahrhunderttagung der Gesellschaft Deutscher Naturforscher und Ärzte zu Leipzig.* Berlin: Junk.

Schwarz, Gerhard. 1994. *Zu Gustav Theodor Fechners Theorie der psychophysischen Bewegung und einigen ihrer naturwissenschaftlichen Entstehungsgründe.* Diplomarbeit, Institut für Psychologie, Technische Universität Berlin.

Schwartze, Peter, Renate Hanitzsch and Frank Rossberg. 1984. "Der Physiologe Ewald Hering." *Wissenschaftliche Zeitschrift der Karl-Marx-Universität Leipzig*, Math.-Naturwiss. Reihe, 35:235–39.

Schweiker, Johann Ev. 1910. "Gustav Theodor Fechner's Gottes- und Sittenlehre." *Historisch-politische Blätter für das katholische Deutschland* 145(4and5):266–88, 323–42.

Séailles, Gabriel. 1925. "La philosophie de Fechner." *Revue Philosophique de la France et de l'Étranger* 100, 50e année:5–47.

Seaman, Francis. 1968. "Mach's Rejection of Atomism." *Journal of the History of Ideas* 29: 381–93.

Searle, John R. 1984. *Minds, Brains and Science.* (The Reith Lectures 1984) London: BBC.

Sengle, Friedrich. 1971. *Biedermeierzeit. Deutsche Literatur im Spannungsfeld zwischen Restauration und Revolution 1815–1848.* Vol. 1: *Allgemeine Voraussetzungen, Richtungen, Darstellungsmittel.* Stuttgart: Metzler.

———. 1972. *Biedermeierzeit. Deutsche Literatur im Spannungsfeld zwischen Restauration und Revolution 1815–1848.* Vol. 2: *Die Formenwelt.* Stuttgart: Metzler.

Seydel, Rudolf. 1866. *Christian Hermann Weiße. Nekrolog.* Leipzig: Breitkopf und Härtel. [Cited according to reprint in Seydel 1887, 84–110.]

———. 1875. "Schelling. Festrede zur Feier des 100jährigen Geburtstages Schellings im akademisch-philosophischen Verein am 27. Januar." [Cited according to reprint in Seydel 1887, 62–83.]

———. 1880. "Fechners 'Tagesansicht.'" *Grenzboten* 39(2):529–44. [Cited according to reprint as "G. Th. Fechner" in Seydel 1887, 111–31.]

———. 1887. *Religion und Wissenschaft. Gesammelte Reden und Abhandlungen.* Breslau: Schottlaender.

Seydler, August. 1893. "Gustav Theodor Fechner." *Véstník Ceské akademie císare Frantiska Josefa pro vedy, slovesnost a umeni* 2(5–8):199–215, 277–90, 367–82, 443–52.

Short, T. L. 1983. "Teleology in Nature." *American Philosophical Quarterly* 20:311–20.

Sieg, Ulrich. 1994. *Aufstieg und Niedergang des Marburger Neukantianismus.* Würzburg: Königshausen and Neumann.

Siegel, Carl. 1913. "Fechners atomistische Naturphilosophie auf phänomenologischer Grundlage." Pp. 301–31 in *Geschichte der deutschen Naturphilosophie.* Leipzig: Akademische Buchhandlung.

Sigwart, Christoph. 1911. *Logik.* 4th ed. arranged by Heinrich Maier. 2 vols. Tübingen: Mohr. [Vol. 2: *Die Methodenlehre.* § 97 b, pp. 542–600: "Die Induction auf psychologischem Gebiete und ihre Voraussetzungen." (Induction in psychology and its presuppositions) 1st ed. 2 vols. Tübingen: Laupp, 1873–1878, 2nd. ed. Tübingen: Mohr, 1895. Translation 1895 as *Logic.* 2 vols. London: Sonnenschein. Reprint 1980. New York: Garland.]

Simon, Herbert A. 1962. "The Architecture of Complexity." *Proceedings of the American Philosophical Society* 106:467–82. [Cited according to Herbert A. Simon 1981. *The Sciences of the Artificial.* 2nd ed. Cambridge, Mass.: MIT Press.]

Simon, Theodor. 1894. *Leib und Seele bei Fechner und Lotze als Vertretern zweier massgebenden Weltanschauungen.* Göttingen: Vandenhoeck and Ruprecht.

———. 1895. "Fechners philosophische Unsterblichkeitslehre." *Christliche Welt* 9(30–31): 699–704, 722–27.

———. 1897. "Die Begründung des Optimismus bei Gustav Theodor Fechner." *Monatshefte der Comenius-Gesellschaft* 6(9–10):285–306.

Singer, Bernard. 1979. "Distribution-Free Methods for Non-Parametric Problems: A Classified and Selected Bibliography." *British Journal of Mathematical and Statistical Psychology* 32:1–60.

Sluga, Hans. 1980. *Gottlob Frege.* (The Arguments of the Philosophers, Nr. 1) London: Routledge.

Smaasen, Willem. 1846. "Vom dynamischen Gleichgewicht der Electricität in einer Ebene oder einem Körper." *Annalen der Physik und Chemie* 69:161–80; and 72:435–49. [Also in *Annales de Chimie* 40:236–47.]

Smart, J. J. C. 1959. "Sensation and Brain Processes." *Philosophical Review* 68:141–56.

Smith, Philip L. 1994. "Fechner's Legacy and Challenge." *Journal of Mathematical Psychology* 38:407–20. [Essay review of Link 1992.]

Snelders, Henricus Adrianus Marie. 1971. "Romanticism and Naturphilosophie and the Inorganic Natural Sciences 1797–1840: An Introductory Survey." *Studies in Romanticism* 9(3):193–215.

———. 1971a. "Point Atomism in Nineteenth Century Germany." *Janus* 58:194–00.

————. 1973. *De invloed van Kant, de Romantiek en de 'Naturphilosophie' op de anorganische naturwetenschappen in Duitsland*. Ph.D. diss. (Proefschrift, dr. in Wiskunde en Naturwetenschappen) Rijksuniversiteit Utrecht.

————. 1994. *Wetenschap en intuitie: Het Duitse romantisch-speculatief natuuronderzoek rond 1800*. Baarn: Ambo.

Snell, Carl. 1874. [Review of Fechner 1873.] *Jenaer Literaturzeitung* Nr. 12 (March 21):164–66.

Sober, Elliott. 1980. "Evolution, Population Thinking and Essentialism." *Philosophy of Science* 47:350–83.

Sommer, Manfred. 1985. *Husserl und der frühe Positivismus*. Frankfurt am Main: Suhrkamp.

————. 1987. *Evidenz im Augenblick. Eine Phänomenologie der reinen Empfindung*. Frankfurt am Main: Suhrkamp. [Pbk. 1996.]

Sommerfeld, Erdmute, Raul Kompass and Thomas Lachmann, eds. 2001. *Fechner Day 2001. Proceedings of the Seventeenth Annual Meeting of the International Society for Psychophysics, University of Leipzig, Germany, 20–23 October 2001*. Lengerich: Pabst.

Speckamp, Hans. 1911. *G. Th. Fechners Ethik im Zusammenhange seines Systems und im Vergleich mit dem englischen Utilitarismus*. Ph.D. diss., phil. Fak., Universität Münster. [Doctoral advisor: Erich Becher.]

Sprink, Walter. 1912. *Spinoza und Fechner. Ein Beitrag zu einer vergleichenden Untersuchung der Lehren Spinozas und Fechners*. Ph.D. diss., phil. Fak., Universität Breslau. [Doctoral advisors: Kühnemann, Stern.]

Sprung, Lothar. 1996. "Gustav Theodor Fechner. 'Mein Leben bietet überhaupt keine denkwürdigen Ereignisse dar.'" Pp. 207–26 in *Die großen Leipziger. 26 Annäherungen*, edited by Vera Hauschild. Frankfurt am Main: Insel.

Sprung, Helga and Lothar Sprung. 1978. "Gustav Theodor Fechner—Wege und Abwege in der Begründung der Psychophysik." *Zeitschrift für Psychologie* 186:429–54.

————. 1980. "Weber—Fechner—Wundt. Aspekte zur Entwicklungsgeschichte einer Wissenschaft, der Psychologie." Pp. 282–301 in *Wilhelm Wundt—progressives Erbe, Wissenschaftsentwicklung und Gegenwart*. (Wissenschaftliche Beiträge der KMU, Reihe Psychologie) Protokoll des internationalen Symposiums Leipzig 1979. Leipzig: Karl-Marx-Universität.

————. 1980a. *Gustav Theodor Fechner und die Entwicklung der Psychologie—Leben, Werk und Wirken in einer "sensiblen Phase" der Entwicklung einer neuen Wissenschaft im 19. Jahrhundert*. Berlin: Urania.

————. 1987. "Gustav Theodor Fechner in der Geschichte der Psychologie—Leben, Werk und Wirkung in der Wissenschaftsentwicklung des 19. Jahrhunderts." *Psychologiehistorische Manuskripte* 1. Leipzig: Karl-Marx-Universität Leipzig, Sektion Psychologie.

————. 1988. "Gustav Theodor Fechner als experimenteller Ästhetiker—Zur Entwicklung der Methodologie und Methodik einer Psychophysik höherer kognitiver Prozesse." Pp. 217–27 in *G. T. Fechner and Psychology*, edited by Josef Brožek and Horst Gundlach. Passau: Passavia.

————. 1997. "Georg Elias Müller (1850–1934). Skizzen zum Leben, Werk und Wirken." *Erinnern und Behalten*. Pp. 338–68 in *Wege zur Erforschung des menschlichen Gedächtnisses*, edited by Gerd Lüer and Uta Lass. (Abhandlungen der Akademie der Wissenschaften in Göttingen, mathem.-physik. Klasse, 3. Folge, Nr. 47) Göttingen: Vandenhoeck.

————. 2000. "Georg Elias Müller and the Beginnings of Modern Psychology." Pp. 70–91 in *Portraits of Pioneers in Psychology*. Vol. 4. Edited by Gregory A. Kimble and Michael Wertheimer. Washington, D.C.: American Psychological Association.

Stach, Petra [see also Lennig, Petra]. 1980. *Zu philosophischen Aspekten des Wirkens von*

Gustav Theodor Fechner. Diplomarbeit. Berlin: Humboldt-Universität zu Berlin, Sektion Marxistisch-leninistische Philosophie.

Stachel, Peter. 1998. "'Ein Kapitel der intellektuellen Entwicklung in Europa': Theoriebildungen in der Wiener Moderne und ihre Wurzeln in den österreichischen Traditionen philosophischen Denkens im 19. Jahrhundert." Pp. 109–76 in *Zwischen Orientierung und Krise. Zum Umgang mit Wissen in der Moderne.* (Studien zur Moderne, Bd. 2) Edited by Sonja Rinofner-Kreidl. Vienna: Böhlau.

Stadler, August. 1878. "Über die Ableitung des psychophysischen Gesetzes." *Philosophische Monatshefte* 14:215–23.

———. 1880. "Das Gesetz der Stetigkeit bei Kant." *Philosophische Monatshefte* 16:577–96.

Stallo, Johann Bernhard. 1848. *General Principles of the Philosophy of Nature with an Outline of Some of Its Recent Developments among the Germans, Embracing the Philosophical Systems of Schelling and Hegel and Oken's System of Nature.* Boston: W. M. Crosby and H. P. Nichols.

Staubermann, Klaus B. 2001. "Tying the Knot: Skill, Judgement and Authority in the 1870s Leipzig Spiritistic Experiments." *The British Journal for the History of Science* 34(1):67–79.

Steinbring, Heinz. 1980. *Zur Entwicklung des Wahrscheinlichkeitsbegriffs. Das Anwendungsproblem in der Wahrscheinlichkeitstheorie aus didaktischer Sicht.* (Materialien und Studien, vol. 18) Bielefeld: Institut für Didaktik der Mathematik der Universität Bielefeld.

Steinmetzler, Johannes. 1956. *Die Anthropogeographie Friedrich Ratzels und ihre ideengeschichtlichen Wurzeln.* (Bonner Geographische Abhandlungen, Heft 19) Bonn: Selbstverlag des Geographischen Instituts.

Stephan, Achim. 1992. "Emergence—A Systematic View on Its Historical Facets." Pp. 25–48 in *Emergence or Reduction? Essays on the Prospects of Nonreductive Physicalism,* edited by Ansgar Beckermann, Hans Flohr and Jaegwon Kim. Berlin: de Gruyter.

———. 1998. "Varieties of Emergence in Artificial and Natural Systems." *Zeitschrift für Naturforschung* 53c (Biosciences):639–56.

———. 1999. *Emergenz. Von der Unvorhersagbarkeit zur Selbstorganisation.* Dresden: Dresden University Press.

Stern, L. William. 1900. "Die psychologische Arbeit des neunzehnten Jahrhunderts, insbesondere in Deutschland." *Zeitschrift für Pädagogische Psychologie und Pathologie* 2:329–52, 413–36. [On Fechner: 341–44.]

———. 1901. "Fechner als Philosoph und Psychophysiker." *Zeitschrift für Pädagogische Psychologie und Pathologie* 3:405–7.

Stern, Peter von. 1967. *Das Leib-Seele-Problem bei Immanuel Hermann Fichte.* Ph.D. diss., phil. Fak., Universität München.

Stevens, Stanley Smith. 1934. "The Volume and Intensity of Tones." *American Journal of Psychology* 46:397–408.

———. 1946. "On the Theory of Scales of Measurement." *Science* 103:677–80. [Cited according to 1960. Pp. 141–49 in *Philosophy of Science,* edited by Arthur Danto and S. Morgenbesser. New York: World Publishing.]

———. 1951. "Mathematics, Measurement, and Psychophysics." Pp. 1–49 in *Handbook of Experimental Psychology,* edited by S. S. Stevens. New York: Wiley.

———. 1957. "On the Psychophysical Law." *Psychological Review* 64:153–81.

———. 1959. "Measurement, Psychophysics and Utility." *Measurement: Definitions and Theories,* edited by C. West Churchman and Philburn Ratoosh. New York: Wiley.

———. 1961. "To Honor Fechner and Repeal His Law." *Science* 133:80–86.

———. 1970. "Neural Events and the Psychophysical Law." *Science* 170:1043–50.

———. 1975. *Psychophysics. Introduction to Its Perceptual, Neural, and Social Prospects.*, edited by Geraldine Stevens. New York: Wiley, 1975. [Reprint 1986, with new introduction by Lawrence E. Marks. New Brunswick, N.J.: Transaction Books.]

Stigler, Stephen M. 1986. *The History of Statistics: The Measurement of Uncertainty before 1900.* Cambridge, Mass.: Harvard University Press.

———. 1987. "The Measurement of Uncertainty in Nineteenth-Century Social Science." Pp. 287–92 in *The Probabilistic Revolution.* Vol. 2: *Ideas in the Sciences.* Edited by Lorenz Krüger, Gerd Gigerenzer, and Mary Morgan. Cambridge, Mass.: MIT Press/Bradford Books.

———. 1999. *Statistics on the Table: The History of Statistical Concepts and Methods.* Cambridge, Mass.: Harvard University Press.

Stöltzner, Michael. 1999. "Vienna Indeterminism: Mach, Boltzmann, Exner." *Synthese* 119:85–111.

———. 2002. "Franz Serafin Exner's Indeterminist Theory of Culture." *Physics in Perspective* 2:267–319.

———. 2003. "Vienna Indeterminism II: From Exner to Frank and von Mises." In *Logical Empiricism: Historical and Contemporary Perspectives*, edited by Paolo Parrini, Wesley C. Salmon, and Merrilee H. Salmon. Pittsburgh.: University of Pittsburgh Press.

Strack, Friedrich. 1994. *Evolution des Geistes: Jena um 1800. Natur und Kunst, Philosophie und Wissenschaft im Spannungsfeld der Geschichte.* Stuttgart: Klett-Cotta.

Stracke, Alexander. 1979. *Zur Entstehung probabilistischer Theoriebildungen 1850–1910.* (Report Wissenschaftsforschung, vol. 17) Bielefeld: B. Kleine.

———. 1981. *Neuere Aspekte zum Prozeß probabilistischer Theoriebildungen. Eine kommentierte Bibliographie 1970–1980.* (Report Wissenschaftsforschung, vol. 21) Bielefeld: B. Kleine.

Stratilescu, Eleonora. 1903. *Die physiologische Grundlage des Seelenlebens bei Fechner und Lotze.* Ph.D. diss., phil. Fak., Friedrich-Wilhelms-Universität Berlin.

Strauss, David Friedrich. 1872. *Der alte und der neue Glaube.* Leipzig: Hirzel. [Cited according to the 12th to 14th. ed. 1895. Bonn: Emil Strauß.]

Strawson, Peter F. 1985. *Skepticism and Naturalism: Some Varieties.* The Woodbridge Lectures 1983. London: Methuen.

Strohl, Jean. 1936. *Lorenz Oken und Georg Büchner. Zwei Gestalten aus der Übergangszeit von Naturphilosophie zu Naturwissenschaft.* Zürich: Corona.

Stubenberg, Leopold. 1986. "Chisholm, Fechner und das Geist-Körper Problem." *Grazer philosophische Studien* 28:187–210.

———. 1997. "Austria vs. Australia: Two Versions of Identity Theory." Pp. 125–46 in *Austrian Philosophy Past and Present. Essays in Honor of Rudolf Haller.* (Boston Studies in the Philosophy of Science, vol. 190.) Edited by Keith Lehrer and Johann Christian Marek. Dordrecht: Kluwer.

Stumpf, Carl. 1896. "Leib und Seele. Rede zur Eröffnung des internationalen Kongresses für Psychologie. München, 4. August 1896." Pp. 65–93 in Stumpf 1910. *Philosophische Reden und Vorträge.* (Beiträge zur Geschichte der Psychologie, vol. 14) Leipzig: Barth. [Reprint 1997. Pp. 154–82 in *Carl Stumpf—Schriften zur Psychologie*, edited by Helga Sprung. Frankfurt am Main: Lang.]

Stumpf, Franz. 1925. *Die Gotteslehre von Hermann Lotze und Gustav Theodor Fechner: Eine vergleichende religionsphilosophische Untersuchung.* Ph.D. diss., phil. Fak., Universität Gießen.

Sturma, Dieter. 1998. "Reductionism in Exile? Herbert Feigl's Identity Theory and the Mind-Body Problem." *Grazer Philosophische Studien* 54:71–87.

Suck, R. 2001. "Measurement, Representational Theory of." Pp. 9442–48 in *International Encyclopedia of the Social and Behavioral Sciences*, edited by Neil J. Smelser and Paul B. Baltes. Amsterdam: Elsevier.

Sully, James. 1877. [Review of Fechner 1876.] *Mind*: 102–8.

Suppe, Frederick, ed. 1977. *The Structure of Scientific Theories*. 2nd ed. Urbana: University of Illinois Press.

Suppes, Patrick. 1980. "Messung." Pp. 415–23 in *Handbuch wissenschaftstheoretischer Begriffe*, edited by Josef Speck. Vol. 2. Göttingen: Vandenhoeck.

Swoboda, Wolfram W. 1974. *The Thought and Work of the Young Ernst Mach and the Antecedents of his Philosophy*. Ph.D. thesis, University of Pittsburgh.

———. 1982. "Physics, Physiology and Psychophysics: The Origins of Ernst Mach's Empiriocriticism." *Rivista di Filosofia* 73(22–23):234–74. [German translation 1988 in *Ernst Mach—Werk und Wirkung*, edited by Rudolf Haller and Friedrich Stadler. Vienna: Hölder.]

Taine, Hippolyte. 1870. *De l'intelligence*. 2 vols. Paris: Hachette. [German translation 1880 from the 3rd ed. by L. Siegfried as *Der Verstand*. 2 vols. Bonn: Emil Strauss.]

[Tannery, Jules.] 1875. "Correspondance. A propos du logarithme des sensations." *La Revue Scientifique* 2e série, 4e année, tome XIV, No 37 (13 mars):876–77. [This article appeared anonymously.]

[———.] 1875a. "La mesure des sensations. Réponses à propos du logarithme des sensations." *La Revue Scientifique* 2e série, 4e année, No 43 (24 avril):1018–20. [This article appeared anonymously.]

———. 1901. [Review of Foucault 1901.] *Bulletin des sciences mathématiques*, 2ème série, 25:101–13.

———. 1912. *Science et Philosophie*. Preface by Émile Borel. Paris: Félix Alcan. Reprint 1924. [Chap. VI "La Psychophysique" is a reprint of Tannery 1875, 1875a, and 1901.]

Tannery, Paul. 1884. "Critique de la Loi de Weber." *Revue philosophique* année 9, tome XVII:15–35.

———. 1886. "A propos de la loi de Weber." *Revue Philosophique*, tome XXI:386–87.

———. 1888. "Psychologie mathématique et psychophysique." [Review of Wernicke 1887, Du Bois-Reymond, Paul 1882, Elsas 1886 and Köhler 1886, among others.] *Revue philosophique* année 13, tome 25:189–97.

Teller, Paul. 1984. "A Poor Man's Guide to Supervenience and Determination." *The Southern Journal of Philosophy* 22(Suppl.):137–67.

Temkin, Owsei. 1950. "German Concepts of Ontogeny and History around 1800." *Bulletin of the History of Medicine* 24:227–46. [Also in Temkin 1977, 373–89.]

———. 1977. *The Double Face of Janus*. Baltimore: Johns Hopkins University Press.

Teo, Thomas. 2002. "Friedrich Albert Lange on Neo-Kantianism, Socialist Darwinism, and a Psychology without a Soul." *Journal of the History of the Behavioral Sciences* 38(3): 285–301.

Tetens, Holm. 1986. "Über Empfindungen. Ein sprachanalytischer Zugang." *Allgemeine Zeitschrift für Philosophie* 11(2):39–48.

Thénard, Louis-Jacques. 1813. *Traité de chimie élémentaire, théorique et pratique*. 4 vols. Paris: Crochand, 1813-16. [6th ed. 5 vols. 1827; 6th ed. 5 vols. 1834–36.]

Thiel, Christian. 1987. "Die Entmaterialisierung der Natur." Pp. 59–67 in *Zum Wandel des Naturverständnisses*, edited by Clemens Burrichter. Paderborn: Schöningh.

Thiel, Manfred. 1982. *Methode*. Vol. 6: *Emerson, Fechner, Feuerbach*. Heidelberg: Elpis.

Thiele, Joachim. 1963. "Ernst-Mach-Bibiliographie." *Centaurus* 8:189–237.

———. 1966. "Briefe von Gustav Theodor Fechner und Ludwig Boltzmann an Ernst Mach." *Centaurus* 11:222–35.

———. 1978. *Wissenschaftliche Kommunikation. Die Korrespondenz Ernst Machs.* Kastellaun: Henn.

Thines, Georges. 1987. "Joseph Delboeuf." *Psychologische Rundschau* 38(4):223.

Thurstone, Louis Leon. 1955. *The Measurement of Values.* Chicago: University of Chicago Press.

Tiberghien, Guy. 1984. *Initiation à la psychophysique.* Paris: Presses Universitaires de France.

Titchener, Edward Bradford. 1905. *Experimental Psychology.* Vol. 2: *Quantitative Experiments.* Part II: *Instructor's Manual.* London: Macmillan, 1905. [Pp. xiii–clxxi "Introduction: The Rise and Progress of Quantitative Psychology," predominantly on Fechner.]

———. 1922. "Mach's 'Lectures on Psychophysics.'" *American Journal of Psychology* 33:213–22.

Tögel, Christfried. 1988. "Fechner und Freuds Traumtheorie." Pp. 131–36 in *G. T. Fechner and Psychology,* edited by Josef Brožek and Horst Gundlach. Passau: Passavia.

Townsend, James T. 1975. "The Mind-Body Equation Revisited." Pp. 200–17 in *Philosophical Aspects of the Mind-Body Problem,* edited by Chung-ying Cheng. Honolulu: University Press of Hawaii.

Toyoda, Tashiyuki. 1997. "Essay on Quetelet and Maxwell. From *La physique sociale* to statistical physics." *Revue des questions scientifiques* 168:279–302.

Turner, R. Steven. 1974. "Helmholtz, Hermann von." Pp. 241–53 in *Dictionary of Scientific Biography.* Vol. 6. New York: Scribner's.

———. 1977. "Hermann von Helmholtz and the Empiricist Vision." *Journal of the History of the Behavioral Sciences* 13:48–58.

———. 1988. "Fechner, Helmholtz, and Hering on the Interpretation of Simultaneous Contrast." Pp. 137–50 in *G. T. Fechner and Psychology,* edited by Josef Brožek and Horst Gundlach. Passau: Passavia.

———. 1993. "Vision Studies in Germany: Helmholtz versus Hering." *Osiris* 8:80–113.

———. 1994. *In the Eye's Mind: Vision and the Helmholtz-Hering Controversy.* Princeton: Princeton University Press.

Ühlein, Herbert O., Jiří Hoskovec and Josef Brožek. 1994. "Deutschsprachige Psychologie in Prag." Pp. 113–22 in *Arbeiten zur Psychologiegeschichte,* edited by Horst Gundlach. Göttingen: Hogrefe.

Ulrici, Hermann. 1874. "Zur Streitfrage des Darwinismus." *Zeitschrift für Philosophie und philosophische Kritik* 65:1–12.

———. 1878. "Psychophysische Fragen und Bedenken." [Review of Fechner 1877 and G. E. Müller 1878.] *Zeitschrift für Philosophie und Philosophische Kritik* 72:281–310.

———. 1883. "Noch einmal die psychophysische Frage." [Review of Fechner 1882.] *Zeitschrift für Philosophie und Philosophische Kritik* 82:267–83.

Vaihinger, Hans. 1875. "Der Akademisch-Philosophische Verein zu Leipzig." *Philosophische Monatshefte* 11:190.

Vermorel, Madelaine and Henri Vermorel. 1986. "Was Freud a Romantic?" *International Review of Psycho-Analysis* 13:15–37.

Vogel, Stephan. 1988. "Fechner und Vierordt: Ein Beispiel für die Begründung und Etablierung der Psychophysik." Pp. 117–29 in *G. T. Fechner and Psychology,* edited by Josef Brožek and Horst Gundlach. Passau: Passavia.

Vogt, Carl. 1847. *Physiologische Briefe für Gebildete aller Stände.* 3 parts. Stuttgart: Cotta.

Volkmann, Wilhelm Fridolin, Ritter von Volkmar. 1856. *Grundriss der Psychologie vom Standpunkte des philosophischen Realismus und nach genetischer Metholde*. Halle: Fricke.

Von [Name], see [Name], Von

Vries, Herman de. 1994. *Remember Gustav Theodor Fechner: A reading-performance at the Royal Botanic Garden Edinburgh at the 16th of August 1992*. Eschenau: H. de Vries.

Vries, Hugo de. 1894. "Über halbe Galton-Curven als Zeichen diskontinuirlicher Variation." *Berichte der Deutschen Botanischen Gesellschaft* 12:197–207.

———. 1898. "Unity in Variability." *University of California Chronicle* 1:329–46.

Waitz, Theodor. 1852. "Der Stand der Parteien auf dem Gebiete der Psychologie." *Allgemeine Monatsschrift für Wissenschaft und Literatur:* 872–88, 1003–26.

Waldrich, Hans-Peter. 1993. *Grenzgänger der Wissenschaft. Hans Driesch, Gustav Theodor Fechner [...]*. München: Kösel.

Walker, Helen M. 1929. *Studies in the History of Statistical Method*. Baltimore: Williams and Wilkins.

Wall, Byron E. 1978. "F. Y. Edgeworth's Mathematical Ethics. Greatest Happiness with the Calculus of Variations." *The Mathematical Intelligencer* 1:177–81.

Ward, James. 1876. "An attempt to interpret Fechner's law." *Mind* 1:452–66.

———. 1902. "Psychology." Pp. 54–70 in *The Encyclopoedia Britannica: The New Volumes. Constituting with the existing vols. of the 9th ed. the 10th ed.* Vol. 8. (Vol. 32 of the 9th ed.) Edinburgh: Adam and Charles Black. [See especially "Relation of Body and Mind: Psychophysical Parallelism," 66–69.]

———. 1911. "Psychology." Pp. 547–604 in *The Encyclopoedia Britannica*. 11th ed. Vol. 22. Cambridge: Cambridge University Press. [Reworked version of Ward 1902. See "Relation of Body and Mind," 600–3.]

Warnke, Camilla. 1998. "Schellings Idee und Theorie des Organismus und der Paradigmawechsel der Biologie um die Wende zum 19. Jahrhundert." *Jahrbuch für Geschichte und Theorie der Biologie* 5:187–234.

Warren, Richard M. 1981. "Measurement of Sensory Intensity." *Behavioral and Brain Sciences* 4:175–223. [With 31 "open peer commentaries" 189–213 and "author's response" 213–23.]

Wasserman, Gerald S., Gary Felsten, and Gene S. Easland. 1979. "The Psychophysical Function: Harmonizing Fechner and Stevens." *Science* 204:85–87.

Watt, Henry. 1913. "Are the Intensity Differences of Sensation Quantitative?" *The British Journal of Psychology* 6:175–83.

Weatherford, Roy. 1982. *Philosophical Foundations of Probability Theory*. London: Routledge.

Weber, Ernst Heinrich. 1834. *De pulsu, resorptione, auditu et tactu. Annotationes anatomicae et physiologicae*. Leipzig: C. F. Koehler.

———. 1835. "Ueber den Tastsinn." *Archiv für Anatomie, Physiologie und wissenschaftliche Medizin:* 152–59.

———. 1846. "Tastsinn und Gemeingefühl." Pp. 481–588 in *Handwörterbuch der Physiologie mit Rücksicht auf physiologische Pathologie*, edited by Rudolph Wagner. Vol. 3, Abt. 2. Braunschweig: Vieweg. [Cited according to Ernst Heinrich Weber 1905. *Tastsinn und Gemeingefühl*. (Ostwald's Klassiker der exakten Wissenschaften, Nr. 149) Edited by Ewald Hering. Leipzig: Wilhelm Engelmann.]

———. 1996. *E. H. Weber on the Tactile Senses*. Edited and translated by Helen E. Ross and David J. Murray. 2nd ed. Hove, East Sussex: Erlbaum/Taylor and Francis, 1996. [Translation of Weber 1834 and 1846.]

Weber, Wilhelm. 1846. "Elektrodynamische Maassbestimmungen insbesondere über ein allgemeines Grundgesetz der elektrischen Wirkung." *Abhandlungen bei Begründung der Königlich Sächsischen Gesellschaft der Wissenschaften:* 209–378. [Cited according to 1893. Pp. 25–214 in *Wilhelm Webers Werke,* edited by Heinrich Weber. Vol. 3, erster Theil. Berlin: Springer, 1893.]

———. 1850. "Zwei Briefe von Wilhelm Weber an G. Th. Fechner über das psychische Maß. Auszugsweise veröffentlicht von Gottlob Friedrich Lipps." [Two letters from December 12, 1850 and January 15, 1851] *Berichte über die Verhandlungen der Königlich Sächsischen Gesellschaft der Wissenschaften,* mathematisch-physische Klasse, 57:388–95. [Cp. Fechner 1860, II, 557f.]

Weinhandl, Ferdinand. 1927. *Die Gestaltanalyse.* Erfurt: Kurt Stenger. [Pp. 159–83 on Fechner.]

Weiss, Georg. 1928. *Herbart und seine Schule.* Munich: Reinhardt.

Weisse, Christian Hermann. 1829. *Über den gegenwärtigen Standpunct der philosophischen Wissenschaften. In besonderer Beziehung auf das System Hegels.* Leipzig: Barth.

———. 1830. *System der Ästhetik als Wissenschaft von der Idee der Schönheit.* 2 vols. Leipzig: Hartmann.

———. 1833. *Die Idee der Gottheit. Eine philosophische Abhandlung. Als wissenschaftliche Grundlegung zur Philosophie der Religion.* Dresden: Grimmer.

———. 1835. *Grundzüge der Metaphysik.* Hamburg: Perthes.

———. 1841. "Ueber die metaphysische Begründung des Raumbegriffs. Antwort an Herrn Dr. Lotze." *Zeitschrift für Philosophie und spekulative Theologie* 8:25–70.

———. 1855. "Ueber die Gränzen des mechanischen Princips der Naturforschung." [Review of Fechner 1855.] *Zeitschrift für Philosophie und philosophische Kritik* 27:97–146, 192–227.

Weizsäcker, Carl Friedrich von. 1972. "Evolution und Entropiewachstum." *Nova Acta Leopoldina* (Edited by Joachim-Hermann Scharf) 37(1):515–30.

Weizsäcker, Viktor von. 1922. "Einleitung." Pp. 5–17 in *Fechner, Tagesansicht und Nachtansicht.* (Frommanns philosophische Taschenbücher, vol. 4) Stuttgart: Frommann.

Wentscher, Else. 1903. *Das Kausalproblem in Lotzes Philosophie.* (Abhandlungen zur Philosophie und ihrer Geschichte, vol. 16) Halle: Max Niemeyer. [Pp. 302–16 on Fechner.]

———. 1911. "Aus Gustav Theodor Fechners Gedankenwelt." *Monatshefte der Comenius-Gesellschaft für Kultur und Geistesleben* 20 (N. F. 3) (1911):49–57.

———. 1921. *Geschichte des Kausalproblems in der neueren Philosophie.* Leipzig: Meiner, 1921.

Wentscher, Max. 1896. *Über physische und psychische Kausalität und das Prinzip des psychophysischen Parallelismus.* Leipzig: J. A. Barth.

———. 1900. "Der psycho-physische Parallelismus in der Gegenwart." *Zeitschrift für Philosophie und philosophische Kritik* 116(1):103–20.

———. 1913. *Hermann Lotze.* Vol. 1: *Lotzes Leben und Werke.* Heidelberg: Winter.

———. 1924. *Fechner und Lotze.* Munich: Reinhardt. [Reprint 1973. Nendeln: Kraus.]

Wenzl, Aloys. 1933. *Das Leib-Seele-Problem im Lichte der neueren Theorien der physischen und seelischen Wirklichkeit.* Leipzig: Meiner.

Wernekke, H. 1901. "Gustav Theodor Fechner." *Psychische Studien* 28(4):193–203.

Wernicke, A. 1883. [Review of Ferdinand August Müller 1882.] *Vierteljahrsschrift für wissenschaftliche Philosophie* 7:213–19.

———. 1886. [Review of Fechner 1882.] *Vierteljahrsschrift für wissenschaftliche Philosophie* 10(2):216–21.

———. 1887. [Review of Ferdinand August Müller 1886.] *Vierteljahrsschrift für wissenschaftliche Philosophie* 11:229–41.

Wetzels, Walter D. 1973. *Johann Wilhelm Ritter: Physik im Wirkungsfeld der deutschen Romantik*. Berlin: de Gruyter.

Whistling, Karl W. 1887. "Gustav Theodor Fechner." *Leipziger Tagblatt* (November 22): 6634. [Courtesy Stadtarchiv Leipzig]

Wiederkehr, Karl Heinrich. 1960. *Wilhelm Webers Stellung in der Entwicklung der Elektrizitätslehre*. Ph.D., diss., math.-nat. Fak., Universität Hamburg. [Doctoral advisor: Schimank.]

———. 1967. *Wilhelm Eduard Weber. Erforscher der Wellenbewegung und der Elektrizität 1804–1891*. Stuttgart: Wissenschaftliche Verlagsgesellschaft.

———. 1994. "Wilhelm Weber und Maxwells elektromagnetische Lichttheorie." *Gesnerus* 51:256–67.

Wiener, Christian. 1892. "Die Empfindungseinheit zum Messen der Empfindungsstärke." *Annalen der Physik und Chemie*, Neue Folge 47:659–70.

Wieser, Johann E. 1881. *Der Spiritismus und das Christenthum*. (Separatabdruck aus der "Zeitschrift für kathol. Theologie." Mit einer Beilage über Dr. G. T. Fechners "Tagesansicht".) Regensburg.

Wille, Bruno. 1905. *Das lebendige All. Idealistische Weltanschauung auf naturwissenschaftlicher Grundlage im Sinne Fechners*. Hamburg/ Leipzig: Leopold Voß.

Wilson, Andrew D. 1997. "Die romantischen Naturphilosophen." Pp. 319–35 in *Die grossen Physiker*. Vol. 1: *Von Aristoteles bis Kelvin*. Munich: Beck.

Windelband, Wilhelm. 1878. "Ueber experimentale Aesthetik." *Im neuen Reich* 8(1):601–16.

———. 1910. "Fechner." *Allgemeine Deutsche Biographie* (Munich) 55:756–63. Reprint 1971. Berlin: Duncker und Humblot.

Winter, H. J. J. 1948. "The Work of G. T. Fechner on the Galvanic Circuit." *Annals of Science* 6:197–205.

Wirth, Johann Ulrich. 1854. "Uebersicht der philosophischen Literatur." [Review of Fechner 1851, see 292–94 and 317–21.] *Zeitschrift für Philosophie und philosophische Kritik* 24:281–321.

Wirth, Wilhelm. 1900. "Der Fechner-Helmholtz'sche Satz über negative Nachbilder und seine Analogien." *Philosophische Studien* 16(1):465–567; 17:311–430. [Also separately as 1900 habilitation thesis, Leipzig: Engelmann.]

———. 1912. *Psychophysik. Darstellung der Methoden der experimentellen Psychologie*. (Handbuch der physiologischen Methodik, vol. 3, Abt. 5. Edited by Robert Tigerstedt) Leipzig: Hirzel.

———. 1929. "Gustav Theodor Fechner. Rede zur Einweihung der Gedenktafel an seinem Geburtshause in Groß-Särchen in der Niederlausitz am 20. Oktober 1929." *Aus der Heimat. Beilage zum Forster Tageblatt*. Edited by Verein für Heimatkunde zu Forst (Lausitz) Nr. 12.

———. 1938. "Gustav Theodor Fechner." Pp. 97–113 in *Sächsische Lebensbilder*, edited by Sächsische Kommission für Geschichte. Vol. 2. Leipzig: O. Leiner.

———. 1938a. "Eine Episode aus G. Th. Fechners Leben vor sechzig Jahren." *Otto Glaunig zum 60. Geburtstag. Festgabe aus Wissenschaft und Bibliothek*. Vol. 2. Leipzig: Richard Hadl. [Passages from Fechner's notebooks of 1877, on Thomas Masaryk's visit with Fechner in Leipzig in 1876 or 1877. See FN, Nachl. 41, 1877, pp. 18–23 of July 13, 1877 and Masaryk's letter to Meinong, in Meinong 1965, 4.]

Wise, M. Norton. 1981. "German Concepts of Force, Energy, and the Electromagnetic Ether: 1845–1880." Pp. 269–307 in *Conceptions of Ether: Studies in the History of Ether Theories, 1740–1900*. Cambridge: Cambridge University Press.

————. 1982. "Atomism and Wilhelm Weber's Concept of Force." Pp. 57–66 in *Atomvorstellungen im 19. Jahrhundert*, edited by Charlotte Schönbeck. Paderborn: F. Schöningh.

————. 1987. "How Do Sums Count? On the Cultural Origins of Statistical Causality." Pp. 395–425 in *The Probabilistic Revolution*. Vol. 1. Edited by Lorenz Krüger, Lorraine J. Daston, and Michael Heidelberger. Cambridge, Mass.: MIT Press/Bradford Books.

————. 1990. "Electromagnetic Theory in the Nineteenth Century." Pp. 342–56 in *Companion to the History of Science*, edited by Robert C. Olby et al. London: Routledge.

Wislicenus, Johannes. 1897. "Rede." Pp. vii–xxi in *Zur fünfzigsten Jubelfeier der Königlich Sächsischen Gesellschaft der Wissenschaften zu Leipzig am 1. Juli 1896. Reden und Register*. Leipzig: Hirzel.

Witting, Hermann. 1990. "Mathematische Statistik." Pp. 781–15 in *Ein Jahrhundert Mathematik 1890–1990*. (Dokumente zur Geschichte der Mathematik, vol. 6.) Edited by Gerd Fischer et al. Braunschweig: Vieweg.

Wittkau-Horgby, Anette. 1998. *Materialismus. Entstehung und Wirkung in den Wissenschaften des 19. Jahrhunderts*. Göttingen: Vandenhoeck.

Wohlgemuth, Joseph. 1924. "Fechner's Seelenlehre und das jüdisch-religiöse Weltbild." *Jeschurun. Monatsschrift für Lehre und Leben im Judentum* 11:477–501; 12:1–34.

Wohlgemuth, Rolf. 1993. *Charles S. Peirce. Zur Begründung einer Metaphysik der Evolution*. Frankfurt am Main: Peter Lang.

Wolters, Gereon. 1988. "Verschmähte Liebe. Mach, Fechner und die Psychophysik." Pp. 103–16 in *G. T. Fechner and Psychology*, edited by Josef Brožek and Horst Gundlach. Passau: Passavia.

————. 1988a. "Atome und Relativität—Was meinte Mach?" Pp. 484–507 in *Ernst Mach— Werk und Wirkung*, edited by Rudolf Haller and Friedrich Stadler. Vienna: Hölder.

Woodruff, A. E. 1962. "Action at a Distance in Nineteenth Century Electrodynamics." *Isis* 53:439–59.

Woodward, William R. 1972. "Fechner's Panpsychism: A Scientific Solution to the Mind-Body Problem." *Journal of the History of the Behavioral Sciences* 8:367–86.

————. 1975. *The Medical Realism of R. Hermann Lotze*. Ph.D. diss., Yale University.

Woodward, William R. and Mitchell G. Ash, eds. 1982. *The Problematic Science: Psychology in Nineteenth Century Thought*. New York: Praeger.

Worcester, Elwood. 1908. *The Living Word*. New York: Moffat, Yard and Co.

Wozniak, Peter. 1998. "The Organisational Outline of the Gymnasia and Technical Schools in Austria and the Beginning of Modern Educational Reform in the Habsburg Empire." Pp. 79–108 in *Zwischen Orientierung und Krise. Zum Umgang mit Wissen in der Moderne*. (Studien zur Moderne, Bd. 2), edited by Sonja Rinofner-Kreidl. Vienna: Böhlau.

Wundt, Wilhelm. 1862. *Beiträge zur Theorie der Sinneswahrnehmung*. Leipzig and Heidelberg: Winter.

————. 1863. *Vorlesungen über die Menschen- und Thierseele*. 2 vols. Leipzig: Voß. [Cited according to reprint 1990. Wolfgang Nitsche. Berlin: Springer. Translation 1998 from the 2nd German ed. by J. E. Creighton and Edward B. Titchener as *Lectures on Human and Animal Psychology*. Bristol: Thoemmes, 1998.]

————. 1875. "La mesure des sensations. Réponses à propos du logarithme des sensations." *La Revue Scientifique* 2e série, 4e année, No 43 (24 avril):1017–18.

————. 1877. "Philosophy in Germany." *Mind* 2(8):493–518.

————. 1880. "Gehirn und Seele." *Deutsche Rundschau* 25:47–72. [Also in Wilhelm Wundt 1885. Pp. 88–126 in *Essays*. Leipzig: Engelmann.]

————. 1883. "Über die Messung psychischer Vorgänge." *Philosophische Studien* 1:251–60. "Weitere Bemerkungen über psychische Messung." 461–71.

————. 1885. "Ueber das Weber'sche Gesetz." *Philosophische Studien* 2(1):1–36.

————. 1887. "Zur Erinnerung an Gustav Theodor Fechner. Worte, gesprochen an seinem Sarge am 21. November 1887." In Johannes Emil Kuntze 1892. Pp. 351–61 in *Gustav Theodor Fechner (Dr. Mises). Ein deutsches Gelehrtenleben.* Leipzig: Breitkopf und Härtel. [Also in *Philosophische Studien* 4 (1888):471–78.]

————. 1894. "Über psychische Causalität und das Princip des psychophysischen Parallelismus." *Philosophische Studien* 10:1–124.

————. 1901. *Gustav Theodor Fechner. Rede zur Feier seines hundertjährigen Geburtstages.* Leipzig: W. Engelmann, 1901. [Cited according to reprint 1914. Pp. 254–343 inWilhelm Wundt, *Reden und Aufsätze.* 2nd ed. Leipzig: A. Kröner.]

————. 1904. "Die Psychologie im Beginn des zwanzigsten Jahrhunderts." *Die Philosophie im Beginn des zwanzigsten Jahrhunderts. Festschrift für Kuno Fischer,* edited by Wilhelm Windelband. Heidelberg: Winter. [Cited according to the 2nd ed. 1907, 1–57. Reprint 1914. Pp. 163–231 in Wilhelm Wundt, *Reden und Aufsätze.* 2nd ed. Leipzig: A. Kröner.]

————. 1908. *Grundzüge der physiologischen Psychologie.* 6th rev. ed., 3 vols. Leipzig: Engelmann. [1st ed., 1 volume 1874. Translation 1904 by Edward B. Titchener (from the 5th German ed. 1902) as *Principles of Physiological Psychology.* London: Swan Sonnenschein and Co. and New York: Macmillan. Reprint 1998. Bristol: Thoemmes.]

————. 1914. "Fechners Tages- und Nachtansicht." Pp. 301–3 in *Sinnliche und übersinnliche Welt.* Leipzig: A. Kröner.

————. 1920. *Erlebtes und Erkanntes.* Stuttgart: A. Kröner.

Zeller, Eduard. 1881. "Über die Messung psychischer Vorgänge." *Philosophische und historische Abhandlungen der Königlichen Akademie der Wissenschaften zu Berlin,* philosophisch-historische Classe, aus dem Jahre II:3–16. [Also 1910. Pp. 488–502 in *Eduard Zellers Kleine Schriften.* Vol. 2. Edited by Otto Leuze. Berlin: Reimer.]

————. 1882. "Einige weitere Bemerkungen über die Messung psychischer Vorgänge." *Sitzungsberichte der Königlich Preussischen Akademie der Wissenschaften zu Berlin,* philosophisch-historische Classe, Stück 15 (March 16):295–305. [Also 1910. Pp. 503–15 in *Eduard Zellers Kleine Schriften.* Vol. 2. Edited by Otto Leuze. Berlin: Reimer.]

Ziegler, Theobald. 1899. *Die geistigen und sozialen Strömungen des neunzehnten Jahrhunderts.* Berlin: Bondi. [Cited according to reprint 1911.]

Zimmermann, Rudolf. 1863. "Die Metaphysik in der Naturwissenschaft." *Oesterreichische Wochenschrift für Wissenschaft, Kunst und öffentliches Leben* 1:481ff. [Cited according to Robert Zimmermann 1870. Pp. 341–47 in *Studien und Kritiken zur Philosophie und Aesthetik.* Vol. 1: *Zur Philosophie. Studien und Kritiken.* Vienna: Wilhelm Braumüller.]

————. 1864. "Ueber philosophische Atomistik." *[Prager] Vierteljahrschrift für praktische Heilkunde* 21(1). [Cited according to Robert Zimmermann 1870. Pp. 347–63 in *Studien und Kritiken zur Philosophie und Aesthetik.* Vol. 1: *Zur Philosophie. Studien und Kritiken.* Vienna: Wilhelm Braumüller.]

Zöllner, (Johann) Karl Friedrich. 1870. "Ueber die Periodicität und heliographische Verbreitung der Sonnenflecken." In Karl Friedrich Zöllner 1881. Pp. 69–82 in *Wissenschaftliche Abhandlungen.* Vol. 4. Leipzig: Staackmann.

————. 1872. *Über die Natur der Kometen. Beiträge zur Geschichte und Theorie der Erkenntniss.* 2nd ed. Leipzig: Engelmann. [3rd ed. 1883 Leipzig: Staackmann and 1886 Leipzig: Engelmann.]

————. 1879. *Transcendentale Physik und die sogenannte Philosophie. Eine deutsche Antwort auf eine "sogenannte wissenschaftliche Frage".* (*Wissenschaftliche Abhandlungen*, vol. 3) Leipzig: Staackmann. [Translation 1880 by Charles Carleton Massey as *Transcendental Physics: An Account of Experimental Investigations from the Scientific Treatises of Johann Carl Friedrich Zöllner.* London: W. H. Harrison. (Also 1888. Boston: Colby and Rich.) Reprint 1976. New York: Arno.]

INDEX OF NAMES

Achelis, Th., 367
Adler, H. E., 10, 13, 367
Adolph, H., 13, 367
Allesch, Ch. G., 11, 333, 367
Ampère, A.-M., 28–29, 50, 141–42, 147, 382
Anderson, D. R., 367
Apel, K., 359–60, 368
Archibald, Th., 343, 368
Armstrong, D. M., 106, 339, 347, 368
Arnheim, R., 333, 368
Arnim, B. von, 329
Ash, M. G., 13, 368, 410, 436
Aubert, H. R., 59, 368
Avenarius, R., 183–84, 264, 267, 358n56, 385

Baer, K. E. von, 293, 368
Bähr, K., 330n94, 368
Bain, A., 169, 179
Balance, W. D. G., 50, 69, 373
Baumgartner, 14, 369, 381
Bauschinger, J., 364n138, 369
Beatty, J., 388
Becher, E., 369, 428
Beckermann, A., 369, 429
Behrens, P. J., 369
Ben-David, J., 319, 369
Bergmann, G., 182
Bergson, H., 12, 210, 352n60, 369–70, 397
Berkeley, G., 77, 124, 370
Bernard, C., 263, 389
Bernfeld, S., 370
Bernhardt, H., 370
Bernoulli, Jak., 305, 311
Bernstein, J., 59, 212, 370
Bertrand, J., 312, 365n147, 370
Berzelius, J., 370
Bessel, F. W., 297, 310, 364n128, 370
Besso, M., 238
Bichat, X., 360n3, 370
Bienaymé, I. J., 362n89
Bieri, P., 338n73, 370
Billroth, 45
Binswanger, L., 358n64, 370
Biot, J.-B., 27–29, 141, 322, 370, 381–82, 387
Bismarck, O., 214

Blackmore, J. T., 14, 345n70, 371
Blumenthal, A. L., 371
Boas, F., 371
Bodio, M. L., 362n89
Bohr, N., 14, 177, 193, 371, 400
Bölsche, W., 12, 261, 357n42, 371
Boltzmann, L., 263, 315–16, 377, 400, 430, 432
Bolzano, B., 44
Bonitz, M., 371, 405
Bonsiepen, W., 14, 371
Boring, E. G., 11, 59, 176, 210, 372, 383, 418, 422
Boscovic, R., 147
Brandes, H. W., 27–28
Brandes, K. W. H., 49
Brandt, R., 372
Brasch, M., 372
Braun, H., 63
Brauns, H.-P., 13, 372–73, 372
Breitkopf, J. G. I. and Ch. G., 43–44, 63, 128, 331n118, 383–84
Brentano, B. (B. von Arnim), 329n85
Brentano, F., 62–63, 69, 181–82, 202, 216, 243–44, 266, 333n185, 334nn190, 197, 335n181, 373
Breuer, J., 62, 182, 266–67, 358n53, 373, 397, 400
Bringmann, N. J., 326
Bringmann, W. G., 10, 50, 69, 373
Broad, C. D., 165
Brockhaus, H., 43
Brown, R., 143, 315
Brown, W., 373
Brozek, J., 10, 373
Bruchmann, K., 373
Bruns, H., 308, 310–13, 364n138, 365n143, 369, 373, 430
Brush, St. G., 317, 374
Buchdahl, G., 343n14, 374
Büchner, L., 56, 151, 154, 163, 167–68, 374, 393
Buck, P., 374
Buek, O., 374
Buggle, F., 374
Bunge, M., 374
Burdach, K. F., 293–96, 298, 362n82, 374

Camerer, (J.) W., 70, 335n232, 374
Caneva, K. L., 14, 375

Cannon, W. B., 263
Carnap, R., 89–90, 167–68, 183, 186–190, 231, 375
Carrier, M., 375
Carriere, M., 375
Carus, C. G., 362n85, 375
Case, Th., 169, 375
Cassirer, E., 183, 356n185, 376
Cauchy, A.-L., 27, 29, 147
Chaitin, G. J., 365n159, 375
Charlier, C. W. L., 364n138
Chisholm, R. M., 336n20, 376
Clausius, R., 309
Cohen, H., 13, 35, 213–32, 235, 353n88, 376
Coleman, W., 376
Collins, R., 319
Coulomb, C. A., 142
Cournot, A.-A., 351n38
Cramér, H., 396
Cramer, K., 376
Cranefield, P. F., 376
Crosland, M., 376
Cummins, R., 376
Cunningham, A., 376, 421
Curd, M. V., 377
Curtius, 53
Czolbe, H., 56
Czuber, E., 312–14, 377

Dalton, J., 143
Darwin, Ch., 168, 173, 181, 250, 252, 258, 260–63, 266, 268, 270, 357n42, 357n43, 373, 396, 400–1, 414, 431–32
Daston, L., 388, 390, 392, 400, 402, 405, 419, 425, 436
Davidson, D., 99, 182, 377, 393
Delboeuf, J. R., 99, 208, 210, 213–14, 352n69, 377, 415, 432
Dennert, E., 377
Descartes, R., 31, 80, 166, 169, 171, 173, 221
Dilthey, W., 179–80
Dittenberger, W., 378
Dittmar, C., 378
Döllinger, I., 293
Dorer, L. M., 378
Dorer, M., 378
Döring, D., 10, 378
Dormoy, E., 365n147
Dove, H. W., 362n89
Driesch, H., 378, 433
Drobisch, M. W., 31, 35, 154, 298, 378, 404
Du Bois-Reymond, E., 119, 163, 225, 231, 259, 263, 285, 291–92, 340n3, 370, 378–80, 431

Du Bois-Reymond, P., 378
Dubos, R., 379
Duncker, G., 369, 373, 379, 391, 393, 396–97, 435

Easland, G. S., 433
Ebbinghaus, H., 180–81, 242, 379, 410, 421
Ebner-Eschenbach, M. von, 266, 400
Ebrecht, A., 11, 333n185, 379
Eccles, J., 429
Edel, G., 379
Edgeworth, F. Y., 332n157, 364n138, 379, 425, 433
Edwards, W., 408
Ehrenfels, Ch. von, 379
Eigen, M., 8, 272, 379
Einstein, A., 78, 143, 177, 238, 315, 372, 379, 386, 396, 400
Elderton, W. P., 379
Ellenberger, H. F., 10, 50, 379
Elliott, E. B., 362n89
Ellis, B., 247, 379, 409
Elsas, A., 229–32, 234, 239, 246, 376, 380, 384, 398, 404, 431
Encke, J. F., 297, 306
Engelhardt, D. von, 380
Engels, F., 25
Erdmann, B., 35, 181, 380, 396
Erdmann, J. E., 25, 380
Eshleman, C. H., 380
Euclid, 67–68, 78
Exner, F., 64, 334n199, 380, 390
Exner, F. S., 316, 334n199, 400, 430
Exner, S., 334n199, 355n180, 380

Falmagne, J.-C., 14, 381
Faraday, M., 50, 141, 374, 382
Fechner, Clara Maria, 322
Fechner, Clementine, 19, 329n85
Fechner, Eduard Clemens, 19, 321
Fechner, Emilie, 9, 19
Fechner, Johanna Dorothea, 321
Fechner, Mathilde, 329n85
Fechner, Samuel Traugott, 19
Feigl, H., 7, 165–67, 174, 182–83, 186, 188–90, 349n76, 367, 385–86, 388, 393, 419, 431
Feitelberg, S., 370
Felsten, G., 433
Feuerbach, L., 25, 39, 47, 432
Feyerabend, P., 3, 385–86
Fichte, I. H., 14, 38–40, 44, 58, 386, 398, 408, 425, 429
Fichte, J. G., 32, 35
Fiedler, J. K., 386

Fine, T. L., 365*n158*, 386
Fischer, G. E., 331
Fischer, J. D., 19, 331
Flach, W., 354*n120*, 386
Flohr, H., 369, 429
Fodor, J. A., 54, 386
Foerster, H. von, 386
Foucault, M., 12, 386, 431
Fraassen, B. C. van, 342*n2*, 387
Frank, M., 89, 387
Frankel, E., 328*n41*, 387
Fraser, A. C., 344*n62*
Frege, G., 330*n106*, 427
Fresnel, A. J., 29
Freud, S., 8, 10–11, 46, 54, 63–64, 87, 182, 249,
      260, 265–66, 268, 334*n197*, 358*n64*, 358*n67*,
      359*n69*, 370, 373, 374, 376, 379, 387, 388, 395,
      406, 415, 421, 432
Freudenberg, G., 387
Freudenreich, H., 387
Friedlein, C., 387
Friedman, M., 354*n144*, 387
Fries, J. F., 58, 371
Fritsch, P., 357*n42*, 387
Fuchs, 208
Fügner, F., 334*n192*, 387
Funkenstein, A., 353*n100*, 387
Fürnrohr, A. E., 387

Galanter, E., 408
Galton, F., 298, 306–7, 315, 362*n89*, 433
Gardner, M. R., 343*n14*, 388
Gauss, C. F., 60, 230, 297–99, 306–7, 310,
      363*n102*, 364*nn128, 131*
Gebhard, W., 47, 327*n9*, 329*n63*, 331*nn121, 128*,
      388
Geiringer, H., 366*n164*, 388
Gerland, G., 159, 249, 261, 267, 359*n77*, 388
Gescheider, G. A., 14, 350*n30*, 35*nn45–46*, 388
Gigerenzer, G., 325*n10*, 388
Gilbert, L. W., 27–28
Glansdorff, P., 272, 419
Gödde, G., 13, 334*n193*, 388
Goethe, J. W. von, 21, 36, 122, 389
Gomperz, Th., 63
Gould, B. A., 362*n89*
Gould, S. J., 327*n20*, 389
Gower, B., 327*n12*, 389
Gräfe, A. von, 67, 322
Grassmann, R., 159
Gregory, Fr., 333*n171*, 346–47*n7*, 389
Gregory, R. L., 350*nn7, 30*, 389
Greiner, R., 420

Grotenfelt, A., 202, 389
Guettler, C., 25, 327*nn15, 25*, 389
Gundlach, H., 10, 13, 325*n11*, 326*nn1–2*,
      333*nn164, 176*, 389
Gutberlet, C., 12, 350*n12*, 389

Haarhaus, J. R., 390
Hacking, I., 359*n83*, 390
Haeckel, E., 12, 25, 61, 159, 168, 260–62, 319,
      326*n35*, 327*n25*, 390, 416
Halbfass, W., 490
Hall, G. S., 70, 335*nn227, 230*, 390
Haller, R., 344*n65*, 390
Hamilton, W., 91, 412
Hanitzsch, R., 426
Hankel, W. G., 333*n164*
Hanle, P. A., 390
Hansen, K., 67
Harms, F., 362*n68*
Härtel, H., 43–44, 63, 66
Hartenstein, G., 35
Hartmann, A., 391
Hartmann, E. von, 13, 62, 65, 68, 159, 319,
      344*n64*, 345*n91*, 376, 391
Hartung, W., 12, 357*n25*, 340*n117*, 357*n25*, 391
Hartungen, H., 391
Harvey, W., 293
Hasenöhrl, F. von, 334*n199*
Hastedt, H., 325*n6*, 339*n104*, 340*n6*, 391
Hatfield, G., 13–14, 329*n65*, 347*n27*, 391
Hauber, K. F., 297
Hausdorff, F., 388, 391
Hegel, G. W. F., 1–2, 8, 21, 23, 38–44, 51, 56, 65,
      92, 121, 148, 150, 285–87, 292–93, 317, 323,
      329*n63*
Heidegger, M., 2
Heinze, M., 330*n106*, 393
Helm, G., 312, 394
Helmholtz, H. von, 14, 29–30, 35, 42, 57–59,
      68–69, 148, 162, 178, 211–13, 218, 224–25,
      231–34, 243, 263, 277, 285, 290–92, 346*n104*,
      354*n145*, 362*n68*, 370, 374, 376, 394–95
Hemecker, W. W., 334*nn191, 193, 194, 201*, 395
Hempel, C. G., 358*n61*
Henderson, L. J., 263, 361*n54*, 395
Hensel, P., 395
Hentschel, K., 358*n61*, 395
Herbart, J. F., 7, 21, 31–35, 41–42, 58, 64, 92, 104,
      115, 148, 150, 154, 156–60, 162–64, 218, 222,
      289, 296, 298, 372–73, 378, 383, 387, 395
Herder, J. G., 293
Hering, E., 14, 64, 182, 216, 224, 243, 248,
      262–64, 266, 369, 374, 377, 395–96

Hermann, D. B., 396
Hermann, I., 396
Herneck, F., 396, 410
Herschel, J. F. W., 120
Hertz, H., 178, 393–95, 416
Herzberg, A., 268, 396
Heuser-Kessler, M.-L., 327*nn*12, *15*, 396–97
Hicks, D., 397
Hiebert, E. N., 397
Hildebrandt, H., 11, 423
Hill, J. A., 397
Hillebrand, F., 373, 397
Hirschmüller, A., 373, 397
Hirzel, Ch. H., 44, 407–8, 430, 435–36
Hochfeld, S., 397
Höffding, H., 177, 182, 193, 397
Hoffmann, D., 10, 398
Höfler, A., 182, 398
Hofmannsthal, H. von, 378
Holbach, P. H. Th. d', 36
Holbein, H., 14, 61, 369
Hollander, K. von, 398
Holmes, F. L., 358*n*54, 359*n*83, 398
Holton, G., 355*n*164, 398
Holzhey, H., 398
Horkheimer, M., 318, 398
Hornstein, G. A., 212, 398
Horstmeier, M., 398
Hoskovec, J., 373, 398, 432
Howard, J. V., 361*n*33, 399
Humboldt, A. von, 36, 56, 370, 394, 396, 399
Hume, D., 77, 86, 94, 336*n*23, 399
Husserl, E., 63, 181, 399, 408, 428

Itelson, G., 399

Jaeger, S., 399, 406
Jaensch, E., 399
James, W., 4, 10–11, 13, 91, 170, 175, 180–81, 187, 211, 268, 323, 348*n*54, 368, 375, 386, 388, 391, 399
Jammer, M., 399
Jardine, N., 376, 421
Jastrow, J., 268, 369, 399, 416
Jean Paul, 36–37, 334*n*204, 399
Jerusalem, W., 182
Jodl, Fr., 182, 400
Jungnickel, Ch., 400

Kaiser, W., 342*n*4, 343*nn*13, *14*, *17*, 358*n*60, 400
Kamlah, A., 362*n*88, 365*n*150, 400

Kamminga, H., 356*n*1, 357*n*46, 400
Kanitscheider, B., 325*n*5, 400
Kann, R., 359*nn*72, *76*, 400
Kant, I., 7, 12–14, 22, 25, 40, 52, 58, 64–65, 68, 92, 138, 140, 143–44, 147, 150, 166–68, 175, 17–79, 181–82, 184–85, 192, 205, 211–18, 223–25, 228–31, 234, 236, 239–41, 243–45, 253, 258–59, 261, 277, 282, 291–92, 296, 357*n*30, 371, 375–76, 380, 387–88, 391, 400–1, 405, 410, 425, 427, 429, 435
Karlik, B., 334*n*199, 366*n*161, 400
Karsten, G., 29
Kaufmann, Al., 400
Kay, A. S., 362*n*74, 400
Kelly, A., 357*n*42, 400
Kendall, M. G., 363*n*123
Kierkegaard, S., 39
Kietz, M., 329*n*85
Kietz, Th., 329*n*85
Kim, J., 165–66, 369, 372, 401, 429
Kirchhoff, G. R., 344*n*58
Klimke, F., 339*n*105, 401
Knight, D. M., 327*n*12, 401
Koenig, E., 309–10, 401
Koffka, K., 268, 401
Köhler, A., 401
Köhler, W., 267, 401
Kohlrausch, R., 142
Köhnke, K. Ch., 14, 214, 401
Kolmogoroff, A. N., 397
Koschnitzke, R., 401
Kössler, K., 401
Kraft, V., 163, 402
Krantz, D. H., 381, 402
Kraus, O., 402
Krengel, U., 402
Kries, J. von, 224–32, 239, 245, 402
Krohn, W., 271–72, 392, 402
Kronenberg, M., 402
Krösche, K., 402
Krueger, L. E., 402, 423
Krug, W. T., 20, 27, 372
Krüger, L., 388, 390, 392–93, 400, 402, 405, 414, 418–19, 425, 430, 436
Kügelgen, W. von, 30, 328*n*59, 403
Kühn, K. G., 20
Kuhn, Th. S., 28, 166, 191, 241, 319, 403
Külpe, O., 308, 403
Kuntze, E., 9
Kuntze, J. E., 9, 21, 38, 41–49, 67, 69–70, 371, 381, 403, 414, 416, 437
Küppers, G., 271, 392, 402

Laitko, H., 398
Lakatos, I., 281
Lalo, Ch., 403
Lamarck, J.-B. de, 268, 270
Lampa, A., 263, 358*n63*, 403
Lamprecht, K., 335*n224*
Lange, F. A., 148, 154, 168, 178, 262, 346*n103*, 362*n91*, 402, 404, 431
Langer, P., 389, 404
Laplace, P. S., 29, 131, 140, 210, 253, 257–58, 261, 283, 289, 297, 308, 311, 425
Lasswitz, K., 12, 13, 231, 326*nn32–33*, 330*n88*, 335*n216*, 354*n141*, 404, 416
Laudan, L., 404
Lauterborn, R., 404
Le Chatelier, H., 358*n63*
Lea, E., 404
Lécuyer, B.-P., 404
Leese, K., 330*nm89, 106*, 405
Lehmann, G., 405
Leibniz, G. W., 55, 103, 106, 113–14, 127, 169, 173, 175, 371, 405
Leisering, B., 405
Lennig, P., 10–11, 13, 329*n85*, 381, 405, 416, 428
Lennon, K., 405
Lenoir, T., 405
Lenz, E., 141
Leplin, J., 405
Lewin, K., 268
Lexis, W., 362*n89*, 365*n147*, 406
Liebe, R., 406
Liebig, J. von, 42
Liebmann, O., 213, 352*n68*, 406
Lindner, G. A., 64, 406
Lipiner, S., 15, 62–63, 387, 391, 405–6, 417
Lipps, G. F., 10, 66, 307, 310–13, 364*n139*, 385, 406–7, 434
Lipps, Th., 364*n139*
Locke, J., 75–77, 124, 407
Lorentz, H. A., 142, 178, 348*n44*
Lotze, R. H., 13, 40–41, 65–66, 104, 115, 138, 151, 154, 159, 181, 212–13, 218, 285, 287–92, 296, 298, 330*n106*, 361*n63*, 381, 386, 390, 398, 402, 407–8, 418, 422, 427, 430, 434, 436
Loveland, D., 365*n159*
Löwith, K., 39, 408
Lowry, R., 408
Lübbe, H., 380, 408
Luce, R. D., 381, 408
Ludwig, F., 408
Ludwig, K., 64, 224–25, 243
Lülmann, C., 408

Lütgert, W., 25, 408
Lyell, C., 258

Mach, E., 4–8, 11, 14, 40, 59, 68, 70, 111, 138, 143, 154–64, 172, 174–75, 182, 187, 192, 205, 218, 231, 234–36, 238–47, 249, 263–64, 267, 286, 309, 315–16, 323, 345*n70*, 346*n95*, 371, 374, 377, 379, 386–87, 390–93, 396–98, 400, 402, 404, 406, 408–10, 426, 430–32, 436
Mahler, G., 15, 63, 369
Marquard, O., 410
Marshall, M. E., 10–11, 13, 193, 381, 403
Martin-Löf, P., 365*n159*
Masaryk, Th., 63, 69, 436
Mattenklott, G., 10, 382, 410
Mausfeld, R., 14, 205, 410
Maxwell, J. C., 50, 142, 419, 432, 435
Mayr, E., 255, 411
McCormmach, R., 400
McGregor, D., 411
Medway, N. L., 373
Meinel, Ch., 411
Meissner, W. W., 411
Melloni, M., 120
Merkel, J., 202, 411
Merz, J. Th., 411
Metzger, W., 267, 411
Metzinger, Th., 412
Meyenn, K. von, 412
Meyer, F. A. E, 412
Meyer, J., 43
Mill, J. St., 91, 155, 279, 412
Minsky, M., 47
Mises (Dr. Mises, pseudonym of Fechner's) 21, 31, 44, 51, 58, 381–84, 389, 403–4, 414, 437
Mises, R. von, 9, 309, 313–17, 365*n158*, 370, 412–13, 430
Möbius, P. J., 49, 70, 261, 413
Moigno, F., 147
Molella, A. Ph., 413
Moleschott, J., 25, 56, 163, 167
Mollweide, K. B., 20–21, 372, 381
Moravec, H., 47, 413
Mühlmann, W. E., 413
Müller, E., 330*n91*
Müller, F. A., 212–17, 240, 353*n118*, 404, 413, 434–35
Müller, G., 413
Müller, G. E., 14, 62, 177, 212–17, 223, 391, 413, 423, 428, 432
Müller, J., 35, 42, 125, 125, 234, 262, 390, 413
Müller, M., 388, 413

Munk, H., 125
Münsterberg, H., 413–14
Murphy, G., 414
Murray, D. J., 11, 13–14, 372, 388, 392, 414–15, 434

Nagel, Th., 414
Natorp, P., 63, 224, 231, 398, 414, 416, 425
Navier, C. L., 29
N-e [anonymous], 357n54
Neumann, F., 50, 373, 414
Neumann, J. von, 177
Neurath, O., 413
Newman, E. B., 415
Newton, I., 23, 78, 139, 149, 164, 238, 250, 374
Nietzsche, F., 63, 388, 408
Nitzschke, B., 373, 415
Noack, L., 38, 333n184, 415

Ockham, W. of, 271
Oelze, B., 10, 415
Oettingen, A. von, 415
Ohm, G., S., 28–29, 382, 391, 400, 415
Oken, L., 23–27, 30, 122, 255–57, 264, 266, 272,
    357n29, 373, 389, 412, 415–16, 419, 426, 430
Ollig, H.-L., 415
Orth, E. W., 415
Ostwald, W., 69, 143, 168, 394–95, 415–16, 433

P., J. von, 327n15, 416
Pander, Ch., 293
Paslack, R., 271, 402
Pastor, W., 13, 404, 416
Paul, Jean 36–37, 334n204, 399, 413, 416
Paul, Jul., 416
Pauli, R., 416
Paulsen, F., 181, 261, 387, 416
Paulsen, J., 416
Pearce Williams, L., 327n12, 416
Pearson, K., 158, 306, 312, 365n148, 416, 420
Peirce, Ch. S., 4, 8, 84, 144, 174, 177, 191, 249,
    260, 268–71, 324, 344n62, 367–69, 374, 392,
    398–99, 413, 416–17, 420, 426, 436
Pelman, C., 417
Penna, A. G., 417
Pernerstorfer, E., 417
Perrin, J., 143
Perry, R. B., 323, 417
Perty, M., 261, 418
Pester, R., 13, 408, 418
Petzoldt, J., 264, 267–68, 358n61, 395, 418
Philippi, E., 418
Picard, É., 418

Pierer, H. A., 43
Piéron, H., 418
Place, U. T., 165, 176
Planck, M., 238–39, 418
Plateau, J. A. F., 30, 216, 244, 403, 418
Plato, J. von, 14, 418
Plätzsch, A., 10, 378
Poggendorff, J. Ch., 50, 332n142
Poisson, S.-D., 29, 147, 298, 305, 308, 382
Polaczek, H. See Geiringer, H.
Popper, K., 106, 419
Porter, Th. M., 14, 368, 388, 419
Preyer, W. Th., 9–10, 70, 262–63, 384–85, 419
Priesner, C., 419
Prigogine, I., 8, 272, 419
Prytulak, L. S., 205, 419
Purkynje, J. E., 331n132
Putnam, H., 54, 78–79, 99, 419

Quetelet, A., 202, 295–98, 306, 362n82, 364n129,
    378, 419, 425, 432

Ranke, K. E., 420
Ratzel, F., 249, 267, 420, 429
Reichenbach, H., 40, 395
Reichlin-Meldegg, K. A. von, 420
Reincke, E., 420
Reininger, R., 182
Rezazadeh-Schafagh, S., 420
Rhode, E., 63
Ribot, Th., 208, 210, 351n52, 380, 420
Richards, E., 421
Rickert, H., 179
Riehl, A., 7, 175–76, 182, 185–87, 189, 261, 396, 421
Riemann, B., 162, 421
Rocke, A. J., 421
Rohracher, H., 421
Roob, H., 404, 422
Rorty, R., 422
Rosenberg, H., 214
Rosenberger, F., 422
Rosencrantz, G., 422
Rosenkranz, K., 422
Rosenzweig, S., 422
Ross, H. E., 11, 14, 414, 422, 434
Rossberg, F., 426
Rothschuh, K. E., 422
Roux, W., 359n78
Royce, J., 312, 422
Rümelin, G., 297, 300, 302, 422
Russell, B., 91, 175, 187, 342n45, 385, 388, 391,
    396, 400, 404, 412, 422

Ryle, G., 165–66

Salmon, W. C., 393, 422, 430
Salomon, 62, 294
Sandkühler, H. J., 393, 422, 424
Särndal, C.-E., 423
Sauer, W., 423
Schaller, J., 56, 154, 423
Scharlau, W., 423
Scheerer, E., 10–11, 13, 383–84, 423
Scheibner, W., 68
Scheler, M., 423
Schelling, F. W. J., 1–2, 4–6, 12, 21–23, 25–27,
    30, 38, 42, 56, 65, 92, 112–15, 121, 141, 148,
    150, 178, 293–95, 327n11, 369, 371, 373,
    379–80, 387–89, 391–92, 397, 399, 412–13,
    424–25, 427, 429, 433
Schleiden, M. J., 25, 42, 58, 120, 298, 383, 420,
    424
Schlick, M., 2, 7, 40, 166–68, 173, 182–89, 242,
    349n69, 349n76, 385, 388, 394–95, 412, 424
Schmid, 362n89
Schmid, E., 400
Schmidt, O., 261, 425
Schmidt, S. J., 386, 402, 425
Schmidt, W., 425
Schmidt, W. de, 353n88, 425
Schmied-Kowarzik, W., 14, 425
Schnabel, F., 425
Schnädelbach, H., 2, 425
Schneider, I., 14, 425
Schnorr, C.-P., 365n159, 425
Schopenhauer, A., 259, 261, 319, 368, 387–88
Schöpf, A., 373
Schramm, M., 425
Schreier, W., 10, 425
Schröder, Ch. and H., 50, 426
Schrödinger, E., 334n199, 390
Schubert, G. H., 25, 380, 426
Schuekarew, A., 426
Schulthess, P., 426
Schulz, L., 426
Schulze, M. G., 35–37
Schumann, Clara 329n85
Schurig, V., 426
Schuster, J., 426
Schwann, Th., 25
Schwartze, P., 426
Schweiker, J. E., 426
Séailles, G., 12, 426
Seaman, F., 426
Searle, J. R., 426

Seebeck, A., 120
Seguin, Ch., 147
Sengle, F., 426
Seydel, R., 12, 115, 427
Seydler, A., 427
Short, T. L., 427
Siegel, C., 12, 427
Sigwart, C., 179, 181
Silberstein, E., 63, 387
Simon, H. A., 271, 427
Simon, Th., 427
Singer, B., 427
Slade, H., 67–68, 323
Sluga, H., 427
Smaasen, W., 259, 427
Smart, J. J. C., 165–66
Smoluchowski, M. von, 143, 315
Snelders, H. A. M., 14, 427
Snell, C., 261, 428
Sober, E., 428
Sommer, M., 428
Speckamp, H., 428
Spinoza, B., 92, 113, 150, 173–74, 177–79, 271,
    338n85, 400, 402, 428
Sprung, H. and L., 10, 13, 406, 428, 430
Stach, P., 405, 428
Stadler, A., 212–15, 217, 222–23, 429, 431
Stadler, Fr., 371, 390, 393, 397, 410, 413, 436
Stallo, J. B., 429
Steffens, H., 22, 412
Steinmetzler, J., 429
Stern, L. W., 429
Stevens, S. S., 7, 11, 202, 205, 355n170, 402, 415,
    419, 429–30, 433
Stigler, St. M., 14, 430
Stracke, A., 430
Strauss, D. Fr., 25, 47, 65, 319
Strawson, P. F., 430
Strohl, J., 430
Strümpell, L., 35
Stubenberg, L., 79, 336n20, 430
Stumpf, C., 70, 179, 181, 218, 400, 430
Suppe, Fr., 431
Suppes, P., 354n146, 431
Swijtink, Z., 388
Swoboda, W. W., 431

Taine, H., 182
Tait, P. G., 69, 394
Tannery, J., 208–10, 213–14, 224, 227–28, 232,
    239, 351n54, 418, 431
Tannery, P., 351n54, 431